Lecture Notes in Computer Science 11688

More information about this series at http://www.springer.com/series/7407

Tuan Phung-Duc · Shoji Kasahara ·
Sabine Wittevrongel (Eds.)

Queueing Theory
and Network Applications

14th International Conference, QTNA 2019
Ghent, Belgium, August 27–29, 2019
Proceedings

 Springer

Editors
Tuan Phung-Duc (ID)
University of Tsukuba
Tsukuba, Japan

Shoji Kasahara (ID)
Nara Institute of Science and Technology
Nara, Japan

Sabine Wittevrongel (ID)
Ghent University
Ghent, Belgium

ISSN 0302-9743   ISSN 1611-3349   (electronic)
Lecture Notes in Computer Science
ISBN 978-3-030-27180-0   ISBN 978-3-030-27181-7   (eBook)
https://doi.org/10.1007/978-3-030-27181-7

LNCS Sublibrary: SL1 – Theoretical Computer Science and General Issues

This Springer imprint is published by the registered company Springer Nature Switzerland AG
The registered company address is: Gewerbestrasse 11, 6330 Cham, Switzerland

Preface

The International Conference on Queueing Theory and Network Applications aims to promote the knowledge and the development of high-quality research on queueing theory and its applications in networks and other related fields covering performance issues on computer and communication systems, job scheduling, blockchain technology, etc. It brings together researchers, scientists, and practitioners from all over the world and offers an open forum to share the latest important research accomplishments and challenging problems in the area of queueing theory and network applications.

This volume contains papers selected and presented at the 14th International Conference on Queueing Theory and Network Applications (QTNA 2019) held during August 27–29, 2019, in Ghent, Belgium.

QTNA 2019 was a continuation of the series of successful QTNA conferences: QTNA 2006 (Seoul, South Korea), QTNA 2007 (Kobe, Japan), QTNA 2008 (Taipei, Taiwan), QTNA 2009 (Singapore), QTNA 2010 (Beijing, China), QTNA 2011 (Seoul, South Korea), QTNA 2012 (Kyoto, Japan), QTNA 2013 (Taichung, Taiwan), QTNA 2014 (Bellingham, USA), QTNA 2015 (Hanoi, Vietnam), QTNA 2016 (Wellington, New Zealand), QTNA 2017 (Qinhuangdao, China), and QTNA 2018 (Tsukuba, Japan).

The conference this year was the first to be held in Europe. We received 49 submissions from 23 countries and areas in five continents: Algeria, Austria, Azerbaijan, Belgium, Brazil, Canada, China, Egypt, Greece, Hong Kong, India, Israel, Italy, Japan, The Netherlands, Poland, Russia, Singapore, South Korea, Sri Lanka, Taiwan, Turkey, and USA. Each submitted paper was peer reviewed by at least three reviewers, and evaluated on the quality, originality, soundness, and significance of its contribution by the members of the Technical Program Committee (TPC) of QTNA 2019 and external reviewers invited by the TPC. After a careful selection, 23 full papers (11+ pages) were accepted for inclusion in this volume of *Lecture Notes in Computer Science* (LNCS) published by Springer.

Furthermore, a number of short papers (6 pages) were selected for presentation at the conference and for inclusion in an electronic version of the conference brochure distributed to all the participants of the conference QTNA 2019.

It was our privilege to invite Professor Mor Harchol-Balter and Professor Johan van Leeuwaarden to give keynote talks at QTNA 2019.

We would like to thank the authors of all the papers appearing in this proceedings for their excellent contribution. Special thanks go to the co-chairs and members of the Program Committee of QTNA 2019 for their time and effort in assuring the quality of the selected papers. We also would like to express our gratitude to the members of the Local Organizing Committee for their hard work throughout the process from planning to holding the conference. Finally, we cordially thank the EasyChair team and Springer for their support in publishing this volume. Thank you all for your contributions to QTNA 2019.

August 2019

Sabine Wittevrongel
Tuan Phung-Duc
Shoji Kasahara

Organization

General Chairs

Sabine Wittevrongel	Ghent University, Belgium
Yutaka Takahashi	Kyoto University, Japan

Program Committee Chairs

Tuan Phung-Duc	University of Tsukuba, Japan
Shoji Kasahara	Nara Institute of Science and Technology, Japan

Program Committee

Herwig Bruneel	Ghent University, Belgium
Wai-Ki Ching	The University of Hong Kong, SAR China
Wanyang Dai	Nanjing University, China
Koen De Turck	CentraleSupélec, France
Ioannis Dimitriou	University of Patras, Greece
Tien Van Do	Budapest University of Technology and Economics, Hungary
Alexander Dudin	Belarusian State University, Belarus
Antonis Economou	University of Athens, Greece
Marco Gribaudo	Politecnico di Milano, Italy
Irina Gudkova	Peoples' Friendship University of Russia, Russia
Qi-Ming He	University of Waterloo, Canada
Ganguk Hwang	Korea Advanced Institute of Science and Technology, South Korea
Yoshiaki Inoue	Osaka University, Japan
Shunfu Jin	Yanshan University, China
Stella Kapodistria	Eindhoven University of Technology, The Netherlands
Shoji Kasahara	Nara Institute of Science and Technology, Japan
Ken'ichi Kawanishi	Gunma University, Japan
Konosuke Kawashima	Tokyo University of Agriculture and Technology, Japan
Jau-Chuan Ke	National Taichung University of Science and Technology, Taiwan
Wojciech Kempa	Silesian University of Technology, Poland
Bara Kim	Korea University, South Korea
Tatsuaki Kimura	Osaka University, Japan
Masahiro Kobayashi	Tokai University, Japan
Achyutha Krishnamoorthy	Cochin University of Science and Technology, India
Ho Woo Lee	Sungkyunkwan University, South Korea

Se Won Lee	Pukyong National University, South Korea
Tony T. Lee	The Chinese University of Hong Kong, SAR China
Bin Liu	Anhui Jianzhu University, China
Zhanyou Ma	Yanshan University, China
Andrea Marin	University of Venice, Italy
Hiroyuki Masuyama	Kyoto University, Japan
Agassi Melikov	Azerbaijan National Academy of Sciences, Azerbaijan
Rein Nobel	Vrije Universiteit Amsterdam, The Netherlands
Toshihisa Ozawa	Komazawa University, Japan
Tuan Phung-Duc	University of Tsukuba, Japan
Wouter Rogiest	Ghent University, Belgium
Poompat Saengudomlert	Bangkok University, Thailand
Zsolt Saffer	Vienna University of Technology, Austria
Yutaka Sakuma	National Defense Academy of Japan, Japan
Yang Woo Shin	Changwon National University, South Korea
Ahmed Tarabia	Damietta University, Egypt
Y. C. Tay	National University of Singapore, Singapore
Miklós Telek	Budapest University of Technology and Economics, Hungary
Nigel Thomas	Newcastle University, UK
Joris Walraevens	Ghent University, Belgium
Jinting Wang	Beijing Jiaotong University, China
Sabine Wittevrongel	Ghent University, Belgium
Hengqing Ye	The Hong Kong Polytechnic University, SAR China
Xue-Ming Yuan	Agency for Science, Research and Technology, Singapore
Dequan Yue	Yanshan University, China
Wuyi Yue	Konan University, Japan
Alexander Zeifman	Vologda State University, Russia
Zhe George Zhang	Western Washington University, USA
Yiqiang Q. Zhao	Carleton University, Canada

Steering Committee

Co-chairs

Bong Dae Choi	Korea University, South Korea
Yutaka Takahashi	Kyoto University, Japan
Wuyi Yue	Konan University, Japan

Members

Hsing Paul Luh	National Chengchi University, Taiwan
Winston Seah	Victoria University of Wellington, New Zealand
Hideaki Takagi	University of Tsukuba, Japan
Y. C. Tay	National University of Singapore, Singapore
Kuo-Hsiung Wang	National Taiwan University, Taiwan

Jinting Wang	Beijing Jiaotong University, China
Dequan Yue	Yanshan University, China
Zhe George Zhang	Western Washington University, USA

Local Organizing Committee

Michiel De Muynck	Ghent University, Belgium
Arnaud Devos	Ghent University, Belgium
Freek Verdonck	Ghent University, Belgium
Sabine Wittevrongel	Ghent University, Belgium

Contents

Strategic Queues

Queueing Networks

Scheduling Policies

Multidimensional Systems

Queueing Models in Applications

Retrial Queues

Retrial Queueing System MMPP/M/1 with Impatient Calls Under Heavy Load Condition

Ekaterina Fedorova[1], Elena Danilyuk[1]([⊠]), Anatoly Nazarov[1], and Agassi Melikov[2]

[1] National Research Tomsk State University, Lenina Avenue, 36, Tomsk, Russia
moiskate@mail.ru, daniluc.elena.yu@gmail.com, nazarov.tsu@gmail.com
[2] Institute of Control Systems, Azerbaijan National Academy of Sciences,
B. Vahabzadeh Street, 9, Baku, Azerbaijan
agassi.melikov@gmail.com

Abstract. In this paper, a single server retrial queue $MMPP/M/1$ with impatient calls is analysed under the heavy load condition. The retrial queue has a dynamical rate of the calls patience depending on the number of calls in the orbit. It is proved that under the heavy load condition the asymptotic characteristic function of the number of calls in the orbit has the gamma distribution with obtained parameters. Also the formula for the system throughput is obtained. Some numerical examples are presented.

Keywords: Retrial queue · Impatient call · Heavy load · Asymptotic analysis

1 Introduction

The classical queueing theory distinguishes two classes of mathematical models: queueing systems with queue and loss systems. But the last decades, a new model appeared – retrial queueing system (or system with repeated calls). Retrial queues are characterized by the feature that an arriving call finding a server busy does not join a queue and does not leave the system immediately, but goes to some virtual place (orbit), then it tries to get service again after some random time.

Retrial queues are widely used for many practical problems in telecommunication networks, mobile networks, computer systems and various daily life situations [1–9]. Retrial queueing systems with impatient customers are mainly applied for analysis of call centers [8–10] where a customer, who cannot connect with an operator, tries later. Obviously, a research of retrial models [6,7,11] is more difficult than a corresponding one without retrial because retrial makes the arrival process more complex. The comprehensive description, the comparison of classical queueing systems and retrial queues and detailed overviews are

© Springer Nature Switzerland AG 2019
T. Phung-Duc et al. (Eds.): QTNA 2019, LNCS 11688, pp. 3–15, 2019.
https://doi.org/10.1007/978-3-030-27181-7_1

contained in books of Artalejo and Gómez-Corral [12], Artalejo and Falin [13], Falin and Templeton [14].

The first retrial model with impatience was considered by Cohen [2]. The $M/M/1$ retrial queue with impatient calls was studied by Falin and Templeton [14]. Also, Yang et al. [15] and Krishnamoorthy et al. [16] performed analysis of the $M/G/1$ model. Retrial queues with Poisson arrivals and the impatience phenomenon were studied in [11,17–20]. In [21] the $M/G/1$ retrial queue with batch arrivals was considered. In the papers above, the impatience is understood as an arriving call joins the orbit with some probability p and leaves the system with the probability $1 - p$.

Studies of some types of retrial queues with MAP (or MMPP) arrivals are known. But they are performed using truncation methods [12,22–24] or matrix methods [25–28] which are applied only for numerical solutions obtaining.

It is known that explicit formulas for stationary distributions in complex retrial queues are derived hard or cannot be obtained at all. But some approximations or asymptotic solutions can be proposed.

In this paper, we use the asymptotic analysis method developed in research [29,30] for different types of queueing systems and networks studies. The principle of the method is a derivation of some asymptotic equations from the systems of equations determined models behaviour, and further getting formulas for asymptotic functions under some limit condition. In previous papers [31,32], we have obtained asymptotic solutions for different types of retrial queues without losses: $M/M/1$, $M/GI/1$ and even $MMPP/M/1$, $MMPP/GI/1$ under the heavy load condition. So in this paper, we generalize our results to models with impatient calls. For comparison, we have also obtained asymptotic solutions for retrial queues $M/M/1$, $MMPP/M/2$, $M/M/N$ with impatient calls, but under another asymptotic condition [33–35].

Performance characteristics for retrial queueing systems under heavy and light loads conditions were also studied by Falin, Anisimov, etc. [36–38]. But for the asymptotic analysis, they use explicit formulas, so they study only Poisson arrivals. Also, work [38] is devoted to the investigation under an "extreme" load (the intensity of primary calls tends to infinity or zero).

The rest of the paper is organized as follows. In Sect. 2, the considered mathematical model is described and the aim of the study is defined. In Sect. 3, the retrial queue is studied under a limit condition of heavy load. The theorem about the gamma distribution of the asymptotic characteristic function is proved and the formula for the system throughput is obtained. In Sect. 4, some numerical examples of the comparison of the asymptotic distributions with exact ones (which are obtained via simulation) are presented. The last section is devoted to some conclusions.

2   Mathematical Model

Consider a single server retrial queueing system $MMPP/M/1$. The system structure is presented in Fig. 1. Primary calls arrive at the system from outside according to Markovian Modulated Poisson Process (MMPP) which is a particular case of Markovian Arrival Process (MAP) defined by matrices $\mathbf{D_0}$ and $\mathbf{D_1}$ [39, 40]. If a primary call finds the server free, it stays here with service time distributed exponentially with a rate μ. Otherwise, the call goes to an orbit, where it stays during random time distributed by the exponential law with a rate σ. After the delay, the call makes an attempt to reach the server again. If the server is free, the call gets the service, otherwise, the call instantly returns to the orbit. From the orbit, calls can leave the system after random time distributed exponentially with a rate α/i depending on the number of calls i in the orbit at this moment.

The arrival process, the service times, the retrial and the impatience times are assumed to be mutually independent.

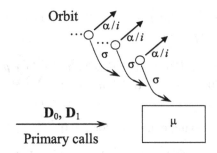

Fig. 1. Retrial queueing system $MMPP/M/1$ with impatient calls

The MMPP underlying process $n(t)$ is a Markov chain with continuous time and finite set of states $n = 1, 2, \ldots, N$.

We introduce a generator of the process $n(t)$ as matrix $\mathbf{Q} = \mathbf{D_0} + \mathbf{D_1}$ with elements q_{mv}, where $m, v = 1, 2, \ldots, N$.

Matrix $\mathbf{D_1}$ is diagonal with elements $\rho\lambda_n$ ($n = 1, 2, \ldots, N$), where λ_n are conditional arrival rates and ρ is a parameter defined below. We denote $\mathbf{\Lambda} = \mathtt{diag}\{\lambda_n\}$. Then the following equality holds $\mathbf{D_1} = \rho\mathbf{\Lambda}$.

Row vector $\mathbf{r}$ is a stationary probability distribution of the underlying process $n(t)$. $\mathbf{r}$ is uniquely determined by the following system

$$\begin{cases} \mathbf{r}\mathbf{Q} = \mathbf{0}, \\ \mathbf{r}\mathbf{e} = 1, \end{cases} \tag{1}$$

where $\mathbf{e} = \{1, 1, \ldots, 1\}^T$ and $\mathbf{0} = \{0, 0, \ldots, 0\}$.

The fundamental rate of MMPP is defined as follows $\lambda = \mathbf{r} \cdot \rho\mathbf{\Lambda} \cdot \mathbf{e}$.

Suppose the system parameters satisfy the following condition

$$\mathbf{r} \cdot \mathbf{\Lambda} \cdot \mathbf{e} = \mu, \tag{2}$$

then the parameter ρ is the system load and it equals $\rho = \lambda/\mu$.

Let $i(t)$ be the number of calls in the orbit and $k(t)$ be the server state

$$k(t) = \begin{cases} 0, & \text{if the server is free,} \\ 1, & \text{if the server is busy.} \end{cases}$$

The problem is to find the probability distribution of the number of calls in the orbit.

The process $i(t)$ is not Markovian, therefore we consider the multidimensional process $\{k(t), n(t), i(t) : t \geq 0\}$ which is a continuous time Markov chain. Denote $P(k, n, i, t) = P\{k(t) = k, n(t) = n, i(t) = i\}$.

The following system of Kolmogorov equations for the stationary distribution $P(k, n, i) = \lim\limits_{t \to \infty} P(k, n, i, t)$ is derived for $i > 0$, $n = \overline{1, N}$.

$$\begin{cases} -(\rho\lambda_n + \alpha + i\sigma - q_{nn})P(0, n, i) + \alpha P(0, n, i+1) + \mu P(1, n, i) \\ \quad + \sum\limits_{v \neq n} P(0, v, i)q_{vn} = 0, \\ -(\rho\lambda_n + \mu + \alpha - q_{nn})P(1, n, i) + \rho\lambda_n P(1, n, i-1) + \rho\lambda_n P(0, n, i) \\ \quad +(i+1)\sigma P(0, n, i+1) + \alpha P(1, n, i+1) + \sum\limits_{v \neq n} P(1, v, i)q_{vn} = 0. \end{cases} \tag{3}$$

Also we have the following equations for $i = 0$

$$\begin{cases} -(\rho\lambda_n - q_{nn})P(0, n, 0) + \alpha P(0, n, 1) + \mu P(1, n, 0) \\ \quad + \sum\limits_{v \neq n} P(0, v, 0)q_{vn} = 0, \\ -(\rho\lambda_n + \mu - q_{nn})P(1, n, 0) + \rho\lambda_n P(0, n, 0) \\ \quad +\sigma P(0, n, 1) + \alpha P(1, n, 1) + \sum\limits_{v \neq n} P(1, v, 0)q_{vn} = 0. \end{cases} \tag{4}$$

Let us introduce row vectors $\mathbf{P}_k(i) = \{P(k, 1, i), P(k, 2, i), \ldots, P(k, N, i)\}$. Then Eqs. (3), (4) have the following matrix form

$$\begin{cases} -\mathbf{P}_0(0)(\rho\Lambda - \mathbf{Q}) + \mu\mathbf{P}_1(0) + \alpha\mathbf{P}_0(1) = \mathbf{0}, \\ -\mathbf{P}_1(0)(\rho\Lambda + \mu\mathbf{I} - \mathbf{Q}) + \mathbf{P}_0(0)\rho\Lambda + \sigma\mathbf{P}_0(1) + \alpha\mathbf{P}_1(1) = \mathbf{0}, \\ -\mathbf{P}_0(i)(\rho\Lambda + (i\sigma + \alpha)\mathbf{I} - \mathbf{Q}) + \mu\mathbf{P}_1(i) + \alpha\mathbf{P}_0(i+1) = \mathbf{0} \quad \text{for } i \geq 0, \\ -\mathbf{P}_1(i)(\rho\Lambda + (\mu + \alpha)\mathbf{I} - \mathbf{Q}) + \mathbf{P}_0(i)\rho\Lambda + \mathbf{P}_1(i-1)\rho\Lambda \\ \quad +\sigma(i+1)\mathbf{P}_0(i+1) + \alpha\mathbf{P}_1(i+1) = \mathbf{0} \quad \text{for } i \geq 0, \end{cases} \tag{5}$$

where $\mathbf{I}$ is the identity matrix.

By $\mathbf{H}_k(u) = \sum\limits_i e^{jui}\mathbf{P}_k(i)$ we denote the partial characteristic functions, where $k = 0, 1$ and $j = \sqrt{-1}$. Then system (5) is rewritten as follows

$$\begin{cases} \mathbf{H}_0(u)(\mathbf{Q} - \rho\Lambda - \alpha\mathbf{I}) + j\sigma\mathbf{H}'_0(u) + \mu\mathbf{H}_1(u) \\ \quad +\alpha e^{-ju}\mathbf{H}_0(u) = \alpha\mathbf{P}_0(0)(e^{-ju} - 1), \\ \mathbf{H}_1(u)(\mathbf{Q} - \rho\Lambda(1 - e^{ju}) - (\mu + \alpha)\mathbf{I}) + \mathbf{H}_0(u)\rho\Lambda - j\sigma e^{-ju}\mathbf{H}'_0(u) \\ \quad +\alpha e^{-ju}\mathbf{H}_1(u) = \alpha\mathbf{P}_1(0)(e^{-ju} - 1). \end{cases} \tag{6}$$

Obviously, system (6) can not be directly solved analytically. To solve system (6) we propose the method of asymptotic analysis under the heavy load condition $\rho \uparrow S$ [31], where S is the throughput of the retrial queue (i.e. a maximum value of the load for the system stationary regime).

3   Asymptotic Analysis Under Heavy Load

The method of asymptotic analysis in queueing theory is the method of research of the equations determining some characteristics of an queueing system under some limit (asymptotic) condition, which is specific for any model and solving problem.

Let us prove the following theorem for the considered retrial queue.

Theorem 1. *Let $i(t)$ be the number of calls in the orbit in the retrial queueing system $MMPP/M/1$ with impatient calls in the stationary regime, then the limit characteristic function $h(u)$ of the process $i(t)$ under the heavy load condition has the gamma distribution form*

$$h(u) = \lim_{\rho \to S} E\left\{ e^{jw(S-\rho)i(t)} \right\} = \left(1 - \frac{jw}{\beta} \right)^{-\gamma},$$

with parameters

$$\beta = \frac{\mu}{Sv\Lambda e + \mu + \alpha}, \quad \gamma = 1 + \frac{\mu}{\sigma}\beta,$$

where $S = 1 + \alpha/\mu$ is the system throughput and the vector $\mathbf{v}$ is a solution of the following system

$$\begin{cases} \mathbf{v}\mathbf{Q} = \mathbf{r}((\alpha + \mu)\mathbf{I} - S\Lambda), \\ \mathbf{v}e = 0, \end{cases}$$

and $E\{X\}$ is mathematical expectation of variable X.

Proof. The proof of the theorem can be divided into two parts: deriving of asymptotic equations and analysis of this equations for getting required characteristics.

Derivation of Asymptotic Equations. First of all, we introduce some notations:

$$\varepsilon = S - \rho, u = \varepsilon w, \mathbf{H}_0(u) = \varepsilon\mathbf{F}_0(w, \varepsilon), \mathbf{H}_1(u) = \mathbf{F}_1(w, \varepsilon), \mathbf{P}_k(0) = \varepsilon\boldsymbol{\pi}_k, \quad (7)$$

where $k = 0; 1$.

So, the heavy load condition is defined as $\rho \uparrow S$ or $\varepsilon \downarrow 0$.

Using (7), we rewrite system (6) as

$$\begin{cases} \varepsilon\mathbf{F}_0(w, \varepsilon)(\mathbf{Q} - (S - \varepsilon)\Lambda - \alpha\mathbf{I}) + j\sigma\dfrac{\partial\mathbf{F}_0(w, \varepsilon)}{\partial w} + \mu\mathbf{F}_1(w, \varepsilon) \\ \quad + \alpha e^{-jw\varepsilon}\varepsilon\mathbf{F}_0(w, \varepsilon) = \alpha\varepsilon\boldsymbol{\pi}_0(e^{-jw\varepsilon} - 1), \\ \mathbf{F}_1(w, \varepsilon)(\mathbf{Q} - (S - \varepsilon)\Lambda(1 - e^{jw\varepsilon}) - (\mu + \alpha)\mathbf{I}) + (S - \varepsilon)\varepsilon\mathbf{F}_0(w, \varepsilon)\Lambda \\ \quad - j\sigma e^{-jw\varepsilon}\dfrac{\partial\mathbf{F}_0(w, \varepsilon)}{\partial w} + \alpha e^{-jw\varepsilon}\mathbf{F}_1(w, \varepsilon) = \alpha\varepsilon\boldsymbol{\pi}_1(e^{-jw\varepsilon} - 1). \end{cases} \quad (8)$$

Let us consider expansions of functions $\mathbf{F}_k(w, \varepsilon)$ in the form

$$\mathbf{F}_k(w, \varepsilon) = \mathbf{F}_k(w) + \varepsilon\mathbf{f}_k(w) + \mathbf{O}(\varepsilon^2), \tag{9}$$

where $\mathbf{F}_k(w) = \lim_{\varepsilon \to 0} \mathbf{F}_k(w, \varepsilon)$, $\mathbf{O}(\varepsilon^2)$ is an infinitesimal value of order ε^2.

Substituting (9) into system (8) and writing equalities for members with equal powers of ε, we obtain the following system of equations

$$\begin{cases} j\sigma\mathbf{F}_0'(w) + \mu\mathbf{F}_1(w) = \mathbf{0}, \\ \mathbf{F}_1(w)(\mathbf{Q} - \mu\mathbf{I}) - j\sigma\mathbf{F}_0'(w) = \mathbf{0}, \\ \mathbf{F}_0(w)(\mathbf{Q} - S\mathbf{\Lambda}) + j\sigma\mathbf{f}_0'(w) + \mu\mathbf{f}_1(w) = \mathbf{0}, \\ jwS\mathbf{F}_1(w)\mathbf{\Lambda} + \mathbf{f}_1(w)(\mathbf{Q} - \mu\mathbf{I}) + S\mathbf{F}_0(w)\mathbf{\Lambda} + j\sigma jw\mathbf{F}_0'(w) \\ \quad -j\sigma\mathbf{f}_0'(w) - \alpha jw\mathbf{F}_1(w) = \mathbf{0}. \end{cases} \tag{10}$$

Then we sum all equations of system (8) and multiply the result by $\mathbf{e}$. After some transformations, we get the following equation

$$\mathbf{F}_1(w, \varepsilon)\left((S - \varepsilon)e^{jw\varepsilon}\mathbf{\Lambda} - \alpha\mathbf{I}\right)\mathbf{e} + j\sigma\frac{\partial\mathbf{F}_0(w, \varepsilon)}{\partial w}\mathbf{e} - \varepsilon\alpha\mathbf{F}_0(w, \varepsilon) = -\varepsilon\alpha(\boldsymbol{\pi}_0 + \boldsymbol{\pi}_1)\mathbf{e}.$$

Substitute expansions (9) and again writing equalities for members with equal powers of ε, we obtain two additional equations

$$\begin{cases} \mathbf{F}_1(w)(S\mathbf{\Lambda} - \alpha\mathbf{I})\mathbf{e} + j\sigma\mathbf{F}_0'(w)\mathbf{e} = 0, \\ \mathbf{F}_1(w)(Sjw - 1)\mathbf{\Lambda}\mathbf{e} + \mathbf{f}_1(w)(S\mathbf{\Lambda} - \alpha\mathbf{I})\mathbf{e} \\ \quad + j\sigma\mathbf{f}_0'(w)\mathbf{e} - \alpha\mathbf{F}_0(w)\mathbf{e} = -\alpha(\boldsymbol{\pi}_0 + \boldsymbol{\pi}_1)\mathbf{e}. \end{cases} \tag{11}$$

Thus, we have four matrix (10) and two scalar asymptotic equations (11).

Analysis of the Equations. The characteristic function of the number of calls in the orbit in considered retrial queue is calculated as

$$H(u) = \mathrm{E}\left\{e^{jui(t)}\right\} = \mathbf{H}_0(u)\mathbf{e} + \mathbf{H}_1(u)\mathbf{e}.$$

Under the heavy load condition $H(u) \approx h(u)$, where $h(u)$ is called the asymptotic characteristic function and is presented in the form

$$h(u) = \lim_{\rho \to S} \mathrm{E}\left\{e^{jw(S-\rho)i(t)}\right\}. \tag{12}$$

Using notations (7), we obtain that

$$h(u) = \mathbf{F}_1\left(\frac{u}{S - \rho}\right)\mathbf{e}.$$

Therefore, it is necessary to obtain the scalar function $\mathbf{F}_1(w)\mathbf{e}$ from Eqs. (10), (11). The derivation includes five steps.

Step 1. Combining the first and the second equations of (10), we get

$$\mathbf{F}_1(w)\mathbf{Q} = \mathbf{0}.$$

Taking into account (1), the function $\mathbf{F}_1(w)$ have the form:

$$\mathbf{F}_1(w) = \mathbf{r}\Phi(w). \tag{13}$$

Hence it is necessary to find the function $\Phi(w)$.

Step 2. From the second equation of (10), we get

$$j\sigma\mathbf{F}_0'(w) = -\mu\mathbf{F}_1(w) = -\mu\mathbf{r}\Phi(w). \tag{14}$$

Substituting (13), (14) into the first equation of (11), it is easy to show that

$$S = \frac{\alpha + \mu}{\mathbf{r}\Lambda\mathbf{e}} = \frac{\alpha + \mu}{\mu}. \tag{15}$$

Therefore, we obtain the formula for the system throughput. The stationary regime of the retrial queue exists if $\rho < 1 + \alpha/\mu$.

Step 3. Summing the third and the fourth equations of (10), we obtain

$$(\mathbf{F}_0(w) + \mathbf{f}_1(w))\mathbf{Q} + j\sigma jw\mathbf{F}_0'(w) + jw\mathbf{F}_1(w)(S\Lambda - \alpha\mathbf{I}) = \mathbf{0}.$$

Take into account formulas (13), (14).

$$(\mathbf{F}_0(w) + \mathbf{f}_1(w))\mathbf{Q} = jw\Phi(w)\mathbf{r}((\mu + \alpha)\mathbf{I} - S\Lambda).$$

Let us suppose that

$$\mathbf{F}_0(w) + \mathbf{f}_1(w) = jw\Phi(w)\mathbf{v}, \tag{16}$$

where $\mathbf{v}$ is a solution of the following equation

$$\mathbf{v}\mathbf{Q} = \mathbf{r}((\mu + \alpha)\mathbf{I} - S\Lambda). \tag{17}$$

Consider Eq. (17). It is matrix with equal ranks of the system matrix and augmented one. Obviously, Eq. (17) has infinitely many solutions. Let us present the general solution of (17) as follows

$$\mathbf{v} = C\mathbf{r} + \mathbf{v}_0,$$

where C is some constant and $\mathbf{v}_0$ is a particular solution (for example, $\mathbf{v}_0\mathbf{e} = 0$). From (16), it follows

$$\mathbf{f}_1(w) = jw\Phi(w)\mathbf{v} - \mathbf{F}_0(w), \tag{18}$$

Step 4. Rewrite the third equation of (10) as

$$j\sigma\mathbf{f}_0'(w) = -\mathbf{F}_0(w)(\mathbf{Q} - S\Lambda) - \mu\mathbf{f}_1(w).$$

Multiply the last equation by $\mathbf{e}$ and substitute expression (18) here.

$$j\sigma\mathbf{f}_0'(w)\mathbf{e} = \mathbf{F}_0(w)(S\Lambda\mathbf{e} + \mu\mathbf{e}) - \mu jw\Phi(w)\mathbf{v}\mathbf{e}. \tag{19}$$

Step 5. Finally, we substitute obtained formulas (13)–(19) in the last equation of system (11).

$$\Phi(w)(Sjw - 1)\mathbf{r}\Lambda\mathbf{e} + jw\Phi(w)(Sv\Lambda\mathbf{e} - \alpha v\mathbf{e}) - \mu jw\Phi(w)v\mathbf{e}$$
$$+(\alpha + \mu)\mathbf{F}_0(w)\mathbf{e} - \alpha\mathbf{F}_0(w)\mathbf{e} = -\alpha(\pi_0 + \pi_1)\mathbf{e}.$$

Differentiating this equation, and taking into account (2), it is easy to obtain the following equation

$$\Phi'(w)\left(-\mu + jw(Sv\Lambda\mathbf{e} - \alpha v\mathbf{e} - \mu v\mathbf{e} + S\mu)\right)$$
$$+j\Phi(w)\left(Sv\Lambda\mathbf{e} - \alpha v\mathbf{e} - \mu v\mathbf{e} + S\mu + \mu^2/\sigma\right) = 0. \tag{20}$$

Let us divide (20) by the expression $Sv\Lambda\mathbf{e} - \alpha v\mathbf{e} - \mu v\mathbf{e} + S\mu$ and introduce denotation

$$\beta = \frac{\mu}{Sv\Lambda\mathbf{e} - \alpha v\mathbf{e} - \mu v\mathbf{e} + S\mu}, \quad \gamma = 1 + \frac{\mu}{\sigma}\beta.$$

So, Eq. (20) is rewritten as

$$\Phi'(w)(\beta - jw) = j\gamma\Phi(w).$$

Clearly, the solution has the form

$$\Phi(w) = C_0\left(1 - \frac{jw}{\beta}\right)^{-\gamma}.$$

From formula (13), we obtain

$$\mathbf{F}_1(w) = \mathbf{r} \cdot C_0\left(1 - \frac{jw}{\beta}\right)^{-\gamma}.$$

Thus, (12) is written as

$$h(u) = C_0\left(1 - \frac{ju}{\beta}\right)^{-\gamma}.$$

It is easy to show that $C_0 = 1$ due to the normalisation requirement.

Note that the parameters β and γ do not depend on the constant C in the solution $\mathbf{v} = C\mathbf{r} + \mathbf{v}_0$. It can be shown by the substituting:

$$\beta = \frac{\mu}{Sv_0\Lambda\mathbf{e} - \alpha v_0\mathbf{e} - \mu v_0\mathbf{e} + S\mu} = \frac{\mu}{Sv_0\Lambda\mathbf{e} + \mu + \alpha}.$$

So, the asymptotic characteristic function of the probability distribution of the number of calls in the orbit $h(u)$ has the form of the gamma distribution characteristic function:

$$h(u) = \left(1 - \frac{ju}{\beta}\right)^{-\gamma}$$

with the inverse scale parameter $\beta = \dfrac{\mu}{Sv\Lambda e + \mu + \alpha}$ and the shape parameter $\gamma = 1 + \dfrac{\mu}{\sigma}\beta$, where the vector $\mathbf{v}$ is a solution of the following system

$$\begin{cases} \mathbf{v}\mathbf{Q} = \mathbf{r}((\alpha + \mu)\mathbf{I} - S\boldsymbol{\Lambda}), \\ \mathbf{v}\mathbf{e} = 0. \end{cases}$$

This completes the proof.

4 Numerical Analysis

Denote the probability distribution function of the gamma distribution as $\Gamma(x)$. Then the discrete probability distribution of the number of calls in the orbit $p(i)$ can be approximated as follows

$$p(i) = \Gamma(i+1) - \Gamma(i).$$

Now we present some numerical examples to demonstrate the applicability area of the obtained results. We perform simulation of the system evolution using software platform ODIS [41], which realizes a discrete-event simulation approach, and we compare statistical results with analytical ones derived in the paper. For the comparison we use Kolmogorov distance between respective distribution functions

$$d = \max_{i \geq 0} \left| \sum_{l=0}^{i} [\tilde{p}(l) - p(l)] \right|.$$

Here, $p(l)$ is a probability distribution calculated using the asymptotic formula and $\tilde{p}(l)$ is an empiric distribution of the number of calls in the orbit obtained as the simulation results of the system evolution. For our purposes, we assume values $d \leq 0.05$ are enough for good accuracy of approximations.

In the example, let the service rate be $\mu = 1$, the retrial rate be $\alpha = 0.1$ and the arrival process be MMPP with 3 states and following parameters

$$\boldsymbol{\Lambda} = \begin{bmatrix} 0.780 & 0 & 0 \\ 0 & 1.014 & 0 \\ 0 & 0 & 1.170 \end{bmatrix}, \quad \mathbf{Q} = \begin{bmatrix} -0.5 & 0.2 & 0.3 \\ 0.1 & -0.3 & 0.2 \\ 0.3 & 0.2 & -0.5 \end{bmatrix}.$$

Table 1. Kolmogorov distances d for various values of the parameter ρ for $\sigma = 10$

ρ	$0.90 \cdot S$	$0.95 \cdot S$	$0.97 \cdot S$
d	0.087	0.044	0.027

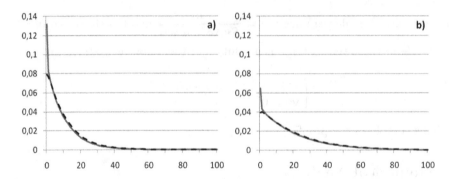

Fig. 2. Comparisons of the asymptotic (dashed line) and the simulation (solid line) distributions for $\sigma = 10$ and: (a) $\rho = 0.90 \cdot S$; (b) $\rho = 0.95 \cdot S$

Table 2. Kolmogorov distances d for various values of the parameter ρ for $\sigma = 2$

ρ	$0.90 \cdot S$	$0.95 \cdot S$	$0.97 \cdot S$
d	0.092	0.046	0.028

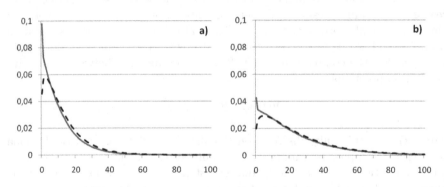

Fig. 3. Comparisons of the asymptotic (dashed line) and the simulation (solid line) distributions for $\sigma = 2$ and: (a) $\rho = 0.90 \cdot S$; (b) $\rho = 0.95 \cdot S$

For these parameters we have $\mathbf{r\Lambda e} = \mu = 1$. Thus, the parameter ρ defines the system load and has values $0 < \rho < S$, where $S = 1 + \alpha/\mu$ is the system throughput.

We vary parameters ρ and σ for demonstrating the application area of asymptotic analysis. The comparison of the distributions is shown in Fig. 2 (for $\sigma = 10$) and Fig. 3 (for $\sigma = 2$). We see that the asymptotic distributions are closer to the simulation for $\rho = 0.95$ (right figures) where probabilities $P(0)$ have less values. Values of Kolmogorov distance are presented in Tables 1, 2.

Note, we obtain the same results of the numerical comparison for different values of service, arrival and impatient parameters.

5   Conclusions

In this way in the paper, the single server retrial queue $MMPP/M/1$ with impatient calls is analysed under the heavy load condition. The retrial queue has non-Poisson arrivals and, as far as we know, such types of retrial queues have not been analytically studied in the literature yet. One more important distinguishing feature of the considered model is that we assume the rate of the calls patience to be dynamical.

We have proved that the asymptotic characteristic function of the number of calls in the orbit has the gamma distribution with parameters (2). By the numerical comparison of the asymptotic distribution and the simulation one, we show a good accuracy of the proposed approximation with the applicability area $\rho \geq 0.95 \cdot S$, where S is the system throughput.

Acknowledgments. The reported study was funded by RFBR according to the research project No. 19-41-703002.

References

1. Wilkinson, R.I.: Theories for toll traffic engineering in the USA. Bell Syst. Tech. J. **35**(2), 421–507 (1956). https://doi.org/10.1002/j.1538-7305.1956.tb02388.x
2. Cohen, J.W.: Basic problems of telephone traffic and the influence of repeated calls. Philips Telecommun. Rev. **18**(2), 49–100 (1957)
3. Elldin, A., Lind, G.: Elementary Telephone Traffic Theory. Ericsson Public Telecommunications, Stockholm (1971)
4. Gosztony, G.: Repeated call attempts and their effect on traffic engineering. Budavox Telecommun. Rev. **2**, 16–26 (1976)
5. Roszik, J., Sztrik, J., Kim, C.: Retrial queues in the performance modelling of cellular mobile networks using MOSEL. Int. J. Simul. **6**, 38–47 (2005)
6. Nazarov, A.A., Kuznetsov, D.Y.: Analysis of Non-Markovian models of communication networks with adaptive protocols of multiple random access. Autom. Remote Control **5**, 789–808 (2001)
7. Choi, B.D., Chang, Y.: Single server retrial queues with priority calls. Mathe. Comput. Modeling **30**, 7–32 (1999)
8. Tran-Gia, P., Mandjes, M.: Modeling of customer retrial phenomenon in cellular mobile networks. IEEE J. Sel. Areas Commun. **15**, 1406–1414 (1997). https://doi.org/10.1109/49.634781
9. Phung-Duc, T., Kawanishi, K.: An efficient method for performance analysis of blended call centers with redial. Asia-Pac. J. Oper. Res. **31**(2), 1–39 (2014). https://doi.org/10.1142/S0217595914400089
10. Aguir, S., Karaesmen, F., Askin, O.Z., Chauvet, F.: The impact of retrials on call center performance. OR Spektrum **26**, 353–376 (2004)
11. Kim, J.: Retrial queueing system with collision and impatience. Commun. Korean Math. Soc. **4**, 647–653 (2010)
12. Artalejo, J.R., Gómez-Corral, A.: Retrial Queueing Systems. A Computational Approach. Springer, Heidelberg (2008). https://doi.org/10.1007/978-3-540-78725-9
13. Artalejo, J.R., Falin, G.I.: Standard and retrial queueing systems: a comparative analysis. Revista Matematica Complutense **15**, 101–129 (2002)

14. Falin, G.I., Templeton, J.G.C.: Retrial Queues. Chapman & Hall, London (1997)
15. Yang, T., Posner, M., Templeton, J.: The M/G/1 retrial queue with non-persistent customers. Queueing Syst. **7**(2), 209–218 (1990)
16. Krishnamoorthy, A., Deepak, T.G., Joshua, V.C.: An M/G/1 retrial queue with non-persistent customers and orbital search. Stoch. Anal. Appl. **23**, 975–997 (2005). https://doi.org/10.1080/07362990500186753
17. Fayolle, G., Brun, M.A.: On a system with impatience and repeated calls. In: Queueing Theory and Its Applications: Liber Amicorum for J.W. Cohen, North Holland, Amsterdam, pp. 283–305 (1998)
18. Martin, M., Artalejo, J.R.: Analysis of an M/G/1 queue with two types of impatient units. Adv. Appl. Probab. **27**, 840–861 (1995). https://doi.org/10.2307/1428136
19. Aissani, A., Taleb, S., Hamadouche, D.: An unreliable retrial queue with impatience and preventive maintenance. In: Proceedings, 15th Applied Stochastic Models and Data Analysis (ASMDA2013), Mataró (Barcelona), Spain, pp. 1–9 (2013)
20. Kumar, M.S., Arumuganathan, R.O.: Performance analysis of single server retrial queue with general retrial time, impatient subscribers, two phases of service and bernoulli schedule. Tamkang J. Sci. Eng. **13**(2), 135–143 (2010)
21. Arrar, N.K., Djellab, N.V., Baillon, J.-B.: On the asymptotic behaviour of M/G/1 retrial queues with batch arrivals and impatience phenomenon. Math. Comput. Modell. **55**, 654–665 (2012). https://doi.org/10.1016/j.mcm.2011.08.039
22. Artalejo, J.R., Pozo, M.: Numerical calculation of the stationary distribution of the main multiserver retrial queue. Ann. Oper. Res. **116**, 41–56 (2002). https://doi.org/10.1023/A:1021359709489
23. Neuts, M.F., Rao, B.M.: Numerical investigation of a multiserver retrial model. Queueing Syst. **7**(2), 169–189 (1990). https://doi.org/10.1007/BF01158473
24. Shin, Y.W., Choo, T.S.: M/M/s queue with impatient customers and retrials. Appl. Math. Model. **33**, 2596–2606 (2009)
25. Dudin, A.N., Klimenok, V.I.: Queueing system $BMAP/G/1$ with repeated calls. Math. Comput. Modell. **30**(3–4), 115–128 (1999). https://doi.org/10.1016/S0895-7177(99)00136-3
26. Gómez-Corral, A.G.: A bibliographical guide to the analysis of retrial queues through matrix analytic techniques. Ann. Oper. Res. **141**, 163–191 (2006). https://doi.org/10.1007/s10479-006-5298-4
27. Diamond, J.E., Alfa, A.S.: Matrix analytical methods for M/PH/1 retrial queues. Stoch. Models **11**, 447–470 (1995). https://doi.org/10.1080/15326349508807355
28. Lopez-Herrero, M.J.: Distribution of the number of customers served in an M/G/1 retrial queue. J. Appl. Probab. **39**(2), 407–412 (2002). https://doi.org/10.1239/jap/1025131437
29. Nazarov, A., Moiseev, A.: Queueing network $MAP - (GI/\infty)^K$ with high-rate arrivals. Eur. J. Oper. Res. **254**, 161–168 (2016). https://doi.org/10.1016/j.ejor.2016.04.011
30. Pankratova, E., Moiseeva, S.: Queueing system MAP/M/∞ with n types of customers. Commun. Comput. Inf. Sci. **487**, 356–366 (2014). https://doi.org/10.1007/978-3-319-13671-4_41
31. Moiseeva, E., Nazarov, A.: Asymptotic analysis of RQ-systems M/M/1 on heavy load condition. In: Proceedings of the IV International Conference Problems of Cybernetics and Informatics, Baku, Azerbaijan, pp. 64–166 (2012)
32. Dudin, A., Nazarov, A., Kirpichnikov, A. (eds.): ITMM 2017. CCIS, vol. 800. Springer, Cham (2017). https://doi.org/10.1007/978-3-319-68069-9

33. Danilyuk, E.Y., Fedorova, E.A., Moiseeva, S.P.: Asymptotic analysis of an retrial queueing system M/M/1 with collisions and impatient calls. Autom. Remote Control **79**(12), 2136–2146 (2018). https://doi.org/10.1134/S0005117918120044
34. Vygovskaya, O., Danilyuk, E., Moiseeva, S.: Retrial queueing system of MMPP/M/2 type with impatient calls in the orbit. In: Dudin, A., Nazarov, A., Moiseev, A. (eds.) ITMM/WRQ-2018. CCIS, vol. 912, pp. 387–399. Springer, Cham (2018). https://doi.org/10.1007/978-3-319-97595-5_30
35. Danilyuk, E., Vygoskaya, O., Moiseeva, S.: Retrial queue M/M/N with impatient customer in the orbit. In: Vishnevskiy, V.M., Kozyrev, D.V. (eds.) DCCN 2018. CCIS, vol. 919, pp. 493–504. Springer, Cham (2018). https://doi.org/10.1007/978-3-319-99447-5_42
36. Falin, G.I.: M/G/1 queue with repeated calls in heavy traffic. Moscow Univ. Math. Bull. **6**, 48–50 (1980)
37. Anisimov, V.V.: Asymptotic analysis of reliability for switching systems in light and heavy traffic conditions. In: Limnios, N., Nikulin, M. (eds.) Recent Advances in Reliability Theory, vol. 8, pp. 119–133. Birkhäuser, Boston (1980). https://doi.org/10.1007/978-1-4612-1384-0_8
38. Stepanov, S.N.: Asymptotic analysis of models with repeated calls in case of extreme load. Probl. Inf. Transm. **29**(3), 248–267 (1993)
39. Neuts, M.F.: A Versatile Markovian point process. J. Appl. Probab. **16**(4), 764–779 (1979). https://doi.org/10.2307/3213143
40. Lucantoni, D.M.: New results on the single server queue with a batch Markovian arrival process. Stoch. Models **7**, 1–46 (1991). https://doi.org/10.1080/15326349108807174
41. Moiseev, A., Demin, A., Dorofeev, V., Sorokin, V.: Discrete-event approach to simulation of queueing networks. Key Eng. Mater. **685**, 939–942 (2016). https://doi.org/10.4028/www.scientific.net/KEM.685.939

Matrix Analytic Solutions for M/M/S Retrial Queues with Impatient Customers

Hsing Paul Luh$^{(\boxtimes)}$ (iD) and Pei-Chun Song

Department of Mathematical Sciences, National Chengchi University,
116 Taipei, Taiwan
slu@nccu.edu.tw

Abstract. In this paper, we investigate the nonhomogeneity of state space for solving retrial queues through the performance of the $M/M/S$ retrial system with impatient customers and S servers that is modeled under quasi-birth-and-death processes with level-dependent transient rates. We derive the analytic solution of multiserver retrial queues with orbit and develop an efficient method to solve this type of systems effectively. The methods proposed are based on nonhomogeneity of the state space although this queueing model was tackled by many researchers before. Under a weaker assumption in this paper, we study and provide the exact expression based on an eigenvector approach. Constructing an efficient algorithm for the stationary probability distribution by the determination of required eigenvalues with a specific accuracy, we develop streamlined matrices of state-balanced equations with the efficient implementation for computation of the performance measures.

Keywords: Quasi-birth-death process · Retrial queues ·
Matrix-geometric method · Eigenvalues

1 Introduction

In this paper, we consider a queueing system with retrials of arrivals following the human behavior that impatience users can abandon the system with certain probability after an unsuccessful retry. Retrial queues have been used to model a phenomenon in modern information and telecommunication systems that blocked customers may retry for service after a certain timeout (See [1,9] and reference therein). Many examples of retrial queues can be found in communication networks nowadays. By an $M/M/S$ model, Do in [5] presents the effect of retrials in data transfers along Internet where in retrial queues a customer who does not receive the allocation of a server joins the orbit and later initiates a request for service. The $M/M/S$ retrial queue has been analyzed by many researchers. However, the stationary probability distribution when the number of servers is larger than two can be only obtained using approximate techniques

Research funds from Taiwan, MOST 107-2221-E-004-007.

T. Phung-Duc et al. (Eds.): QTNA 2019, LNCS 11688, pp. 16–33, 2019.
https://doi.org/10.1007/978-3-030-27181-7_2

(e.g., [1,6]). Our goal in this study is to develop an effective method with a closed-form solution for solving this type of problems in which the number of servers is big.

The modeling of repeated attempts has been the subject of numerous investigations in queueing systems. In [7], it explains that two functional blocks are typically distinguished in models which consider retrials: a block that accommodates the servers and possibly a waiting queue, and a block where users that retry are accommodated, usually called a retrial orbit. Because the retrial rate among customers depends on the number of customers in the retrial orbit, when it is modeled by a Quasi-Birth-and-Death approach, it assumably shall build a nonhomogeneous and infinite state space. The term of nonhomogeneity of state space is used to describe the state transition probability, or the probability of increments/decrements, is not homogeneous instead it depends on the state over the studied system. When the state homogeneity condition does not hold for the case of multiserver retrial queues, the absence of closed-form solutions for the main performance characteristics is ineluctable. Either the finite truncated or generalized truncated methods may be used to replace the original infinite state space by a solvable state space, that is, a model where steady state probabilities can be computed. In this paper with an eigenvector approach, we investigate a computationally solvable with infinite state space to tackle this problem.

Falin and Templeton in [9] present necessary and sufficient conditions for ergodicity of the retrial queues with $M/M/S$. Falin in [8] presents an approximation which is based on the truncation of the state space at a sufficiently large level related to the number of customers in the orbit. Another approximation based on the homogenization of the model was pioneered by Neuts and Rao in [13], where the $M/M/S$ retrial queue is approximated by the multiserver retrial queue with the total retrial rate that does not depend on the number of customers in the orbit as long as the orbit contains the number of customers greater than the specified value N. Note that the discussion for the choice of N is presented in the book in [1] on retrial queues. With this assumption, the stationary probabilities of the $M/M/S$ retrial queue can be estimated by any algorithm of [3,12] based on the matrix-geometric method (MGM). Domenech-Benlloch et al. [7] consider a multiserver retrial queue with the impatient phenomenon of customers waiting in the orbit. They propose two different generalized truncated methods (called HM1 and HM2) based on the homogenization of the state space when beyond the number of customers in the retrial orbit. The steady-state probabilities of the multiserver retrial queue with impatient customers are approximated with a modified retrial queue where the retrial rate beyond a certain level only depends on the conditional mean value of the number of customers in the orbit. They in [7] also compared their methods with other well-known algorithms that belong to different categories in [2], showing that the proposed HM2 method outperforms previous approaches from the aspect of accuracy at the price of increasing computation cost. Based on the HM2 algorithm, Do et al. in [6] propose an approximation that first obtains the conditional mean value $E[L|L \geq N]$ of the number L of customers in the orbit under the condition $L \geq N$ which is the simple function of both the single eigenvalue and N where N is suggested in [7].

Our contributions allow an efficient computation for the stationary probability distribution and the performance measures. The research direction is to evaluate the maximal eigenvalue of $\mathbf{R}$ without actually having computed $\mathbf{R}$. Instead, we adopted an approach based on the nonhomogeneity of the state space and provide an efficient method with the time complexity of only $O(S)$ to compute the eigenvectors of matrix $\mathbf{R}$. Then, we develop simplified equations that allow the efficient implementation of the computation of the performance measures. With a given precision level $\epsilon > 0$, we may construct an efficient computation algorithm for solving the stationary probability distribution, which guarantees a specific accuracy for the computation of performance measures.

The paper is organized as follows. A nonhomogeneous quasi-birth-and-death queueing model with orbit for impatient customers is constructed in Sect. 2. Matrix analytic derivation is presented in Sect. 3. The algorithmic solution procedures are described in Sect. 4. In Sect. 5, numerical test examples are presented for comparison with the results in [6].

2   A Queueing System with Orbit for Impatient Customers

In the system under study, we consider a queueing system with S servers. Customers arrive according to a Poisson process with rate λ and upon encountering an available server, request an exponentially distributed service time with rate μ. Without loss of generality, assume that each customer occupies one resource unit. When a new request finds all servers occupied, it joins the retrial orbit immediately. There is an infinite capacity for the retrial orbit with a random service discipline. After a random time that is exponentially distributed of rate γ this customer retries, being a successful retrial if it finds a free server. Otherwise, the customer leaves the system with probability p or returns to the retrial orbit with probability $1 - p$ independently of the occupancy rate of the servers and start the retrial procedure again. Conventionally, denote by ρ the average load of the system.

The model considered here can be represented as a bi-dimensional continuous-time Markov chain (MC) whose state space is defined by the number of customers in the retrial orbit and the number of customers being served, constituting a Level Dependent Quasi-Birth-and-Death Process (LDQBD). In QBD related literature, the term level refers to a set of states with the same first coordinate. Consider a retrial queueing model with S homogeneous servers and impatient customers. Let a random variable $J(t)$ represent the number of occupied servers at time t, $0 \leq J(t) \leq S$. When $J(t) = S$, a customer joins the orbit in order to wait and retry. Let $L(t)$ be the number of customers in the orbit waiting for retrial at time t. Each customer retries with rate γ. A retrying customer either leaves the queue with probability p if all servers are busy upon the retrial or rejoins the orbit with probability $1 - p$. Note that a time between subsequent retrials of a specific customer follows the exponential distribution with parameter γ. The main characteristics of this model are its infinite state space $(L(t), J(t))$ and also its space of nonhomogeneity produced by the fact that the retrial rate depends on the number of customers in the retrial orbit.

Suppose the system is stable and $\lim_{t\to\infty} Pr(L(t), J(t))$ exists. This system can be represented by two-dimensional continuous-time Markov chain (CTMC) $X = \{L(t), J(t)\}$ with state space $\{0, 1, ...\} \times \{0, 1, ..., S\}$. We will use a two dimensional state space description where state (ℓ, j) denotes that the number of customers in orbit equals ℓ ($\ell = 0, 1, 2, \cdots$) and that j ($j = 0, 1, 2, \ldots, S$) servers are busy. Hence, the total effective retrial rate is $\ell\gamma$ when $L = \ell$. The infinitesimal generator of this process has an infinite block tridiagonal structure $\mathbf{Q}$ defined in (1). Let $m = S + 1$ and $\mathbf{e}_i$ be the row vector in which the ith component is 1, 0 elsewhere, $i = 1, 2, \cdots, m$. Denote by $\mathbf{e}_i^t$ the transpose of the vector $\mathbf{e}_i$. Construct a discrete-time and nonhomogeneously infinitesimal generator for X as the following $\mathbf{Q}$,

$$\mathbf{Q} \triangleq \begin{bmatrix} \mathbf{Q}_1^{(0)} & \mathbf{Q}_0^{(0)} & 0 & & \cdots \\ \mathbf{Q}_2^{(1)} & \mathbf{Q}_1^{(1)} & \mathbf{Q}_0^{(1)} & 0 & \cdots \\ 0 & \mathbf{Q}_2^{(2)} & \mathbf{Q}_1^{(2)} & \mathbf{Q}_0^{(2)} & \cdots \\ & & \cdots & \cdots & \cdots \end{bmatrix} \tag{1}$$

where

$$\mathbf{Q}_2^{(\ell)} = \ell\mathcal{B} \text{ for } \ell = 1, 2, \cdots$$
$$\mathbf{Q}_1^{(\ell)} = \mathcal{A} - D^{\mathcal{A}} - \mathbf{Q}_0^{(\ell)} - D^{(\ell)}, \text{ for } \ell = 0, 1, 2, \cdots$$
$$\mathbf{Q}_0^{(\ell)} = \lambda\mathbf{e}_m^t\mathbf{e}_m \quad \text{for } \ell = 0, 1, 2, \cdots$$

and

$$\mathcal{A} = \begin{bmatrix} 0 & \lambda & 0 & & & \\ \mu & 0 & \lambda & 0 & & \\ 0 & 2\mu & 0 & \lambda & & \\ 0 & \cdots & & & \cdots & \\ 0 & & (S-1)\mu & 0 & \lambda \\ & & & S\mu & 0 \end{bmatrix}_{m\times m}$$

$$\mathcal{B} = \begin{bmatrix} 0 & \gamma & 0 & & \\ 0 & 0 & \gamma & & \\ \vdots & & & \ddots & \\ 0 & & & & \gamma \\ 0 & & & 0 & \gamma p \end{bmatrix}_{m\times m}$$

and $D^{(\cdot)}$ denotes a diagonal matrix with the diagonal elements defined as

$$D^{\mathcal{A}}(i, i) = \sum_{k=0}^{S} \mathcal{A}(i, k), \quad i = 0, 1, 2, \ldots, S$$
$$D^{(\ell)}(i, i) = \sum_{k=0}^{S} \mathbf{Q}_2^{(\ell)}(i, k), \quad \ell = 0, 1, 2, \ldots$$
$$D^{(0)} = 0.$$

3   LDQBD Model Formulation

Let the steady-state probabilities of X be denoted by $\pi_{\ell,j} = \lim_{t \to \infty} \Pr(L(t) = \ell, J(t) = j)$. Define the row vector $\boldsymbol{\pi}_\ell = [\pi_{\ell,0}, ..., \pi_{\ell,s}]$, $\ell = 0, 1, 2, \dots$. Throughout the paper, we adhere to the convention, unless stated otherwise, that probability vectors are row vectors. $\mathbf{Q}$ in (1) is an irreducible stochastic matrix, its steady state probability vector associated to it is denoted by $\boldsymbol{\pi}$, and we partition it as $\boldsymbol{\pi} = (\boldsymbol{\pi}_0, \boldsymbol{\pi}_1, \boldsymbol{\pi}_2....)$, where $\boldsymbol{\pi}_\ell$, $\ell \geq 0$, is an m-vector. Being a stationary probability distribution, $\boldsymbol{\pi}$ satisfies $\boldsymbol{\pi}\mathbf{Q} = \mathbf{0}$ and $\boldsymbol{\pi}\mathbf{1} = 1$, where $\mathbf{0}$ is a zero matrix and $\mathbf{1}$ is a column of all 1.

An MC is said to be positive recurrent if the mean time to return to each state for the first time after leaving it is finite. In infinite QBD MCs, this requires that the drift to higher level states be smaller than the drift to lower level states. To preserve the stability of the system, i.e., the existence of the steady state probability distribution, the MC is assumed aperiodic as well.

Theorem 1 ([10]). *If the LDQBD process with a transition rate matrix given by (1) is irreducible, aperiodic, and positive recurrent, then there exist matrices* $\{\mathbf{R}^{(\ell)} : \ell \geq 1\}$ *such that*

$$\boldsymbol{\pi}_{(\ell+1)} = \boldsymbol{\pi}_\ell \mathbf{R}^{(\ell+1)}, \ \ell \geq 0$$

where the sequence $\{\mathbf{R}^{(\ell)}\}$ *is the minimal nonnegative solution of the set of equations given by*

$$\mathbf{Q}_0^{(\ell)} + \mathbf{R}^{(\ell+1)}\mathbf{Q}_1^{(\ell+1)} + \mathbf{R}^{(\ell+1)}\mathbf{R}^{(\ell+2)}\mathbf{Q}_2^{(\ell+2)} = \mathbf{0}, \ \ell \geq 0 \qquad (2)$$

The proof may be found in [10].

Let $|| \cdot ||_1$ denote a matrix norm by

$$||\mathbf{Z}||_1 = \max_{1 \leq j \leq m} \sum_{i=1}^{m} |[\mathbf{Z}]_{ij}|$$

where $\mathbf{Z}$ is an $m \times m$ matrix with its element $[\mathbf{Z}]_{ij}$ at the ith row and the jth column. With an extension of Theorem 1, we claim the following fact.

Corollary 1. *If* $\mathbf{Q}$ *is irreducible then for any* $\epsilon > 0$, *there exists a number* K *such that for all* $n > K$ *we have* $\frac{1}{n} \frac{||\mathbf{Q}_1^{(n)} - \mathbf{Q}_1^{(n+1)}||_1}{||\mathbf{Q}_1^{(n+1)}||_1} < \epsilon$, *if and only if* $\boldsymbol{\pi}_n \mathbf{1} < \epsilon$.

To extend the result from a stable queueing system, we have the following lemma in general.

Lemma 1. *If* $\mathbf{Q}$ *and* $\mathbf{A}^{(n)}$ *for* $n > K$ *are irreducible, where* $\mathbf{A}^{(n)} = \mathbf{Q}_0^{(n)} + \mathbf{Q}_1^{(n)} + \mathbf{Q}_2^{(n)}$, *then* $\mathbf{Q}$ *is positive recurrent if and only if* $\mathbf{p}(\mathbf{Q}_0^{(n)} - \mathbf{Q}_2^{(n)})\mathbf{1} < 0$, *where* $\mathbf{p}$ *satisfies* $\mathbf{p}\mathbf{A}^{(n)} = \mathbf{0}$ *and* $\mathbf{p}\mathbf{1} = 1$, *for all* $n > K$.

The proof could be done with a similar homogeneous case and can be found in [11].

Since we know for every $\ell > 0$

$$|[\mathbf{Q}_1^{(\ell)}]_{ii}| > \sum_{\substack{j=1 \\ j\neq i}}^{m} |[\mathbf{Q}_1^{(\ell)}]_{ij}| \text{ for all } i = 1, 2, \ldots, m,$$

it implies that $\mathbf{Q}_1^{(\ell)}$ is invertible. Note $\sup_i |[\mathbf{Q}_1^{(\ell)}]_{ii}| < \infty$ and

$$|[\mathbf{Q}_1^{(\ell)}]_{ii}| > \sum_{j=1}^{m} |[\mathbf{Q}_2^{(\ell)}]_{ij}| \text{ for all } i = 1, 2, \ldots, m.$$

By observation, given a fixed n, if the maximal eigenvalue of $(x\mathbf{Q}_2^{(n+2)}(\mathbf{Q}_1^{(n+1)})^{-1}) < 1$, then $(\mathbf{Q}_1^{(n+1)} + x\mathbf{Q}_2^{(n+2)})$ is invertible. Define $\mathbf{I}$ as the identity matrix. Consider (2), for $n > K$, and for $0 < x < 1$, claim that matrix $(\mathbf{Q}_1^{(n+1)} + x\mathbf{Q}_2^{(n+2)})$ is invertible. Because of $(\mathbf{Q}_1^{(n+1)} + x\mathbf{Q}_2^{(n+2)}) = (\mathbf{I} + x\mathbf{Q}_2^{(n+2)}(\mathbf{Q}_1^{(n+1)})^{-1})\mathbf{Q}_1^{(n+1)}$. If $||x\mathbf{Q}_2^{(n+2)}(\mathbf{Q}_1^{(n+1)})^{-1}||_1 < 1$, then $(\mathbf{I} + x\mathbf{Q}_2^{(n+2)}(\mathbf{Q}_1^{(n+1)})^{-1})$ exists. We write $(\mathbf{I} + x\mathbf{Q}_2^{(n+2)}(\mathbf{Q}_1^{(n+1)})^{-1})^{-1}$ as a power series in $x\mathbf{Q}_2^{(n+2)}(\mathbf{Q}_1^{(n+1)})^{-1}$. This gives

$$(\mathbf{Q}_1^{(n+1)} + x\mathbf{Q}_2^{(n+2)})^{-1} = (\mathbf{Q}_1^{(n+1)})^{-1}(\mathbf{I} + x\mathbf{Q}_2^{(n+2)}(\mathbf{Q}_1^{(n+1)})^{-1})^{-1}$$

$$= (\mathbf{Q}_1^{(n+1)})^{-1}\sum_{k=0}^{\infty}(-1)^k(x\mathbf{Q}_2^{(n+2)}(\mathbf{Q}_1^{(n+1)})^{-1})^k$$

Note $\mathbf{Q}_0^{(n)}$ is rank-1 since $\mathbf{Q}_0^{(n)} = \lambda \mathbf{e}_m^t \mathbf{e}_m$. Define

$$h^{(n)}(x) \triangleq -\lambda \mathbf{e}_m(\mathbf{Q}_1^{(n+1)} + x\mathbf{Q}_2^{(n+2)})^{-1}\mathbf{e}_m^t.$$

Moreover, $h^{(n)}(x)$ can be written as

$$h^{(n)}(x) = \sum_{k=0}^{\infty} c_k x^k, \quad 0 < x < 1,$$

where $c_k = (-1)^{k+1}\mathbf{e}_m(\mathbf{Q}_1^{(n+1)})^{-1}[\mathbf{Q}_2^{(n+2)}(\mathbf{Q}_1^{(n+1)})^{-1}]^k \lambda \mathbf{e}_m^t$.

Lemma 2. *Given* $(\mathbf{Q}_1^{(n+1)} + x\mathbf{Q}_2^{(n+2)})$ *is invertible, we have* $h^{(n)}(x) = x$, *such that* $0 < x < 1$ *for* $n > K$.

Proof. Under the stability in (2), we assume that

$$\mathbf{R}^{(n)} \to \mathbf{R}, \text{ for } n > K,$$

and there exists $0 < x < 1$ such that

$$\det[\mathbf{Q}_0^{(n)} + x\mathbf{Q}_1^{(n+1)} + x^2\mathbf{Q}_2^{(n+2)}] = 0.$$

Consider the following characteristic polynomial

$$
\begin{aligned}
&\det[\mathbf{Q}_0^{(n)} + x\mathbf{Q}_1^{(n+1)} + x^2\mathbf{Q}_2^{(n+2)}] \\
&= \det[\mathbf{Q}_0^{(n)} + x(\mathbf{Q}_1^{(n+1)} + x\mathbf{Q}_2^{(n+2)})] \\
&= \det[\mathbf{Q}_0^{(n)}(\mathbf{Q}_1^{(n+1)} + x\mathbf{Q}_2^{(n+2)})^{-1} + x\mathbf{I}]\det[\mathbf{Q}_1^{(n+1)} + x\mathbf{Q}_2^{(n+2)}].
\end{aligned}
$$

Since $\det[\mathbf{Q}_1^{(n+1)} + x\mathbf{Q}_2^{(n+2)}] \neq 0$, finding a zero of the characteristic polynomial is equivalent to finding a zero in the following equation,

$$
\det[\mathbf{Q}_0^{(n)}(\mathbf{Q}_1^{(n+1)} + x\mathbf{Q}_2^{(n+2)})^{-1} + x\mathbf{I}] = 0,
$$

$$
\det[1 + \mathbf{e}_m(\mathbf{Q}_1^{(n+1)} + x\mathbf{Q}_2^{(n+2)})^{-1}\frac{1}{x}\lambda\mathbf{e}_m^t] = 0,
$$

$$
-\mathbf{e}_m(\mathbf{Q}_1^{(n+1)} + x\mathbf{Q}_2^{(n+2)})^{-1}\mathbf{e}_m^t\lambda - x = 0,
$$

$$
h^{(n)}(x) - x = 0. \qquad\qquad\qquad\square
$$

Lemma 3. *When* $\mathbf{Q}_0^{(n)} = \lambda\mathbf{e}_m^t\mathbf{e}_m$ *is rank-1, there exists uniquely* x *satisfying* $x = h^{(n)}(x)$, $0 < x < 1$.

Proof. First, $[\mathbf{Q}_1^{(n)}]_{ii} < 0$ defined by (1), for all i, we have $c_0 = (-1)\mathbf{e}_m$ $(\mathbf{Q}_1^{(n)})^{-1}\lambda\mathbf{e}_m^t > 0$ and $c_1 = \mathbf{e}_m(\mathbf{Q}_1^{(n)})^{-1}(\mathbf{Q}_2^{(n)}(\mathbf{Q}_1^{(n)})^{-1})\lambda\mathbf{e}_m^t > 0$. By induction on k, we have $c_k > 0$, for all $k \geq 0$. The function is thus increasing for $0 < x < 1$, and $h(0) = c_0 = -\mathbf{e}_m(\mathbf{Q}_1^{(n)})^{-1}\lambda\mathbf{e}_m^t > 0$. In addition, one may check that $h'(x) > 0$, and $h''(x) > 0$. Second, claim $h(1) = 1$ in the following arguments. By Lemma 1 with $\mathbf{pA}^{(n)} = \mathbf{0}$ for $n > K$, we have $-\mathbf{p}(\mathbf{Q}_1^{(n)} + \mathbf{Q}_2^{(n)}) = \mathbf{pQ}_0^{(n)} = \lambda\mathbf{pe}_m^t\mathbf{e}_m$, implying that $\mathbf{p} = -\lambda\mathbf{pe}_m^t\mathbf{e}_m(\mathbf{Q}_1^{(n)} + \mathbf{Q}_2^{(n)})^{-1}$. Multiplying $\mathbf{e}_m^t$ from right on both sides, it gives $\mathbf{pe}_m^t = -\mathbf{pe}_m^t\mathbf{e}_m(\mathbf{Q}_1^{(n)} + \mathbf{Q}_2^{(n)})^{-1}\mathbf{e}_m^t\lambda$. Thus, it produces $\lambda\mathbf{e}_m(\mathbf{Q}_1^{(n)} + \mathbf{Q}_2^{(n)})^{-1}\mathbf{e}_m^t = -1$. Hence, we have that $h^{(n)}(1) = 1$. It is now clear that there is a unique solution of $x = h^{(n)}(x)$ between 0 and 1 when $n > K$. $\qquad\square$

Hence, x can be found when $n > K$ in the following equation,

$$
h^{(n)}(x) = x. \tag{3}
$$

Corollary 2. *Suppose* $\mathbf{Q}_0^{(\ell)}$ *is rank-1. A fixed point of (3) is an eigenvalue of* $\mathbf{R}$ *and* $\boldsymbol{\xi}$ *is the corresponding left eigenvector.*

The proof is straightforward.

Theorem 2. *If* $\mathbf{Q}_0^{(\ell)}$ *is rank-1 and* $\mathbf{Q}_2^{(\ell)}$ *is nonsingular, then for any* $\epsilon > 0$ *there exists a* K *such that* $\sum_{k=n}^{\infty}\pi_k < \epsilon$ *for* $n > K$ *and solve a fixed point for* $h^{(K)}(x) = x$.

The proof is easily obtained by the arguments provided in Lemma 2. For further discuss explicitly, denote that the fixed point is σ such that $h^{(K)}(\sigma) = \sigma$.

Define

$$\boldsymbol{\xi}^{(\ell)}(x) \triangleq -\lambda \mathbf{e}_m (\mathbf{Q}_1^{(\ell+1)} + x \mathbf{Q}_2^{(\ell+2)})^{-1}.$$

Since $\mathbf{Q}_0^{(\ell)} = \lambda \mathbf{e}_m^t \mathbf{e}_m$ is rank-1, let $\mathbf{R}^{(\ell)} = \mathbf{e}_m^t \boldsymbol{\xi}^{(\ell)}$. Suppose x_ℓ is an eigenvalue of $\mathbf{R}^{(\ell)}$ and the corresponding left eigenvector is $\boldsymbol{\xi}^{(\ell)}$, that is $\boldsymbol{\xi}^{(\ell)} \mathbf{R}^{(\ell)} = x_\ell \boldsymbol{\xi}^{(\ell)} = \boldsymbol{\xi}^{(\ell)} \mathbf{e}_m^t \boldsymbol{\xi}^{(\ell)}$. Thus, it implies $x_\ell = \boldsymbol{\xi}^{(\ell)} \mathbf{e}_m^t$. Similarly, $x_{\ell+1}$ and $\boldsymbol{\xi}^{(\ell+1)}$ is an eigenpair associated with $\mathbf{R}^{(\ell+1)}$. Consider (2) again and let it be written as

$$\mathbf{Q}_0^{(\ell)} + \mathbf{e}_m^t \boldsymbol{\xi}^{(\ell+1)} (\mathbf{Q}_1^{(\ell+1)} + \mathbf{e}_m^t \boldsymbol{\xi}^{(\ell+2)} \mathbf{Q}_2^{(\ell+2)}) = 0.$$

Multiplying $\mathbf{e}_m$ on both sides of the equation above, we have

$$\mathbf{e}_m \mathbf{Q}_0^{(\ell)} + \mathbf{e}_m \mathbf{e}_m^t \boldsymbol{\xi}^{(\ell+1)} (\mathbf{Q}_1^{(\ell+1)} + \mathbf{e}_m^t \boldsymbol{\xi}^{(\ell+2)} \mathbf{Q}_2^{(\ell+2)}) = 0$$
$$\lambda \mathbf{e}_m + \boldsymbol{\xi}^{(\ell+1)} (\mathbf{Q}_1^{(\ell+1)} + \mathbf{e}_m^t \boldsymbol{\xi}^{(\ell+2)} \mathbf{Q}_2^{(\ell+2)}) = 0.$$

Finally, we have

$$\boldsymbol{\xi}^{(\ell+1)} = -\lambda \mathbf{e}_m (\mathbf{Q}_1^{(\ell+1)} + \mathbf{e}_m^t \boldsymbol{\xi}^{(\ell+2)} \mathbf{Q}_2^{(\ell+2)})^{-1} \quad \text{for } \ell = 0, 1, \cdots, K-1. \quad (4)$$

Because $\boldsymbol{\xi}^{(\ell+1)}$ is a function of x and $\boldsymbol{\xi}^{(\ell+1)} \mathbf{e}_m^t = x_{\ell+1}$, we may decide $x_{\ell+1}$ by $\boldsymbol{\xi}^{(\ell+1)}(x) \mathbf{e}_m^t$ with assigning a x, $0 < x < 1$.

Consider $h^{(\ell)}(x) = \boldsymbol{\xi}^\ell(x) \mathbf{e}_m^t$ again but we are going to use it with $\boldsymbol{\xi}^{(\ell)}$ for $0 < \ell < K$.

Theorem 3. *If the Markov chain is positive recurrent, we have $0 < x < 1$ and $h^{(\ell)}(1) > 1$, for $0 < \ell < K$.*

Proof. From (4) we know

$$\mathbf{e}_m (\mathbf{Q}_1^{(\ell+1)} + (\mathbf{Q}_2^{(\ell+2)})^{-1} \mathbf{e}_m^t < \mathbf{e}_m (\mathbf{Q}_1^{(\ell+1)} + (\mathbf{Q}_2^{(\ell+1)})^{-1} \mathbf{e}_m^t,$$

and

$$-\mathbf{e}_m (\mathbf{Q}_1^{(\ell+1)} + (\mathbf{Q}_2^{(\ell+2)})^{-1} \mathbf{e}_m^t > -\mathbf{e}_m (\mathbf{Q}_1^{(\ell+1)} + (\mathbf{Q}_2^{(\ell+1)})^{-1} \mathbf{e}_m^t = 1.$$

Thus we have $h^{(\ell)}(1) > 1$. $\qquad \square$

Then it is shown in Theorem 3 that under certain irreducibility conditions, the value of the $h(x)$ in lies $(0,1)$, which may efficiently be solved and will be expressed by $h^{(\ell)}(x)$.

4 Deficient Matrix Approaches

In this section, we will focus on an efficient approach by taking into account π_0 at the boundary state when it solves the stationary probability π. We rewrite the system state balance equations as

$$
\left[\pi_0, \pi_1, \pi_2, \cdots\right]
\begin{bmatrix}
\mathbf{Q}_1^{(0)} & \mathbf{Q}_0^{(0)} & 0 & & 1 \\
\mathbf{Q}_2^{(1)} & \mathbf{Q}_1^{(1)} & \mathbf{Q}_0^{(1)} & \ddots & 1 \\
0 & \mathbf{Q}_2^{(2)} & \mathbf{Q}_1^{(2)} & \ddots & \vdots \\
& & \ddots & & \vdots
\end{bmatrix}
= [0, 0, \cdots, 0, 1].
$$

4.1 Boundary Equations and Eigenvector Approaches

Let $\mathbf{T} = \mathbf{Q}_1^{(0)} + \mathbf{e}_m^t \boldsymbol{\xi}^{(1)} \mathbf{Q}_2^{(1)}$ and $\boldsymbol{\xi}^{(0)} = -\lambda \mathbf{e}_m [(\mathbf{T} + \mathbf{e}_m^t \mathbf{e}_m)^{-1} - (\mathbf{T} + a\mathbf{e}_m^t \mathbf{e}_m)^{-1}]$. We consider a general structure of $\mathbf{T}$ in the following lemma.

Lemma 4. *Let $\mathbf{T}$ be an $m \times m$ matrix with rank $m - 1$, and $\mathbf{T}$ has no zero row or column. There exists a rank-1 matrix $\mathbf{P}$ that satisfies $\mathbf{TP} = \mathbf{PT} = 0$, which is determined uniquely only up to a constant.*

Proof. Let $S = m - 1$. Without losing of generality, we may assume $\{\mathbf{e}_1 \mathbf{T}, \mathbf{e}_2 \mathbf{T}, \cdots, \mathbf{e}_S \mathbf{T}\}$ are linearly independent. Since $\text{rank}(\mathbf{T}) = S$, there exist some constants $c_1, c_2, \cdots, c_S$ which are not all zero such that

$$
c_1 \mathbf{e}_1 \mathbf{T} + c_2 \mathbf{e}_2 \mathbf{T} + \cdots + c_S \mathbf{e}_S \mathbf{T} = \mathbf{e}_m \mathbf{T}. \tag{5}
$$

To prove $c_1, c_2, \cdots, c_S$ are uniquely determined. Suppose there are other numbers $d_1, d_2, \cdots, d_S$ such that $d_1 \mathbf{e}_1 \mathbf{T} + d_2 \mathbf{e}_2 \mathbf{T} + \cdots + d_S \mathbf{e}_S \mathbf{T} = \mathbf{e}_m \mathbf{T}$ which is also an expression of $\mathbf{e}_m \mathbf{T}$. By subtracting one from another, we obtain $(c_1 - d_1)\mathbf{e}_1 \mathbf{T} + (c_2 - d_2)\mathbf{e}_2 \mathbf{T} + \cdots + (c_S - d_S)\mathbf{e}_S \mathbf{T} = \mathbf{0}$. According to our assumption of linear independence of $\{\mathbf{e}_1 \mathbf{T}, \mathbf{e}_2 \mathbf{T}, \cdots, \mathbf{e}_S \mathbf{T}\}$, we have $c_i - d_i = 0$, $1 \leq i \leq S$. Hence, $c_i - d_i = 0$ for $1 \leq i \leq S$ and the expression is determined uniquely.

From (5), we have $(c_1, c_2, \cdots, c_S, -1)\mathbf{T} = 0$ which implies

$$
\begin{bmatrix}
c_1 & c_2 & c_3 & \cdots & c_S & -1 \\
c_1 & c_2 & c_3 & \cdots & c_S & -1 \\
& & \cdots\cdots\cdots & & \\
c_1 & c_2 & c_3 & \cdots & c_S & -1
\end{bmatrix}
\mathbf{T} = \mathbf{0}
$$

Similarly, suppose $\{\mathbf{T}\mathbf{e}_1^t, \mathbf{T}\mathbf{e}_2^t, \cdots, \mathbf{T}\mathbf{e}_S^t\}$ are linearly independent and there exist some constants $a_1, a_2, \cdots, a_S$ which are not all zero such that $a_1 \mathbf{T}\mathbf{e}_1^t + a_2 \mathbf{T}\mathbf{e}_2^t + \cdots + a_S \mathbf{T}\mathbf{e}_S^t = \mathbf{T}\mathbf{e}_m^t$. It implies that

$$
\mathbf{T}
\begin{bmatrix}
a_1 & a_1 & a_1 & \cdots & a_1 \\
a_2 & a_2 & a_2 & \cdots & a_2 \\
& & \cdots\cdots\cdots & \\
a_S & a_S & a_S & \cdots & a_S \\
-1 & -1 & -1 & \cdots & -1
\end{bmatrix}
= \mathbf{0}
$$

Based on two cases described, $\mathbf{P}$ may be written as

$$\mathbf{P} = c \times \begin{bmatrix} a_1c_1 & a_2c_1 & a_3c_1 & \cdots & a_Sc_1 & -c_1 \\ a_1c_2 & a_2c_2 & a_3c_2 & \cdots & a_Sc_2 & -c_2 \\ & & \cdots\cdots\cdots & & & \cdots \\ -a_1 & -a_2 & -a_3 & \cdots & -a_S & 1 \end{bmatrix}$$

where c is an arbitrary real number. This means $\mathbf{P}$ is uniquely determined up to a constant. It concludes that $\mathbf{TP} = \mathbf{PT} = \mathbf{0}$. □

Define $\mathbf{E}$ with a nonzero row vector $\mathbf{b}$ as follows

$$\mathbf{E} \overset{\Delta}{=} \mathbf{e}_m^t \mathbf{b}.$$

Consider the matrix $\mathbf{EPE}$ with its element at the ith row and the jth column, i.e., $[\mathbf{EPE}]_{i,j}$. We write

$$[\mathbf{EPE}]_{i,j} = \sum_{k=1}^{m} [\mathbf{E}]_{i,k} [\mathbf{PE}]_{k,j}$$
$$= \sum_{k=1}^{m} \sum_{\ell=1}^{m} [\mathbf{E}]_{i,k} [\mathbf{P}]_{k,\ell} [\mathbf{E}]_{\ell,j}$$
$$= \sum_{k=1}^{m} [\mathbf{E}]_{i,k} [\mathbf{P}]_{k,m} [\mathbf{E}]_{m,j}$$

Since $\mathbf{E}$ is a rank-1 matrix with the first S rows of zeros, we know that

$$\text{If } i \neq m, [\mathbf{EPE}]_{i,j} = 0$$
$$\text{If } i = m, [\mathbf{EPE}]_{m,j} = [\mathbf{EP}]_{m,m} [\mathbf{E}]_{m,j}$$

This implies $\mathbf{EPE} = [\mathbf{EP}]_{m,m} \times \mathbf{E}$, so we choose a proper r with $\mathbf{P}$ such that $[\mathbf{EP}]_{m,m} = 1$ and $\mathbf{EPE} = \mathbf{E}$. It is easy to see that by given a one may choose a proper r such that $\mathbf{EPE} = a\,\mathbf{E}$.

Suppose the first S columns or rows of $\mathbf{T}$ are linear independent. It produces for $a \neq 0$ that $\mathbf{T} + a\mathbf{e}_m^t\mathbf{b}$ is full rank and $(\mathbf{T} + a\,\mathbf{e}_m^t\mathbf{b})^{-1}$ exists. Suppose

$$(\mathbf{T} + a\,\mathbf{e}_m^t\mathbf{b})^{-1} \overset{\Delta}{=} \mathbf{P}_a + \mathbf{W}_a$$

where $\mathbf{P}_a$ satisfies $\mathbf{TP}_a = \mathbf{P}_a\mathbf{T} = \mathbf{0}$ and $\mathbf{W}_a$ denotes a matrix for the remainders with respect to $(\mathbf{T} + a\,\mathbf{e}_m^t\mathbf{b})^{-1}$.

Lemma 5. *Let* $(\mathbf{T} + a\mathbf{E})^{-1} = \mathbf{P}_a + \mathbf{W}_a$, *where* $\mathbf{TP}_a = \mathbf{P}_a\mathbf{T} = \mathbf{0}$ *and* $\mathbf{EP}_a\mathbf{E} = \mathbf{E}$, *then* $\mathbf{EW}_a = \mathbf{W}_a\mathbf{E} = \mathbf{0}$.

Proof. Since $\mathbf{TP}_a = \mathbf{P}_a\mathbf{T} = \mathbf{0}$, we may choose a proper r such that $\mathbf{EP}_a\mathbf{E} = \mathbf{E}$. Consider

$$(-\mathbf{E})[(\mathbf{T}+a\mathbf{E})^{-1}(\mathbf{T}+a\mathbf{E})] + \mathbf{E} = \mathbf{0}$$
$$(-\mathbf{E})(\mathbf{T}+a\mathbf{E})^{-1}\mathbf{T} - \mathbf{E}(\mathbf{T}+a\mathbf{E})^{-1}\mathbf{E} + \mathbf{E} = \mathbf{0}$$

$$\Rightarrow$$

$$-\mathbf{E}(\mathbf{T}+a\,\mathbf{E})^{-1}\mathbf{T}$$
$$= \mathbf{E}(\mathbf{T}+a\,\mathbf{E})^{-1}\mathbf{E} - \mathbf{E}$$
$$= \mathbf{EP}_a\mathbf{E} + \mathbf{EW}_a\mathbf{E} - \mathbf{E}$$
$$= \mathbf{EP}_a\mathbf{E} - \mathbf{E} + \mathbf{EW}_a\mathbf{E}$$
$$= \mathbf{EW}_a\mathbf{E} \text{ with a properly chosen } c.$$

On the other hand, we have a similar derivation in the following,

$$-\mathbf{E}(\mathbf{T}+a\,\mathbf{E})^{-1}\mathbf{T}$$
$$= -\mathbf{E}(\mathbf{P}_a + \mathbf{W}_a)\mathbf{T}$$
$$= -\mathbf{EW}_a\mathbf{T}$$

From the two expressions above, it gives that

$$\mathbf{EW}_a\mathbf{E} = -\mathbf{EW}_a\mathbf{T}$$
$$\mathbf{EW}_a(\mathbf{T}+\mathbf{E}) = \mathbf{0}$$
$$\mathbf{EW}_a = \mathbf{0}$$

For $(\mathbf{T}+\mathbf{E})$ is of the full rank, we can similarly acquire $\mathbf{W}_a\mathbf{E} = \mathbf{0}$.   □

Lemma 6. *Let $\mathbf{T}$ be an $m \times m$ matrix with rank $m - 1$, has no zero row or column, and $\mathbf{b}$ be a row vector satisfies that $\mathbf{T} + \mathbf{e}_m^t\mathbf{b}$ is full rank, then $[(\mathbf{T}+\mathbf{e}_m^t\mathbf{b})^{-1} - (\mathbf{T}+a\,\mathbf{e}_m^t\mathbf{b})^{-1}]\mathbf{T} = \mathbf{0}$ for all $a \neq 0$.*

Proof. Recall $\mathbf{E} = \mathbf{e}_m^t\mathbf{b}$ and consider

$$\mathbf{I} = (\mathbf{T}+\mathbf{E})(\mathbf{P}_1 + \mathbf{W}_1) = \mathbf{EP}_1 + \mathbf{TW}_1 \quad \text{with } a = 1,$$
$$\mathbf{I} = (\mathbf{T}+a\mathbf{E})(\frac{1}{a}\mathbf{P}_1 + \mathbf{W}_a) = \mathbf{EP}_1 + \mathbf{TW}_a$$

Combining the equations above we have

$$\mathbf{T}(\mathbf{W}_1 - \mathbf{W}_a) = \mathbf{0}.$$

Similarly, we have

$$(\mathbf{W}_1 - \mathbf{W}_a)\mathbf{T} = \mathbf{0}.$$

If $\mathbf{W}_1 = \mathbf{W}_a$, then we are done. Otherwise, it means $\mathbf{W}_1 - \mathbf{W}_a = \beta\mathbf{P}_1$ for some constant $\beta \neq 0$ by Lemma 4, then it produces

$$\beta\mathbf{EP}_1 = \mathbf{E}(\mathbf{W}_1 - \mathbf{W}_a) = \mathbf{0}.$$

Since $\mathbf{EP}_1 \neq \mathbf{0}$, we have $\beta = 0$, implying $\mathbf{W}_1 = \mathbf{W}_a$.   □

From Lemma 12, $\mathbf{e}_m[(\mathbf{T}+\mathbf{e}_m^t\mathbf{b})^{-1} - (\mathbf{T}+a\,\mathbf{e}_m^t\mathbf{b})^{-1}]\mathbf{T} = \mathbf{0}$, it provides a way to obtain $\boldsymbol{\xi}^{(0)}$, because $\mathbf{Q}_1^{(0)} + \mathbf{R}^{(1)}\mathbf{Q}_2^{(1)}$ plays the role of $\mathbf{T}$ in the equation. So we can set

$$\boldsymbol{\xi}^{(0)} = -\lambda\mathbf{e}_m[(\mathbf{Q}_1^{(0)} + \mathbf{e}_m^t\boldsymbol{\xi}^{(1)}\mathbf{Q}_2^{(1)} + \mathbf{e}_m^t\mathbf{b})^{-1} - (\mathbf{Q}_1^{(0)} + \mathbf{e}_m^t\boldsymbol{\xi}^{(1)}\mathbf{Q}_2^{(1)} + a\mathbf{e}_m^t\mathbf{b})^{-1}]. \quad (6)$$

for any $a \neq 0$, and any row vector $\mathbf{b}$ such that $\text{rank}(\mathbf{Q}_1^{(0)} + \mathbf{e}_m^t\boldsymbol{\xi}^{(1)}\mathbf{Q}_2^{(1)} + \mathbf{e}_m^t\mathbf{b}) = m$.

In our case, we set $\mathbf{b} = \mathbf{e}_m$ in (6).

4.2 LU Decomposition Approaches

In order to reduce the time complexity of matrix multiplication and inversion, we will adapt LU decomposition for computing an $m \times m$ matrix, namely

$$\mathbf{T} = \begin{bmatrix} [\mathbf{T}]_{0,0} & [\mathbf{T}]_{0,1} \\ [\mathbf{T}]_{1,0} & [\mathbf{T}]_{1,1} & [\mathbf{T}]_{1,2} \\ 0 & [\mathbf{T}]_{2,1} & [\mathbf{T}]_{2,2} & [\mathbf{T}]_{2,3} \\ \vdots & 0 & \ddots & \ddots & \ddots \\ \vdots & \vdots & & \ddots & \ddots & \ddots & \ddots \\ \vdots & \vdots & & & \ddots & \ddots & \ddots & \ddots \\ 0 & 0 & 0 & 0 & \cdots & [\mathbf{T}]_{S-1,S-2} & [\mathbf{T}]_{S-1,S-1} & [\mathbf{T}]_{S-1,S} \\ 0 & [\mathbf{T}]_{S,1} & [\mathbf{T}]_{S,2} & [\mathbf{T}]_{S,3} & \cdots & [\mathbf{T}]_{S,S-2} & [\mathbf{T}]_{S,S-1} & [\mathbf{T}]_{S,S} \end{bmatrix} \quad (7)$$

with its all diagonal elements are nonzero. Denote by $[\mathbf{T}]_{i,j}$ the element at the $(i+1)$th row and the $(j+1)$th column of matrix $\mathbf{T}$.

Let $\mathbf{L}$ and $\mathbf{U}$ be component matrices of LU decomposition of $\mathbf{T}$, where $\mathbf{L}$ is a lower triangular matrix and $\mathbf{U}$ is an unit upper triangular matrix, then $\mathbf{L}$ and $\mathbf{U}$ can be expressed as

$$\mathbf{L} = \begin{bmatrix} [\mathbf{L}]_{0,0} \\ [\mathbf{L}]_{1,0} & [\mathbf{L}]_{1,1} \\ 0 & [\mathbf{L}]_{2,1} & [\mathbf{L}]_{2,2} \\ \vdots & 0 & \ddots & \ddots \\ \vdots & \vdots & & \ddots & \ddots & \ddots \\ 0 & 0 & \cdots & 0 & [\mathbf{L}]_{S-1,S-2} & [\mathbf{L}]_{S-1,S-1} \\ 0 & [\mathbf{L}]_{S,1} & [\mathbf{L}]_{S,2} & \cdots & [\mathbf{L}]_{S,S-2} & [\mathbf{L}]_{S,S-1} & [\mathbf{L}]_{S,S} \end{bmatrix} \quad (8)$$

$$\mathbf{U} = \begin{bmatrix} 1 & [\mathbf{U}]_{0,1} & 0 & 0 & \cdots & \cdots & 0 \\ & 1 & [\mathbf{U}]_{1,2} & 0 & \cdots & \cdots & 0 \\ & & 1 & [\mathbf{U}]_{2,3} & 0 & \cdots & 0 \\ & & & \ddots & \ddots & \ddots & \vdots \\ & & & & 1 & [\mathbf{U}]_{S-2,S-1} & 0 \\ & & & & & 1 & [\mathbf{U}]_{S-1,S} \\ & & & & & & 1 \end{bmatrix} \quad (9)$$

From (8), (9), and $\mathbf{T} = \mathbf{LU}$, we can write the following equations.

$$\begin{cases} [\mathbf{T}]_{0,0} = [\mathbf{L}]_{0,0} \\ [\mathbf{T}]_{0,1} = [\mathbf{L}]_{0,0}[\mathbf{U}]_{0,1} \\ [\mathbf{T}]_{i,i-1} = [\mathbf{L}]_{i,i-1}, \quad 1 \le i \le S-1 \\ [\mathbf{T}]_{i,i} = [\mathbf{L}]_{i,i} + [\mathbf{L}]_{i,i-1}[\mathbf{U}]_{i-1,i}, \quad 1 \le i \le S-1 \\ [\mathbf{T}]_{i,i+1} = [\mathbf{L}]_{i,i}[\mathbf{U}]_{i,i+1}, \quad 1 \le i \le S-1 \\ [\mathbf{T}]_{S,1} = [\mathbf{L}]_{S,1} \\ [\mathbf{T}]_{S,i} = [\mathbf{L}]_{S,i} + [\mathbf{L}]_{S,i-1}[\mathbf{U}]_{i-1,i}, \quad 2 \le i \le S \end{cases}$$

which induces the Algorithm 1 with time complexity $\mathcal{O}(S)$.

Algorithm 1: LU decomposition

[Step 1] $[\mathbf{L}]_{0,0} = [\mathbf{T}]_{0,0}$, $[\mathbf{U}]_{0,1} = \frac{[\mathbf{T}]_{0,1}}{[\mathbf{L}]_{0,0}}$

[Step 2]   Compute recursively for $i = 1, 2, \cdots, S-1$

$$[\mathbf{L}]_{i,i-1} = [\mathbf{T}]_{i,i-1}$$

$$[\mathbf{L}]_{i,i} = [\mathbf{T}]_{i,i} - [\mathbf{L}]_{i,i-1}[\mathbf{U}]_{i-1,i}$$

$$[\mathbf{U}]_{i,i+1} = \frac{[\mathbf{T}]_{i,i+1}}{[\mathbf{L}]_{i,i}}$$

[Step 3] $[\mathbf{L}]_{S,1} = [\mathbf{T}]_{S,1}$

[Step 4]   Compute recursively for $i = 2, 3, \cdots, S$

$$[\mathbf{L}]_{S,i} = [\mathbf{T}]_{S,i} - [\mathbf{L}]_{S,i-1}[\mathbf{U}]_{i-1,i}$$

Lemma 7. *If $\mathbf{T}$ is of the special form as described in (7) and $\mathbf{L}, \mathbf{U}$ are the component matrices of LU decomposition of $\mathbf{T}$, then the last row of $\mathbf{T}^{-1}$ is the same as the last row of $\mathbf{L}^{-1}$, i.e. $\mathbf{e}_m \mathbf{T}^{-1} = \mathbf{e}_m \mathbf{L}^{-1}$.*

Proof. It is easy to check that $\mathbf{U}^{-1}$ is modified to an upper triangular matrix with all diagonal elements of one by Gaussian elimination.

This implies that the last row of $\mathbf{U}^{-1}$ is $\mathbf{e}_m \mathbf{U}^{-1} = [0, 0, \cdots, 0, 1] = \mathbf{e}_m$

Since $\mathbf{T}^{-1} = (\mathbf{LU})^{-1} = \mathbf{U}^{-1}\mathbf{L}^{-1}$, by multiply $\mathbf{e}_m$ on both sides, we obtain

$$\mathbf{e}_m \mathbf{T}^{-1} = \mathbf{e}_m \mathbf{U}^{-1}\mathbf{L}^{-1} = \mathbf{e}_m \mathbf{L}^{-1}$$

Hence, we have the result.   □

Now, define $\mathbf{e}_m \mathbf{T}^{-1} = \mathbf{e}_m \mathbf{L}^{-1} = [\ell_0, \ell_1, \ell_2, \cdots, \ell_S]$. Since $\mathbf{e}_m \mathbf{L}^{-1}\mathbf{L} = \mathbf{e}_m$, we obtain the following equation, i.e.,

$$\begin{cases} \ell_0[\mathbf{L}]_{0,0} + \ell_1[\mathbf{L}]_{1,0} = 0 \\ \ell_i[\mathbf{L}]_{i,i} + \ell_{i+1}[\mathbf{L}]_{i+1,i} + \ell_S[\mathbf{L}]_{S,i} = 0, \quad 1 \le i \le S-2 \\ \ell_{S-1}[\mathbf{L}]_{S-1,S-1} + \ell_S[\mathbf{L}]_{S,S-1} = 0 \\ \ell_S[\mathbf{L}]_{S,S} = 1. \end{cases}$$

Algorithm 2: Compute $\mathbf{e}_m^t \mathbf{L}^{-1}$

[Step 1] $\ell_S = \frac{1}{[\mathbf{L}]_{S,S}}$

[Step 2] $\ell_{S-1} = -\frac{[\mathbf{L}]_{S,S-1}}{[\mathbf{L}]_{S-1,S-1}}$

[Step 3]   Compute recursively for $i = S-2, S-1, \cdots, 1$

$$\ell_i = -\frac{[\mathbf{L}]_{i+1,i}}{[\mathbf{L}]_{i,i}} \ell_{i+1} - \frac{[\mathbf{L}]_{S,i}}{[\mathbf{L}]_{i,i}} \ell_S$$

[Step 4] $\ell_0 = -\frac{[\mathbf{L}]_{1,0}}{[\mathbf{L}]_{0,0}}$

which induces the Algorithm 2 with time complexity $\mathcal{O}(S)$.

Thus, replacing $\mathbf{T}$ by $(\mathbf{Q}_1^{(K)} + \sigma\mathbf{Q}_2^{(K)})$ or $(\mathbf{Q}_1^{(\ell)} + \mathbf{e}_m\boldsymbol{\xi}^{(\ell+1)}\mathbf{Q}_2^{(\ell+1)})$, the time complexity of solving $\boldsymbol{\xi}^{(\ell)}$ can be reduced from $\mathcal{O}(S^3)$ to $\mathcal{O}(S)$.

In summary, we present a general computing procedure for obtaining $\boldsymbol{\pi}$ in the following.

Algorithm 3: Computing stationary probabilities $\boldsymbol{\pi}$

[Step 0]   Let K be determined in Corollary 1 by a preset $\epsilon > 0$

[Step 1]   solve $x = -\lambda\mathbf{e}_m(\mathbf{Q}_1^{(K)} + x\mathbf{Q}_2^{(K)})^{-1}\mathbf{e}_m^t$
and $\boldsymbol{\xi}^{(K)} = -\lambda\mathbf{e}_m(\mathbf{Q}_1^{(K)} + \sigma\mathbf{Q}_2^{(K)})^{-1}, \sigma = x$

[Step 2] $\boldsymbol{\xi}^{(k)} = -\lambda\mathbf{e}_m(\mathbf{Q}_1^{(k)} + \mathbf{e}_m^t\boldsymbol{\xi}^{(k+1)}\mathbf{Q}_2^{(k+1)})^{-1}, \quad k = K-1, K-2, \cdots, 1$

[Step 3] $\boldsymbol{\xi}^{(0)} = -\lambda\mathbf{e}_m[(\mathbf{Q}_1^{(0)} + \mathbf{e}_m^t\boldsymbol{\xi}^{(1)}\mathbf{Q}_2^{(1)} + \mathbf{e}_m^t\mathbf{e}_m)^{-1} - (\mathbf{Q}_1^{(0)} + \mathbf{e}_m^t\boldsymbol{\xi}^{(1)}\mathbf{Q}_2^{(1)} + a\mathbf{e}_m^t\mathbf{e}_m)^{-1}]$

[Step 4] $\sigma_k = \boldsymbol{\xi}^{(k)}\mathbf{e}_m^t, \quad k = 0, 1, \cdots, K-1$

[Step 5] $s_k \stackrel{\Delta}{=} \prod_{i=0}^{k-1}\sigma_i, \quad s_0 = 1, \quad k = 0, 1, \cdots, K$

[Step 6] $\phi = (\sum_{k=0}^{K-1} s_k\boldsymbol{\xi}^{(k)}\mathbf{1} + \frac{s_K}{1-\sigma}\boldsymbol{\xi}^{(K)}\mathbf{1})^{-1}$

[Step 7] $\boldsymbol{\pi}_k = \phi s_k\boldsymbol{\xi}^{(k)}, \quad 0 \le k \le K$

[Step 8] $\boldsymbol{\pi}_k = (\sigma)^{k-K}\boldsymbol{\pi}_K, \quad k \ge K+1$.

The expected number of customers in the retrial orbit L_q can be determined by

$$L_q = \sum_{i=0}^{K-1} i\boldsymbol{\pi}_i\mathbf{1} + \sum_{i=K}^{\infty} i\sigma^{i-K}\boldsymbol{\pi}_K\mathbf{1} = \sum_{i=0}^{K-1} i\boldsymbol{\pi}_i\mathbf{1} + \boldsymbol{\pi}_K\mathbf{1}\{\frac{K}{1-\sigma} + \frac{\sigma}{(1-\sigma)^2}\}.$$

Define the effective retrial rate and the effective service rate as E_r and E_s, respectively. The performance measures are expressed by

$$E_r = \sum_{k=1}^{\infty} k\gamma\boldsymbol{\pi}_k\mathbf{1} = L_q\gamma,$$

$$E_s = \sum_{i=1}^{S}\sum_{k=1}^{\infty} i\boldsymbol{\pi}_k\mathbf{e}_i^t\mu.$$

5   Numerical Experiments

By our model, the computational effort of the suggested approach in Algorithm 3 is significantly reduced while the numerical stability associated with the computational procedure is controlled under a preset precision level. We will conduct numerical experiments on PC with Intel(R) Core(TM) i5-5200U CPU @ 2.20GHz for the proposed method with the test problems appeared in [6]. With the help of stationary distribution $\pi = (\pi_0, \pi_1, \pi_2, \ldots)$ of $M/M/S$ with impatient customers, we can compute the expected number of customers in the retrial queue and the expected number of customers at each level by treating ρ and S as the decision variables, e.g., $\lambda = \rho \times S \times \mu$. In experiments, we first use the test problems in [6] and comparing the results by computing N_{ret} which is denoted by L_q in our model, the blocking probability P_b, the delayed service probability P_{ds}, and the nonservice probability P_{ns}, i.e.,

$$P_b = \sum_{i=0}^{K-1} \pi_i e_m^t + \pi_K e_m^t \frac{1}{1-\sigma}$$

$$P_{ds} = \{L_q - \sum_{i=1}^{K} i\pi_i e_m^t - \pi_K e_m^t (\frac{K}{1-\sigma} + \frac{\sigma}{(1-\sigma)^2})\}\gamma/\lambda$$

$$P_{ns} = p\gamma\{\sum_{i=0}^{K} i\pi_i e_m^t + \pi_K e_m^t (\frac{K}{1-\sigma} + \frac{\sigma}{(1-\sigma)^2})\}/\lambda.$$

Referring to [6], the default value of number of examples are set as $S = 50$, 100, 200, 500, and 1000, $\mu = 1/180$, $\gamma = 0.01$, $p = 0,2$, $\epsilon = 10^{-5}$, respectively. We confirm the computing procedure and robustness of Algorithm 3. The performance measures are presented in particular for $S = 500$, $\rho = 1$, $\mu = 1/180$, $\gamma = 0.01$ in Fig. 1.

The main purpose of this paper is the development of an eigenpair approach that results in an efficient method to effectively solve retrial systems with customer impatience. This novel method is a continuing effort inspired in the previous research papers. The proposed algorithm depends on a series of eigenvalues and eigenvectors for nonhomogeneous QBD. The computational complexity is much lower because it only needs to solve an eigenvalue once and the remaining probabilities are attained by substitution. According to our experiments over 100 test problems including $S = 2000$, we found the computational complexity depends on ρ which confirms the observation in [6]. In specific, our test problems illustrate the relationship K which is denoted by N in [6] among other system parameters in Fig. 1. Therefore, we choose K by the following rule in our case rather than using Corollary 1 which only provides a rough upper bound in general.

Observation: There exists a $f(\rho)$ such that $\ln(K)$ is proportional to $f(\rho)\ln(S)$ where $f(\rho)$ may be written as

$$f(\rho) = \{ \begin{matrix} 0.3, & \text{if } \rho < 0.9 \\ 1.4\rho, & \text{if } \rho \geq 0.9. \end{matrix}$$

Table 1. Computational Time for $\lambda = \rho(\mu S), \mu = 1/180, \gamma = 0.01, p = 0.2$

ρ	$S = 50$		$S = 100$		$S = 200$		$S = 500$		$S = 1000$	
	K	Time (s)	K	Time (s)	K	Time (s)	K	Time (s)	K	Time (s)
0.4	10	0.001	10	0.001	10	0.004	10	0.010	10	0.021
	20	0.001	20	0.001	20	0.006	20	0.016	20	0.016
	30	0.001	30	0.003	30	0.006	30	0.014	30	0.020
	40	0.001	40	0.005	40	0.008	40	0.014	40	0.030
0.8	30	0.002	30	0.003	30	0.005	30	0.016	30	0.033
	40	0.002	40	0.003	40	0.007	40	0.026	40	0.033
	50	0.004	50	0.004	50	0.008	50	0.026	50	0.047
	60	0.003	60	0.005	60	0.009	60	0.03	60	0.047
1.0	50	0.009	80	0.008	120	0.033	200	0.097	350	0.288
	75	0.011	100	0.009	160	0.033	250	0.102	400	0.339
	100	0.01	120	0.009	200	0.039	300	0.124	450	0.352
	125	0.014	140	0.012	240	0.05	350	0.157	500	0.396
1.4	100	0.009	200	0.015	300	0.05	800	0.279	1000	0.937
	150	0.013	250	0.02	400	0.061	900	0.332	1500	1.368
	200	0.014	300	0.018	500	0.07	1000	0.35	2000	1.648
	250	0.01	350	0.029	600	0.084	1100	0.378	2500	2.146

Although this method is one of the generalized truncated methods, we believe our method can be used in many cases in engineering problems where the matrix $\mathbf{Q}_0^{(\ell)}$ has only one non-zero row of which examples are found in [4,14,15]. We expect that this method will outperform the previous proposals in terms of accuracy for the most common performance parameters used in retrial systems and under a wide range of scenarios in applications (Tables 1 and 2).

Table 2. Computational Time for $\lambda = \rho(\mu S), \mu = 10, \gamma = 1.6, p = 0.15$

ρ	$S = 100$		$S = 200$		$S = 500$		$S = 1000$		$S = 2000$	
	K	Time (s)	K	Time (s)	K	Time (s)	K	Time (s)	K	Time (s)
0.8	60	0.003	50	0.01	50	0.021	40	0.027	40	0.060
0.9	170	0.01	150	0.031	100	0.009	90	0.07	80	0.124
0.95	320	0.02	350	0.055	320	0.11	270	0.19	190	0.287
1.0	560	0.043	810	0.15	1510	0.507	2390	1.853	3760	6.119

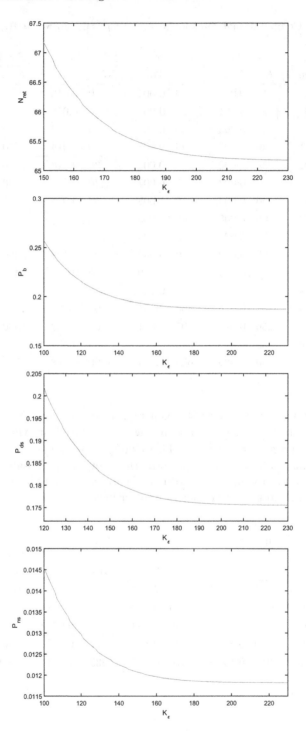

Fig. 1. Stats v.s. K for $S = 500, \rho = 1.0, \mu = 1/180, \gamma = 0.01$

References

1. Artalejo, J.R., Gómez-Corral, A.: Retrial Queueing Systems. Springer, Heidelberg (2008). https://doi.org/10.1007/978-3-540-78725-9
2. Artalejo, J.R., Pozo, M.: Numerical calculation of the stationary distribution of the main multiserver retrial queue. Ann. Oper. Res. **116**(1–4), 41–56 (2002)
3. Bini, D.A., Favati, P., Meini, B.: Matrix-Analytic Methods in Stochastic Models. Springer, New York (2013). https://doi.org/10.1007/978-1-4614-4909-6
4. Bright, L., Taylor, P.G.: Calculating the equilibrium distribution in level dependent quasi-birth-and-death processes. Stoch. Models **11**(3), 497–525 (1995)
5. Do, T.V.: Solution for a retrial queueing problem in cellular networks with the fractional guard channel policy. Math. Comput. Modell. **53**(11–12), 2058–2065 (2011)
6. Do, T.V., Do, N.H., Zhang, J.: An enhanced algorithm to solve multiververer retrial queueing systems with impatient customers. Comput. Ind. Eng. **65**, 719–728 (2013). https://doi.org/10.1016/j.cie.2013.04.008
7. Domenech-Benlloch, M.J., Gimenez-Guzman, J.M., Pla, V., Martinez-Bauset, J., Casares-Giner, V.: Generalized truncated methods for an efficient solution of retrial systems. Mathe. Probl. Eng. (2008). http://eudml.org/doc/55480
8. Falin, G.I.: On sufficient conditions for ergodicity of multichannel queueing systems with repeated calls. Adv. Appl. Prob. **16**, 447–448 (1984)
9. Falin, G.I., Templeton, J.G.C.: Retrial Queues. Chapman & Hall, London (1997)
10. Kharoufeh, J.P.: Wiley Encyclopedia of Operations Research and Management Science (2011). https://doi.org/10.1002/9780470400531.eorms0460
11. Latouche, G., Ramaswami, V.: Introduction to Matrix Analytic Methods in Stochastic Modelling. ASA & SIAM. SIAM, Philadelphia (1999)
12. Neuts, M.F.: Matrix-Geometric Solutions in Stochastic Models. The John Hopkins University Press, Baltimore (1981)
13. Neuts, M.F., Rao, B.M.: Numerical investigation of a multiserver retrial model. Queueing Syst. **7**, 169–189 (1990)
14. Ramaswami, V., Latouche, G.: A General Class of Markov processes with explicit matrix-geometric solution. OR Spectrum **8**, 209–218 (1986)
15. Tijms, H.C., van Vuuren, D.J.: Markov Processes on a semi-infinite strip and the geometric tail algorithm. Ann. Oper. Res. **113**, 133–140 (2002)

A Coupling-Based Analysis
of a Multiclass Retrial System
with State-Dependent Retrial Rates

Evsey Morozov[1,2] and Taisia Morozova[2]

[1] Institute of Applied Mathematical Research Karelian Research Centre RAS Center,
Petrozavodsk, Russia
emorozov@karelia.ru
[2] Petrozavodsk State University, Petrozavodsk, Russia
tiamorozova@mail.ru

Abstract. We study a multiclass single-server retrial system with independent Poisson inputs and the *state-dependent* retrial rates. Meeting busy server, a new class-i customer joins orbit i. Orbit i is working as a FIFO-type queueing system, in which the top customer retries to occupy server. The retrial times are exponentially distributed with a rate depending on the current configuration of the binary states of all orbits, *idle* or *non-idle*. We present a new coupling-based proof of the necessary stability conditions of this retrial system, found earlier in the paper [17]. The key ingredient of the proof is a coupling of the processes of retrials with the corresponding independent Poisson processes. This result allows to apply classic property PASTA in the following performance analysis. A few numerical results verifying stability conditions of a 3-class system are included as well.

Keywords: Retrial system · Multiclass coupled-orbit queues · Stability condition · Coupling · Regeneration · Simulation

1 Introduction

In this work, we provide a coupling-based proof of the necessary stability conditions of a single-server retrial multiclass queueing systems with state-dependent retrial rates, called *coupled orbit queues* [7–9]. The new proof allows to apply the key property PASTA ("Poisson arrivals see time average") for equating time-average and customer-average limits and, as a result, to obtain the corresponding stationary probabilities. This in turn allows to obtain the necessary stability condition by an analogy with analysis of queueing Markov processes by means of Kolmogorov equations.

We assume that there are N classes of customers following independent Poisson inputs. Meeting server busy, a class-i customer joins a virtual class-i orbit and then attempts to occupy server for transmission. Under *coupled orbit queues*

© Springer Nature Switzerland AG 2019
T. Phung-Duc et al. (Eds.): QTNA 2019, LNCS 11688, pp. 34–50, 2019.
https://doi.org/10.1007/978-3-030-27181-7_3

policy, the retrial rate of a given orbit depends on the *binary states* of other orbits. By this reason (and not to confuse it with the *coupling method* applied in the present work), we call such a system also the system with *state-dependent retrial rates*. This setting is well-motivated by modelling wireless multiple access systems, in particular, relay-assisted cognitive cooperative systems in which users transmit packets to a common destination node, and the orbit queues play a role of relay nodes [23]. In general, such models are motivated by increase the impact of wireless interference, see details in [7–9]. This work is devoted mainly to a new proof of the necessary stability conditions of the coupled orbit queues, and by this reason we will not further motivate this setting. Instead, the interested readers, besides the mentioned above works [7–9], can see papers [14,21,22], where further references can be found as well.

The service times of customers are assumed to have general class-dependent distribution. In a few previous works [16–18], a regenerative performance analysis of the single-server system with state-dependent retrial rates has been developed as well as the numerical verification of the stability conditions of such models was performed. In work [16], where the regenerative performance and stability analysis of coupled orbit queues has been applied for the first time, the retransmission rate of each orbit i had only two values: μ_i^*, if *all other orbits* are idle, and μ_i when at least one orbit $j \neq i$ is not idle. In the following works [17,18] the approach suggested in [16] has been extended to much more general model in which retrial rate of a given orbit in general depends on each possible *configuration* of other orbits. By configuration we mean a fixed set of the binary states of orbits: *busy or idle*.

In the framework of the Markovian models of retrial systems, performance analysis is typically performed using Kolmogorov equations describing the detailed dynamics of the process, for instance see [4,13]. In this regard we mention the fundamental works [1–3,11,12], devoted to analysis and bibliography of the retrial systems.

However, to analyze non-Markov models, we have applied the *regenerative approach* and local balance equations in works [16–18], to obtain some steady-state characteristics, bounds and stability conditions of the models with state-dependent retrial rates.

In this research we again apply the regenerative approach and balance equations for each orbit to develop an *alternative proof* of the necessary stability stationary and performance analysis of the corresponding models. This unified approach is based on the *coupling arguments* and the property PASTA [5]. A challenging problem of the performance analysis of such a model (and the retrial models in general) is a complicated structure of the successful attempts of orbital customers. The key ingredient of the approach proposed in this work is the *coupling* of the process of retrial attempts from each orbit with an *independent Poisson process*. This then allows to apply the property PASTA to the process of successful attempts to equate the limiting fraction of customers which meet server busy and the (limiting) fraction of the busy time of server. This novel approach is the *main contribution* of the present research, and it has a promising potential in stability and performance analysis of a wide class of the retrial systems.

In previous works, both necessary and sufficient stability conditions of the systems with state-dependent retrial rates have been established [8,16,17]. It is worth mentioning that all performed experiments indicated that sufficient condition seems to be excessive and the necessary stability condition solely is stability criterion. In this regard, we now numerically verify the necessary stability condition for each orbit separately. As a result, we establish that violation of the sufficient stability condition indeed leads to instability of the system (for a set of the system parameters). The idea used to select the corresponding set of parameters (and described in Sect. 5 after formula (25)) is another contribution of the present research.

As a by-product of our analysis, we obtain the coupling-based proof of some steady-state results for the multiclass retrial system with class-dependent constant retrial rates, *independent* of the states of other orbits [15], see Remark 1 at the end of Sect. 4.

The paper is organized as follows. In Sect. 2 we describe the model under consideration. In Sect. 3, preliminary results are shortly given, including previous proof of the necessary stability conditions. Section 4 contains the new coupling-based proof of these conditions. In Sect. 5, the proof of sufficient stability conditions is outlined. Finally, in Sect. 6, we present simulation results for $N = 3$ classes of customers to verify stability conditions, including symmetric system, when all orbits behave similarly. For the first time we demonstrate examples in which violation of the necessary stability conditions implies instability of the orbits. We could not construct such examples in previous works [17,18], and the new examples are based on a delicate analysis of the mutual influence of the orbits.

2   Model Description

We study a single-server retrial queueing model with N classes of customers following independent Poisson inputs. Class-i customers have input rate λ_i, $1 \leq i \leq N$. Alternatively, we may assume a Poisson input with rate $\lambda = \sum_{i=1}^{N} \lambda_i$, and a new arrival is class-i customer with the probability $p_i =: \lambda_i/\lambda$, $1 \leq i \leq N$. Let $\{t_n, n \geq 1\}$ be the arrival instants of the input with rate λ, with interarrival times $\tau_n = t_{n+1} - t_n$. Throughout the paper, we omit the corresponding serial index to denote a generic element of an independent identically distributed (iid) sequence. Thus generic interarrival time τ is exponentially distributed with parameter λ. It is assumed that service times of class-i customers, $\{S_n^{(i)}, n \geq 1\}$ are iid with service rate $\gamma_i = 1/\mathsf{E}S^{(i)}$, $1 \leq i \leq N$. We stress that (generic) service time $S^{(i)}$ has *general class-dependent* distribution. If a class-i customer meets server busy, then it joins (virtual) FIFO orbit queue i. In this context, discipline FIFO means that the head customer from orbit i retries until he occupies server, while other customers blocked in orbit i are waiting "in a queue". The attempts from orbit i follow exponential distribution with a rate *depending on the current status of other orbits: idle or non-idle.*

For the following analysis we need definitions from [17]. For each i, we define the set $\mathcal{G}(i)$ of N-dimensional vectors,

$$J_n^{(i)} = \{j_{n,1}^{(i)}, \ldots, j_{n,i-1}^{(i)}, 1, j_{n,i+1}^{(i)}, \ldots, j_{n,N}^{(i)}\},$$

with binary components $j_{n,k}^{(i)} \in \{0,1\}$, if $k \neq i$, while the ith component $j_{n,i}^{(i)} = 1$ always. We assume that $\mathcal{G}(i) = \{J_n^{(i)}, 1 \leq n \leq 2^{N-1}\}$ is the ordered set (say, in the lexicographical order), where index n denotes the nth element of this set. Each vector $J_n^{(i)}$ is called *configuration* and has the following interpretation: if the kth orbit is non-idle, then we put $j_{n,k}^{(i)} = 1$, otherwise, $j_{n,k}^{(i)} = 0$. For a given configuration $J_n^{(i)}$, we denote $\mu_n^{(i)}$ the retransmission rate of orbit i. Thus, the set $\mathcal{G}(i)$ contains all possible different configurations of the orbits "observed from orbit i", and each configuration from this set describes a fixed set of the states of orbits: busy $= 1$ or idle $= 0$, provided orbit i is *busy*. Also we denote the set of rates

$$\mathcal{M}_i = \{\mu_n^{(i)} : J_n^{(i)} \in \mathcal{G}(i)\}, \tag{1}$$

of all configurations belonging to $\mathcal{G}(i)$, $1 \leq i \leq N$. In general, different configurations have different retransmission rates but it is possible that, for given i, the set $\mathcal{M}_i$ contains repetitive elements, say $\mu_k^{(i)} = \mu_l^{(i)}$. It means that the retrial rate of orbit i is insensitive to the switching between configurations $J_k^{(i)}$ and $J_l^{(i)}$.

Our analysis is based on the *regenerative approach* [5]. To describe the *regenerative structure* of the retrial system, we define the basic stochastic processes. Let $N_i(t)$ be the number of class-i customers blocked in orbit i and $W_i(t)$ be the remaining workload in orbit i, at instant t^-, $1 \leq i \leq N$. In other words, $W_i(t)$ is the time which is required to serve all class-i customers blocked at instant t. Denote $N(t) = \sum_i N_i(t)$ the number of customers in all orbits at instant t. Let $Q(t) = 1$ if server be busy at instant t^-, and $Q(t) = 0$, otherwise. Denote

$$X(t) = N(t) + Q(t), \ t \geq 0 \quad \text{and} \quad X(t_n) = X_n, \ n \geq 1.$$

We will consider the zero initial state when the 1st customer arrives in the idle system at instant $t_1 = 0$. It is well-known that the process $X = \{X(t), t \geq 0\}$ regenerates when a new customer *meets an empty system* [5]. More exactly, regenerations are defined as follows:

$$T_{n+1} = \inf\left(t_k > T_n : X_k = 0\right), \ n \geq 0, \ T_0 := 0.$$

After each regeneration, the process starts anew independently of the pre-history. It implies that the "fragments" of the paths between regenerations, that is $\{X(t) : T_n \leq t < T_{n+1}\}$, $n \geq 0$, are iid random elements [5,24]. The regeneration periods $T_{n+1} - T_n$ are iid as well, distributed as a random variable T. The regenerative process X is called *positive recurrent* if $\mathsf{E}T < \infty$, and, when the input is Poisson (as in our setting), positive recurrence implies *stationarity* of the system, that is the existence of the stationary distribution of $X(t)$ as $t \to \infty$.

3   Preliminary Results

Before to present a new proof of the necessary stability conditions of the system, we give results obtained in [17], which are used in the following analysis. Moreover, we recall in brief the proof of the necessary stability conditions found in [17] and in [16] (for a less general model), to demonstrate the difference between these proofs. Note that the new proof can also be applied to reprove stability conditions of a two-way communication system with multiple classes of incoming customers and outgoing calls, obtained in [15].

Denote $A_i(t)$ the number of class-i arrivals in the interval of time $[0, t)$. Then the work which is required to process all class-i customers arriving in the interval $[0, t)$ is defined as

$$V_i(t) = \sum_{n=1}^{A_i(t)} S_n^{(i)}, \ 1 \le i \le N. \tag{2}$$

Also denote $S(t)$ the remaining service time of a customer at instant t^- ($S(t) = 0$, if the server is free). Denote $B_i(t)$ the time when the server is occupied by class-i customers, in the interval $[0, t]$; then $\sum_i B_i(t) = B(t)$ is the total busy time of server. Finally, denote the traffic intensity for each class, and the summary traffic intensity, respectively,

$$\rho_i = \frac{\lambda_i}{\gamma_i}, \quad \rho = \sum_{i=1}^{N} \rho_i.$$

It has been shown in [16,17], using the balance equations

$$V_i(t) = \sum_{i=1}^{N} W_i(t) + S(t) + B_i(t), \ i = 1, \ldots, N,$$

the Strong Law of Large Numbers (SLLN), and representation (2), that the stationary probability the *server is occupied by a class-i customer* is obtained as the following limit with probability (w.p.) 1,

$$\mathsf{P}_b^{(i)} = \lim_{t \to \infty} \frac{B_i(t)}{t} = \rho_i, \ 1 \le i \le N. \tag{3}$$

Then the stationary *busy probability of the server* is

$$\mathsf{P}_b = \sum_i \mathsf{P}_b^{(i)} = \rho. \tag{4}$$

Denote the indicator $I_k^{(i)} = 1$, if the kth class-i customer joins orbit i, and $I_k^{(i)} = 0$, otherwise. Let $A_i^{(0)}(t)$ be the number of customers joining orbit i, and $D_i(t)$ the number of customers leaving orbit i, respectively, in the interval $[0, t)$. (Note that $D_i(t)$ is the number of the successful attempts from orbit i). The proof of the stability conditions in [17] is based on the asymptotic analysis of the following balance relations,

$$A_i^{(0)}(t) = N_i(t) + D_i(t), \ 1 \le i \le N, \tag{5}$$

between the input to orbit i and the output from orbit i, in the interval $[0, t)$. In this analysis developed in [16,17], to find the limit $D_i(t)/t$ as $t \to \infty$, we have applied the following approach. Define, in interval $[0, t]$, the time when the system has configuration $J_n^{(i)}$ and server is free,

$$T_n^{(i)}(t) = \int_0^t 1(J_i(u) = J_n^{(i)}, \mathcal{I}(u) = 1)du,$$

where 1 is the indicator function, the random variable $J_i(t)$ represents the current configuration of the system, "observed" from the non-idle orbit i, at instant t; $\mathcal{I}(u) = 1$, if the server is idle at instant u, and $\mathcal{I}(u) = 0$ otherwise. Then

$$T_0^{(i)}(t) = \sum_{n:J_n^{(i)} \in \mathcal{G}(i)} T_n^{(i)}(t),$$

is the time, in the interval $[0, t]$, when *server is idle and orbit i is busy*. In the papers [16,17], we defined (in slightly another notation) the Poisson process $\hat{D}_n^{(i)}(t)$ of retrials from orbit i provided that the retrial system has configuration $J_n^{(i)}$ during whole interval $[0, t]$. Evidently, then

$$\frac{\hat{D}_i(T_n^{(i)}(t))}{T_n^{(i)}(t)} \to \mu_n^{(i)}, \ t \to \infty, \tag{6}$$

while the *time-average* limit,

$$\lim_{t \to \infty} \frac{T_n^{(i)}(t)}{t} =: \mathsf{P}_n^{(i)}, \tag{7}$$

represents the stationary probability that the *system has configuration $J_n^{(i)}$ and server is idle*. Thus,

$$\sum_{n:J_n^{(i)} \in \mathcal{G}(i)} \mathsf{P}_n^{(i)} = \mathsf{P}_0^{(i)} \tag{8}$$

is the stationary probability that *server is idle and orbit i is busy*. The further analysis in [16,17] is based on the replacement of the relations (5) by the following *stochastic equality*

$$A_i^{(0)}(t) = \sum_{k=1}^{A_i(t)} I_k^{(i)} =_{st} N_i(t) + \sum_{n:J_n^{(i)} \in \mathcal{G}(i)} \hat{D}_i(T_n^{(i)}(t)), \tag{9}$$

and using (6)–(8).

4 Coupling-Based Proof of the Necessary Stability Condition

In this section, we present the *new proof* of Theorem 1 below, which as we mention in Sect. 3, has been proved in [17] by another way. Before to explain why we present the new proof, we recall the *classic* property PASTA [25]. This property states that the long-run *fraction of Poisson arrivals* which meet the system in a particular state equals the long-run *fraction of time* the system is in that state. In the proof in [17], using Eq. (9), we equate the limiting *fraction of customers* joining orbit i, and the limiting *fraction of time* when server is busy (and orbit i is busy). However the limit (7) in general has not such interpretation because the process of the attempts from orbit in general *is not Poisson*. This is the main reason why we develop a new proof in the present work. (Note that, to justify the approach applied in [16,17], it seems promising to use the so-called *conditional PASTA* [10]). To overcome the mentioned difficulty, we use a coupling to connect the process of retrials from a given orbit with an *independent Poisson process*. In turn, this construction further allows to apply the property PASTA, to establish the equality of the corresponding limits. Also note that, in the new proof, we replace the last term in (9) by a sum of indicators by analogy with representation of $A_i^{(0)}(t)$, see (13), (14) below. In our opinion, this new proof can be useful to analyze a wide class of the retrial systems. (An example is a general multicalss system with constant retrial rates studied in the paper [19]; another example is the multiclass system with outgoing calls considered in [15]).

Introduce the *maximal* and the *minimal* possible retrial rate from orbit i, respectively,

$$\hat{\mu}_i = \max_{n:J_n^{(i)} \in \mathcal{G}(i)} \mu_n^{(i)}, \ \mu_i^0 = \min_{n:J_n^{(i)} \in \mathcal{G}(i)} \mu_n^{(i)}, \ 1 \leq i \leq N.$$

Theorem 1. If the system under consideration is positive recurrent, then,

$$\rho \leq \min_{1 \leq i \leq N} \left[\frac{\hat{\mu}_i}{\lambda_i + \hat{\mu}_i} \right]. \tag{10}$$

Proof. As in [17], using the renewal theory, we find that w.p.1,

$$\lim_{t \to \infty} \frac{1}{t} A_i^{(0)}(t) = \lambda_i \rho, \ 1 \leq i \leq N. \tag{11}$$

Denote $\hat{D}_n^{(i)} = \{\hat{D}_n^{(i)}(t), \ t \geq 0\}$ the Poisson process with rate $\mu_n^{(i)}$, that is

$$\mathsf{P}(\hat{D}_n^{(i)}(t) = k) = e^{-\mu_n^{(i)}t} \frac{[\mu_n^{(i)}t]^k}{k!}, \ k \geq 0; \ t \geq 0, \ 1 \leq i \leq N. \tag{12}$$

We may assume that this process is generated by the attempts from (non-idle) orbit i if the system has permanent configuration $J_n^{(i)}$. For each i, introduce the family of *independent* processes

$$\mathcal{D}_i = \{\hat{D}_n^{(i)} : J_n^{(i)} \in \mathcal{G}(i)\},$$

and let $\{z_n^{(i)}(k), \, k \geq 1\}$ be the instants of the process $\hat{D}_n^{(i)}$. It is assumed that the families $\{\mathcal{D}_i, \, i = 1, \ldots, N\}$ are independent as well. Denote $D_i(t)$ the number of actual attempts from orbit i in interval $[0, \, t)$. Now, using a coupling, we connect the process $D_i = \{D_i(t), \, t \geq 0\}$ with the family of the Poisson processes $\mathcal{D}_i$. At instant $t = 0$, we start to sample all Poisson processes $\hat{D}_n^{(i)}$ from the family $\mathcal{D}_i$, and this procedure is simultaneously performed for all orbits, that is for all families $\{\mathcal{D}_i, \, i = 1, \ldots, N\}$. From now on we fix an arbitrary i. Then, the first configuration, say $J_n^{(i)}$, appears when the 1st class-i customer is blocked on orbit i at some instant $v_{1,n}^{(i)}$. (Recall that the zero initial state is assumed, while configurations are defined for non-idle orbit i). At the instant $v_{1,n}^{(i)}$ we replace the remaining exponential time (in the interval covering instant $v_{1,n}^{(i)}$) in the process $\hat{D}_n^{(i)}$ by a new independent exponential variable with rate $\mu_n^{(i)}$. (This procedure is called *resampling*). Note that by construction, a switching between configurations happens if and only if an orbit switches between idle/busy or busy/idle states. Then, after instant $v_{1,n}^{(i)}$, we *synchronize* (take identical) exponential intervals in both processes $D_n^{(i)}$ and $\hat{D}_n^{(i)}$, that is synchronize the process of retrials from orbit i with the corresponding Poisson process. (Namely this synchronization is called a *coupling* [5]). This synchronization lasts until the current configuration $J_n^{(i)}$ switches, at some instant $v_{2,n}^{(i)}$, to the next configuration $J_k^{(i)}$, say. Then, at the instant $v_{2,n}^{(i)}$, we interrupt the current interval in the process $\hat{D}_k^{(i)}$ and synchronize the following intervals in process $\hat{D}_k^{(i)}$ and process D_i of real attempts, and so on. Analogously, this procedure is applied to all other orbits. Thus, by construction, the instants of the actual retrials D_i constitute a *subsequence* of the renewal points of the corresponding processes from the family $\mathcal{D}_i$ *with resampling*. We emphasize that the resampling does not change distribution of the Poisson processes $\mathcal{D}_i$, and we will keep the notation $\hat{D}_n^{(i)}$ for the resampled process and the notation $\{z_n^{(i)}(k), \, k \geq 1\}$ for its instants. Now, for each i, denote

$$Q(z_n^{(i)}(k)) = Q_n^{(i)}(k), \; J_i(z_n^{(i)}(k)) = J_n^{(i)}(k), \; n \geq 1.$$

Then, in particular, the equality

$$1(Q_n^{(i)}(k) = 0, \; J_n^{(i)}(k) = J_n^{(i)}) = 1$$

means that the kth instant of the Poisson process $\hat{D}_n^{(i)}$ is the instant of the *successful attempt* of a class-i orbital customer, which occurs while configuration $J_n^{(i)}$ takes place. (If $1(Q_n^{(i)}(k) = 1, \; J_n^{(i)}(k) = J_n^{(i)}) = 1$, that is server is busy, then the attempt is *unsuccessful* and indeed can be ignored). Then the number of successful attempts $D_n^{(i)}(t)$ from orbit i in the interval $[0, t)$, when the system has configuration $J_n^{(i)}$, is defined as

$$D_n^{(i)}(t) = \sum_{k=1}^{\hat{D}_n^{(i)}(t)} \mathbb{1}(Q_n^{(i)}(k) = 0,\ J_n^{(i)}(k) = J_n^{(i)}). \tag{13}$$

Consequently, the number of departures from orbit i within this interval is

$$D_i(t) = \sum_{n: J_n^{(i)} \in \mathcal{G}(i)} D_n^{(i)}(t),\ 1 \le i \le N. \tag{14}$$

Recall that $\mathsf{P}_n^{(i)}$ is the stationary probability that *server is idle and system has configuration* $J_n^{(i)}$, obtained as *time-average* limit (7). On the other hand, by construction of the process $\hat{D}_n^{(i)}(t)$, we can apply the property PASTA and see that the following *event-average* limit coincides with the limit (7):

$$\lim_{t \to \infty} \frac{\sum_{k=1}^{\hat{D}_n^{(i)}(t)} \mathbb{1}(Q_n^{(i)}(k) = 0,\ J_n^{(i)}(k) = J_n^{(i)})}{\hat{D}_n^{(i)}(t)} = \mathsf{P}_n^{(i)}. \tag{15}$$

Thus, by (12), (15), it follows that, for $1 \le i \le N$, as $t \to \infty$,

$$\frac{D_n^{(i)}(t)}{t} = \frac{\sum_{k=1}^{\hat{D}_n^{(i)}(t)} \mathbb{1}(Q_n^{(i)}(k) = 0,\ J_n^{(i)}(k) = J_n^{(i)})}{\hat{D}_n^{(i)}(t)} \cdot \frac{\hat{D}_n^{(i)}(t)}{t} \to \mathsf{P}_n^{(i)} \mu_n^{(i)}. \tag{16}$$

Now, by (14), (16), we obtain

$$\lim_{t \to \infty} \frac{D_i(t)}{t} = \sum_{n: J_n^{(i)} \in \mathcal{G}(i)} \mu_n^{(i)} \mathsf{P}_n^{(i)},\ 1 \le i \le N. \tag{17}$$

It remains to note that, in the positive recurrent case, $N_i(t) = o(t)$, $t \to \infty$ [24], and that, by (5), (11), (17), the following relation holds for each i:

$$\lambda_i \mathsf{P}_b = \lambda_i \rho = \sum_{n: J_n^{(i)} \in \mathcal{G}(i)} \mu_n^{(i)} \mathsf{P}_n^{(i)}. \tag{18}$$

Note now that the stationary probability $\mathsf{P}_0^{(i)}$ that *server is idle and orbit i is busy*, defined in (8), connects, for each i, with the stationary *idle server* probability P_0 as follows:

$$\mathsf{P}_0 = \mathsf{P}_0^{(i)} + \mathsf{P}_{00}^{(i)}, \tag{19}$$

where $\mathsf{P}_{00}^{(i)}$ is the stationary probability that *both server and orbit i are idle*. Denote π_0 the stationary probability that the *system is completely empty*.

By the positive recurrence, it then follows from the regeneration theory [5] and from PASTA that

$$\pi_0 = \frac{1}{ET} > 0.$$

Evidently, $\pi_0 \leq \mathsf{P}_{00}^{(i)}$, and then it follows from (19) that $\mathsf{P}_0 > \mathsf{P}_0^{(i)}$. Finally, by (18), the following strict inequality holds,

$$\lambda_i \rho \leq \hat{\mu}_i \mathsf{P}_0^{(i)} < \hat{\mu}_i \mathsf{P}_0 = \hat{\mu}_i (1 - \rho), \ 1 \leq i \leq N, \tag{20}$$

implying (10).

By (18), we also have the opposite inequality

$$\mathsf{P}_0^{(i)} \leq \frac{\lambda_i}{\mu_i^0} \rho, \ 1 \leq i \leq N.$$

Then, by (19), we obtain the following lower bound for the probability $\mathsf{P}_{00}^{(i)}$:

$$\mathsf{P}_{00}^{(i)} \geq \mathsf{P}_0 - \frac{\lambda_i}{\mu_i^0} \rho = 1 - \rho(1 + \frac{\lambda_i}{\mu_i^0}), \ 1 \leq i \leq N. \tag{21}$$

Remark 1. If the retrial rates are insensitive to the configurations, that is $\mu_n^{(i)} \equiv \mu_i$ for all $J_n^{(i)}$, then the system becomes a conventional *constant retrial rates multiserver system* [20]. Hence, Theorem 1 can be applied to reprove the necessary stability conditions of such a system found in [15]. Moreover, in this case, the first inequality in (20) becomes equality, implying the following explicit expressions (also see [16,17]):

$$\mathsf{P}_0^{(i)} = \frac{\lambda_i}{\mu_i} \rho, \ 1 \leq i \leq N. \tag{22}$$

5 Discussion of Sufficient Stability Condition

The complete proof of the sufficient stability condition is indeed quite similar to that is given for the multiclass retrial system with constant (state-independent) retrial rates in [19]. By this reason, in this section, we only outline the proof of the sufficient stability condition of the system with the state-dependent retrial rates.

For each i, introduce the iid sequence $\{S_n^{(i)}, \ n \geq 1\}$, the service times of class-i customers. Denote the indicator $I_n^{(i)} = 1$, if the nth arrival belongs to class i, and $I_n^{(i)} = 0$, otherwise, that is $\mathsf{E}I_n^{(i)} = p_i, \ 1 \leq i \leq N$. Then, the service time of the nth arrival can be represented as

$$S_n = \sum_{i=1}^{N} I_n^{(i)} S_n^{(i)}, \ n \geq 1, \tag{23}$$

where the sequences $\{I_n^{(i)}\}$ and $\{S_n^{(i)}\}$ are independent. In the retrial model under consideration, the server has an idle time after each departure. It allows us to construct a *dominating buffered system*, denoted $\hat{\Sigma}$, as follows. Using coupling, we take the same interarrival times (the same input flow). Then, in the system $\hat{\Sigma}$, for each class-i customer, we assign, additionally to the original service time, an extra exponential service time $\zeta_i = \min(\tau, \xi_i)$, where ξ_i is exponential random variable with the (minimal) parameter μ_i^0. In other words, ξ_i is the distance between the *"slowest"* attempts from orbit i. The exponential variable ζ_i has parameter $\lambda + \mu_i^0$ and is an upper bound of the idle time of server after departure, provided orbit i is not empty. Then it follows that the (generic) service time $\hat{S}^{(i)}$ in the system $\hat{\Sigma}$ satisfies the following stochastic equality $\hat{S}^{(i)} =_{st} S^{(i)} + \zeta_i$, $i = 1, \ldots, N$. Further, using the corresponding coupling and an induction, we show that (i) it is possible to serve customers in both systems in the *same order*, and (ii) that a customer leaves the system $\hat{\Sigma}$ *not earlier* than the same customer in the original system [19]. This monotonicity property implies that the workload (remaining work) in the original system is dominated by the workload in the system $\hat{\Sigma}$. In turns, it means that the positive recurrence of the system $\hat{\Sigma}$ implies positive recurrence of the original system. On the other hand, a standard *negative drift* condition for the buffered system $\hat{\Sigma}$ to be stable (positive recurrent) is $\mathsf{E}\hat{S} < \mathsf{E}\tau$. Recall that $\mathsf{E}S^{(i)} = 1/\gamma_i$, $p_i = \lambda_i/\lambda$. Then, after some algebra and using (23), we obtain the negative drift condition as follows

$$\rho + \max_{1 \le i \le N} \frac{\lambda}{\mu_i^0 + \lambda} < 1. \tag{24}$$

We rewrite this *sufficient stability* condition in the following form

$$\rho < \min_i \left(\frac{\mu_i^0}{\lambda + \mu_i^0} \right), \tag{25}$$

to compare it with the necessary condition (10).

It is worth mentioning that in all previous experiments given in [17,18], simulations have shown that the orbits remain stable if the necessary stability condition (10) holds, while the sufficient stability condition (25) is violated. On the other hand, the presence of the maximal retrial rates in (20) indicates that condition (10) *cannot be stability criterion*. To motivate this conclusion, we assume that the rate $\hat{\mu}_i$ is achieved for a configuration which *rarely happens*, while orbit i is heavily loaded. Then, one can expect that orbit i, working as a queuing system with capacity less than $\hat{\mu}_i$ most of time, cannot process incoming traffic, implying instability of orbit i. In the next section, this observation is used to select such retrial rates for which sufficient condition (25) is false, but condition (10) still holds, and it implies instability of the orbits. Moreover, it was detected in [17] that if condition (25) is violated, while condition (10) holds, then both stability and instability of separate orbits may happen. To "localize" the behaviour of the orbits, it seems reasonable to check stability of orbit i depending on whether the *local* condition

$$\rho \le \frac{\hat{\mu}_i}{\lambda_i + \hat{\mu}_i}, \tag{26}$$

is fulfilled for orbit i, or not. In the next section, we present numerical results confirming these observations.

6   Simulation Results

In this section, we present the results of a few numerical experiments to illustrate the statements given above. We simulate the *system with 3 classes of customers* and estimate each orbit size $N_i(t)$. In all experiments we simulate 600 arrivals and then average orbit sizes over 30 independent sample paths. In all figures, the black, grey and (grey) dotted curve corresponds to the 1st, 2nd and 3rd orbit, respectively, and the axis t counts the number of "events" (arrivals, departures, attempts) used in the *discrete-event simulation algorithm*.

Define the measure expressing the proximity between the load and the upper bound present in the local necessary condition (26) for orbit i:

$$\Delta_i = \frac{\hat{\mu}_i}{\lambda_i + \hat{\mu}_i} - \rho, \ i = 1, 2, 3. \tag{27}$$

Recall that, for the system with three orbits, the capacity of each set $\mathcal{M}_i$ and $\mathcal{G}(i)$ equals 4 [17]. Following the comments in previous section, we now choose the retrial system parameters in such a way to get Δ_i rather small.

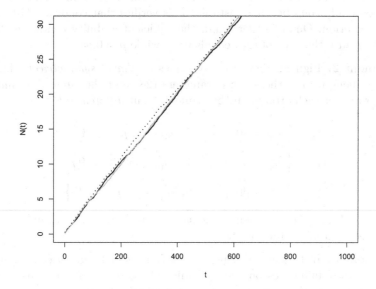

Fig. 1. Condition (25) is violated, condition (10) holds, all symmetrical orbits are unstable.

Experiment 1. Figure 1 demonstrates the dynamics of the *symmetric* orbits for the following input and service rates,

$$\lambda_i = 2, \ \gamma_i = 7, \ i = 1, 2, 3, \tag{28}$$

and the following retrial rates

$$\mathcal{M}_1 = \left\{ \mu_{00}^1 = 30, \ \mu_{10}^1 = 5, \ \mu_{01}^1 = 5, \ \mu_{11}^1 = 5 \right\},$$

$$\mathcal{M}_2 = \left\{ \mu_{00}^2 = 30, \ \mu_{10}^2 = 5, \ \mu_{01}^2 = 5, \ \mu_{11}^2 = 5 \right\},$$

$$\mathcal{M}_3 = \left\{ \mu_{00}^3 = 30, \ \mu_{10}^3 = 5, \ \mu_{01}^3 = 5, \ \mu_{11}^3 = 5 \right\}. \tag{29}$$

Notation μ_{kj}^i reflects the status of a fixed configuration. For instance, $\mu_{00}^1 = 30$ is the retrial rate of (non-idle) orbit 1 when orbits 2 and 3 are idle; $\mu_{10}^2 = 5$ is the rate of orbit 2, when orbit 1 is busy and orbit 3 is idle, and so on. For the selected parameters, all orbits are *symmetrical* and have similar behaviour. (For more detail on symmetric orbits see [18]). With these parameters we also have $\rho = 0.85$, and the r.h.s. of condition (10) equals 0.93. As a result, the necessary stability condition (26) holds true for each i and

$$\Delta_i = 0.08, \ \ i = 1, 2, 3. \tag{30}$$

The maximal rate $\hat{\mu}_i = 30$ corresponds to the configuration when, for each orbit i, both other orbits are *empty*. But the latter event seems to be "rare" because Δ_i in (30) are small and thus all orbits are expected to be heavily loaded. As Fig. 1 shows, all orbits become unstable. This verifies that condition (10) is not stability criterion. On the other hand, the sufficient stability condition (25) is violated because the r.h.s of (25) equals 0.71, while $\rho = 0.85$.

Experiment 2. Figure 2 shows the dynamics of *almost symmetrical* orbits. In this experiment we use the same parameters (28) and the *same maximal and minimal* rates for each orbit as in Experiment 1, but *different other retrial rates*:

$$\mathcal{M}_1 = \left\{ \mu_{00}^1 = 30, \ \mu_{10}^1 = 5, \ \mu_{01}^1 = 8, \ \mu_{11}^1 = 10 \right\},$$

$$\mathcal{M}_2 = \left\{ \mu_{00}^2 = 30, \ \mu_{10}^2 = 10, \ \mu_{01}^2 = 5, \ \mu_{11}^2 = 9 \right\},$$

$$\mathcal{M}_3 = \left\{ \mu_{00}^3 = 30, \ \mu_{10}^3 = 5, \ \mu_{01}^3 = 15, \ \mu_{11}^3 = 25 \right\}. \tag{31}$$

In this case all calculations remain the same as in Experiment 1, and we again observe the instability of all orbits.

Experiment 3. In this experiment, we verify how violation of condition (26) for *given orbit* i influences on the dynamics of *this* orbit. In this case we use parameters $\lambda_i \equiv 1, \ \gamma_i \equiv 4$, and *non-symmetrical* retrial rates

$$\mathcal{M}_1 = \left\{ \mu_{00}^1 = 2, \ \mu_{10}^1 = 2, \ \mu_{01}^2 = 20, \ \mu_{11}^1 = 1 \right\},$$

$$\mathcal{M}_2 = \left\{ \mu_{00}^2 = 7, \ \mu_{10}^2 = 5, \ \mu_{01}^2 = 5, \ \mu_{11}^2 = 3 \right\},$$

$$\mathcal{M}_3 = \left\{ \mu_{00}^3 = 10, \ \mu_{10}^3 = 9, \ \mu_{01}^3 = 5, \ \mu_{11}^3 = 5 \right\}.$$

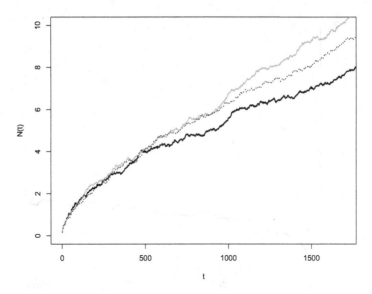

Fig. 2. Condition (25) is violated, condition (10) holds, all (non-symmetrical) orbits are unstable.

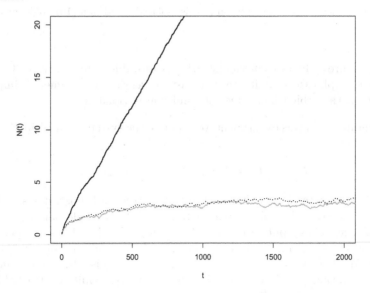

Fig. 3. Condition (25) is violated, condition (26) holds for orbits 2,3 and is violated for orbit 1.

Then the load coefficients are $\rho_i \equiv 0.25$, $\rho = 0.75$, while the r.h.s. of (25) is 0.25. Hence the sufficient condition (25) is violated. On the other hand, the r.h.s. of (26) is $0.66, 0.88, 0.90$ for orbit $i = 1, 2, 3$, respectively. Because $0.66 < \rho = 0.75 < 0.88$, then condition (26) is violated for orbit 1 and satisfied for orbits

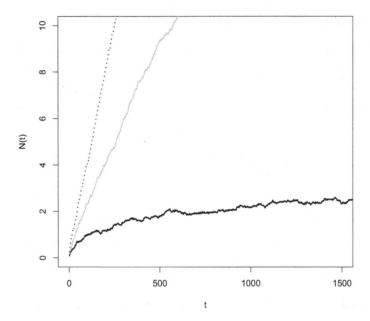

Fig. 4. Weibull service time: condition (25) is violated, condition (10) holds, orbits are unstable.

2 and 3. Figure 3 shows that the 1st orbit is unstable, while orbits 2,3 remain stable. We emphasize that it was not easy to select the parameters implying instability of the orbit when sufficient condition is violated.

Experiment 4. In this experiment we simulate the system with Weibull service time distribution

$$F(x) = 1 - e^{-(\frac{x}{k})^{\beta}}, \; x \geq 0; \; k > 0, \; \beta > 0,$$

with scale parameter k and shape parameter β. Figure 4 illustrates simulation results for the case when shape parameters $\beta_i = 2, 2, 3$, scale parameters $k_i = 0.5, 0.1, 0.1$, and the input rates $\lambda_i = 1, 2, 3$, for orbit $i = 1, 2, 3$, respectively. The retrial rates remain the same as in (29). As a result, we obtain that the r.h.s of (25) equals 0.54, $\rho = 0.88$ and the r.h.s. of (10) is 0.91. Thus, sufficient stability condition (25) is violated but the necessary condition (10) holds. As Fig. 4 shows, in this case all orbits are unstable.

In summary, the experiments confirm that stability criterion of the retrial system under consideration is located in the interval

$$\left[\min_{i} \frac{\mu_i^0}{\lambda + \mu_i^0}, \; \min_{i} \frac{\hat{\mu}_i}{\lambda_i + \hat{\mu}_i} \right],$$

in which stability/instability of orbit i may depend on whether condition (26) is fulfilled for this orbit or not.

7   Conclusion

We consider a single-server N-class retrial system with independent Poisson inputs and the state-dependent retrial rates (also called the system with *coupled orbits*). A class-i customer, meeting busy server, joins an infinite capacity orbit i, $1 \leq i \leq N$. The head customer in orbit i attempts to occupy server after an exponentially distributed time with a rate depending on the current configuration of the binary states of all orbits, idle/non-idle. We give a new proof of the necessary stability conditions of this system obtained earlier in the paper [17]. The new proof is based on a coupling of the processes of attempts from the orbits with independent Poisson processes, allowing to apply classic property PASTA. We simulate a 3-class system to verify stability conditions.

Acknowledgements. The study was carried out under state order to the Karelian Research Centre of the Russian Academy of Sciences (Institute of Applied Mathematical Research KRC RAS). The research of is partly supported by Russian Foundation for Basic Research, projects 18-07-00147, 18-07-00156, 19-07-00303.

References

1. Artalejo, J.R.: Accessible bibliography on retrial queues. Math. Comput. Modell. **30**, 1–6 (1999)
2. Artalejo, J.R., Gomez-Corral, A.: Retrial Queueing Systems: A Computational Approach. Springer, Heidelberg (2008). https://doi.org/10.1007/978-3-540-78725-9
3. Artalejo, J.R., Phung-Duc, T.: Markovian retrial queues with two way communication. J. Ind. Manag. Optim. **8**, 781–806 (2012)
4. Artalejo, J.R., Phung-Duc, T.: Single server retrial queues with two way communication. Appl. Math. Model. **37**, 1811–1822 (2013)
5. Asmussen, S.: Applied Probability and Queues. Wiley, New York (1987)
6. Avrachenkov, K., Morozov, E., Steyaert, B.: Sufficient stability conditions for multi-class constant retrial ratesystems. Queueing Syst. **82**(12), 149–171 (2016). https://doi.org/10.1007/s11134-015-9463-9
7. Dimitriou, I.: A two class retrial system with coupled orbit queues. Prob. Engin. Infor. Sc. **31**(2), 139–179 (2017)
8. Dimitriou, I.: A queueing system for modeling cooperative wireless networks with coupled relay nodes and synchronized packet arrivals. Perform. Eval. **114**, 16–31 (2017). https://doi.org/10.1016/j.peva.2017.04.002
9. Dimitriou, I.: Modeling and analysis of a relay-assisted cooperative cognitive network. In: Thomas, N., Forshaw, M. (eds.) ASMTA 2017. LNCS, vol. 10378, pp. 47–62. Springer, Cham (2017). https://doi.org/10.1007/978-3-319-61428-1_4
10. van Doorn, E.A., Regterschot, G.J.K.: Conditional PASTA. Oper. Res. Lett. **7**(5), 229–232 (1988)
11. Falin, G.I., Templeton, J.G.D.: Retrial Queues. Chapman & Hall, London (1997)
12. Falin, G.I.: A survey of retrial queues. Queueing Syst. **7**, 127–167 (1990)
13. Kim, J., Kim, B.: A survey of retrial queueing systems. Ann. Oper. Res. **247**(1), 3–36 (2016)
14. Mitola, J., Maguire, G.: Cognitive radio: making software radios more personal. IEEE Pers. Commun. **6**(4), 13–18 (1999)

15. Morozov, E., Phung-Duc, T.: Regenerative analysis of two-way communication orbit-queue with general service time. In: Takahashi, Y., Phung-Duc, T., Wittevrongel, S., Yue, W. (eds.) QTNA 2018. LNCS, vol. 10932, pp. 22–32. Springer, Cham (2018). https://doi.org/10.1007/978-3-319-93736-6_2

16. Morozov, E., Dimitriou, I.: Stability analysis of a multiclass retrial system with coupled orbit queues. In: Reinecke, P., Di Marco, A. (eds.) EPEW 2017. LNCS, vol. 10497, pp. 85–98. Springer, Cham (2017). https://doi.org/10.1007/978-3-319-66583-2_6

17. Morozov E., Morozova T.: Analysis of a generalized retrial system with coupled orbits. In: Proceeding of the 23rd Conference of FRUCT Assosiation, pp. 253–260 (2018)

18. Morozov, E., Morozova, T., Dimitriou, I.: Simulation of multiclass retrial system with coupled orbits. In: Proceedings of the First International Workshop on Stochastic Modeling and Applied Research of Technology Petrozavodsk, Russia, 21–25 September 6–16. CEUR-WS Proceedings (2018)

19. Morozov, E., Rumyantsev, A., Dey, S., Deepak, T.G.: Performance Analysis and stability of multiclass orbit queue with constant retrial rates and balking. (submitted for special issue Performance Evaluation)

20. Avrachenkov, K., Morozov, E., Steyaert, B.: Sufficient stability conditions for multi-class constant retrial rate systems. Queueing Syst. **82**(1–2), 149–171 (2016)

21. Moutzoukis, E., Langaris, C.: Non-preemptive priorities and vacations in a multiclass retrial queueing system. Stoch. Models **12**(3), 455–472 (1996)

22. Pappas, N., Kountouris, M., Ephremides, A., Traganitis, A.: Relay-assisted multiple access with full-duplex multi-packet reception. IEEE Trans. Wireless Commun. **14**, 3544–3558 (2015)

23. Sadek, A., Liu, K., Ephremides, A.: Cognitive multiple access via cooperation: protocol design and performance analysis. IEEE Trans. Inf. Theory **53**(10), 3677–3696 (2007)

24. Smith, W.L.: Regenerative stochastic processes. Proc. Roy. Soc. Ser. A **232**, 6–31 (1955)

25. Wolff, R.W.: Poisson arrivals see time averages. Oper. Res. **30**, 223–231 (1982)

Stability Conditions of a Multiclass System with NBU Retrials

Evsey Morozov[1,2] and Ruslana Nekrasova[1,2(✉)]

[1] Institute of Applied Mathematical Research Karelian Research Centre RAS,
Petrozavodsk, Russia
emorozov@karelia.ru, ruslana.nekrasova@mail.ru
[2] Petrozavodsk State University, Petrozavodsk, Russia

Abstract. We consider a multiclass multiserver retrial queuing system with classical retrial discipline: the customers, meeting server busy, are blocked on the corresponding (virtual) orbit and then retry to occupy server independently. The retrial times have general class-dependent distributions. The input process is renewal and a new arrival is class-i customer with a given probability p_i. We exploit a regenerative structure of a basic process describing the dynamics of the system to establish stability conditions. More exactly, we show that, provided the retrial times belong to the New-Better-Than-Used class, the convenient requirement that the mean load (traffic intensity) is less than the number of servers, is the stability criterion of the model. A few numerical results are included which, in particular, show that this condition ensures stability of the system with the New-Worse-Than-Used Weibull retrial times as well.

Keywords: Retrial system · Stability condition · NBU retrials ·
Multiclass queues · Multiple servers

1 Introduction

In this paper, we study an m-server retrial queueing system with classical retrial discipline and K classes of customers. In this system, a class-i customer which meets all servers busy, is blocked on the corresponding (virtual) class-i orbit and then makes attempts until finds a server idle. To establish the necessary stability conditions, we allow general class-dependent retrial times distributions. To prove the sufficient stability conditions, the retrial times are assumed to be independent identically distributed (i.i.d.) with class-dependent distribution belonging to a subclass of the *New-Better-Than-Used* (NBU) distributions. The classical retrial discipline means that all orbital customers make the attempts independently. The arrivals follow a renewal process with rate λ, and a new arrival belongs to class i with a given probability p_i and has service rate μ_i. In this research, we show that if the retrial time of each class has the NBU distribution, then the well-known stability criterion of the buffered multiserver multiclass system, the traffic intensity is less than the number of servers, is indeed the stability criterion

© Springer Nature Switzerland AG 2019
T. Phung-Duc et al. (Eds.): QTNA 2019, LNCS 11688, pp. 51–63, 2019.
https://doi.org/10.1007/978-3-030-27181-7_4

of the retrial model as well. A few ideas of the proof are similar to that have been used in the closely related paper [5], and omit the corresponding steps focusing on the new aspects of the proof.

This system has a regenerative structure, and we use this fact to develop stability analysis of the system using a characterization of the limiting remaining renewal time in the process generated by the regenerations of a basic process describing the dynamics of the system [4, 6, 12].

The retrial systems play an increasing role to model the dynamics of the modern wireless telecommunication systems. Another area of the application of the retrial queues is *call centres* in which customers, who meet an operator busy, call later again until find operator idle. Because these *orbital* or blocked customers are not present in the system, they can not occupies operator immediately when it becomes idle. As a result, an idle time of the operator (server) after each departure appears, and this effect reduces the capacity of the system. However, as it has been shown in a number of papers, for instance, [2, 5, 6], under classic retrial policy, the *lost capacity* becomes negligible as the number of the blocked customers increases. As we mentioned above, by this reason the classical retrial discipline has been called *asymptotically work-conserving* in the paper [6].

The existing literature devoted to analysis of retrial queues is vast, and we point out here only a few important sources of the bibliography in this area, [7, 8], and the basic books [9, 11]. We also mention a recent survey paper [1] devoted to retrial queues, which contains many performance results and pays a lot of attention to stability conditions. An interested reader can find further references in the mentioned above works.

In the most of the works, the retrial times are assumed to be exponential, and, in many cases, it allows to construct a suitable Markov process describing the dynamics of the system, to study stability conditions and find stationary performance measures in an explicit form, see for instance, [1, 13, 14, 17].

To the best of our knowledge, the paper [2] is one of the first works where the non-exponential retrial times are considered. This notable paper is devoted to a detailed stability analysis and rate of convergence in a general single-server single-class retrial queue in various settings. Then, the regenerative approach has been applied to extend stability analysis to a multiserver single-class system with classical retrial discipline [6]. More recently, stability of a single-class multiserver retrial system with classical discipline and *generally distributed* retrial times has been analysed by the fluid limit approach in the paper [15]. This analysis reduces the original stochastic model to its deterministic analogue. The dynamics of this deterministic model then allows to establish stability/instability of the original model. However, the analysis in [15], being highly complicated, is applied to the single-class retrial system only and it seems to be quite difficult, if possible, to apply it in the analysis of the multiclass retrial system.

By this reason, our goal in this work is to extend to non-exponential retrials the regenerative approach, which has been successfully applied earlier to analyze stability of general retrial queueing systems (both multi-server and multiclass) with the renewal input general class-dependent service times and exponential class-dependent retrial times [5,6]. In this work, we first give the necessary stability condition of the system, using the analysis developed in [5]. Since the proof is *independent of the retrial times* and similar to the proof that has given in [5], we only give some comments for easy understanding. Then, using the mean drift analysis, we prove that the same condition is also sufficient for the stability of the system with class-dependent NBU retrial times. The proof of sufficient stability conditions of such a general model is the *main contribution* of this work. It is worth mentioning that the main idea of the proof is that we widely use the NBU property to replace the remaining retrial times of the blocked customers by the new copies of the retrial times. The replacement of the remaining retrial times by the new retrial times makes the new system "more loaded" and allows to apply the approach from [5]. The regenerative approach is highly effective, using negative drift condition to "verify" stability of a basic queueing process (the summary orbit size in our case) at the *high levels*. Namely by this reason the obtained (sufficient) stability condition coincides with the *stability criterion* of a classical buffered multiclass, multiserver system. The insensitivity of the stability of the system to the retrial times is confirmed by the *independence of the stability criterion on the retrial rates*.

The paper is organized as follows. In Sect. 2, description of the model and its regenerative property are given. In Sect. 3, we formulate and outline the proof of the necessary stability condition for generally distributed retrial times (Theorem 1). Then, the proof of the sufficient stability condition for the NBU retrials is presented (Theorem 2). It is intuitive that the stability condition indeed must be true for general retrial times. To illustrate this intuition, in Sect. 4, we demonstrate the positive results of a few simulation experiments for a two-class single-server system in which retrial times have class-dependent *New-Worse-Than-Used* Weibull distribution. Moreover, we present also the dynamics of the queues in the corresponding two-class single-server *buffered system* with the same input and the same service times as in the original system, to demonstrate the influence of the retrial rates on the orbit sizes.

2 Description of the Model

We consider an m-server, bufferless K-class retrial queueing system with *classical retrial policy*. The customers arrive at the instants $\{t_n, \, n \geq 1\}$, which constitute a renewal input with the independent identically distributed (i.i.d.) interarrival times $\tau_n := t_{n+1} - t_n$, $n \geq 1$, with a finite mean $\mathsf{E}\tau =: 1/\lambda$. (Here τ is a generic interarrival time, and in what follows we omit serial index to denote a generic element of an i.i.d. sequence). Assume the service times of class-i customers, $\{S_n^{(i)}, \, n \geq 1\}$, are i.i.d. with corresponding rate $\mu_i = 1/\mathsf{E}S^{(i)}$, $1 \leq i \leq K$. A new arrival belongs to class-i with a given probability p_i regardless of other arrivals and the state of the system. Thus, the arrival rate of class-i customers is $\lambda_i = \lambda \cdot p_i$, $1 \leq i \leq K$. We also denote class-i traffic intensity (load) and summary load, respectively,

$$\rho_i := \lambda_i/\mu_i, \quad \rho := \sum_{i=1}^{K} \rho_i. \tag{1}$$

A new class-i customer meeting the system (all servers) busy, joins the i-th infinite capacity virtual orbit and attacks server after a (generic) retrial time $\xi^{(i)}$. The classical retrial policy means that the orbital customers make attempts independently. If the retrial times are exponential, then the retrial rate from each orbit is proportional to the orbit size, and it is the most studied case in the literature, see [5,7,11]. (In general, there are various forms of a dependence between the orbit size and retrial rate, see for instance, [8].)

For general retrial times, a connection between orbit size and a "retrial rate" (which must be correctly redefined in this case) becomes less evident. However it is intuitive, that the bigger orbit is, then the retrial rate is more intensive as well. This informal observation indicates how to investigate the stability of such a system. We assume that the orbit size increases unlimitedly, in which case the orbit size approaches to the queue size in a conventional buffered queueing system.

The main distinctive feature of the system we consider in this research is that we assume a wider class of retrial times than exponential. Namely, to establish the necessary stability conditions, we allow general class-dependent retrial distributions, while to prove the sufficient conditions, we assume, that the retrial times have a class-dependent NBU distribution. The distribution F of a random variable $\xi \geq 0$ is called NBU if, for each $x, y \geq 0$, its *tail* satisfies the inequality

$$\mathsf{P}(\xi > x + y | \xi > y) \leq \mathsf{P}(\xi > x). \tag{2}$$

Denote by $\nu(t)$ the number of busy servers and by $N^{(i)}(t)$ the number of customers in the i-th orbit, at instant t^-. Denote the summary orbit size by $N(t) = \sum_{i=1}^{K} N^{(i)}(t)$, and let

$$X(t) = N(t) + \nu(t), \ t \geq 0; \quad X_n := X(t_n), \ n \geq 1. \tag{3}$$

The basic process $X := \{X(t), t \geq 0\}$ is *regenerative*, with the regeneration instants

$$T_{n+1} = \inf_{k} \left(t_k > T_n : X_k = 0 \right), \qquad T_0 = 0, \ n \geq 0, \tag{4}$$

the i.i.d. regeneration periods $T_{n+1} - T_n$ and generic period T. We restrict analysis by the *zero initial state* in which case the 1st period $T_1 =_{st} T$ (stochastically). We call the process X (and the retrial system) *positive recurrent* if the mean $\mathsf{E}T < \infty$. (If $T_1 \neq_{st} T$, then we additionally require $T_1 < \infty$ with probability (w.p.) 1). The positive recurrence is the key requirement to establish stability [3], so in what follows we use the terms "stability" and "positive recurrence" as synonyms.

Define the remaining regeneration time at instant t as

$$T(t) := \min_{k}(T_k - t : T_k - t > 0), \ t \geq 0.$$

In the further analysis we use the following asymptotic result from the renewal theory [4]: if

$$T(t) \not\Rightarrow \infty, \ t \to \infty, \tag{5}$$

then $\mathsf{E}T < \infty$. (Here $\Rightarrow$ denotes the convergence in probability). The property (5) means that there exist a deterministic sequences of instances $z_k \to \infty$ and some constants $C, \varepsilon \in (0, \infty)$ such that

$$\inf_k \mathsf{P}(T(z_k) \le C) \ge \varepsilon. \tag{6}$$

3 Stability Conditions

It has been proved in the [5], that the requirement $\rho < m$ is stability criterion of a less general m-server system with class-dependent exponential retrials.

A careful analysis of the proof in [5] shows that the necessary stability condition for the new setting can be established exactly as in [5] because the proof is indeed *independent of the retrial distributions*. Thus, we formulate below this statement for the system with general service times and give only a few comments to the proof for easy understanding. Denote Σ the original retrial system.

Theorem 1. *Assume that the system Σ with generally distributed retrial times is positive recurrent and condition*

$$\max_{1 \le i \le K} \mathsf{P}(\tau > S^{(i)}) > 0 \tag{7}$$

holds. Then

$$\rho < m. \tag{8}$$

To prove, we apply the following balance equation, connecting the work $V(t) = \sum_{i=1}^{K} \sum_{n=1}^{A_i(t)} S_n^{(i)}$, arrived in the system in the interval $[0, t)$, where $A_i(t)$ is the number of class-i arrivals, with the departed work $D(t)$ (in the same interval) and the remaining work $R(t)$ (in the servers and in orbits) at the instant t:

$$V(t) = R(t) + D(t) = R(t) + mt - I(t), \ t \ge 0, \tag{9}$$

where $I(t)$ is the summary idle time of all servers in $[0, t]$. By positive recurrence, $R(t) = o(t)$, $t \to \infty$, with w.p. 1 [18]. Moreover, using condition (7) and regeneration theory, one can easily show that w.p.1, $\lim_{t\to\infty} I(t)/t > 0$, while by the Strong Law of Large Numbers, $V(t)/t \to \rho$, $t \to \infty$, and then the statement of Theorem 1 follows.

Remark 1. We emphasize again that the proof of Theorem 1 is independent of the retrial time distributions. Also note that the statement of Theorem 1 holds for an *arbitrary initial state* $X(0)$.

Remark 2. As in [5], one can show that the stationary *busy probability* of an arbitrary server equals

$$P_b = \frac{\rho}{m}. \tag{10}$$

However, the proof of the sufficiency of conditions (7), (8) for the stability is a more challenging problem. Below we resolve this problem for a subclass of the NBU retrials. For each i, denote $\xi^{(i)}$ class-i generic retrial time, and let F_i be its distribution function with the mean $E\xi^{(i)} < \infty$, $1 \leq i \leq K$. Now define the distribution F_0 as follows: for each $x \geq 0$,

$$F_0(x) = \min_{1 \leq i \leq K} F_i(x). \tag{11}$$

Denote $\xi^{(0)}$ the random variable with distribution F_0. Because $F_0(x) \leq F_i(x)$, it then follows that the *stochastic ordering* holds [16]:

$$\xi^{(0)} \geq_{st} \xi^{(i)}. \tag{12}$$

Now we formulate and prove the following sufficient stability conditions.

Theorem 2. *Assume condition (8) holds, the system Σ has zero initial state and the interarrival time τ is unbounded, that is, for each $x \geq 0$,*

$$P(\tau > x) > 0. \tag{13}$$

Moreover, assume that each function F_i is NBU distribution such, that

$$\inf(x : F_i(x) > 0) = 0, \ 1 \leq i \leq K. \tag{14}$$

Then the system is stable, $ET < \infty$.

Note that condition (13) implies (7), and that by (14), the retrial time $\xi^{(i)}$ takes arbitrary small value with a positive probability.

    First of all, we outline the main steps of the proof. We will focus on the application of NBU property because it is the key new point of the analysis, while the remaining steps follow the analysis presented in [5]. First, using condition (8), we establish the negative drift of the orbit size process. This step is especially laborious, and to establish it, we show that the idle time of the server between a departure and the initiation of next service goes to zero as the orbit size increases. It is intuitive because, as the orbit size grows, the retrial attempts become more frequent, the idle time decreases and the retrial systems approaches conventional buffered $GI/GI/m$ system (for which (8) is the stability criterion). It then follows that the orbit size can not increases unlimitedly and, as a result, visits a bounded set infinitely often. In the last step, using condition (7), we show that starting in the bounded set, the process X reaches zero state (regeneration point) with a positive probability within a finite interval. In other words, the property (5) holds and the system is positive recurrent.

Proof. Denote $S_i(t)$ the remaining service time in server i at instant t^- ($S_i(t) = 0$ if the server is idle). Then the summary idle time of all servers in interval $[0, t_n]$ is defined as

$$I_n = \int_0^{t_n} \sum_{i=1}^m 1(S_i(u) = 0)du, \ n \geq 1, \tag{15}$$

where 1 is the indicator function. As in [5] we show that, see (8),

$$\liminf_{n \to \infty} \mathsf{E}\left(\frac{I_n}{n}\right) \geq (m - \rho)\mathsf{E}\tau =: \varepsilon > 0. \tag{16}$$

Denote by $\Delta_n = I_{n+1} - I_n$, the summary idle time in the interval $[t_n, t_{n+1}]$, and assume that $\mathsf{E}\Delta_n \to 0$. Then, there exists $n_0 = n_0(\varepsilon) > 1$ such that $\mathsf{E}\Delta_n \leq \varepsilon/2$ for $n \geq n_0$, implying

$$\mathsf{E}I_n = \sum_{k=1}^{n_0-1} \mathsf{E}\Delta_k + \sum_{k=n_0}^{n-1} \mathsf{E}\Delta_k \leq o(n) + \frac{\varepsilon}{2}(n - n_0), \ n \geq n_0. \tag{17}$$

It is easy to see that it contradicts (16), and thus $\mathsf{E}\Delta_n \not\to 0$. Hence there exist a deterministic subsequence $n_k \to \infty$ and a constant $\varepsilon_0 > 0$ such that

$$\inf_k \mathsf{E}\Delta_{n_k} \geq \varepsilon_0. \tag{18}$$

Denote $N_n = N(t_n)$, $D(t_n) = D_n$ and, for arbitrary constants $d, d_0 > 0$, write

$$\begin{aligned} \mathsf{E}\Delta_n &= \mathsf{E}[\Delta_n, N_n \leq d + d_0] + \mathsf{E}[\Delta_n, N_n > d + d_0, D_n > d_0] \\ &\quad + \mathsf{E}[\Delta_n, N_n > d + d_0, D_n \leq d_0]. \end{aligned} \tag{19}$$

Using the independence between N_n and the next interarrival time τ_n, one can obtain the following upper bound for the 1st summand in (19):

$$\mathsf{E}[\Delta_n, N_n \leq d + d_0] \leq m\mathsf{E}\tau \mathsf{P}(N_n \leq d + d_0). \tag{20}$$

Following [5], we consider m i.i.d. sequences of the i.i.d. variables $\{\eta_n^{(j)}\}$, $j = 1, \ldots, m$, where each $\eta_n^{(j)}$ is distributed as the *shortest* service time

$$\eta_n^{(j)} =_{st} \eta := \min_{1 \leq i \leq K} S^{(i)}, \quad \mathsf{E}\eta < \infty.$$

Then we construct a superposition of the zero-delayed renewal processes generated by the sequences $\{\eta_n^{(j)}\}$,

$$M_j(\tau_n) := \inf\left(k \geq 1 : \eta_1^{(j)} + \cdots + \eta_k^{(j)} \geq \tau_n\right), \tag{21}$$

and denote

$$M(\tau_n) = \sum_{j=1}^m M_j(\tau_n), \quad M(0) = m. \tag{22}$$

(The process $M_j(\tau_n)$ represents the number of renewals in interval $[t_n, t_{n+1})$ generated by the shortest service times in server j). Because there are idle times ("gap") between the departures in the original retrial system, then the summary number of departures, D_n, in the interval $[t_n, t_{n+1})$ is upper bounded as follows:

$$D_n \leq_{st} M(\tau_n), \ n \geq 1.$$

This bound allows to construct the following upper bound for the 2nd term in (19):

$$\mathsf{E}[\Delta_n, \ N_n > d + d_0, \ D_n > d_0] \leq ma\mathsf{P}(M(a) > d_0) + m\mathsf{E}[\tau; \tau > a], \quad (23)$$

where $a \geq 0$ is an arbitrary constant.

While details of previous discussion can be found in [5] and do not depend on the retrial time distribution, the next step, where we construct an upper bound of the last term in (19), is new and critically depends on the NBU property. Note that the remaining retrial time of a class-i (orbital) customer, provided the attained retrial time $\geq x$, denoted $\hat{\xi}^{(i)}(x)$, has the tail distribution

$$\mathsf{P}(\hat{\xi}^{(i)}(x) > y) = \mathsf{P}(\xi^{(i)} > x + y | \xi^{(i)} > x) = \frac{1 - F_i(x+y)}{1 - F_i(x)}, \ x \geq 0, \ y \geq 0.$$

Fix for a moment an arbitrary instant t_n, and assume that $N_n = d$, for some $d \geq 1$. Denote $\zeta_n(d)$ the time since the instant t_n until the 1st retrial attempt after t_n occurs, if $N_n \geq d$. By the independence of the retrials and by (12), we have the following upper bound

$$\mathsf{P}(\zeta_n(d) > x) \leq \left[\mathsf{P}(\xi^{(0)} > x)\right]^d \leq \mathsf{P}(\xi^{(0)} > x), \quad (24)$$

which is independent of n. It implies the upper bound of the expectation

$$\mathsf{E}\zeta_n(d) = \int_0^\infty \mathsf{P}(\zeta_n(d) > x) dx \leq \int_0^\infty \left[\mathsf{P}(\xi^{(0)} > x)\right]^d dx. \quad (25)$$

Note that, for each x,

$$\left[\mathsf{P}(\xi^{(0)} > x)\right]^d \to 0, \ d \to \infty, \quad (26)$$

and that

$$\mathsf{E}\xi^{(0)} = \int_0^\infty \mathsf{P}(\xi^{(0)} > x) dx < \infty.$$

It means that the function $\mathsf{P}(\xi^{(0)} > x)$ is integrable and (26) implies convergence in mean (by Lebesgue's Dominated Convergence Theorem [10]):

$$\lim_{d \to \infty} \mathsf{E}\zeta_n(d) \leq \lim_{d \to \infty} \int_0^\infty \left[\mathsf{P}(\xi^{(0)} > x)\right]^d dx = 0. \quad (27)$$

In the event $\{N_n > d + d_0, D_n \le d_0\}$, there are at most m initial idle periods of the servers and at most d_0 customers depart the system, and thus the orbit size remains grater than d during interval $[t_n, t_{n+1})$. Then, by (12), the mean idle period (of any server) after each departure is upper bounded by $E\zeta_n(d)$, and the number of the idle periods in this interval is upper bounded by $m + d_0$. It gives the following upper bound for the last term in (19):

$$E[\Delta_n, N_n > d + d_0, D_n \le d_0] \le (m + d_0) E\zeta_n(d). \tag{28}$$

Note that $E\tau < \infty$ and $EM(a) < \infty$. Now we first take a in (23) such that $mE[\tau; \tau > a] \le \varepsilon_0/4$ (see (18)), then select $d_0 = d_0(a)$ in such a way that $ma P(M(a) > d_0) \le \varepsilon_0/4$, and, applying convergence (27), take $d = d(d_0)$ so large, that $(m + d_0) E\zeta_n(d) \le \varepsilon_0/4$, see (28). Now we assume that the summary orbit size N_n grows unlimitedly in probability,

$$N_n \Rightarrow \infty, \qquad n \to \infty, \tag{29}$$

and show that it implies a contradiction. Indeed, by (29), we can take such n_1, that $mE\tau P(N_n \le d + d_0) \le \varepsilon_0/4$ for all $n \ge n_1$. Now, picking together all the upper bounds found above, we obtain from (19), that $E\Delta_n < \varepsilon_0$ for $n \ge n_1$, that contradicts (18). Thus, $N_n \not\Rightarrow \infty$, and there exist a deterministic subsequence of the arrival instances $t_{n_i} \to \infty$ as $n_i \to \infty$, and constants $\delta > 0$, $C < \infty$ such that

$$\inf_i P(N_{n_i} \le C) \ge \delta. \tag{30}$$

The following arguments are similar to that have been used in [5], and the unboundness of τ plays here a key role (see (13)). Denote W_n the summary remaining work in all servers at the instant t_n^- (that is, $W_n \le R(t_n)$, see (9)). Because the sequence $\{W_n\}$ is tight, then there exist a constant C_0 such, that, see (30),

$$\inf_i P(N_{n_i} \le C, W_{n_i} \le C_0) \ge \frac{\delta}{2}. \tag{31}$$

It is easy to see that each customer can be served within interval with the length $Z := \max_{1 \le i \le K} S^{(i)} + \xi^{(0)}$ (maximal service time plus maximal orbital delay). By $EZ < \infty$, there are constants $a_0 < \infty$, $\epsilon > 0$, such, that $P(Z \le a_0) \ge \epsilon$, and then, on the event $\{N_{n_i} \le C, W_{n_i} \le C_0\}$, the system becomes idle in the interval $[t_{n_i}, t_{n_i} + C_0 + aC]$ with a probability which is not less than ϵ^C, *provided the external customers do not arrive since the instant t_{n_i}*. Now, because $E\tau < \infty$ and by (13), there are constants $A < \infty$, $\varepsilon' > 0$ such that

$$P(C_0 + a_0 C < \tau_{n_i} \le A) \ge \varepsilon', $$

where we take into account that the probability of the event $\{C_0 + a_0 C < \tau_{n_i} \le A\}$ is independent of i. By (31), it now follows that a new customer meets completely idle system in the interval $[t_{n_i}, t_{n_i} + A]$ with a probability which is bounded from below by the constant $\delta/2\, \epsilon^C \varepsilon' > 0$. This implies (5), and the positive recurrence follows.

4   Simulation Results

It is intuitive that the NBU retrials assumption in Theorem 2 is rather technical requirement related to the proof, and condition (8) must be indeed stability criterion for general retrial times. (In this regard we mention the paper [15]). To verify this assumption, below we describe a few experiments with a two-class single-server system with the *New-Worse-Than-Used* (NWU) retrial times. Recall that a random variable ξ has a NWU distribution if, for each $x, y \geq 0$

$$\mathsf{P}(\xi > x + y | \xi > y) \geq \mathsf{P}(\xi > x). \tag{32}$$

We assume $K = 2$ classes of customers, following independent *Poisson inputs*, in which service times are *exponential*, and class-i retrial times $\xi^{(i)}$ have *Weibull distribution* with the scale parameter 1 and the shape parameter w_i:

$$F_i(x) = 1 - \exp(-x^{w_i}), \qquad i = 1, 2. \tag{33}$$

Weibull distribution is NWU one, if $w_i \leq 1$, while $w_i \geq 1$ corresponds to the NBU distribution, including exponential if $w_i = 1$ (with the equality in (32)). The mean retrial time for class i is

$$\mathsf{E}\xi^{(i)} = \Gamma\left(\frac{1}{w_i} + 1\right), \tag{34}$$

where Γ denotes Gamma function. If $0 < w_i \leq 1$ then class-i retrial rate decreases monotonically,

$$1 \geq \gamma_i = \frac{1}{\mathsf{E}\xi^{(i)}} \to 0, \ w_i \to 0.$$

(Note that, for the NBU Weibull, γ_i is not monotone and simulation results are less illustrative). We present the dynamics of orbits for a few values of the retrial rates γ_1, γ_2, under condition $\rho = \rho_1 + \rho_2 < 1$. We consider 10000 arrivals and take average over 100 independent paths of the orbit sizes $N_n^{(1)}$, $N_n^{(2)}$.

In the 1st experiment, we apply parameters $\lambda_1 = 0.2$, $\lambda_2 = 1.2$, $\mu_1 = 1$, $\mu_2 = 2$, giving the probabilities $p_1 = 0.14$, $p_2 = 0.86$, and the traffic intensities $\rho_1 = 0.2$, $\rho_2 = 0.6$, $\rho = 0.8$. The parameters of Weibull distribution are, respectively, $w_1 = 0.25$, $w_2 = 0.9$, implying the retrial rates $\gamma_1 = 0.042$, $\gamma_2 = 0.95$. The results are presented in Fig. 1. We observe that both orbits are stable. However $\rho_1 < \rho_2$, the 1st orbit size strongly dominates: $N_n^{(1)} \gg N_n^{(2)}$ (we keep the same notation for the sample mean and for the orbit size). It is explained by relation $\gamma_1 \ll \gamma_2$ (2nd orbit attempts are much more intensive than the 1st). This observation indicates that the orbit rate plays a significant role in the behaviour of orbits.

We emphasize that the stable dynamics of the orbits is intuitive and confirms our conjecture that stability criterion holds for a wider class of retrial distributions, because the retrial times in this experiment *do not satisfy NBU* assumption.

The virtual orbits in the system Σ are similar to the *queues* in the conventional infinite buffer system (denoted $\hat{\Sigma}$), where a waiting customer *immediately*

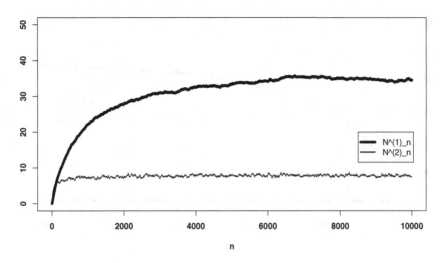

Fig. 1. Orbit dynamics, light load: $\rho_1 = 0.2$, $\rho_2 = 0.6$, $\gamma_1 = 0.042$, $\gamma_2 = 0.95$.

enters server as it becomes idle. The difference with the retrial system, where an idle time exists after each departure, must lead to a dominance of the summary orbit size in the system Σ over the summary queue size in the system $\hat{\Sigma}$ (provided other corresponding parameters remain equal).

In the following experiment we consider a 2-class FIFO infinite-buffer single-server system $\hat{\Sigma}$ with the same Poisson inputs and the same service times $\{S_n^{(i)}\}$, $i = 1, 2$; $n \geq 1$. Note, that the service order in the system Σ is not FIFO in general. Denote by $Q_n^{(i)}$ the number of class-i customers in the buffer of $\hat{\Sigma}$ at the instant t_n. We compare the sample means of orbits with the sample means of the queues in $\hat{\Sigma}$. We use the following parameters $\lambda_1 = 0.23$, $\lambda_2 = 0.46$, $\mu_1 = 0.5$, $\mu_2 = 1$. Thus, $\rho_1 = \rho_2 = 0.46$, $\rho = 0.92$. In the system Σ, we take Weibull retrials with parameters $w_1 = 0.9$, $w_2 = 0.25$ (and the corresponding orbit rates $\gamma_1 = 0.95$, $\gamma_2 = 0.042$). Figure 2 illustrates stability of both orbits and a dominance of orbit 2: $N_n^{(2)} \gg N_n^{(1)}$. In more detail (in another time scale) simulation results describing *both systems* are presented in Fig. 3. This figure demonstrates the dynamics of both orbits and both queues simultaneously. It is expected and easy to see that $N_n^{(1)} + N_n^{(2)} \gg Q_n^{(1)} + Q_n^{(2)}$, but it is rather surprising that the queue size $Q_n^{(1)}$ dominates the orbit $N_n^{(1)}$.

These observations allow to conclude that the varying of retrial rates can be used to optimize the system performance. In particular, orbits can be used to realize *priority-like policy*, when customers of some classes capture server, breaking FIFO discipline. Such a policy seems to be more flexible than the conventional *preemptive* or *non-preemptive* priority and needs a further comparative study.

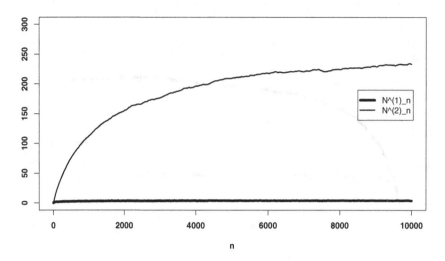

Fig. 2. Orbit dynamics: $\rho_1 = \rho_2 = 0.46$, $\gamma_1 = 0.95$, $\gamma_2 = 0.042$.

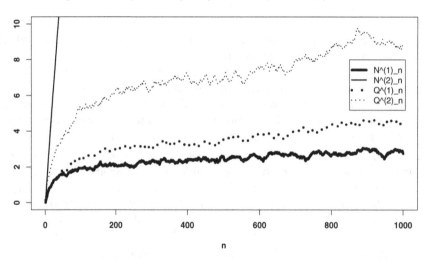

Fig. 3. Orbit dymanics and queue size: $\rho_1 = \rho_2 = 0.46$, $\gamma_1 = 0.95$, $\gamma_2 = 0.042$.

5   Conclusion

In this research, we consider the stability analysis of a general multiclass multiserver retrial system with non-exponential retrial times and classical retrial policy. In this system, an arriving customer who meets all servers busy joins the corresponding (virtual) orbit and then attempts to occupy server again independently of other blocked customers. Both service time and retrial times are class-dependent. While the distributions of service times are general, it is assumed that the distributions of the retrial times belong to a subclass of the so-called

New-Better-Than-Used (NBU) distributions. We develop the regenerative analysis to obtain sufficient stability conditions of the system. This analysis is also based on the coupling method and the stochastic ordering. The numerical examples are given to show that the stability conditions hold for a wider class of the retrial times distributions.

Acknowledgement. The study was carried out under state order to the Karelian Research Centre of the Russian Academy of Sciences (Institute of Applied Mathematical Research KRC RAS). The research is partly supported by Russian Foundation for Basic Research, projects 18-07-00147, 18-07-00156, 19-07-00303.

References

1. Kim, J., Kim, B.: A survey of retrial queueing systems. Ann. Oper. Res. **247**, 1–34 (2015)
2. Altman, E., Borovkov, A.A.: On the stability of retrial queues. Queueing Syst. **26**, 343–363 (1997)
3. Asmussen, S.: Applied Probability and Queues, 2nd edn. Springer, New York (2003). https://doi.org/10.1007/b97236
4. Feller, W.: An Introduction to Probability Theory, vol. 2. Wiley, New York (1971)
5. Morozov, E., Phung-Duc, T.: Stability analysis of a multiclass retrial system with classical retrial policy. Perform. Eval. **112**, 15–26 (2017)
6. Morozov, E.: A multiserver retrial queue: regenerative stability analysis. Queueing Syst. **56**, 157–168 (2007)
7. Artalejo, J.R.: Accessible bibliography on retrial queues. Math. Comput. Modell. **30**, 1–6 (1999)
8. Artalejo, J.R.: A classified bibliography of research on retrial queues: progress in 1990–1999. Top **7**, 187–211 (1999)
9. Artalejo, J.R., Gomez-Corral, A.: Retrial Queueing Systems: A Computational Approach. Springer, Heidelberg (2008). https://doi.org/10.1007/978-3-540-78725-9
10. Billingsley, P.: Convergence of Probability Measures, 2nd edn. Wiley, New York (1999)
11. Falin, G.I., Templeton, J.G.D.: Retrial Queues. Chapman and Hall, London (1997)
12. Morozov, E., Delgado, R.: Stability analysis of regenerative queues. Autom. Remote Control **70**, 1977–1991 (2009)
13. Artalejo, J.R., Phung-Duc, T.: Markovian retrial queues with two way communication. J. Ind. Manag. Optim. **8**, 781–806 (2012)
14. Artalejo, J.R., Phung-Duc, T.: Single server retrial queues with two way communication. Appl. Math. Model. **37**, 1811–1822 (2013)
15. Kang, W.: Fluid limits of many-server retrial queues with non-persistent customers. Queueing Syst. **79**(2), 183–219 (2015)
16. Muller, A., Stoyan, D.: Comparison Methods for Stochastic Models and Risks. Wiley Series in Probability and Statistics. Wiley, Chichester (2002)
17. Sakurai, H., Phung-Duc, T.: Two-way communication retrial queues with types of outgoing calls. TOP **23**, 466–492 (2015)
18. Smith, W.L.: Regenerative stochastic processes. Proc. Roy. Soc. A **A**(232), 6–31 (1955)

Controlled M/M/1-RQ System with Randomized Acceptance from the Orbit

Andrei V. Zorine$^{(\boxtimes)}$ (ID)

Lobachevsky State University of Nizhni Novgorod,
23 Gagarina prospekt, Nizhni Novgorod 603950, Russian Federation
zoav1602@gmail.com
http://www.itmm.unn.ru

Abstract. A retrial queuing system with Poisson input, exponential service times, and exponential in-orbit times is studied. The server implements a randomized acceptance policy: a retrial request from the orbit is accepted by the server with a probability depending on the number in the orbit. Such a policy can be used to give preference in service to newcomers. The queuing process is modelled as a continuous-time countable Markov process. For its embedded jump chain a sufficient condition for the existence of the stationary mode is proved. The proof is based on the iterative dominating technique. Further, a least unloading cost problem is considered. Given a partition of the state space into an "empty orbit" set, "sparsely populated orbit" set, and a prohibited "crowded orbit" set, the conditional expectation and the conditional variance for the cost of the first passage time with prohibited set (also known as the "Chung functional") are used to pose the optimization problem. Results of a numerical investigation of the structure of optimal policies are presented.

Keywords: Retrial queue · Randomized acceptance for service · Stationary mode · Iterative dominating approach · Chung functionals

1 Introduction

Active waiting in an orbit contrary to classical waiting queues is attracting increasing attention of researchers in queuing theory. A new customer occupies the server if the server is idle upon arrival, or joins an 'orbit' and stays there, making retrial attempts to occupy the server at (random) times. For a bibliographical reviews, as well as for some interesting derived schemes, see e.g., [1–7]. Retrial queues are widely used to model telecommucation and computer networks. Modern telecommunication networks follow complex control protocols. Mathematical study of static protocols, dynamic protocols, and adaptive protocols to resolve conflicts in multiple access networks can be found in [8,9]. At the same time, optimal control problems for queuing systems with retrial customers are less studied. For example, in [10], the control policy consists in selecting the threshold for the number of customers to terminate the server vacation.

© Springer Nature Switzerland AG 2019
T. Phung-Duc et al. (Eds.): QTNA 2019, LNCS 11688, pp. 64–76, 2019.
https://doi.org/10.1007/978-3-030-27181-7_5

Since different control policies may essentially change the system dynamics, the development of methods for obtaining stationarity conditions for retrial queueing models is important [11–13]. The most popular method to establish stationarity condition uses a drift criteria. As a rule a stationarity condition for classical retrial queueing systems involves arrival rates of the input flow together with mean service times. On the other hand, one can't exclude a possibility that controlled retrial queueing systems exist whose stationarity conditions depends on control policies. Obviously, it is fruitful to learn that the stationarity conditions for a particular retrial queuing system are independent of the selected control policy.

The majority of researchers consider average operation cost as an objective function [14,15]. But we would like to draw attention to another kind of objective functions arising from stochastic least time (least cost) of hitting a state subspace by a random process. For discrete-time Markov chains, functionals of this kind were proposed in [16], an application to traffic control problem can be found in [17]. Original Chung functionals in the sense of [16] were defined as the number of steps it takes a discrete-time Markov chain to reach a given set with a given prohibet set (taboo set). In [18,19] Chung functionals were also defined for continuous-time Markov processes together with an additive cost functional, and they were applied to the study of a priority queueing system with feedback customers, setup times, and input flows modulated by a two-state random environment.

In the present work a retrial queueing system is investigated in which a request from the orbit is accepted by an idle server with a given probability. Such a policy can be used to give preference to newcomers. A control policy is determined by an infinite sequence of acceptance probabilities which can depend of the number in the orbit in quite an irregular way. A mathematical model for this queueing system recently has been considered in [21]. However, a new approach to establish sufficient conditions for the existence of the stationary probability distribution will be demonstrated which for a wider class of processes than studied in [21,22]. Also, the structure of a policy which unloads the orbit at the least cost is revealed by means of numerical study.

2   The Queueing Model and the Stationarity Condition

Consider a retrial queueing system of $M|M|1 - RQ$ type with randomized acceptance from the orbit. To be precise, let us assume that the input flow is Poisson with intensity $\lambda > 0$, service times are i.i.d. exponentially distributed with the mean μ^{-1}. If an arriving customer finds the server busy it joins the orbit and waits there, emitting service requests with i.i.d. exponentially distributed intervals with the mean γ^{-1}. When a request from the orbit comes to the idle server it's accepted with probability p_x where x is the number in the orbit.

Let us denote by $\Gamma^{(0)}$ the idle server state, by $\Gamma^{(1)}$ the busy server state. Let $\Gamma(t) \in \Gamma = \{\Gamma^{(0)}, \Gamma^{(1)}\}$ be the server state at time t, $t \geqslant 0$. Let $\kappa(t)$ be the number in the orbit at time t. Under the above assumptions, the process

$$\{(\Gamma(t), \kappa(t)), t \geqslant 0\} \tag{1}$$

is a continuous-time Markov process, and the Kolmogorov equations for the probabilities $Q_{j,x}(t) = \mathbf{P}(\Gamma(t) = \Gamma^{(j)}, \kappa(t) = x)$ are

$$\frac{d}{dt}Q_{0,x}(t) = -(\lambda + x\gamma p_x)Q_{0,x}(t) + \mu Q_{1,x}(t),$$

$$\frac{d}{dt}Q_{1,x}(t) = \lambda Q_{0,x}(t) + (x+1)\gamma p_{x+1}Q_{0,x+1}(t) - (\lambda + \mu)Q_{1,x}(t) \tag{2}$$

$$+ \lambda Q_{1,x-1}(t), \qquad x = 0, 1, \dots$$

It's a quasi-birth-and-death process [20] with space state

$$\{(\Gamma^{(j)}, x) \colon j = 0, 1, \ x = 0, 1, \dots\}$$

and with infinitesimal generator $\tilde{Q}$, given by

$$\tilde{Q} = \begin{pmatrix} N_0 & \Lambda & 0 & 0 & \dots \\ M_1 & N_1 & \Lambda & 0 & \dots \\ 0 & M_2 & N_2 & \Lambda & \dots \\ \vdots & \vdots & \vdots & \vdots & \ddots \end{pmatrix}$$

where

$$\Lambda = \begin{pmatrix} 0 & 0 \\ 0 & \lambda \end{pmatrix}, \quad N_x = \begin{pmatrix} -\lambda - x\gamma p_x & \lambda \\ \mu & -\lambda - \mu \end{pmatrix}, \quad M_x = \begin{pmatrix} 0 & x\gamma p_x \\ 0 & 0 \end{pmatrix},$$

and the states ordering is such that j varies first, x varies last, i.e.

$$(\Gamma^{(0)}, 0), \ (\Gamma^{(1)}, 0), \ (\Gamma^{(0)}, 1), \ (\Gamma^{(1)}, 1), \ (\Gamma^{(0)}, 2), \ (\Gamma^{(1)}, 2), \quad \dots.$$

Besides the dependence on x, the block matrices N_x and M_x depend on the acceptance probabilities p_x, $x = 0, 1, \dots$ in an irregular way. In particular, if the limit $\lim_{x \to \infty} x p_x$ exists (either finite or infinite), then the chain is a multidimensional asymptotically quasi-Toeplitz Markov chain [22]. But in general we don't assume the limit exists. The Markov process with similar generator matrix was studied in [21]. In particular, a closed-form solution for the stationary probability distribution was giver there.

Let $\tau_1, \tau_2, \dots$ be the jump instantst of the Markov process (1) and $\tau_0 = 0$. Let us consider the server state $\tilde{\Gamma}_i = \Gamma(\tau_i + 0)$ and the number in the orbit $\tilde{\kappa}_i = \kappa(\tau_i + 0)$ after the i-th jump. It was proved in the theory of time-homogeneous Markov processes that a two-variate sequence

$$\{(\tilde{\Gamma}_i, \tilde{\kappa}_i); i = 0, 1, \dots\} \tag{3}$$

is an homogeneous Markov chain.

If $p_x = 0$ for a particular x then the states $(\Gamma^{(j)}, w)$ with $w < x$ are transient. Then the chain (3) is transient when $p_x = 0$ for infinitely many x's. Thus, to proceed we assume that $p_x > 0$ for all $x = 1, 2, \ldots$. Then all the states here constitute a single class of communicating aperiodic states.

Put $\underline{p} = \liminf\limits_{x \to \infty} x p_x \leqslant \infty$. The central claim of this section is the following theorem.

Theorem 1. *If either*

(1) $\underline{p} = \infty$ and $\lambda \mu^{-1} < 1$, or
(2) $0 < \underline{p} < \infty$ and $\lambda \mu^{-1} < \gamma q (\lambda + \gamma q)^{-1}$
    then there exists the unique stationary probability distribution for the Markov chain (3).

The proof is based on the iterative dominating approach [19] and consists of several lemmas. Set $Q_i(j, x) = \mathbf{P}(\tilde{\Gamma}_i = \Gamma^{(j)}, \tilde{\kappa}_i = x)$, $j = 0, 1$ and $x = 0, 1, \ldots$. Let us introduce partial probability generating functions

$$F_{j,i}(z) = \sum_{x=0}^{\infty} z^x Q_i(j, x), \quad |z| \leqslant 1. \tag{4}$$

Lemma 1. *The partial probability generating functions $F_{0,i}(z)$, $F_{1,i}(z)$, $i = 0, 1, \ldots$ satisfy the following functional equations:*

$$F_{0,i+1}(z) = \frac{\mu}{\lambda + \mu} F_{1,i}(z), \tag{5}$$

$$F_{1,i+1}(z) = \frac{\lambda z}{\lambda + \mu} F_{1,i}(z) + \sum_{x=0}^{\infty} z^x Q_i(0, x) \frac{x \gamma p_x + \lambda z}{(\lambda + x \gamma p_x) z}. \tag{6}$$

Proof. From the infinitesimal generator $\tilde{Q}$ we find the Chapman–Kolmogorov equations for the stationary probability distribution of Markov chain (3):

$$Q_{i+1}(0, x) = Q_i(1, x) \cdot \frac{\mu}{\lambda + \mu}, \tag{7}$$

$$Q_{i+1}(1, 0) = Q_i(0, 1) \cdot \frac{\gamma p_1}{\lambda + \gamma p_1} + Q_i(0, 0), \tag{8}$$

$$Q_{i+1}(1, x) = Q_i(0, x) \cdot \frac{\lambda}{\lambda + x \gamma p_x} + Q_i(0, x + 1) \cdot \frac{(x + 1) \gamma p_{x+1}}{\lambda + (x + 1) \gamma p_{x+1}} + Q_i(1, x - 1) \cdot \frac{\lambda}{\mu + \lambda}. \tag{9}$$

Multiplying Eq. (7) by z^x and summing for $x = 0, 1, \ldots$ we get (5). Multiplying Eq. (9) by z^x, summing for $x = 1, 2, \ldots$ and adding (8) we get (6). This proves the Lemma.

Lemma 2. *If $F_{0,i}(z)$, $F_{1,i}(z)$ are analytic functions of a complex variable z in a disk $|z| < 1 + \varepsilon$, $\varepsilon > 0$ and $p_x > 0$ for $x = 1, 2, \ldots$ then $F_{0,i+1}(z)$, $F_{1,i+1}(z)$ can be analytically continued to the disk $|z| < 1 + \varepsilon$.*

Proof. Let us assume that the series $F_{0,i}(z)$, $F_{1,i}(z)$ converge at z_0 ($1 < z_0 < 1 + \varepsilon$). It suffices then to prove convergence of $F_{1,i+1}(z)$ at z_0.

A function

$$f(q) = \frac{\gamma q + \lambda z_0}{(\gamma q + \lambda) z_0} = \frac{1}{z_0} \left(1 + \frac{\lambda(z_0 - 1)}{\gamma q + \lambda} \right)$$

is a monotonously decreasing function of $q > 0$. So, since $0 < xp_x \leqslant x$, we have

$$1 = f(0) > f(xp_x) \geqslant f(x) = \frac{x\gamma + \lambda z}{(x\gamma + \lambda)z_0} \to \frac{1}{z_0}$$

from above.

One has

$$\sum_{x=0}^{\infty} Q_i(0, x) z_0^x \frac{x\gamma p_x + \lambda z_0}{(x\gamma p_x + \lambda)z_0} < \sum_{x=0}^{x_0-1} Q_i(x, 0) z_0^x = F_{1,i}(z_0).$$

The right-hand sides of equations from Lemma 1 analytic in the disk $|z| < 1+\varepsilon$. Then the partial probability generating functions in the left-hand sides can be analyticall continued into the disk as well. The Lemma is proved.

Put

$$F_i(z) = F_{0,i}(z) + F_{1,i}(z), \qquad i = 0, 1, \ldots.$$

Lemma 3. *If $\underline{p} = \infty$ and $\lambda\mu^{-1} < 1$ then there exists an $\varepsilon > 0$ such that $|F_i(z)|$, $i = 0$, 1, $\ldots$, are uniformely bounded inside the disk $|z| \leqslant 1 + \varepsilon$ if $F_0(z)$ is analytic in the disk.*

Proof. Let $F_{0,0}(z) = 1$ and $F_{1,0}(z) = 0$. We will demonstrate that one can find $\varepsilon > 0$ such that $|F_{1,i}(z)|$, $i = 0, 1, \ldots$, are uniformly bounded inside the disk $|z| \leqslant 1 + \varepsilon$. Then from Eq. (5) we deduce uniform boundness of $|F_{0,i}(z)|$, $i = 0$, $1, \ldots$ inside the disk.

Pick a $\varepsilon' > 0$ and $z' > 1$ such that

$$\varepsilon' < 1 + \rho - \rho z' - \frac{1}{z'}, \qquad \rho z' < 2 + 2\rho, \tag{10}$$

$$\varepsilon' < 1 + \rho + \rho z' - \frac{1}{z'} \tag{11}$$

where $\rho = \lambda/\mu$. Under assumption that $xp_x \to \infty$ as $x \to \infty$, there exists $x_0 = x_0(\varepsilon', z')$ such that

$$0 \leqslant \frac{x\gamma p_x + \lambda z'}{(x\gamma p_x + \lambda)z'} \leqslant \frac{1}{z'} + \varepsilon'$$

for all $x \geqslant x_0$. Then

$$F_{1,i+1}(z') = \frac{\lambda z'}{\lambda + \mu} F_{1,i}(z') + \sum_{x=0}^{\infty} (z')^x Q_i(0,x) \frac{x\gamma p_x + \lambda z'}{(\lambda + x\gamma p_x)z'}$$

$$= \frac{\lambda z'}{\lambda + \mu} F_{1,i}(z') + \sum_{x=0}^{x_0-1} (z')^x Q_i(0,x) \frac{x\gamma p_x + \lambda z'}{(\lambda + x\gamma p_x)z'} + \sum_{x=x_0}^{\infty} (z')^x Q_i(0,x) \frac{x\gamma p_x + \lambda z'}{(\lambda + x\gamma p_x)z'}$$

$$\leqslant \frac{\lambda z'}{\lambda + \mu} F_{1,i}(z') + \left(\frac{1}{z'} + \varepsilon'\right) \sum_{x=x_0}^{\infty} (z')^x Q_i(0,x) + \sum_{x=0}^{x_0-1} (z')^x \frac{x\gamma p_x + \lambda z'}{(\lambda + x\gamma p_x)z'}.$$

Let

$$B^* = \sum_{x=0}^{x_0-1} (z')^x \frac{x\gamma p_x + \lambda z'}{(\lambda + x\gamma p_x)z'}.$$

Since

$$\sum_{x=x_0}^{\infty} (z')^x Q_i(0,x) \leqslant F_{0,i}(z')$$

independently of ε' and z', recalling also (5) and setting $\rho = \lambda/\mu$ we get

$$F_{1,i+1}(z') \leqslant \frac{\rho z'}{1+\rho} F_{1,i}(z') + \left(\frac{1}{z'} + \varepsilon'\right) \cdot \frac{1}{1+\rho} \cdot F_{1,i-1}(z') + B^*.$$

Let us introduce a dominating sequence

$$F_{i+2}^* = \frac{\rho z'}{1+\rho} F_{i+1}^* + \left(\frac{1}{z'} + \varepsilon'\right) \frac{1}{1+\rho} F_i^* + B^*, \tag{12}$$
$$F_i^* \geqslant F_{1,i}(z').$$

and prove its convergence. Equation (12) in matrix form is

$$\begin{pmatrix} F_{i+2}^* \\ F_{i+1}^* \end{pmatrix} = \begin{pmatrix} \dfrac{\rho z'}{1+\rho} & \left(\dfrac{1}{z'} + \varepsilon'\right) \dfrac{1}{1+\rho} \\ 1 & 0 \end{pmatrix} \cdot \begin{pmatrix} F_{i+1}^* \\ F_i^* \end{pmatrix} + \begin{pmatrix} B^* \\ 0 \end{pmatrix}. \tag{13}$$

Set

$$A = \begin{pmatrix} \dfrac{\rho z'}{1+\rho} & \left(\dfrac{1}{z'} + \varepsilon'\right) \dfrac{1}{1+\rho} \\ 1 & 0 \end{pmatrix}$$

The dominating sequence (12) is convergent when all eigenvalues of the matrix A have moduli below one. The eigenvalues θ_1, θ_2 of the matrix A are the roots of the equation

$$\theta^2 - \frac{\rho z'}{1+\rho} \theta - \left(\frac{1}{z'} + \varepsilon'\right) \frac{1}{1+\rho} = 0. \tag{14}$$

They are the real numbers

$$\theta_1 = \frac{1}{2}\left(\frac{\rho z'}{1+\rho} + \sqrt{\left(\frac{\rho z'}{1+\rho}\right)^2 + \frac{4}{1+\rho} \cdot \left(\frac{1}{z'} + \varepsilon'\right)}\right) > 0,$$

$$\theta_2 = \frac{1}{2}\left(\frac{\rho z'}{1+\rho} - \sqrt{\left(\frac{\rho z'}{1+\rho}\right)^2 + \frac{4}{1+\rho}\left(\frac{1}{z'} + \varepsilon'\right)}\right) < 0.$$

Firstly, $|\theta_1| < 1$ is equivalent to

$$\sqrt{\left(\frac{\rho z'}{1+\rho}\right)^2 + \frac{4}{1+\rho}\left(\frac{1}{z'}+\varepsilon'\right)} < 2 - \frac{\rho z'}{1+\rho}. \tag{15}$$

Inequality (15) holds if

$$\frac{1}{z'}+\varepsilon' < 1+\rho-\rho z', \qquad 2 - \frac{\rho z'}{1+\rho} > 0$$

and these follow from (10).

Secondly, $|\theta_2| < 1$ is when

$$\sqrt{\left(\frac{\rho z'}{1+\rho}\right)^2 + \frac{4}{1+\rho}\left(\frac{1}{z'}+\varepsilon'\right)} < 2 + \frac{\rho z'}{1+\rho}, \tag{16}$$

or

$$\frac{1}{z'}+\varepsilon' < 1+\rho+\rho z'.$$

The latter inequality is true because of (11).

So, the sequence $\{F_i^*, i = 0, 1, \ldots\}$ dominating $\{F_{1,i}(z), i = 0, 1, \ldots\}$ is convergent and bounded from above by some constant M. Then the series $F_{1,i}(z)$ are convergent in the disk $|z| \leqslant z'$ for all $i = 0, 1, \ldots$ and are uniformly bounded in modulus by the same M (i.e., $\varepsilon = z' - 1$).

Lemma 4. *If $0 < \underline{p} < \infty$ and $\lambda\mu^{-1} < \gamma p(\lambda + \gamma p)^{-1}$ then there exists an $\varepsilon > 0$ such that $|F_i(z)|$, $i = 0, 1, \ldots$, are uniformely bounded inside the disk $|z| \leqslant 1+\varepsilon$ if $F_0(z)$ is analytic in the disk.*

Proof. The proof goes along the proof of Lemma 3. When $0 < \underline{p} < 0$, there exists $x_0 = x_0(z',\varepsilon')$ such that $\underline{p} - \varepsilon' < x p_p$ and

$$\frac{x\gamma p_x + \lambda z'}{(x\gamma p_x + \lambda)z'} \leqslant \frac{1}{z'}\left(1 + \frac{\lambda(z'-1)}{\gamma\underline{p} + \lambda}\right) + \varepsilon'$$

for all $x \geqslant x_0$. On this account, we have

$$F_{1,i+1}(z') \leqslant \frac{\rho z'}{1+\rho}F_{1,i}(z') + \left(\frac{1}{z'}\left(\frac{\gamma\underline{p} + \lambda z'}{\gamma\underline{p} + \lambda}\right) + \varepsilon'\right)\cdot\frac{1}{1+\rho}\cdot F_{1,i-1}(z') + B^*.$$

We introduce a dominating sequence

$$F_{i+2}^* = \frac{\rho z'}{1+\rho}F_{i+1}^* + \left(\frac{1}{z'}\left(\frac{\gamma\underline{p} + \lambda z'}{\gamma\underline{p} + \lambda}\right) + \varepsilon'\right)\frac{1}{1+\rho}F_i^* + B^*,$$
$$F_i^* \geqslant F_{1,i}(z')$$

and, as above, consider the root of a corresponding equation

$$\theta^2 - \frac{\rho z'}{1+\rho}\theta - \left(\frac{\gamma q + \lambda z'}{z'(\gamma q + \lambda)} + \varepsilon'\right)\frac{1}{1+\rho} = 0$$

in place of (14). The roots lie inside the unit circle if $z' > 1$ and $\varepsilon' > 0$ satisfy inequalities

$$\varepsilon' < z'\Big(\frac{\gamma\underline{p}}{\gamma\underline{p} + \lambda} + \rho\Big) - \frac{\gamma\underline{p}}{\gamma\underline{p} + \lambda} - \rho(z')^2 = (z' - 1)\Big(\frac{\gamma\underline{p}}{\rho(\gamma\underline{p} + \lambda)} - z'\Big), \quad (17)$$

$$\rho z' < 2 + 2\rho, \tag{18}$$

$$\varepsilon' < z'\Big(\frac{\gamma\underline{p}}{\gamma\underline{p} + \lambda} + \rho\Big) - \frac{\gamma\underline{p}}{\gamma\underline{p} + \lambda} + \rho(z')^2. \tag{19}$$

The condition $\rho < \gamma\underline{p}(\gamma\underline{p} + \lambda)^{-1}$ guarantees that inequalities (17)–(19) have a solution. The finish the proof of the Lemma as in Lemma 3.

Proof (of Theorem 1). Assume for a moment that the hypothesis on λ, μ, and p_x, $x = 1, 2, \ldots$ holds but no stationary probability distribution exists. Then, independently of the initial probability distribution of the Markov chain (3),

$$\lim_{i \to \infty} Q_i(j, x) = 0, \qquad j = 0, 1, \quad x = 0, 1, \ldots.$$

This in turn means that mathematical expectations $\mathbf{E}\tilde{\kappa}_i$, $i = 0, 1, \ldots$ grow to infinity. By Lemmas 3 and 4, there's a constant M^* and an initial probability distribution such that functions $F_i(z)$ are analytic in an open disk $|z| < 1 + \varepsilon$ for some $\varepsilon > 0$. But for some small $\delta > 0$,

$$\mathbf{E}\tilde{\kappa}_i = \lim_{z \to 1} F_i'(z) = \frac{1}{2\pi\sqrt{-1}} \int_{|z-1|=\delta} \frac{F(z)}{(z - 1)^2}\, dz \leqslant \frac{M^*}{\delta}.$$

The contradiction proves the theorem.

Theorem 2. *Under assumtions of Theorem 1, the CTMC (1) has a unique stationary probability distribution.*

Proof. Since, under assumptions of Theorem 1, the embedded jump chain is positive recurrent, we may choose any state and consider cycles made by visits to this state. The continous-time Markov process (1) becomes a regenerative process then, and it has the limiting probability distribution if the mean cycle length is finite. But it is, indeed, since the mean sojourn times of the continous-time Markov process (1) are bounded by

$$\max\big\{(\lambda + \mu)^{-1}, (\lambda + \gamma p_1)^{-1}, (\lambda + 2\gamma p_2)^{-1}, \ldots\big\} < \lambda^{-1}.$$

Finally, the limiting distribution is also a stationary one.

The meaning of Theorems 1, 2 consists in easily verifiable sufficient conditions for stationarity. For example, the conditions are fulfilled whenn all p_x are bounded from below by a positive constant p_*, $p_x \geqslant p_* > 0$, and $\rho < 1$. Another example is

$$\rho < \frac{\gamma}{\lambda + \gamma}, \qquad p_x = \begin{cases} x^{-1} & \text{if } x \text{ is odd,} \\ p_* & \text{if } x \text{ is even,} \end{cases}$$

here

$$\liminf_{x \to \infty} x p_x = 1, \qquad \limsup_{x \to \infty} x p_x = \infty.$$

The latter example is not covered by results from [21]. Besides that it's worth mentioning that if all p_x are bounded from below by some $p_* > 0$, the stationarity condition is free of the acceptance probabilities p_x, $x = 1, 2, \ldots$. This allows to solve optimization problems (like the one in the following section) without taking special care about the stationary regime existence at different parameters values.

3   A Problem of Acceptance Probabilities Optimization with Respect to Unloading Costs

Denote by $q(\Gamma^{(j)}, x; \Gamma^{(l)}, w)$ the element of the matrix $\tilde{Q}$, corresponding to the transition from the state $(\Gamma^{(j)}, x)$ to a different state $(\Gamma^{(l)}, w)$, and by $-q(\Gamma^{(j)}, x)$ the diagonal element of $\tilde{Q}$ corresponding to the state $(\Gamma^{(j)}, x)$. Let S be the state space of the process (3). Suppose that it's partitioned into mutually disjoint subsets S_-, S_0, and S_+, such that $S_0 \neq \varnothing$, $S_+ \neq \varnothing$. We call S_- a taboo set, S_0 a critical set, and S_+ a final set. Denote by $f(j, x)$ the probability of reaching the set S_+ from an initial state $(\Gamma^{(j)}, x)$ without visits to the taboo set S_- by the process (3), and call it the taboo-probability (of taboo first passage). The taboo probabilities $f(j, x)$, $(\Gamma^{(j)}, x) \in S_0$, can be found as the minimal nonnegative solution to the linear algebraic system

$$q(\Gamma^{(j)}, x) f(j, x) = \sum_{(\Gamma^{(l)}, w) \in S_+} q(\Gamma^{(j)}, x; \Gamma^{(l)}, w)$$

$$+ \sum_{\substack{(\Gamma^{(l)}, w) \in S_0, \\ (\Gamma^{(l)}, w) \neq (\Gamma^{(j)}, x)}} q(\Gamma^{(j)}, x; \Gamma^{(l)}, w) f(l, w), (\Gamma^{(j)}, x) \in S_0 \qquad (20)$$

Denote by η the number of jumps it takes the process (3) to reach S_+ from a state $(\Gamma^{(j)}, x)$ without visiting the taboo set S_-. Furthermore, let the cost $c(\gamma, x)$ of sojourn at a state (γ, x) be given for the process (3). Then the total sojourn cost ζ over all customers up to the visit of S_+ is given by

$$\zeta = \sum_{i=0}^{\eta-1} c(\tilde{\Gamma}_i, \tilde{\kappa}_i)(\tau_{i+1} - \tau_i).$$

It follows from the general theory of Markov processes that a three-dimensional sequence $\{(\tilde{\Gamma}_i, \tilde{\kappa}_i, \tau_{i+1} - \tau_i); i = 0, 1, \ldots\}$ is a homogeneous Markov chain, and that for every $t_0 > 0$, $t_1 > 0$, ..., $t_{i+1} > 0$, $j_0, j_1, \ldots, j_{i+1} \in \{0, 1\}$, $w_0, w_1,$ ..., $w_{i+1} \in \{1, 2, \ldots, n\}$, $(j_{i+1}, w_{i+1}) \neq (j_i, w_i)$, one has

$$\mathbf{P}\left(\{\tilde{\Gamma}_{i+1} = \Gamma^{(j_{i+1})}, \tilde{\kappa}_{i+1} = w_{i+1}, \tau_{i+2} - \tau_{i+1} < t_{i+1}\} \mid H\right)$$
$$= (1 - e^{-q(\Gamma^{(j_{i+1})}, w_{i+1})t_{i+1}}) \frac{q(\Gamma^{(j_i)}, w_i; \Gamma^{(j_{i+1})}, w_{i+1})}{q(\Gamma^{(j_i)}, w_i)},$$

where

$$H = \bigcap_{l=0}^{i} \{\tilde{\Gamma}_l = \Gamma^{(j_l)}, \tilde{\kappa}_l = w_l, \tau_{l+1} - \tau_l < t_l\}.$$

This together with methods in [18] leads to the following claim.

Theorem 3. *For all* $(\Gamma^{(j)}, x) \in S_0$

$$\mathbf{E}(\zeta \mid \{\eta < \infty, \tilde{\Gamma}_0 = \Gamma^{(j)}, \tilde{\kappa}_0 = x\}) = \frac{L^{(1)}(j, x)}{f(j, x)}$$

$$\mathbf{Var}(\zeta \mid \{\eta < \infty, \tilde{\Gamma}_0 = \Gamma^{(j)}, \tilde{\kappa}_0 = x\}) = \frac{L^{(2)}(j, x)}{f(j, x)} - \left(\frac{L^{(1)}(j, x)}{f(j, x)}\right)^2,$$

where $L^{(1)}(j, x)$, $L^{(2)}(j, x)$ *are the solution of a finite linear algebraic system of equations*

$$L^{(1)}(j, x) = c(\Gamma^{(j)}, x) \frac{f(j, x)}{q(\Gamma^{(j)}, x)} + \sum_{\substack{(\Gamma^{(l)}, w) \in S_0, \\ (\Gamma^{(l)}, w) \neq (\Gamma^{(j)}, x)}} \frac{q(\Gamma^{(j)}, x; \Gamma^{(l)}, w)}{q(\Gamma^{(j)}, x)} L^{(1)}(l, w),$$

$$L^{(2)}(j, x) = 2c(\Gamma^{(j)}, x) \frac{L^{(1)}(j, x)}{q(\Gamma^{(j)}, x)} + \sum_{\substack{(\Gamma^{(l)}, w) \in S_0, \\ (\Gamma^{(l)}, w) \neq (\Gamma^{(j)}, x)}} \frac{q(\Gamma^{(j)}, x; \Gamma^{(l)}, w)}{q(\Gamma^{(j)}, x)} L^{(2)}(l, w),$$

by iterations with zero initial condition.

To formulate the optimal unloading problem, choose an integer n and set

$$S_+ = \{(\Gamma^{(0)}, 0), (\Gamma^{(1)}, 0)\},$$
$$S_0 = \{(\Gamma^{(0)}, 1), (\Gamma^{(1)}, 1), (\Gamma^{(0)}, 2), (\Gamma^{(1)}, 2), \ldots, (\Gamma^{(0)}, n), (\Gamma^{(1)}, n)\},$$
$$S_- = \{(\Gamma^{(0)}, n+1), (\Gamma^{(1)}, n+1), (\Gamma^{(0)}, n+2), (\Gamma^{(1)}, n+2), \ldots\}.$$

Let $c(\gamma, x) = x$. Then $c(\tilde{\Gamma}_i, \tilde{\kappa}_i)(\tau_{i+1} - \tau_i) = \tilde{\kappa}_i(\tau_{i+1} - \tau_i)$ is the total sojourn time over all customers during the time interval $[\tau_i, \tau_{i+1})$, and ζ is the total sojourn time over all customers up to the first visit to S_+. One natural optimization problem is to minimize the mean costs $\mathbf{E}(\zeta \mid \{\eta < \infty, \tilde{\Gamma}_0 = \Gamma^{(j)}, \tilde{\kappa}_0 = x\})$ for each $(\Gamma^{(j)}, x) \in S_0$. The overcome the multi-objective nature of the problem let us introduce a convolution of objectives with equal weights:

$$g_1(p_1, p_2, \ldots, p_n) = \frac{1}{2n} \sum_{\substack{j \in \{0,1\} \\ x \in \{1,2,\ldots,n\}}} \frac{L^{(1)}(j, x)}{f(j, x)},$$

and setup an optimization problem

$$g_1(p_1, p_2, \ldots, p_n) \to \min_{(p_1,\ldots,p_n)\in[p^*,1]^n} . \tag{21}$$

The average variance

$$g_2(p_1, p_2, \ldots, p_n) = \frac{1}{2n} \sum_{\substack{j\in\{0,1\} \\ x\in\{1,2,\ldots,n\}}} \left(\frac{L^{(2)}(j,x)}{f(j,x)} - \left(\frac{L^{(1)}(j,x)}{f(j,x)}\right)^2 \right)$$

may characterize the risk (as in financial mathematics). We also can study the risk minization problem

$$g_2(p_1, p_2, \ldots, p_n) \to \min_{(p_1,\ldots,p_n)\in[p^*,1]^n} . \tag{22}$$

For instance, for $n = 1$ we have

$$g_1(p_1) = \frac{2\mu^2 + 6\lambda\mu + p_1\gamma\mu + 3\lambda^2 + p_1\gamma\lambda}{2(\lambda + \mu)(\lambda^2 + p_1\gamma(\lambda + \mu))},$$

$$g_1'(p_1) = -\frac{\gamma(\lambda^2 + 3\lambda\mu + \mu^2)}{(\lambda^2 + p_1\gamma(\lambda + \mu))^2} < 0,$$

$$g_2(p_1) = \frac{A}{2(\lambda + \mu)^2(\lambda^2 + p_1\gamma(\lambda + \mu))^2},$$

$$A = 3\lambda^4 + 16\lambda^3\mu + 22\lambda^2\mu^2 + 12\lambda\mu^3 + 2\mu^4$$
$$+2p_1\gamma\lambda(2\mu^2 + 3\lambda\mu + \lambda^2) + 2p_1^2\gamma^2(\lambda + \mu)^2,$$

$$g_2'(p_1) = -\frac{2\gamma(\mu^3 5\lambda\mu^2 + 6\lambda^2\mu + p_1\gamma\lambda\mu + \lambda^3)}{(\lambda^2 + p_1\gamma(\lambda + \mu))^3} < 0.$$

Obviously, the solutions to optimization problems (21), (22) are $p_1 = 1$.

When $n \geqslant 2$, both optimization problems become analytically untractable. A script was written in the Octave programming language [23] to build and solve equations from Theorem 3, and also an heuristic search was done, firstly among threshold policies of the form ($i_0 = \overline{0, n}$)

$$p_i = 1, \ i \leqslant i_0; \quad p_i = p^* > 0, \ i > i_0 \tag{23}$$

and, secondly, using the Octave's built-in successive quadratic programming solver for nonlinear optimization with several random initial positions in the hypercube $[p^*, 1]^n$.

Our experiments demonstrate, that $\rho = \lambda/\mu$ plays an important role for the structure of the optimal acceptance probabilities. When ρ is small, the best conditional mean unloading cost and the best conditional variance of the unloading cost were with $p_i = 1$ for all $i = 1, 2, \ldots, n$, i.e. the best control is to always accept requests from the orbit. But when ρ approaches 1, a threshold policy becomes much more efficient. Let us consider the following numerical example.

Let $N = 15$, $p^* = 10^{-4}$, $\lambda = 0.5$, $\gamma = 1.0$, and $\mu = 1.5$. Then the optimal choice is (23) with $i_0 = N$ (i.e., always accept), $g_1(1, \ldots, 1) = 140.78$. When $\lambda = 0.9$, the optimal choice with respect to the quantity $g_1(\cdot)$ is is (23) with $i_0 = 12$, $g_1 = 132.66$. And the optimal choice with respect to the quantity $g_2(\cdot)$ is

$$p_1 = \ldots = p_5 = 1, \quad p_6 = 0.457, \quad p_7 = \ldots = p_{15} = 10^{-4},$$

the value of the objective is 4785.58 (compare it to $g_2(1, \ldots, 1) = 13141.08$). Finally, when $\lambda = 1.4$, the threshold policy with $i_0 = 5$ still minimizes the conditional mean unloading cost, while the threshold policy with $i_0 = 3$ minimizes the conditional variance.

4   Conclusion

This paper discusses a new optimization problem for retrial queues. The stationarity condition for a retrial queue with randomized acceptance from the orbit given here extends known results for the classical retrial M/M/1 queue. Numerical experiments demonstrate that a retrial queue can unload with less total waiting time (without getting too crowded) if the acceptance probabilities have a threshold form.

References

1. Falin, G.I., Templeton, J.G.C.: Retrial Queues. CRC Press, Boca Raton (1997)
2. Koba, E.V.: Stability conditions for some typical retrial queues. Cybern. Syst. Anal. **44**(1), 100–103 (2005)
3. Artalejo, J.R., Gómez-Corral, A.: Retrial Queueing Systems. Springer, Heidelberg (2008). https://doi.org/10.1007/978-3-540-78725-9
4. Lakatos, L.: Cycling-waiting systems. Cybern. Syst. Anal. **19**(2), 176–180 (2010)
5. Nazarov, A.A., Sudyko, E.A.: Method of asymptotic semiinvariants for studying a mathematical model of a random access network. Probl. Inform. Transm. **46**(1), 86–102 (2010)
6. Nazarov, A., Sztrik, J., Kvach, A.: A survey of recent results in finite-source retrial queues with collisions. In: Dudin, A., Nazarov, A., Moiseev, A. (eds.) ITMM/WRQ -2018. CCIS, vol. 912, pp. 1–15. Springer, Cham (2018). https://doi.org/10.1007/978-3-319-97595-5_1
7. Kim, J., Kim, B.: A survey of retrial queueing systems. Ann. Oper. Res. **247**(1), 3–36 (2016)
8. Nazarov, A.A., Yurevich, N.M.: A study of a network with a dynamic ALOHA random multiple acces protocol. Autom. Control Comput. Sci. **29**(6), 39–44 (1995)
9. Nazarov, A.A., Kuznetsov, D.Y.: Analysis of a communication network governed by an adaptive random multiple access protocol in critical load. Probl. Inform. Transm. **40**(3), 243–253 (2004)
10. Artalejo, J.R.: Analysis of an $M/G/1$ queue with constant repeated attempts and server vacations. Comput. Oper. Res. **24**(6), 493–504 (1997)
11. Artalejo, J.R., Dudin, A.N., Klimenok, V.I.: Stationary analysis of a retrial queue with preemptive repeated attempts. Oper. Res. Lett. **28**(4), 173–180 (2001)

12. Kernane, T.: Conditions for stability and instability of retrial queueing systems with general retrial times. Stat. Probab. Lett. **78**, 3244–3248 (2008)

13. Nazarov, A.A., Sadyko, E.A.: Nonergodicity of mathematic model of a random access network. Vestnik of SibGAU **36**(3), 62–65 (2011)

14. Kitaev, M.Yu., Rykov, V.V.: Controlled queueing systems. CRC Press, Boca Raton (1995)

15. Sennott, L.: Stochastic Dynamic Programming and The control of Queueing Systems. Wiley, Hoboken (1999)

16. Fedotkin, M.A.: Algebraic properties of Chung functionals of homogeneous Markov chains with countable state space (Russian). Dokl. Akad. Nauk SSSR **1**, 43–46 (1976)

17. Golysheva, N.M., Fedotkin, M.A.: Cyclic control of conflicting flows under conditions of the birth and death of critical size queues. Autom. Remote Control **51**(4), 479–484 (1990)

18. Zorine, A.V.: Minimization of unloading cost for an exponential time-sharing queueing process. Tomsk State Univ. J. Control Comput. Sci. **4**(17), 55–63 (2011). (in Russian)

19. Zorine, A.V.: On ergodicity conditions in a polling model with Markov modulated input and state-dependent routing. Queue. Syst. **76**(2), 223–241 (2014)

20. Neuts, M.: Matrix-Geometric Solutions in Stochastic Models: An Algorithmic Approach. Dover publications, Inc., Mineola (1981)

21. Dudin, A., Nazarov, A.: On a tandem queue with retrials and losses and state dependent arrival, service and retrial rates. Int. J. Oper. Res. **29**(2), 170–182 (2017)

22. Klimenok, V., Dudin, A.: Multi-dimensional asymptotically quasi-Toeplitz Markov chains and their application in queueing theory. Queue. Syst. **54**, 245–259 (2006)

23. GNU Octave Homepage. http://www.gnu.org/software/octave. Accessed 14 May 2019

Analysis of Retrial Queues for Cognitive Wireless Networks with Sensing Time of Secondary Users

Kohei Akutsu[1] and Tuan Phung-Duc[2(✉)]

[1] Graduate School of Systems and Information Engineering,
University of Tsukuba, 1-1-1 Tennodai, Tsukuba, Ibaraki, Japan
s1820435@s.tsukuba.ac.jp
[2] Faculty of Engineering, Information and Systems, University of Tsukuba,
1-1-1 Tennodai, Tsukuba, Ibaraki, Japan
tuan@sk.tsukuba.ac.jp

Abstract. This paper considers a cognitive radio network system which has two classes of users that we call Primary Users and Secondary Users. We model this system using a single server retrial queueing model. In the conventional retrial queue, if a customer cannot find an idle channel, he joins the orbit and retries to occupy a channel. The new feature of our model is that every arriving Secondary User first enters the orbit. He starts to sense the channels and tries to find an idle channel. For this model, we obtain explicit expressions for the joint generating functions of the state of the server and the number of Secondary Users in the orbit. In addition, we derive the necessary stability condition. We obtain the distribution of the number of retrials by a Secondary User using simulation. Beside, we consider the multiserver model for which we obtain the average of the number of Secondary Users in the orbit and the distribution of the number of retrials by simulation.

Keywords: Retrial queues · Interruption · Number of retrials · Cognitive wireless networks · Number of sensings

1 Introduction

In recent years, Internet traffic has been explosively increasing due to the increase of smart phones, tablet computers and other electronic devices. As a result, wireless resource, i.e. radio frequencies are shortage [1]. Thus, we need technologies that improve the efficiency of a wireless spectrum. Cognitive radio is considered as one of promising technologies that can solve the bandwidth shortage problem. Cognitive wireless technology enables unlicensed users to utilize frequencies that were originally allocated to licensed users [1]. There are two types of cognitive radio technologies; frequency sharing type and heterogeneous type [2]. In the former, wireless terminals search for unused frequencies or idle time of licensed bandwidth and cognitively use them to communicate. On the other hand, in

© Springer Nature Switzerland AG 2019
T. Phung-Duc et al. (Eds.): QTNA 2019, LNCS 11688, pp. 77–91, 2019.
https://doi.org/10.1007/978-3-030-27181-7_6

the latter, the mobile terminals dynamically choose their optimal communication method according to the congestion level and connection situations. In this paper, we focus on the frequency sharing cognitive radio networks. In these systems, there exists two classes of users that we call Primary Users (PUs) and Secondary Users (SUs). PUs are referred to as the licensed users that own the bandwidth and SUs use the bandwidth when PUs are not present.

The sensing mechanism of SUs is similar to the retrial behavior of customers in retrial queues. In retrial queues, an arriving customer that cannot occupy a server upon arrival is blocked and joins a virtual waiting room called orbit from which the customer retries again after some random time until being served. From that point of view, our queueing model is close to some retrial queues in the literature. The main difference of our model is that every SU in our model needs to sense the state of the servers, meaning that each SU first enters the sensing pool (orbit in retrial queue literature) spending sensing time before occupying a server.

There are previous studies concerning retrial queues. Keilson et al. [3] analyzes $M/G/1/1$ retrial queue, deriving performance measures of the system. Falin [4,5] derives the waiting time distribution of customers in the orbit while Falin [6] derives the limiting theorem for the waiting time of customers under heavy traffic. Falin [7] focuses on the number of retrials, deriving the average, the variance and the probability distribution. In these models, there is only one class of users. In our model in this paper, there are two classes of users, i.e. PUs and SUs. SUs must sense before entering the server and a SU on transmission may be interrupted upon the arrival of a PU.

Some closely related works concerning cognitive radio networks are as follows. Konishi et al. [8] consider cognitive radio networks in which SUs request for a random number channels simultaneously, analyzing the performance measures such as the probability that SUs cannot enter the servers and the probability that SUs are forced to terminate their transmissions. The former probability is called blocking probability. In this model, if the number of idle servers is less than that a SU requires, the SU is blocked. In [8], the sensing mechanism is not taken into account.

Salameh et al. [9] propose a queueing model for cognitive radio networks with reactive-decision spectrum handoff. By limiting the number of simultaneously sensing SUs, the authors derive some performance measures such as the blocking probability, the average of the number of blockings, and the time to complete the transmission. In [9], a SU continuously senses the servers one by one until he finds an idle one. Furthermore, a SU on transmission may be interrupted upon the arrival of a PU. The interrupted SU senses again until he finds an idle server. In this model, they assume that the size of the orbit is limited.

Retrial queues with two-way communication presented by Artalejo and Phung-Duc [10,11], Sakurai and Phung-Duc [12] are also closely related to our model. In the models with two-way communications, the server makes out-going calls when he is idle. In contrast, in our model PUs may interrupt SUs in transmission.

In this paper, we consider a cognitive radio network system and model it using a single server retrial queue. The new feature of our model is that every arriving SU enters the orbit. They start to sense the server until they find an idle server. For this model, we obtain explicit expressions for the joint generating function of the state of the server and the number of SUs in the orbit. In addition, we derive the necessary stability condition. We find the distribution of the number of retrials (sensings) of SUs by simulation. Besides, we consider an extended model with multiple servers (multiserver). We obtain the average of the number of SUs in the orbit and the distribution of the number of retrials of SUs for this model by simulation.

The rest of this paper is organized as follows. In Sect. 2, we present an overview of the system and the setting of the model. In Sect. 3, we present the analysis of the model deriving the performance measures. Section 4 shows some numerical results for the performance measures. In Sect. 5, we expand the single server model to the multiserver one and show some numerical examples. Finally, in Sect. 6, we conclude our paper.

2 An Overview of the System and the Setting of the Model

2.1 An Overview of the System

In the system in this paper, we consider two classes of users, PUs and SUs. They share channels and PUs have preemptive priorities over SUs. Their behaviors differ in the following two points. First, every SU must sense before occupying a channel. Second, when a PU arrives at a channel and he cannot find an idle one, a SU on transmission is interrupted. The interrupted SU returns to the sensing pool and starts to sense again.

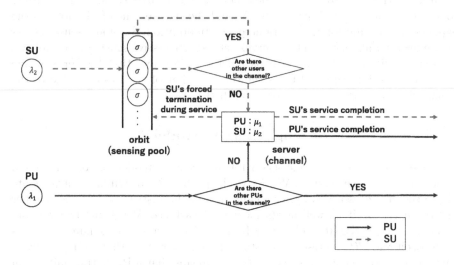

Fig. 1. Our model's flowchart.

In the sequel, we explain PU and SU's behaviors in the system. PUs perform the following procedures;

1. A PU arrives at the system.
2. If the PU finds an idle channel, he starts to transmit. If all the channels are occupied by other PUs, the newly arriving PU leaves the system. In case some channels are used by SUs, the ongoing SU in one of these channels is interrupted and the PU occupies the channel.
3. The PU finishes the transmission and leaves the system.

Furthermore, SUs perform the following procedures;

1. A SU arrives at the system.
2. The SU starts to sense channels.
3. When the SU finishes sensing and finds an idle channel, the SU starts to transmit. If the SU finishes sensing and cannot find an idle channel, he senses again.
4. When a PU arrives at a channel which a SU is using and there are no other idle channels, the SU is interrupted and returns to the sensing pool again.
5. If the SU finishes his transmission, he leaves the system.

Figure 1 represents an overview of the system as a flowchart.

2.2   The Setting of the Model

In Sect. 2.1, we described an overview of the system with multiple channels. In this section, we make necessary assumptions in this model. Note that in this paper, a channel has the same meaning as a server. We can represent this system with a retrial queueing model. PUs arrive at servers according to a Poisson process with parameter λ_1. The service time of PUs follows the exponential distribution with parameter μ_1. On the other hand, SUs arrive at servers according to a Poisson process with parameter λ_2 and the service time of SUs follows the exponential distribution with parameter μ_2. In addition, SUs must sense servers before arriving at servers. The interval between successive sensing times follows the exponential distribution with parameter σ. We note that the service time distribution of interrupted SUs is the same as that of the fresh SU due to the memoryless property of the exponential distribution.

3   The Analysis of Single Server Model

In this section, we use a single server retrial queueing model for cognitive radio networks with single channel. We derive the probability distribution of the state of the server, necessary stability condition and the distribution of the number of SUs in the orbit as well as its mean and variance. We define two random variables; $N(t) \in \mathbb{N} \cup \{0\}$ and $C(t) \in \{0, 1, 2\}$. $N(t)$ expresses the number of SUs in the orbit and $C(t)$ represents the state of the server. Note that $C(t) = 0$ means that there is no user in the server; $C(t) = 1$ means that a PU is transmitting in the server and $C(t) = 2$ means that a SU is transmitting in the server. Under these settings, $\{(C(t), N(t))|t \geq 0\}$ becomes a continuous time Markov chain. We assume that this chain is stable and that the input processes of PUs and SUs and their service time are mutually independent. Subsequently, we define the joint probability mass function of $\{C(t) = i, N(t) = j\}$ as $\pi_{i,j}$. Note that we assume that the chain is in steady state, therefore it is not necessary to take into account the time. We set the partial generating functions of the state of the server and the number of SUs in the orbit as $\Pi_i(z) = \sum_{j=0}^{\infty} \pi_{i,j} z^j$ for $i \in \{0, 1, 2\}$.

3.1   The Probability Distribution of the State of the Server

In this section, we set up the balance equations for states $(0, j)$, $(1, j)$ and $(2, j)$ $(j \in \mathbb{N} \cup \{0\})$ and derive the probability distribution of the state of the server. The balance equations read as follows.

$$
\begin{cases}
(\lambda_1 + \lambda_2 + j\sigma)\pi_{0,j} = \lambda_2 \pi_{0,j-1} + \mu_1 \pi_{1,j} + \mu_2 \pi_{2,j}, \\
(\lambda_2 + \mu_1)\pi_{1,j} = \lambda_1 \pi_{0,j} + \lambda_1 \pi_{2,j-1} + \lambda_2 \pi_{1,j-1}, \\
(\lambda_1 + \lambda_2 + \mu_2)\pi_{2,j} = (j+1)\sigma\pi_{0,j+1} + \lambda_2 \pi_{2,j-1},
\end{cases}
$$

where $\pi_{i,-1} = 0$ for $i \in \{0, 1, 2\}$. Multiplying these equations by z^j and summing up over $j \in \mathbb{N} \cup \{0\}$, we obtain

$$
\begin{cases}
(\lambda_1 + \lambda_2)\Pi_0(z) + \sigma z \Pi_0'(z) = \lambda_2 z \Pi_0(z) + \mu_1 \Pi_1(z) + \mu_2 \Pi_2(z), & (1) \\
(\lambda_2 + \mu_1)\Pi_1(z) = \lambda_1 \Pi_0(z) + \lambda_1 z \Pi_2(z) + \lambda_2 z \Pi_1(z), & (2) \\
(\lambda_1 + \lambda_2 + \mu_2)\Pi_2(z) = \sigma \Pi_0'(z) + \lambda_2 z \Pi_2(z). & (3)
\end{cases}
$$

By (2) and $\Pi_0(1) + \Pi_1(1) + \Pi_2(1) = 1$, we obtain

$$
\Pi_1(1) = \frac{\lambda_1}{\mu_1 + \lambda_1}, \tag{4}
$$

$$
\Pi_0(1) + \Pi_2(1) = \frac{\mu_1}{\mu_1 + \lambda_1}. \tag{5}
$$

Transforming (2) and (3) and rearranging the result yields

$$
\Pi_1(z) = \frac{\lambda_1 \Pi_0(z) + \lambda_1 z \Pi_2(z)}{\lambda_2 + \mu_1 - \lambda_2 z}, \tag{6}
$$

$$
\Pi_2(z) = \frac{\sigma \Pi_0'(z)}{\lambda_1 + \lambda_2 + \mu_2 - \lambda_2 z}. \tag{7}
$$

Substituting (6) and (7) into (1), we have

$$
\frac{\Pi_0'(z)}{\Pi_0(z)} = \gamma(z), \tag{8}
$$

where
$$\gamma(z) := \frac{\lambda_2(\lambda_1 + \lambda_2 + \mu_1 - \lambda_2 z)(\lambda_1 + \lambda_2 + \mu_2 - \lambda_2 z)}{\sigma(\mu_2(\lambda_2 + \mu_1) - \lambda_2(\lambda_1 + \lambda_2 + \mu_1 + \mu_2)z + z^2\lambda_2^2)}.$$

Substituting (8) in terms of $\Pi_0'(z)$ into (7) and setting $z = 1$ in the result, we obtain

$$\Pi_2(1) = \frac{\lambda_2(\lambda_1 + \mu_1)}{\mu_2\mu_1 - \lambda_1\lambda_2 - \lambda_2\mu_1}\Pi_0(1). \tag{9}$$

Solving (5) and (9), we have

$$\Pi_0(1) = \frac{\mu_2\mu_1 - \lambda_1\lambda_2 - \lambda_2\mu_1}{\mu_2(\mu_1 + \lambda_1)}, \tag{10}$$

$$\Pi_2(1) = \frac{\lambda_2}{\mu_2}. \tag{11}$$

(4), (10) and (11) represent the probability distribution of the state of the server. Note that the sensing rate, σ, is not appeared in the probability of the state of the server.

3.2   Necessary Stability Condition

In Sect. 3.1, we can obtain the probability distribution of the state of the server. We assume that this system is in steady state. If this assumption satisfies, there exists a stationary distribution. Therefore we get

$$\Pi_1(1) + \Pi_2(1) < 1.$$

From this, we obtain the following condition as the necessary stability condition.

$$\frac{\lambda_1}{\mu_1 + \lambda_1} + \frac{\lambda_2}{\mu_2} < 1, \tag{12}$$

$$\Leftrightarrow \frac{\lambda_2}{\mu_2} < \frac{\mu_1}{\mu_1 + \lambda_1}. \tag{13}$$

Here we consider this condition. In (12), the first term and the second term in the left side respectively equal $\Pi_1(1)$ and $\Pi_2(1)$. And, by (4), we realize that the right side of (13) represents the probability that a PU is not transmitting. Noting this and the fact that the left side of (13) represents SU's traffic intensity, (13) means that SU's traffic intensity is smaller than the probability that PU is not transmitting. Finally, the sensing rate, σ, did not arise in this condition.

3.3   Joint Distributions of the Number of SUs in the Orbit

In this section, we derive partial generating functions of the number of SUs in the orbit. Solving (8) in terms of $\Pi_0(z)$, we obtain the following result.

$$\Pi_0(z) = \Pi_0(1)\exp(\int_1^z \gamma(u)du). \tag{14}$$

Letting z_1 and z_2 ($z_1 < z_2$) denote the poles of $\gamma(z)$, we can represent $\gamma(z)$ below.

$$\gamma(z) = \frac{A}{z_1 - z} + \frac{B}{z_2 - z}, \tag{15}$$

where

$$z_1 = \frac{\lambda_1 + \lambda_2 + \mu_1 + \mu_2 - \sqrt{(\lambda_1 + \lambda_2 + \mu_1 + \mu_2)^2 - 4\mu_2(\lambda_2 + \mu_1)}}{2\lambda_2},$$

$$z_2 = \frac{\lambda_1 + \lambda_2 + \mu_1 + \mu_2 + \sqrt{(\lambda_1 + \lambda_2 + \mu_1 + \mu_2)^2 - 4\mu_2(\lambda_2 + \mu_1)}}{2\lambda_2},$$

$$A = \frac{\lambda_2}{\sigma} \frac{(\lambda_1 + \lambda_2 + \mu_2 - \lambda_2 z_1)(\lambda_1 + \lambda_2 + \mu_2 - \lambda_2 z_1)}{\lambda_2^2(z_2 - z_1)},$$

$$B = \frac{\lambda_2}{\sigma} \frac{(\lambda_1 + \lambda_2 + \mu_2 - \lambda_2 z_2)(\lambda_1 + \lambda_2 + \mu_2 - \lambda_2 z_2)}{\lambda_2^2(z_1 - z_2)}.$$

Differentiating (6) and (7) with respect to z and substituting $z = 1$, we have

$$\Pi_1'(1) = \frac{\mu_1\{\lambda_1\Pi_0'(1) + \lambda_1\Pi_2(1) + \lambda_1\Pi_2'(1)\} + \lambda_1\lambda_2\{\Pi_0(1) + \Pi_2(1)\}}{\mu_1^2}, \tag{16}$$

$$\Pi_2'(1) = \frac{\sigma\Pi_0''(1)(\lambda_1 + \mu_2) + \lambda_2\sigma\Pi_0'(1)}{(\lambda_1 + \mu_2)^2}. \tag{17}$$

$\Pi_1'(1)$ represents the average of the number of SUs in the orbit when a PU is transmitting and $\Pi_2'(1)$ represents the average of the number of SUs in the orbit when a SU is transmitting. Substituting $z = 1$ into (8) and substituting (10) into the result, we obtain

$$\Pi_0'(1) = \frac{\lambda_2(\lambda_1 + \mu_2)}{\sigma\mu_2}, \tag{18}$$

$$\Pi_0''(1) = \gamma'(1)\Pi_0(1) + \gamma(1)\Pi_0'(1). \tag{19}$$

Combining (16), (17) and (18), we obtain the average of the number of SUs in the orbit,

$$E[N] = \frac{\Pi_0''(1)\sigma}{\lambda_1 + \mu_2} + \frac{\lambda_2^3}{\mu_1\mu_2(\lambda_1 + \mu_2)} + \frac{\lambda_2^2(\lambda_1 + \mu_1 + \mu_2)}{\mu_1\mu_2(\lambda_1 + \mu_2)}$$
$$+ \lambda_2\left\{\frac{1}{\sigma} + \frac{\lambda_1^2}{\sigma\mu_1\mu_2} + \frac{\Pi_0''(1)\sigma\mu_2}{\mu_1(\lambda_1 + \mu_2)^2} + \lambda_1\left(\frac{1}{\mu_1(\lambda_1 + \mu_1)} + \frac{1}{\sigma\mu_2} + \frac{1}{\sigma\mu_1} + \frac{\Pi_0''(1)\sigma}{\mu_1(\lambda_1 + \mu_2)^2}\right)\right\}.$$

Finally, we derive generating functions of the joint stationary probability of the state of the server and the number of SUs in the orbit. Substituting (15) into (14), we obtain the partial generating function when nobody is transmitting.

$$\Pi_0(z) = \Pi_0(1)\left(\frac{z_1 - 1}{z_1 - z}\right)^A \left(\frac{z_2 - 1}{z_2 - z}\right)^B. \tag{20}$$

Using (6), (7) and (20), we obtain $\Pi_1(z)$ and $\Pi_2(z)$ in detail.

4 Numerical Results

In this section, when PU and SU's arrival rates or SU's retrial rate change, we discuss how the mean and the variance of the number of SUs in the orbit change, using figures. Generally, in the cognitive radio systems, it is likely that the transmission time of SUs is shorter than that of PUs. Therefore, we set $\mu_1 = 4$ and $\mu_2 = 20$ in all the numerical experiments.

4.1 The Average and Variance of the Number of SUs in the Orbit

First, we consider how the average and the variance of the number of SUs in the orbit change under the stability condition. Figure 2 shows that the average and the variance of the number of SUs in the orbit against the arrival rate of PUs for various values of the sensing parameter. We observe from Fig. 2 that the average and the variance of the number of SUs in the orbit decrease with the increase of sensing parameter σ. We can interpret this change below; the higher the arrival rate of PUs is, the higher the probability of PU's existence in the server is, too. As a result, it is difficult for a SU to occupy the server and the number of SUs in the orbit increases. Figure 3 shows the average and the variance of the number of SUs in the orbit against the arrival rate of SUs for various values of the sensing parameter. Similar to Fig. 2, we observe from Fig. 3 that the average and the variance of the number of SUs in the orbit decrease with the increase of sensing parameter σ and increase with the arrival rate of SUs. We also interpret this

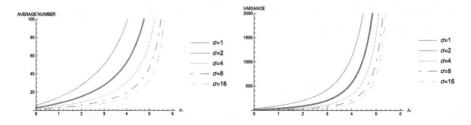

Fig. 2. The change of the average and the variance of the number of SUs in the orbit when the arrival rate of PUs changes, where $\mu_1 = 4$ and $\mu_2 = 20$.

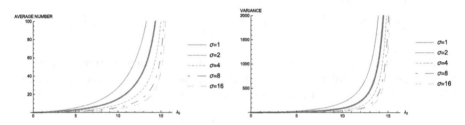

Fig. 3. The change of the average and the variance of the number of SUs in the orbit when the arrival rate of SUs changes, where $\mu_1 = 4$ and $\mu_2 = 20$.

change below; the higher the arrival rate of SUs is, the higher the probability of SU's presence in the server is, too. As a result, it is difficult for a SU to occupy the server and the number of SUs in the orbit increases.

4.2 The Number of Retrials for SUs

In the sequel, we consider the distribution of the number of retrials of SUs under the stability condition. Note that Fig. 4 is obtained by simulation. The left side of Fig. 4 shows the distribution of the number of retrials of SUs against the arrival rate of SUs (λ_2). We observe that the distribution of the number of retrials is greatly affected by the arrival rate of SUs. The right side of Fig. 4 shows the distribution of the number of retrials of SUs for different retrial rates (σ). We observe that the distribution of the number of retrials for SUs is more insensitive to the change of retrial rate than the change of the arrival rate of SUs. These results are natural because the arrival rate is used in stability condition and the retrial rate is not used in it.

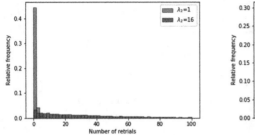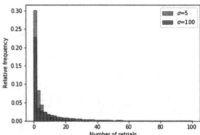

Fig. 4. The distribution of the number of retrials of SUs for different λ_2 or different σ.

5   Multiserver Model

5.1   The Setting of the Model

In the above section, we considered a single server retrial queueing model. In this section, we extend this model to a model with c identical servers. Note that in [9], the size of the orbit is limited, however, in our model, the size of the orbit is unlimited. Similar to the single server model, PUs arrive at the servers according to a Poisson process with parameter λ_1. The service time of PUs follows the exponential distribution with parameter μ_1. On the other hand, SUs arrive at the servers according to a Poisson process with parameter λ_2 and the service time of SUs follows the exponential distribution with parameter μ_2. As with our single server model, SUs must sense servers before arriving at the servers. The sensing time follows the exponential distribution with parameter σ. The behaviors of PUs and SUs are the same as that in the single server model. Compared with the single server model, we make a change in the definitions of random variables. The reason is that we have to consider the number of idle servers. We define

$N(t), C_1(t)$ and $C_2(t)$ as the number of SUs in the orbit, the number of PUs on transmission and the number of SUs on transmission. Under these settings, $\{(N(t), C_1(t), C_2(t))|t \geq 0\}$ forms a continuous time Markov chain. We assume that the input process of PUs and SUs and their service time are mutually independent. Under these settings, we perform simulations.

5.2   The Average of the Number of SUs in the Orbit

First, we consider the average of the number of SUs in the orbit with the change of the arrival rate of PUs when the number of the servers $c = 1, 2, 3, 4$ and 5. The left hand side of Fig. 5 shows the change of the average of the number of SUs in the orbit with the change of the arrival rate of PUs. This result shows that when the number of servers is small, a little change in the number of servers greatly affects the average of the number of SUs in the orbit. However, the change is not significant if the number of servers is large.

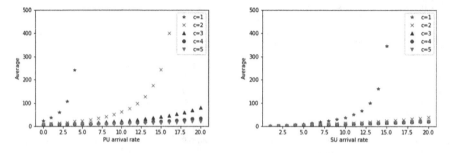

Fig. 5. The average of the number of SUs in the orbit with the change of the arrival rate of PUs or the arrival rate of SUs when $c = 1, 2, 3, 4$ and 5.

Next, we consider the average of the number of SUs in the orbit with the change of the arrival rate of SUs when the number of the servers $c = 1, 2, 3, 4$ and 5. The right hand side of Fig. 5 shows the change of the average of the number of SUs in the orbit with the change of the arrival rate of SUs. As with the left side of Fig. 5, this result shows that when the number of servers is small, a little change in the number of servers greatly affects the average of the number of SUs in the orbit. However, the change is not significant if the number of servers is large.

5.3   Distribution of the Number of Retrials for SUs

In the sequel, we consider the distribution of the number of retrials of SUs with the change of the arrival rate of SUs or the retrial rate of SUs when the number of the servers $c = 3, 4$ and 5. Figures 6, 7 and 8 show the distribution of the number of retrials for several values of λ_2 and σ. As with the single server model, these results show that the distribution of the number of retrials for SUs is more insensitive to the change of retrial rate than the change of the arrival rate of SUs.

5.4   Stability Condition for Multiserver Model

In Sect. 3.2, we derived the necessary stability condition for a single server. From the results of that stability condition, we consider the stability condition for our multiserver model. First, we consider the stability condition in the $M/M/c$ queueing model. In this model, we assume that users arrive at servers according to a Poisson process with parameter λ and their service time follows the exponential distribution with parameter μ. Then, we can derive the model's stability condition as follows.

$$\frac{\lambda}{c\mu} < 1 \Leftrightarrow \frac{\lambda}{\mu} < c.$$

This condition means that the number of available servers is greater than the mean of the number of customers in the servers. Next, we consider the stability condition for our single server model again. In (13), we realize that the left side represents the probability that SUs are transmitting, and the right side represents the probability that PUs are not transmitting. And we can interpret that the right side of (13) represents the expected value of the number of servers that PUs are not transmitting. Therefore, we conjecture that the stability condition for our multiserver model as follows.

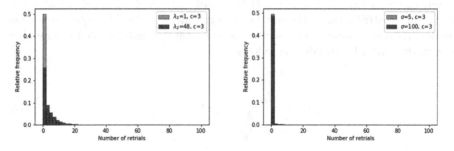

Fig. 6. The distribution of the number of retrials of SUs for different λ_2 or different σ when $c = 3$.

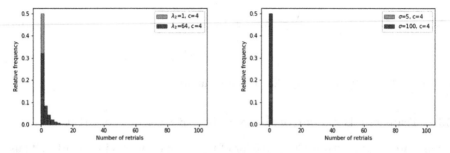

Fig. 7. The distribution of the number of retrials of SUs for different λ_2 or different σ when $c = 4$

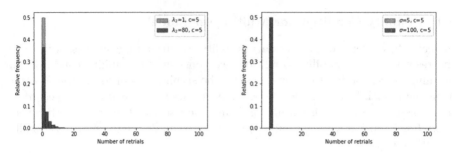

Fig. 8. The distribution of the number of retrials of SUs for different λ_2 or different σ when $c = 5$.

$$\frac{\lambda_2}{\mu_2} < \sum_{i=0}^{c}(c-i)\pi_i, \tag{21}$$

where π_i represents the steady state probability that the number of transmitting PUs equals i. We explain the meaning of (21). The left side represents the offered load of SUs. On the other hand, the right side represents the expected value of the number of servers that PUs are not transmitting, namely the expected value of the number of servers that are available for SUs. And, because of the setting of our model, it follows that the stationary distribution of PUs is identical to the stationary distribution of the number of customers of the $M/M/c/c$ queueing model. Therefore, we can represent the steady state probability that the number of transmitting PUs equals i as follows.

$$\pi_i = \frac{\dfrac{\lambda_1^i}{i!\mu_1^i}}{\sum_{k=0}^{c}\dfrac{\lambda_1^k}{k!\mu_1^k}}, i = 0, 1, \cdots, c. \tag{22}$$

Substituting it into (21) and transforming the result, we obtain

$$\frac{\lambda_2}{\mu_2} < c - \frac{\lambda_1}{\mu_1}\frac{\sum_{k=0}^{c-1}\dfrac{\lambda_1^k}{k!\mu_1^k}}{\sum_{k=0}^{c}\dfrac{\lambda_1^k}{k!\mu_1^k}}. \tag{23}$$

In the next section, we verify (23) using simulation.

5.5   Transient Behavior of the Number of SUs in the Orbit

Finally, we show the transient behavior of the number of SUs in the orbit. Thus, we perform simulation of the transition of the number of SUs in the orbit in the range. We set the parameters with $\lambda_1 = 1, \mu_1 = 4, \mu_2 = 20$ and $\sigma = 1$. Therefore, from (23), we can estimate the parameter of the boundary of stability region for

λ_2 for each c. When $c = 3$, the parameter of the boundary of stability region is $\lambda_2 \fallingdotseq 55$. In the same way, we obtain the parameters of the boundary of stability region are $\lambda_2 \fallingdotseq 75$ when $c = 4$ and $\lambda_2 \fallingdotseq 95$ when $c = 5$. Based on these parameters, we obtain Figs. 10 and 12 with the change of SU arrival rate by simulation. These figures show that our estimation is reasonable.

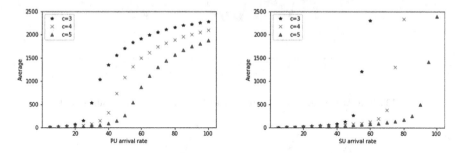

Fig. 9. The average of the number of SUs in the orbit with the change of the arrival rate of PUs or the arrival rate of SUs when $c = 3, 4$ and 5, where $\lambda_1 = 1, \lambda_2 = 8, \mu_1 = 4, \mu_2 = 20$ and $\sigma = 1$.

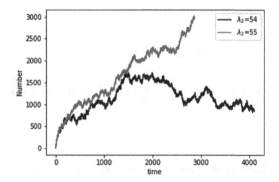

Fig. 10. The transition of the number of SUs in the orbit when $c = 3$.

5.6   Remarks on the Simulations

In all the simulations concerning the average of the number of SUs in the orbit and the distribution of the number of retrials of SUs, we consider 50 sample paths of the underlying Markov chains and take the average of these 50 sample paths. In these cases, in each sample path, we consider 1,100,000 jumps and we discard the first 100,000 jumps so as to consider the stationary behavior. For the graphs of the average of the number of SUs in the orbit, we plot the average of the 50 sample paths and in the graphs of the distributions we plot the histogram based on these 50 sample paths. In the simulation concerning the transition of the number of SUs in the orbit, we consider 1 sample path of the underlying Markov chains. We simulate 5,000,000 jumps.

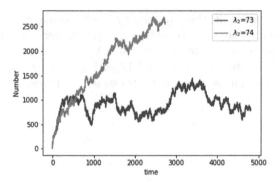

Fig. 11. The transition of the number of SUs in the orbit when $c = 4$.

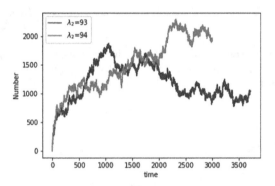

Fig. 12. The transition of the number of SUs in the orbit when $c = 5$.

6   Conclusion

In this paper, we have considered cognitive radio network systems and model them using a single server retrial queueing model and a multiserver queueing model. The feature of our model is that every arriving SU enters the orbit. They start to sense servers until they find an idle server. In single server model, we obtain explicit expressions for the partial generating functions. In addition, we derive the necessary stability condition. We find the distribution of the number of retrials of SUs for various values by simulation. In multiserver model, we obtain the average of the number of SUs in the orbit and the distribution of retrials of SUs for various values by simulation.

In the future work, we plan to derive the stability condition for the multi-server model and perform the stationary analysis under the stability condition.

Acknowledgments. The research of Tuan Phung-Duc was supported in part by JSPS KAKENHI Grant Number 18K18006.

References

1. Cognitive radio, efficiently utilizing white-space for communication. Web, 30 November 2018. (in Japanese), https://cend.jp/user/headline_technology/20180319.html
2. What is cognitive radio? Indispensable technology for realizing ubiquitous. Web, 30 November 2018. (in Japanese), https://www.sbbit.jp/article/cont1/19899
3. Keilson, J., Cozzolino, J., Young, H.: Service system with unfilled requests repeated. Oper. Res. **6**, 1126–1137 (1968)
4. Falin, G.I.: Aggregate arrival of customers in a one-line system with repeated calls. Ukrainian Math. J. **28**, 437–440 (1975)
5. Falin, G.I.: Waiting time in a single-channel queueing system with repeated calls. Moscow Univ. Comput. Math. Cybern. **4**, 83–87 (1977)
6. Falin, G.I.: An M/G/1 system with repeated calls in heavy traffic. Moscow Univ. Comput. Math. Cybern. **6**, 48–50 (1980)
7. Falin, G.I.: On the waiting-time process in a single-line queue with repeated calls. J. Appl. Probab. **23**, 185–192 (1986)
8. Konishi, Y., Masuyama, H., Kasahara, S., Takahashi, Y.: Performance analysis of dynamic spectrum handoff scheme with variable bandwidth demand of secondary users for cognitive radio networks. Wireless Netw. **19**, 607–617 (2013)
9. Salameh, O., De Turck, K., Bruneel, H., Blondia, C., Wittevrongel, S.: Analysis of secondary user performance in cognitive radio networks with reactive spectrum handoff. Telecommun. Syst. **65**, 539–550 (2017)
10. Artalejo, J.R., Phung-Duc, T.: Single server retrial queues with two way communication. Appl. Math. Model. **37**(4), 1811–1822 (2013)
11. Artalejo, J.R., Phung-Duc, T.: Markovian retrial queues with two way communication. J. Ind. Manage. Optim. **8**(4), 781–806 (2012)
12. Sakurai, H., Phung-Duc, T.: Two-way communication retrial queues with multiple types of outgoing calls. TOP **23**(2), 466–492 (2015)

Controllable Queues

$M/M/1$ Queue with Controllable Service Rate

Zsolt Saffer[1(✉)], Karl Grill[1], and Wuyi Yue[2]

[1] Institute of Statistics and Mathematical Methods in Economics,
Vienna University of Technology, Vienna, Austria
{zsolt.saffer,karl.grill}@tuwien.ac.at
[2] Department of Intelligence and Informatics, Konan University,
Kobe 658-8501, Japan
yue@konan-u.ac.jp

Abstract. In this paper we consider an $M/M/1$ queue, in which the customer service rate is allowed to be increased and decreased by a fixed value at each customer service completion. These changes in service rate are controlled by probabilities depending on the actual number of customers and the actual service rate. The dependency on the number of customers follows a specific power form, while the dependency on actual service rate is general and independent of the dependency on the number of customers.

We describe how to compute the partial stationary distribution of the service rate values when the system is empty. Based on it we provide a computational procedure for computing the stationary probability vectors of the number of customers in the system. We derive also the vector probability generating function and the vector mean of the stationary number of customers.

We establish a methodology which utilizes the specific structure of the model. This methodology inherits some element from the stationary analysis of the standard QBD model and provides a first order, forward algorithm for computing the stationary probability vectors of the number of customers in the system.

Keywords: Queueing theory · Control of queues ·
State dependent service rate · Variable service rate

1 Introduction

Controlling the service capacity of a queueing model supports achieving the optimal operational setting according to the requirements of the application scenario, like reaching an optimal resource allocation or the optimum of a cost model. The customer service time is determined by the capacity demand of the customer together with the actually available service capacity. However in regular queueing system the capacity demand of the customer and the service capacity of the service are not modeled separately, instead a customer service time is modeled.

© Springer Nature Switzerland AG 2019
T. Phung-Duc et al. (Eds.): QTNA 2019, LNCS 11688, pp. 95–111, 2019.
https://doi.org/10.1007/978-3-030-27181-7_7

The change of the available service capacity can be modeled on several ways. Perhaps the simplest service capacity model is to allow changeable number of servers [1]. Similar ways of modeling service capacity are to enable variable service rate in queueing models with Markovian service time [2] or applying speed scaling [3]. In these models the service capacity determines also the customer service time.

Another way of modeling service capacity is to see it as time slice, the fraction of time allocated for service, like in the case of processor time sharing [4]. This can be realized e.g. by time limited polling models [5] or time limited vacation models [6]. Recently Bruneel et al. modeled the service capacity in discrete-time queueing models as an individual random variable from slot to slot, while the service demand is described by another random variable. They investigated such models under different settings in a series of papers [7–9].

In general controlling the service capacity is motivated by providing an automated mechanism to achieve an optimal performance, like e.g. minimal queue length or minimal operating cost. The control of service capacity is modeled in several works as dependency of the service capacity on the state of the system. One early work on such service rate control is [10], which investigates a specific dependency of the service rate on the number of customers in the system. According to this dependency the service rate is proportional to the power of the number of customers. An $M/M/s$ system with variable number of servers is studied in the paper [11].

Another way of controlling the service capacity is to let increase or decrease the number of servers depending on the actual state of the system. In the paper [12] a queue with exponentially distributed service time and a specific control schema is investigated, in which the number of active servers increases or decreases when the queue length changes by $\pm k$ customers. The control of the number of servers.

In applications, like manufacturing systems, in which the processing speed follows the processing demand, a control mechanism can be used in order to keep the number of incoming items on conveyor belt limited. Such an adaptation mechanism can be realized by enabling the increase and decrease of the service capacity on probabilistic way after each processing step depending on the number of not processed items. A power form family of probabilistic dependency of increase and decrease the service capacity ensures the lower (higher) probability of decrementing (incrementing) the service capacity in case of higher number of items present in the system. Such a service capacity control mechanism is investigated in the multi-server queue with constant service time of [13], in which the concurrently served customers are served in synchronized manner. The increase and decrease of the number of active servers are controlled by probabilities depending on the actual number of customers and the number of active servers.

In this paper, motivated by the above application scenario, we investigate an $M/M/1$ queue, in which the customer service rate is allowed to be increased and decreased by a fixed value at each customer service completion. The changes in service rate are controlled by probabilities depending on the actual number of customers and the actual service rate. The dependency on the number of customers follows a power form, while the dependency on actual service rate is general and independent of the dependency on the number of customers. The stationary joint distribution of the number of customers and the actual service rate is described in vector formalism, where the vector $\mathbf{p}_n$ describes the partial distribution when the number of customers is $n \geq 0$.

We give a description for computing $\mathbf{p}_0$. Based on it we provide a computational procedure for computing the stationary probability vectors $\mathbf{p}_n$, $n \geq 1$. We derive also the vector probability generating function (PGF) and the vector mean of the stationary number of customers in terms of $\mathbf{p}_0$. This queueing system can be described by a special case of level dependent QBD having special structure on block matrix level. We establish a methodology which utilizes this specific structure of the model. Thus the methodology inherits some elements from the stationary analysis of the standard quasi-birth-and-death (QBD) model. For computing the stationary distribution in terms of the stationary probability vectors $\mathbf{p}_n$, $n \geq 1$ a first order, forward algorithm is provided. Moreover we provide an additional condition, under which the forward recursion does not require any matrix inversion in the iteration steps.

The rest of the paper is organized as follows. Section 2 gives the model description. In Sect. 3 preliminary results are provided. In Sect. 4 we provide the recursion for computing the stationary distribution. The vector PGF and the vector mean of the stationary number of customers are derived in Sect. 5. Finally, Sect. 6 deals with the computational aspects of the determination of the stationary distribution.

2   Model and Notation

2.1   Modell Description

We consider a queue with controllable capacity, in which the capacity control is realized by changeable rate of customer service. The customers arrive according to a Poisson process with parameter λ. The model has one server and infinite buffer. The customer service times are i.i.d. and exponentially distributed. The service rate, μ^* can only be an integer multiple of μ, more precisely $\mu^* = \mu, 2 * \mu, \ldots M * \mu$. We call $\frac{\mu^*}{\mu}$ as the multiplication factor of the service rate. The service rate can be changed only at service completion epochs. It can be incremented by μ, decremented by μ or kept changed with some probabilities depending on both the number of customers in the system and the actual service rate. We assume that the customer arrival process, the customer service times and the change of the service rate are mutually independent.

Let $q(n, m)$ be the probability of decrementing the service rate by μ at the end of a customer service, when there are $n = 1, \ldots$ customers in the system and the actual service rate is $m * \mu$, $m = 2, \ldots, M$. Similarly let $r(n, m)$ stand for the probability of incrementing the service rate by μ at the end of a customer service, when there are $n = 1, \ldots$ customers in the system and the actual service rate is $m * \mu$, $m = 1, \ldots, M - 1$. These probabilities are expressed in the form

$$q(n, m) = \alpha \gamma^n r_m, \quad n = 1, \ldots, m = 2, \ldots, M,$$
$$r(n, m) = \alpha - q(n, m) = \alpha \left(1 - \gamma^n r_m\right), \quad n = 1, \ldots, m = 1, \ldots, M - 1, \quad (1)$$

where $0 \le r_m \le 1$, for $m = 1, \ldots, M$, $0 \le \alpha \le 1$, and $0 \le \gamma < 1$.

These formulas provide a control on the probabilities of decrementing, incrementing and keeping unchanged the multiplication factor of the service rate in terms of parameters r_m-s, α and γ. The construction of the expressions in (1) ensures that $0 \le r(n, m) \le 1$ and $0 \le q(n, m) \le 1$, i.e. they are probabilities. The parameters r_m-s implement a general dependency on the actual service rate and, as it can be seen from the formulas, it is independent of the dependency on the number of customers.

The parameter γ counts for the control depending on the number of customers. The higher the number of customers, the lower (higher) the probability of decrementing (incrementing) the multiplication factor of the service rate. The parameter α determines the probability of no change in the multiplication factor of the service rate and this probability equals to $1 - \alpha$.

For the matrix function $\mathbf{M}(z)$ the notation $\mathbf{M}^{(n)}(z)$ stands for the n-th derivative of $\mathbf{M}(z)$ with respect to z, for $n \ge 1$. Similarly $\mathbf{M}^{(n)}$ denotes the n-th derivative of $\mathbf{M}(z)$ with respect to z, at $z = 0$ for $n \ge 1$, i.e. $\frac{d^n}{dz^n} \left(\mathbf{M}(z) \right) |_{z=0}$. For $n = 0$ $\mathbf{M}^{(n)}(z)$ and $\mathbf{M}^{(n)}$ denotes $\mathbf{M}(z)$ and $\mathbf{M}(0)$, respectively. When $\mathbf{M}$ is a matrix then $\lambda_{max}(\mathbf{M})$ denotes its eigenvalue with the highest absolute value.

2.2   Formulating a Markov Chain

Let the number of customers in the system and the multiplication factor of the service rate be denoted at time $t > 0$ by $N(t)$ and $M(t)$, respectively. Then the process $(N(t); M(t); t \ge 0)$ is a continuous-time bivariate Markov chain (CTMC). The infinitesimal generator matrix of this CTMC, $\mathbf{Q}$ can be given in terms of block matrices. The outer index of matrix $\mathbf{Q}$ is the number of customers (n), while the index inside of the block matrices represents the multiplication factor of the service rate (m). The matrix $\mathbf{Q}$ can be expressed as

$$\mathbf{Q} = \begin{pmatrix} \mathbf{B}^0 & \mathbf{C} & \mathbf{0} & \mathbf{0} \cdots \\ (\mathbf{A} - \gamma^1 \mathbf{T}) & \mathbf{B} & \mathbf{C} & \mathbf{0} \cdots \\ \mathbf{0} & (\mathbf{A} - \gamma^2 \mathbf{T}) & \mathbf{B} & \mathbf{C} \cdots \\ \mathbf{0} & \mathbf{0} & (\mathbf{A} - \gamma^3 \mathbf{T}) & \mathbf{B} \cdots \\ \vdots & \vdots & \vdots & \vdots & \ddots \end{pmatrix}. \quad (2)$$

The block matrices $\mathbf{A}$, $\mathbf{T}$, $\mathbf{C}$, $\mathbf{B}^0$ and $\mathbf{B}$ are given by

$$\mathbf{A} = \mathbf{S}\mathbf{A}^*, \qquad\qquad \mathbf{B}^0 = -\mathbf{C},$$
$$\mathbf{T} = \mathbf{S}\mathbf{T}^*. \qquad\qquad \mathbf{B} = -(\mathbf{S} + \mathbf{C}).$$

$$S = \begin{pmatrix} \mu & 0 & 0 & \cdots \\ 0 & 2\mu & 0 & \cdots \\ \vdots & \vdots & \vdots & \ddots \\ 0 & 0 & \cdots & M\mu \end{pmatrix}. \qquad A^* = \begin{pmatrix} 1-\alpha & \alpha & 0 \cdots & 0 & 0 \\ 0 & 1-\alpha & \alpha \cdots & 0 & 0 \\ \vdots & \vdots & \vdots \ddots & \vdots & \vdots \\ 0 & 0 & 0 \cdots & 1-\alpha & \alpha \\ 0 & 0 & 0 \cdots & 0 & 1 \end{pmatrix},$$

$$T^* = \begin{pmatrix} -\alpha r_1 & \alpha r_1 & 0 \cdots & 0 & 0 & 0 \\ -\alpha r_2 & 0 & \alpha r_2 \cdots & 0 & 0 & 0 \\ \vdots & \vdots & \vdots \ddots & \vdots & \vdots & \vdots \\ 0 & 0 & 0 \cdots & -\alpha r_{M-1} & 0 & \alpha r_{M-1} \\ 0 & 0 & 0 \cdots & 0 & -\alpha r_M & \alpha r_M \end{pmatrix}. \qquad C = \begin{pmatrix} \lambda & 0 & 0 & \cdots \\ 0 & \lambda & 0 & \cdots \\ \vdots & \vdots & \vdots & \ddots \\ 0 & 0 & \cdots & \lambda \end{pmatrix}.$$

Remark 1. We remark here that matrix A^* is stochastic, matrix T is singular and hence the row sums of matrices $A + B + C$ are 0. Moreover matrix A is non-singular.

From now on we call the model in short form as controllable service rate queue. Furthermore we assume that the controllable service rate queue is stable.

2.3   Relaxing the Structure of the Block Matrices

Without loss of focusing on the controllable service rate queue, from now on we consider a somewhat more general queueing model having the generator matrix's structure like (2), i.e. matrices C, A, T and B may not have to have the forms as specified above. This is because the solution of the queueing model depends mainly on the structure of the generator matrix on block matrix level. We assume that the model is stable, the row sums of matrices $A + B + C$ are 0 as well as matrix A is non-singular.

3   Preliminary Results

In this section we investigate the properties of the matrices $\frac{d^n}{dz^n}(A + Bz + Cz^2)\,|_{z=0}$ for $n \geq 0$, which we need in the subsequent sections. We introduce notations for these matrices as follows

$$X_n = \frac{d^n}{dz^n}(A + Bz + Cz^2)\,|_{z=0}, \quad n \geq 0.$$

3.1   Second Order Recursions

Lemma 1. *The matrices X_n can be computed from the following second order recursion*

$$X_n = -\left(n X_{n-1} BA^{-1} + n(n-1) X_{n-2} CA^{-1}\right), \quad n \geq 2, \tag{3}$$

where

$$X_0 = A^{-1} \text{ and } X_1 = -A^{-1}BA^{-1}. \tag{4}$$

Proof. We consider a complex valued matrix function $\mathbf{f}(z)$ on values for z, for which the inverse of matrix $\mathbf{f}(z)$ exists as well as both $\mathbf{f}(z)$ and $\mathbf{f}^{-1}(\mathbf{z})$ are analytic. Then we have

$$\mathbf{f}^{-1}(z)\mathbf{f}(z) = \mathbf{I}. \tag{5}$$

Taking the n-th derivative of (5) yields

$$(\mathbf{f}^{-1})^{(n)}(z)\mathbf{f}(z) + \sum_{k=0}^{n-1} \binom{n}{k} (\mathbf{f}^{-1})^{(k)}(z)\mathbf{f}^{(n-k)}(z) = \mathbf{0}, \quad n \geq 1,$$

from which

$$(\mathbf{f}^{-1})^{(n)}(z) = - \left(\sum_{k=0}^{n-1} \binom{n}{k} (\mathbf{f}^{-1})^{(k)}(z)\mathbf{f}^{(n-k)}(z) \right) \mathbf{f}^{-1}(z), \quad n \geq 1.$$

follows. Applying it to

$$\mathbf{f}(z) = \left(\mathbf{A} + \mathbf{B}z + \mathbf{C}z^2 \right), \quad |z| \leq 1.$$

and setting $z = 0$ leads to

$$(\mathbf{f}^{-1})^{(1)} = -\mathbf{f}^{-1}(0)\mathbf{f}^{(1)}\mathbf{f}^{-1}(0). \tag{6}$$

and

$$\begin{aligned}(\mathbf{f}^{-1})^{(n)} = - \Bigg(&\binom{n}{n-1}(\mathbf{f}^{-1})^{(n-1)}\mathbf{f}^{(1)} \\ &+ \binom{n}{n-2}(\mathbf{f}^{-1})^{(n-2)}\mathbf{f}^{(2)} \Bigg) \mathbf{f}^{-1}(0), \quad n \geq 2.\end{aligned} \tag{7}$$

The statements come by applying $\mathbf{f}^{-1}(0) = \mathbf{A}^{-1}$, $\mathbf{f}^{(1)} = \mathbf{B}$ and $\mathbf{f}^{(2)} = 2\mathbf{C}$ as well as the notation $\mathbf{X}_n = (\mathbf{f}^{-1})^{(n)}$ for $n \geq 0$ in (6) and (7). □

We define matrices $\mathbf{Y}_n$ for $n \geq 0$ by means of the following relations

$$\mathbf{X}_n = n!\mathbf{Y}_n, \quad n \geq 0. \tag{8}$$

Lemma 2. *The matrices $\mathbf{Y}_n$ can be computed recursively as*

$$\mathbf{Y}_n = - \left(\mathbf{Y}_{n-1}\mathbf{B}\mathbf{A}^{-1} + \mathbf{Y}_{n-2}\mathbf{C}\mathbf{A}^{-1} \right), \quad n \geq 2, \tag{9}$$

where

$$\mathbf{Y}_0 = \mathbf{A}^{-1} \text{ and } \mathbf{Y}_1 = -\mathbf{A}^{-1}\mathbf{B}\mathbf{A}^{-1}. \tag{10}$$

Proof. The lemma comes by applying (8) in (3) and (4). □

3.2 Explicit Formula for Matrices $\mathbf{Y}_n$

In this subsection we give an explicit expression for matrices $\mathbf{Y}_n$. We define a corresponding QBD by setting $\gamma = 0$ in (2) leading to

$$\mathbf{Q} = \begin{pmatrix} \mathbf{B}^0 & \mathbf{C} & \mathbf{0} & \mathbf{0} \cdots \\ \mathbf{A} & \mathbf{B} & \mathbf{C} & \mathbf{0} \cdots \\ \mathbf{0} & \mathbf{A} & \mathbf{B} & \mathbf{C} \cdots \\ \mathbf{0} & \mathbf{0} & \mathbf{A} & \mathbf{B} \cdots \\ \vdots & \vdots & \vdots & \vdots & \ddots \end{pmatrix}. \tag{11}$$

Theorem 1. *If corresponding QBD is stable then the matrices $\mathbf{Y}_n$ can be expressed in explicit form as*

$$\mathbf{Y}_n = \mathbf{Y}_0 \sum_{i=0}^{n} \mathbf{V}^i \mathbf{R}^{n-i}, \quad n \geq 1, \tag{12}$$

where matrix $\mathbf{R}$ is the minimal non-negative solution of the quadratic matrix equation

$$\mathbf{R}^2 \mathbf{A} + \mathbf{R}\mathbf{B} + \mathbf{C} = \mathbf{0} \tag{13}$$

and

$$\mathbf{V} = -\mathbf{B}\mathbf{A}^{-1} - \mathbf{R}. \tag{14}$$

Proof. Substracting $\mathbf{Y}_{n-1}\mathbf{R}$ from both sides of (9) results in

$$\mathbf{Y}_n - \mathbf{Y}_{n-1}\mathbf{R} = \mathbf{Y}_{n-1}\left((-\mathbf{B}\mathbf{A}^{-1}) - \mathbf{R}\right) + \mathbf{Y}_{n-2}(-\mathbf{C}\mathbf{A}^{-1}), \quad n \geq 2. \tag{15}$$

We make the conjenture

$$\mathbf{Y}_{n-1}\left((-\mathbf{B}\mathbf{A}^{-1}) - \mathbf{R}\right) + \mathbf{Y}_{n-2}(-\mathbf{C}\mathbf{A}^{-1}) = (\mathbf{Y}_{n-1} - \mathbf{Y}_{n-2}\mathbf{R})\,\mathbf{V}, \quad n \geq 2. \tag{16}$$

In order to show that this conjenture holds we will determine the matrices $\mathbf{R}$ and $\mathbf{V}$ without getting any contradiction.

From (16) we conclude

$$\begin{aligned} \mathbf{V} &= -\mathbf{B}\mathbf{A}^{-1} - \mathbf{R}, \\ -\mathbf{R}\mathbf{V} &= -\mathbf{C}\mathbf{A}^{-1}. \end{aligned} \tag{17}$$

Combining the above equations and rearranging leads to a quadratic matrix equation for $\mathbf{R}$ as

$$\mathbf{R}^2 + \mathbf{R}\mathbf{B}\mathbf{A}^{-1} + \mathbf{C}\mathbf{A}^{-1} = \mathbf{0}. \tag{18}$$

Multiplying (18) by matrix $\mathbf{A}$ from right gives

$$\mathbf{R}^2 \mathbf{A} + \mathbf{R}\mathbf{B} + \mathbf{C} = \mathbf{0}. \tag{19}$$

The quadratic matrix equation could have more solutions. Any solution of this equation together with the corresponding matrix $\mathbf{V}$, specified by (17), satisfy the conjecture and hence any second order sequence $\mathbf{Y}_n$, $n \geq 2$ determined from (15) and (16) is a solution of (9). However (9) has only one solution for a given initial matrices $\mathbf{Y}_0$ and $\mathbf{Y}_1$. It follows that any solution pair $\mathbf{R}$ and the corresponding matrix $\mathbf{V}$ leads to the same solution for the sequence $\mathbf{Y}_n$, $n \geq 0$. If the corresponding QBD is stable then there exists at least a minimal non-negative solution of this equation [14]. Therefore we select $\mathbf{R}$ as the minimal non-negative solution of this equation.

The second order sequence $\mathbf{Y}_n$, $n \geq 2$ can be determined from (15) and (16) recursively as follows. Combining these two equations gives

$$\mathbf{Y}_n - \mathbf{Y}_{n-1}\mathbf{R} = (\mathbf{Y}_{n-1} - \mathbf{Y}_{n-2}\mathbf{R})\,\mathbf{V}, \quad n \geq 2.$$

Solving it recursively and replacing the index n by k leads to

$$\mathbf{Y}_k - \mathbf{Y}_{k-1}\mathbf{R} = (\mathbf{Y}_1 - \mathbf{Y}_0\mathbf{R})\,\mathbf{V}^{k-1}, \quad k \geq 2.$$

In fact the above relation holds also for $k = 1$. Multiplying it by $\mathbf{R}^{n-k}$ from right and summing from $k = 1$ up to n results in

$$\mathbf{Y}_n - \mathbf{Y}_0\mathbf{R}^n = (\mathbf{Y}_1 - \mathbf{Y}_0\mathbf{R}) \sum_{i=0}^{n-1} \mathbf{V}^i \mathbf{R}^{n-i-1}, \quad n \geq 1. \qquad (20)$$

Observe from (10) that

$$\mathbf{Y}_1 = \mathbf{Y}_0(-\mathbf{BA}^{-1}). \qquad (21)$$

Applying (21) and the first relation of (17) in (20) and rearranging gives

$$\mathbf{Y}_n - \mathbf{Y}_0\mathbf{R}^n = \mathbf{Y}_0 \left(-\mathbf{BA}^{-1} - \mathbf{R}\right) \sum_{i=0}^{n-1} \mathbf{V}^i \mathbf{R}^{n-i-1} = \mathbf{Y}_0 \mathbf{V} \sum_{i=0}^{n-1} \mathbf{V}^i \mathbf{R}^{n-i-1}$$

$$= \mathbf{Y}_0 \sum_{i=0}^{n-1} \mathbf{V}^{i+1} \mathbf{R}^{n-(i+1)} = \mathbf{Y}_0 \sum_{i=1}^{n} \mathbf{V}^i \mathbf{R}^{n-i}, \quad n \geq 1. \qquad (22)$$

The statement of the theorem comes by further rearranging of (22). $\qquad\square$

3.3   First Order Recursions for Matrices $\mathbf{Y}_n$

Based on the explicit expression for matrices $\mathbf{Y}_n$ also first order recursions can be given for them. We will use these recursions in the next sections.

Corollary 1. *The following first order recursions hold for matrices $\mathbf{Y}_n$*

$$\mathbf{Y}_n = \mathbf{Y}_{n-1}\mathbf{R} + \mathbf{A}^{-1}\mathbf{V}^n, \quad n \geq 1, \qquad (23)$$

$$\mathbf{Y}_n = \mathbf{A}^{-1}\mathbf{VAY}_{n-1} + \mathbf{A}^{-1}\mathbf{R}^n, \quad n \geq 1. \qquad (24)$$

Proof. The first statement can be obtained by rearranging (12) as

$$\mathbf{Y}_n = \mathbf{Y}_0 \sum_{i=0}^{n} \mathbf{V}^i \mathbf{R}^{n-i} = \mathbf{Y}_0 \left(\sum_{i=0}^{n-1} \mathbf{V}^i \mathbf{R}^{n-1-i} \right) \mathbf{R} + \mathbf{Y}_0 \mathbf{V}^n$$
$$= \mathbf{Y}_0 \mathbf{Y}_{n-1} \mathbf{R} + \mathbf{A}^{-1} \mathbf{V}^n.$$

Similarly the second statement can be obtained also by rearranging (12) leading to

$$\mathbf{Y}_n = \mathbf{Y}_0 \sum_{i=0}^{n} \mathbf{V}^i \mathbf{R}^{n-i} = \mathbf{Y}_0 \sum_{i=1}^{n} \mathbf{V}^i \mathbf{R}^{n-i} + \mathbf{Y}_0 \mathbf{R}^n$$
$$= \mathbf{Y}_0 \mathbf{V} \sum_{i=1}^{n} \mathbf{V}^{i-1} \mathbf{R}^{(n-1)-(i-1)} + \mathbf{Y}_0 \mathbf{R}^n$$
$$= \mathbf{Y}_0 \mathbf{V} \mathbf{Y}_0^{-1} \left(\mathbf{Y}_0 \sum_{i=0}^{n-1} \mathbf{V}^i \mathbf{R}^{n-1-i} \right) + \mathbf{Y}_0 \mathbf{R}^n = \mathbf{A}^{-1} \mathbf{V} \mathbf{A} \mathbf{Y}_{n-1} + \mathbf{A}^{-1} \mathbf{R}^n.$$

□

4   Stationary Solution

In this section we derive a recursive relation for the stationary probability vectors. Based on the preliminary result we reformulate this recursion into an alternative form, which enables to guess the stationary solution. Then we establish a theorem for determining the stationary probability vectors.

4.1   Relation for the PGF of the Stationary Probability Vectors

Let $\mathbf{p}$ be the $1 \times \infty$ stationary probability vector of the number of customers in the system. Partitioning $\mathbf{p}$ in the form $\mathbf{p} = (\mathbf{p}_0, \mathbf{p}_1, \ldots)$, where $\mathbf{p}_i$, $i \geq 0$ are $1 \times M$ stationary probability vectors, enables to expand the system of equations $\mathbf{p}\mathbf{Q} = \mathbf{0}$ as

$$(\mathbf{p}_0, \mathbf{p}_1, \mathbf{p}_2, \mathbf{p}_3, \ldots) \begin{pmatrix} B^0 & \mathbf{C} & \mathbf{0} & \mathbf{0} \cdots \\ (\mathbf{A} - \gamma^1 \mathbf{T}) & \mathbf{B} & \mathbf{C} & \mathbf{0} \cdots \\ \mathbf{0} & (\mathbf{A} - \gamma^2 \mathbf{T}) & \mathbf{B} & \mathbf{C} \cdots \\ \mathbf{0} & \mathbf{0} & (\mathbf{A} - \gamma^3 \mathbf{T}) & \mathbf{B} \cdots \\ \vdots & \vdots & \vdots & \vdots \quad \ddots \end{pmatrix} = \mathbf{0}. \tag{25}$$

Let us multiply (25) by the column hypervector

$$\begin{pmatrix} \mathbf{I} \\ z\mathbf{I} \\ z^2\mathbf{I} \\ \vdots \end{pmatrix},$$

from right, where $\mathbf{I}$ is the $M \times M$ identity matrix and $|z| \leq 1$ is a complex value. This results in

$$
(\mathbf{p}_0, \mathbf{p}_1, \mathbf{p}_2, \mathbf{p}_3, \ldots)
\begin{pmatrix}
\mathbf{B}^0 & \mathbf{C} & \mathbf{0} & \mathbf{0} \cdots \\
(\mathbf{A} - \gamma^1 \mathbf{T}) & \mathbf{B} & \mathbf{C} & \mathbf{0} \cdots \\
\mathbf{0} & (\mathbf{A} - \gamma^2 \mathbf{T}) & \mathbf{B} & \mathbf{C} \cdots \\
\mathbf{0} & \mathbf{0} & (\mathbf{A} - \gamma^3 \mathbf{T}) & \mathbf{B} \cdots \\
\vdots & \vdots & \vdots & \vdots \qquad \ddots
\end{pmatrix}
\begin{pmatrix}
\mathbf{I} \\
z\mathbf{I} \\
z^2\mathbf{I} \\
\vdots
\end{pmatrix}
= \mathbf{0}.
$$

Collecting the terms according to matrices $\mathbf{A}$, $\mathbf{T}$, $\mathbf{B}$ and $\mathbf{C}$ yields

$$
\begin{aligned}
& \left(\mathbf{p}_1 + \mathbf{p}_2 z + \mathbf{p}_3 z^2 + \ldots\right) \mathbf{A} - \left(\mathbf{p}_1 \gamma + \mathbf{p}_2 \gamma^2 z + \mathbf{p}_3 \gamma^3 z^2 + \ldots\right) \mathbf{T} \\
& + \left(\mathbf{p}_0 + \mathbf{p}_1 z + \mathbf{p}_2 z^2 + \ldots\right) \mathbf{B} + \mathbf{p}_0 \left(\mathbf{B}^0 - \mathbf{B}\right) \\
& + \left(\mathbf{p}_0 z + \mathbf{p}_1 z^2 + \mathbf{p}_2 z^3 + \ldots\right) \mathbf{C} = \mathbf{0}.
\end{aligned}
\tag{26}
$$

The vector PGF of the stationary number of customers is defined as

$$
\mathbf{p}(z) = \sum_{n=0}^{\infty} \mathbf{p}_n z^n, \quad |z| \leq 1.
$$

Using it in (26) and rearranging it gives

$$
\begin{aligned}
& \mathbf{p}(z)\frac{1}{z}\mathbf{A} - \mathbf{p}_0 \frac{1}{z}\mathbf{A} - \mathbf{p}(\gamma z)\frac{1}{z}\mathbf{T} + \mathbf{p}_0 \frac{1}{z}\mathbf{T} + \mathbf{p}(z)\mathbf{B} + \mathbf{p}_0 \left(\mathbf{B}^0 - \mathbf{B}\right) \\
& + \mathbf{p}(z)z\mathbf{C} = \mathbf{0}.
\end{aligned}
\tag{27}
$$

Multiplying (27) by z and further rearranging leads to the PGF relation

$$
\begin{aligned}
& \mathbf{p}(z) - \mathbf{p}(\gamma z)\mathbf{T}\left(\mathbf{A} + \mathbf{B}z + \mathbf{C}z^2\right)^{-1} \\
& = \mathbf{p}_0 \left((\mathbf{A} - \mathbf{T}) + (\mathbf{B} - \mathbf{B}^0)z\right)\left(\mathbf{A} + \mathbf{B}z + \mathbf{C}z^2\right)^{-1}.
\end{aligned}
\tag{28}
$$

4.2   Recursive Relations for the Stationary Probability Vectors

Proposition 1. *The following recursive relations hold for the stationary probability vectors in the stable controllable service rate queue*

$$
\mathbf{p}_n - \sum_{i=0}^{n} \gamma^i \mathbf{p}_i \mathbf{T}\mathbf{Y}_{n-i} = \mathbf{p}_0 \left((\mathbf{A} - \mathbf{T})\mathbf{Y}_n + (\mathbf{B} - \mathbf{B}^0)\mathbf{Y}_{n-1}\right). \quad n \geq 1
\tag{29}
$$

Proof. Taking the n-th derivative of (28) with respect to z for $n = 1, \ldots$ and setting $z = 0$ gives

$$
n!\mathbf{p}_n - \sum_{i=0}^{n} \binom{n}{i} i! \gamma^i \mathbf{p}_i \mathbf{T}\mathbf{X}_{n-i} = \mathbf{p}_0 \left((\mathbf{A} - \mathbf{T})\mathbf{X}_n + n(\mathbf{B} - \mathbf{B}^0)\mathbf{X}_{n-1}\right), \quad n \geq 1.
\tag{30}
$$

The proposition comes by substituting (8) into (30) and rearranging it. $\square$

Proposition 2. *An alternative form of the recursive relations for the stationary probability vectors in the stable controllable service rate queue can be given as*

$$\mathbf{p}_n - \sum_{i=1}^{n} \gamma^i \mathbf{p}_i \mathbf{T} \mathbf{Y}_{n-i} = \mathbf{p}_0 \mathbf{R}^n - \mathbf{p}_0 \left(\mathbf{R}\mathbf{A} + \mathbf{B}^0 \right) \mathbf{Y}_{n-1}. \tag{31}$$

Proof. Observe that for $i = 0$ the term in the sum of the l.h.s of (29) equals to $\mathbf{p}_0 \mathbf{T} \mathbf{Y}_n$. This enables the following rearrangement of (29)

$$\mathbf{p}_n - \sum_{i=1}^{n} \gamma^i \mathbf{p}_i \mathbf{T} \mathbf{Y}_{n-i} = \mathbf{p}_0 \left(\mathbf{A}\mathbf{Y}_n + (\mathbf{B} - \mathbf{B}^0)\mathbf{Y}_{n-1} \right). \quad n \geq 1 \tag{32}$$

The r.h.s of (30) can be rearranged by applying (24) and the first relation of (17) as

$$\mathbf{p}_0 \left(\mathbf{A}\mathbf{Y}_n + (\mathbf{B} - \mathbf{B}^0)\mathbf{Y}_{n-1} \right) = \mathbf{p}_0 \left(\mathbf{V}\mathbf{A}\mathbf{Y}_{n-1} + \mathbf{R}^n + (\mathbf{B} - \mathbf{B}^0)\mathbf{Y}_{n-1} \right)$$
$$= \mathbf{p}_0 \mathbf{R}^n + \mathbf{p}_0 \left(-\mathbf{B} - \mathbf{R}\mathbf{A} + (\mathbf{B} - \mathbf{B}^0) \right) \mathbf{Y}_{n-1}$$
$$= \mathbf{p}_0 \mathbf{R}^n - \mathbf{p}_0 \left(\mathbf{R}\mathbf{A} + \mathbf{B}^0 \right) \mathbf{Y}_{n-1} \tag{33}$$

The proposition comes by applying (33) in (32). □

4.3   Solution in Terms of Stationary Probability Vectors

The structure of the equations in Proposition 2 suggests the following conjecture.

$$\mathbf{p}_0 \left(\mathbf{R}\mathbf{A} + \mathbf{B}^0 \right) = \mathbf{0}. \tag{34}$$

$$\mathbf{p}_n - \sum_{i=1}^{n} \gamma^i \mathbf{p}_i \mathbf{T} \mathbf{Y}_{n-i} = \mathbf{p}_0 \mathbf{R}^n, \quad n \geq 1.$$

This leads to the next theorem.

Theorem 2. *The stationary probability vectors in the stable controllable service rate queue can be determined from the following equations*

$$\mathbf{p}_0 \left(\mathbf{R}\mathbf{A} + \mathbf{B}^0 \right) = \mathbf{0}. \tag{35}$$

$$\mathbf{p}_n = \left(\mathbf{p}_0 \mathbf{R}^n + \sum_{i=1}^{n-1} \gamma^i \mathbf{p}_i \mathbf{T} \mathbf{Y}_{n-i} \right) \left(\mathbf{I} - \gamma^n \mathbf{T}\mathbf{A}^{-1} \right)^{-1}, \quad n \geq 1. \tag{36}$$

Proof. We show that the probability vectors determined by (34) satisfy the partitioned form of the equilibrium equation, (25). For $n = 1$ (34) gives

$$\mathbf{p}_1 - \gamma \mathbf{p}_1 \mathbf{T} \mathbf{Y}_0 = \mathbf{p}_0 \mathbf{R}. \tag{37}$$

Multiplying (37) by $\mathbf{A}$ from right and rearranging it leads to

$$\mathbf{p}_1 (\mathbf{A} - \gamma\mathbf{T}) - \mathbf{p}_0\mathbf{RA} = 0. \tag{38}$$

Summing (38) with the first equation of (34) yields

$$\mathbf{p}_1 (\mathbf{A} - \gamma\mathbf{T}) + \mathbf{p}_0\mathbf{B}^0 = 0,$$

which is the first partitioned equilibrium equation.

For $n \geq 2$ we apply (34) for n, $n-1$ and $n-2$, $n \geq 2$ after each other, which leads to

$$\mathbf{p}_n - \sum_{i=1}^{n}\gamma^i\mathbf{p}_i\mathbf{TY}_{n-i} = \mathbf{p}_0\mathbf{R}^n.$$

$$\mathbf{p}_{n-1} - \sum_{i=1}^{n-1}\gamma^i\mathbf{p}_i\mathbf{TY}_{n-i-1} = \mathbf{p}_0\mathbf{R}^{n-1}.$$

$$\mathbf{p}_{n-2} - \sum_{i=1}^{n-2}\gamma^i\mathbf{p}_i\mathbf{TY}_{n-i-2} = \mathbf{p}_0\mathbf{R}^{n-2}.$$

Multiplying the first, second and third equation by matrix $\mathbf{A}$, $\mathbf{B}$ and $\mathbf{C}$ from right, respectively and summing up them yields

$$\mathbf{p}_n\mathbf{A} + \mathbf{p}_{n-1}\mathbf{B} + \mathbf{p}_{n-2}\mathbf{C} - \sum_{i=1}^{n-2}\gamma^i\mathbf{p}_i\mathbf{T} (\mathbf{Y}_{n-i}\mathbf{A} + \mathbf{Y}_{n-i-1}\mathbf{B} + \mathbf{Y}_{n-i-2}\mathbf{C})$$
$$- \gamma^{n-1}\mathbf{p}_{n-1}\mathbf{T} (\mathbf{Y}_0\mathbf{B} + \mathbf{Y}_1\mathbf{A}) - \gamma^n\mathbf{p}_n\mathbf{TY}_0\mathbf{A}$$
$$= \mathbf{p}_0\mathbf{R}^{n-2} (\mathbf{C} + \mathbf{RB} + +\mathbf{R}^2\mathbf{A}) . \tag{39}$$

It follows from (9) that $\mathbf{Y}_{n-i}\mathbf{A} + \mathbf{Y}_{n-i-1}\mathbf{B} + \mathbf{Y}_{n-i-2}\mathbf{C} = 0$ for $i \leq n - 2$. Substituting it, (9) and (21) into (39) and rearranging gives

$$\mathbf{p}_n\mathbf{A} + \mathbf{p}_{n-1}\mathbf{B} + \mathbf{p}_{n-2}\mathbf{C} - \gamma^{n-1}\mathbf{p}_{n-1}\mathbf{T} (\mathbf{Y}_0\mathbf{B} + \mathbf{Y}_0(-\mathbf{BA}^{-1})\mathbf{A}) - \gamma^n\mathbf{p}_n\mathbf{T}$$
$$= \mathbf{p}_n\mathbf{A} + \mathbf{p}_{n-1}\mathbf{B} + \mathbf{p}_{n-2}\mathbf{C} - \gamma^n\mathbf{p}_n\mathbf{T} = 0.$$

Further rearrangement results in

$$\mathbf{p}_n (\mathbf{A} - \gamma^n\mathbf{T}) + \mathbf{p}_{n-1}\mathbf{B} + \mathbf{p}_{n-2}\mathbf{C} = 0,$$

which is the n-th partitioned equilibrium equation for $n \geq 2$. According to the condition of the theorem the model is stable and hence the solution of the equilibrium equations is unique. It follows that probability vectors determined by (34) are the stationary probability vectors of the model. The second equation of the theorem comes by rearranging the second equation of the conjecture (34).   □

Remark 2. If the model is stable then the second relation of (34) determine the probability vector $\mathbf{p}_n$ (from a known $\mathbf{p}_0$) uniquely. It follows that under stability the matrices $(\mathbf{I} - \gamma^n\mathbf{TA})$ for $n \geq 1$ are non-singular, otherwise the probability vectors $\mathbf{p}_n$ were determined only up to a multiplication constant, i.e. not uniquely.

Theorem 2 provides a recursive way of computing the stationary probability vectors. The steps of computational procedure can be summarized as follows:

1. Determination of matrix $\mathbf{R}$ as the minimal non-negative solution of the quadratic equation (19) by means of iterative computation.
2. Computation of probability vector $\mathbf{p}_0$ up to a multiplication constant as a solution of the homogenous system of linear equations (35).
3. Recursive computation of the probability vectors $\mathbf{p}_n$, for $n \geq 1$ from the relation (36).
4. Determination of the normalization constant from $\sum_n \mathbf{p}_n \mathbf{e} = 1$ and adjusting the stationary probability vectors $\mathbf{p}_n$, $n \geq 0$ accordingly.

5   The Stationary PGF and Mean of the Number of Customers

In this section we derive the stationary PGF of the number of customers. Based on them we provide an expression for the stationary mean of the number of customers and also give a sufficient condition for the stability of the system.

5.1   The Stationary PGF of the Number of Customers

Theorem 3. *The stationary PGF of the number of customers in the stable controllable service rate queue is given in terms of $\mathbf{p}_0$ by*

$$\mathbf{p}(z) = \mathbf{p}_0 \sum_{n=0}^{\infty} \left((\mathbf{A} - \mathbf{T}) + (\mathbf{B} - \mathbf{B}^0) \gamma^n z \right) \left(\mathbf{A} + \mathbf{B} \gamma^n z + \mathbf{C} \gamma^{2n} z^2 \right)^{-1}$$
$$\times \overleftarrow{\prod}_{j=0}^{n-1} \mathbf{T} \left(\mathbf{A} + \mathbf{B} \gamma^j z + \mathbf{C} \gamma^{2j} z^2 \right)^{-1}, \tag{40}$$

where $\overleftarrow{\prod}_{i=0}^{n-1}$ stands for multiplying from left with increasing index and the empty product is 0.

Proof. Rearranging (28) leads to

$$\mathbf{p}(z) = \mathbf{p}(\gamma z)\mathbf{T} \left(\mathbf{A} + \mathbf{B}z + \mathbf{C}z^2 \right)^{-1}$$
$$+ \mathbf{p}_0 \left((\mathbf{A} - \mathbf{T}) + (\mathbf{B} - \mathbf{B}^0)z \right) \left(\mathbf{A} + \mathbf{B}z + \mathbf{C}z^2 \right)^{-1}.$$

This relation can be solved for $\mathbf{p}(z)$ by recursive substitution of $z = \gamma^n z$, for $n = 0, 1, \ldots$, since $\gamma < 1$, which leads to

$$\mathbf{p}(z) = \lim_{n \to \infty} \mathbf{p}(\gamma^n z) \overleftarrow{\prod}_{n=0}^{\infty} \mathbf{T} \left(\mathbf{A} + \mathbf{B} \gamma^n z + \mathbf{C} \gamma^{2n} z^2 \right)^{-1}$$
$$+ \mathbf{p}_0 \sum_{n=0}^{\infty} \left((\mathbf{A} - \mathbf{T}) + (\mathbf{B} - \mathbf{B}^0) \gamma^n z \right) \left(\mathbf{A} + \mathbf{B} \gamma^n z + \mathbf{C} \gamma^{2n} z^2 \right)^{-1}$$
$$\times \overleftarrow{\prod}_{j=0}^{n-1} \mathbf{T} \left(\mathbf{A} + \mathbf{B} \gamma^j z + \mathbf{C} \gamma^{2j} z^2 \right)^{-1}.$$

The first term vanishes due to due to $\gamma < 1$. $\square$

5.2 The Stationary Mean of the Number of Customers

Corollary 2. *The stationary mean of the number of customers in the stable controllable service rate queue is given in terms of* $\mathbf{p_0}$ *by*

$$
\sum_{n=1}^{\infty} n\mathbf{p}_n = \mathbf{p_0} \sum_{n=0}^{\infty} \left(((\mathbf{B} - \mathbf{B}^0)\gamma^n) \left(\mathbf{A} + \mathbf{B}\gamma^n + \mathbf{C}\gamma^{2n} \right)^{-1} \overleftarrow{\prod_{j=0}^{n-1}} \mathbf{T} \left(\mathbf{A} + \mathbf{B}\gamma^j + \mathbf{C}\gamma^{2j} \right)^{-1} \right.
$$
$$
+ \mathbf{p_0} \sum_{n=0}^{\infty} \left((\mathbf{A} - \mathbf{T}) + (\mathbf{B} - \mathbf{B}^0)\gamma^n \right) \frac{d}{dz} \left(\left(\mathbf{A} + \mathbf{B}\gamma^n z + \mathbf{C}\gamma^{2n} z^2 \right)^{-1} \right) |_{z=1}
$$
$$
\times \overleftarrow{\prod_{j=0}^{n-1}} \mathbf{T} \left(\mathbf{A} + \mathbf{B}\gamma^j + \mathbf{C}\gamma^{2j} \right)^{-1}
$$
$$
+ \mathbf{p_0} \sum_{n=0}^{\infty} \left((\mathbf{A} - \mathbf{T}) + (\mathbf{B} - \mathbf{B}^0)\gamma^n \right) \left(\mathbf{A} + \mathbf{B}\gamma^n + \mathbf{C}\gamma^{2n} \right)^{-1}
$$
$$
\times \sum_{k=0}^{n-1} \left(\overleftarrow{\prod_{j_1=0}^{k-1}} \mathbf{T} \left(\mathbf{A} + \mathbf{B}\gamma^{j_1} + \mathbf{C}\gamma^{2j_1} \right)^{-1} \mathbf{T} \frac{d}{dz} \left(\left(\mathbf{A} + \mathbf{B}\gamma^n z + \mathbf{C}\gamma^{2n} z^2 \right)^{-1} \right) |_{z=1} \right.
$$
$$
\left. \times \overleftarrow{\prod_{j_2=k+1}^{n-1}} \mathbf{T} \left(\mathbf{A} + \mathbf{B}\gamma^{j_2} + \mathbf{C}\gamma^{2j_2} \right)^{-1} \right) \tag{41}
$$

Proof. The statement of the corollary is obtained by taking the first derivative of (40) with respect to z and setting $z = 1$. □

5.3 Sufficient Condition of the Stability

Lemma 3. *The stability condition of the corresponding QBD is given by*

1. *Its generator matrix is irreducible.*
2. *It has negative drift, in other words*

$$
\boldsymbol{\pi}\mathbf{Ce} - \boldsymbol{\pi}\mathbf{Ae} < 0, \tag{42}
$$

where vector $\boldsymbol{\pi}$ is the stationary probability vector of the corresponding QBD's generator matrix and it is determined uniquely by

$$
\boldsymbol{\pi}\left(\mathbf{A} + \mathbf{B} + \mathbf{C} = 0\right) \quad and \quad \boldsymbol{\pi}\mathbf{e} = 1. \tag{43}
$$

Proof. The statement can be derived from the stability condition of the analogous discrete-time QBD, see in [14]. □

Corollary 3. *A sufficient condition of the stability of the controllable service rate queue is given by,*

1. *Setting $\gamma = 0$ in matrix $\mathbf{Q}$ results in an irreducible generator matrix.*
2. *$\boldsymbol{\pi}\mathbf{Ce} - \boldsymbol{\pi}\mathbf{Ae} < 0$, where $\boldsymbol{\pi}$ is given by (43).*
3.

$$
\sum_{n=0}^{\infty} \left((\mathbf{A} - \mathbf{T}) + (\mathbf{B} - \mathbf{B}^0)\gamma^n \right) \left(\mathbf{A} + \mathbf{B}\gamma^n + \mathbf{C}\gamma^{2n} \right)^{-1}
$$
$$
\times \overleftarrow{\prod_{j=0}^{n-1}} \mathbf{T} \left(\mathbf{A} + \mathbf{B}\gamma^j + \mathbf{C}\gamma^{2j} \right)^{-1} < \infty. \tag{44}
$$

Proof. The system is stable if $\sum_{n=0}^{\infty} \mathbf{p}_n < \infty$. Setting $z = 1$ in (40) gives a stability condition as

$$\mathbf{p}_0 \sum_{n=0}^{\infty} \left((\mathbf{A} - \mathbf{T}) + (\mathbf{B} - \mathbf{B}^0)\gamma^n \right) \left(\mathbf{A} + \mathbf{B}\gamma^n + \mathbf{C}\gamma^{2n} \right)^{-1} \times$$

$$\overleftarrow{\prod}_{j=0}^{n-1} \mathbf{T} \left(\mathbf{A} + \mathbf{B}\gamma^j + \mathbf{C}\gamma^{2j} \right)^{-1} < \infty, \tag{45}$$

The first criterion of the corollary means that the generator of the corresponding QBD is irreducible. Similarly the second one ensures that the corresponding QBD has negative drift. Hence the first two criteria together with Lemma 3 ensures the stability of the corresponding QBD, in which case the matrix R, the minimal non-negative solution of the equation (19) and the probability vector $\mathbf{p}_0$ determined from (34) exist (see in [14]). Hence the above stability condition reduces to the third criterion of the corollary. □

6 Computational Considerations

Although the Theorem 2 determines the solution in terms of stationary probability vectors, it is computationally extensive as it requires all the previously determined probability vectors in each iteration step. In this section we give a first order forward recursion for determining the stationary probability vectors, which requires only the lastly determined probability vector and a temporary vector variable in each iteration step. Additionally we provide an additional condition, under which the matrix inversions in the iteration steps of the forward recursion can be replaced by sums.

Theorem 4. *The stationary probability vectors in the stable controllable service rate queue can be computed from the following first order recursion*

$$\mathbf{p}_n = (\mathbf{p}_{n-1}\mathbf{R} + \mathbf{z}_{n-1}) \left(\mathbf{I} - \gamma^n \mathbf{T}\mathbf{A}^{-1} \right)^{-1}, \quad n \geq 1, \tag{46}$$

$$\mathbf{z}_n = \left(\mathbf{z}_{n-1} + \gamma^n \mathbf{p}_n \mathbf{T}\mathbf{A}^{-1} \right) \mathbf{V}, \quad n \geq 1, \tag{47}$$

where $\mathbf{p}_0$ is given by (35) and $\mathbf{z}_0 = \mathbf{0}$.

Proof. We apply the second relation of (34) for n and $n - 1$, for $n \geq 1$, which leads to

$$\mathbf{p}_n - \sum_{i=1}^{n} \gamma^i \mathbf{p}_i \mathbf{T} \mathbf{Y}_{n-i} = \mathbf{p}_0 \mathbf{R}^n$$

$$\mathbf{p}_{n-1} - \sum_{i=1}^{n-1} \gamma^i \mathbf{p}_i \mathbf{T} \mathbf{Y}_{n-i-1} = \mathbf{p}_0 \mathbf{R}^{n-1} \tag{48}$$

Multiplying the second relation of (48) by $\mathbf{R}$ from right and subtracting it from the first one gives

$$\mathbf{p}_n - \mathbf{p}_{n-1}\mathbf{R} - \sum_{i=1}^{n-1} \gamma^i \mathbf{p}_i \mathbf{T} \left(\mathbf{Y}_{n-i} - \mathbf{Y}_{n-i-1}\mathbf{R} \right) - \gamma^n \mathbf{p}_n \mathbf{T}\mathbf{Y}_0 = \mathbf{0}. \tag{49}$$

Applying (23) in (49) and performing rearrangements yields

$$\mathbf{p}_n \left(\mathbf{I} - \gamma^n \mathbf{TA}^{-1}\right) = \mathbf{p}_{n-1}\mathbf{R} + \sum_{i=1}^{n-1} \gamma^i \mathbf{p}_i \mathbf{TA}^{-1}\mathbf{V}^{n-i}. \tag{50}$$

We define the vector sequence $\mathbf{z}_n$ as

$$\mathbf{z}_n = \sum_{i=1}^{n} \gamma^i \mathbf{p}_i \mathbf{TA}^{-1}\mathbf{V}^{n+1-i}, \quad n \geq 0. \tag{51}$$

The second relation of the statement can be obtained by rearranging (51) as

$$\mathbf{z}_n = \sum_{i=1}^{n} \gamma^i \mathbf{p}_i \mathbf{TA}^{-1}\mathbf{V}^{n+1-i} = \left(\sum_{i=1}^{n-1} \gamma^i \mathbf{p}_i \mathbf{TA}^{-1}\mathbf{V}^{n-i}\right)\mathbf{V} + \gamma^n \mathbf{p}_n \mathbf{TA}^{-1}\mathbf{V}$$

$$= \left(\mathbf{z}_{n-1} + \gamma^n \mathbf{p}_n \mathbf{TA}^{-1}\right)\mathbf{V}, \quad n \geq 1.$$

Observe that the last term on r.h.s. of (51) equals to $\mathbf{z}_{n-1}$. Substituting $\mathbf{z}_{n-1}$ into (51) and rearranging it gives the first relation of the statement.   □

If additional criterion fulfills then the matrix inversions in the iteration steps of (46) can be replaced by sums.

Lemma 4. *If matrix $\mathbf{A}$ is non-singular and $\lambda_{max}(\mathbf{TA}^{-1}) < 1$ then*

$$\left(\mathbf{I} - \gamma^n \mathbf{TA}^{-1}\right)^{-1} = \sum_{\ell=0}^{\infty} \left(\gamma^n \mathbf{TA}^{-1}\right)^{\ell}, \quad n \geq 1. \tag{52}$$

Proof. If $\lambda_{max}(\mathbf{TA}^{-1}) < 1$ then $\sum_{\ell=0}^{\infty} \left(\gamma^n \mathbf{TA}^{-1}\right)^{\ell} < \infty$ due to $\lambda_{max}(\gamma^n \mathbf{TA}^{-1}) = \gamma^n \lambda_{max}(\mathbf{TA}^{-1}) < 1$.   □

Based on Theorem 4, a first order recursion can be established to compute the stationary probability vectors. The computational steps of the procedure can be summarized as follows:

1. Determination of matrix $\mathbf{R}$ as the minimal non-negative solution of the quadratic equation (19) by means of iterative computation.
2. Initializations
   - Initialize $\mathbf{z}_0 = \mathbf{0}$.
   - Initialize $\mathbf{p}_0$ by computating the solution of the homogenous system of linear equations (35).
3. Recursive computation of the probability vectors $\mathbf{p}_n$ based on the relations (46) and (47), for $n \geq 1$
   - Compute $\left(\mathbf{I} - \gamma^n \mathbf{TA}^{-1}\right)^{-1}$.
   - Compute $\mathbf{p}_n$.
   - Compute $\mathbf{z}_n$.
4. Determination of the normalization constant from $\sum_n \mathbf{p}_n \mathbf{e} = 1$ and adjusting the stationary probability vectors $\mathbf{p}_n$, $n \geq 0$ accordingly.

References

1. Phung-Duc, T.: Exact solutions for M/M/c/setup queues. Telecommun. Syst. **64**(2), 309–324 (2017)
2. Yajima, M., Phung-Duc, T.: Batch arrival single server queue with variable service speed and setup time. Queueing Syst. **86**(3–4), 241–260 (2017)
3. Phung-Duc, T., Rogiest, W., Wittevrongel, S.: Single server retrial queues with speed scaling: analysis and performance evaluation. J. Ind. Manage. Optim. **13**(4), 1927–1943 (2017)
4. Kleinrock, L., Muntz, R.: Processor sharing queueing models of mixed scheduling disciplines for time shared systems. J. ACM **19**, 464–482 (1972)
5. Saffer, Z., Telek, M.: Analysis of globally gated Markovian limited cyclic polling model and its application to uplink traffic in the IEEE 802.16 network. J. Ind. Manage. Optim. (JIMO) **7**(3), 677–697 (2011)
6. Leung, K.K., Eisenberg, M.: A single-server queue with vacations and non-gated time-limited service. Perform. Eval. **12**(2), 115–125 (1991)
7. Bruneel, H., Rogiest, W., Walraevens, J., Wittevrongel, S.: On queues with general service demands and constant service capacity. In: Norman, G., Sanders, W. (eds.) QEST 2014. LNCS, vol. 8657, pp. 210–225. Springer, Cham (2014). https://doi. org/10.1007/978-3-319-10696-0_17
8. Bruneel, H., Wittevrongel, S., Claeys, D., Walraevens, J.: Discrete-time queues with variable service capacity: a basic model and its analysis. Ann. Oper. Res. **239**(2), 359–380 (2016)
9. De Muynck, M., Bruneel, H., Wittevrongel, S.: Delay analysis of a queue with general service demands and correlated service capacities. In: Takahashi, Y., Phung-Duc, T., Wittevrongel, S., Yue, W. (eds.) QTNA 2018. LNCS, vol. 10932, pp. 64–85. Springer, Cham (2018). https://doi.org/10.1007/978-3-319-93736-6_5
10. Conway, R.W., Maxwell, W.L.: A queueing model with state dependent service rate. J. Ind. Eng. **12**, 132–136 (1961)
11. Lia, H., Yang, T.: Queues with a variable number of servers. EJOR **124**(3), 615–628 (2000)
12. Mazalov, V., Gurtov, A.: Queuing system with on-demand number of servers. Math. Applicanda 40(2) **15/56**, 1–12 (2012)
13. Saffer, Z., Grill, K., Yue, W.: Controllable capacity queue with synchronous constant service time and loss. In: Takahashi, Y., Phung-Duc, T., Wittevrongel, S., Yue, W. (eds.) QTNA 2018. LNCS, vol. 10932, pp. 51–63. Springer, Cham (2018). https://doi.org/10.1007/978-3-319-93736-6_4
14. Neuts, M.F.: Matrix-Geometric Solutions in Stochastic Models: An Algorithmic Approach. The John Hopkins University Press, Baltimore (1981)

A Single Server Queue
with Workload-Dependent Service
Speed and Vacations

Yutaka Sakuma[1]($\boxtimes$) (iD), Onno Boxma[2], and Tuan Phung-Duc[3] (iD)

[1] Department of Computer Science, National Defense Academy of Japan,
Yokosuka-shi, Japan
sakuma@nda.ac.jp

[2] Department of Mathematics and Computer Science,
Eindhoven University of Technology, Eindhoven, The Netherlands
o.j.boxma@tue.nl

[3] Department of Policy and Planning Sciences,
University of Tsukuba, Tsukuba-shi, Japan
tuan@sk.tsukuba.ac.jp

Abstract. In modern data centers, the trade-off between processing speed and energy consumption is an important issue. Motivated by this, we consider a queueing system in which the service speed is a function of the workload, and in which the server switches off when the system becomes empty, only to be activated again when the workload reaches a certain threshold. For this system we obtain the steady-state workload distribution. We use this result to choose the activation threshold such that a certain cost function, involving holding costs and activation costs, is minimized.

Keywords: Single server queue ·
Workload-dependent service speed and vacations ·
Steady-state workload distribution · Level crossing technique ·
Power-saving mechanisms · Cost optimization

1 Introduction

In this paper we consider an $M/G/1$-type queueing system with the following two special features: (i) the service speed is not constant, but a function of the workload, and (ii) the server switches off when the system becomes empty, only to be activated again when the workload reaches a certain threshold. In the remainder of this introduction we successively provide a motivation for this study, present a detailed model description, discuss related literature and give an overview of the rest of the paper.

© Springer Nature Switzerland AG 2019
T. Phung-Duc et al. (Eds.): QTNA 2019, LNCS 11688, pp. 112–127, 2019.
https://doi.org/10.1007/978-3-030-27181-7_8

1.1 Motivation

Cloud service has become ubiquitous in our modern information society. Most Internet users are familiar with some cloud service such as Dropbox, Slack, Google drive etc. These services are supported by data centers where thousands of servers are available, consuming a large amount of energy. Thus, it is crucial to have mechanisms balancing energy consumption and performance for users. While energy saving is very important, most data centers are still designed for peak traffic of users. As a result, in the off-peak period, most servers are idle but still consume about 60% of their peak energy consumption [9,13]. One simple idea is to use an ON-OFF control that automatically adjusts the number of active servers according to the workload. In addition, dynamic scaling techniques such as frequency scaling or voltage scaling enable individual computers to adjust their processing speed in accordance with their workload.

These automatic scaling techniques have the advantage of balancing performance and energy consumption. Because the energy consumption is a monotonic function of the processing speed, less energy is consumed when the system is less congested. When the workload is high, the processing speed is scaled up and thus, the delay performance is improved. At the single computer (server, CPU) level, on the other hand, energy could be saved by adjusting the processing speed of a server according to its own workload. These considerations, featuring the important trade-off between processing speed and energy consumption, motivate the analysis and optimization of queueing systems where the server capacity is dynamically changed according to the workload.

Apart from the interest in power-saving computer systems, queues with variable service speed also naturally arise in service systems with human servers. In particular, in service systems such as call centers, staff numbers are scheduled to meet the demands of customers. Also a human server may speed up when the workload is large, and may spend more time on a job when the workload is small.

In this paper, we propose and analyze a queueing model that features two power-saving mechanisms. The speed of the server is scaled according to the workload in the system. Moreover, the server is turned off when the system is empty and is activated again once the workload reaches a certain threshold. We obtain the distribution for the stationary workload in the system and its mean. We also formulate an optimization problem.

1.2 Model Description

The model under consideration is an $M/G/1$ queue with two special features (cf. Fig. 1): (i) when the server is active and the amount of work present equals $x > 0$, the server works at speed $r(x)$, and (ii) when the workload has dropped to zero, the server becomes inactive ("takes a vacation") and remains inactive until the workload has reached some level $M > 0$, after which it immediately resumes service. We denote the rate of the Poisson arrival process by λ, and the i.i.d. (independent, identically distributed) service requirements by $B_1, B_2, \ldots,$

with distribution $B(\cdot)$ and Laplace-Stieltjes transform (LST) $\beta(s)$. B will denote a generic service requirement. For much of the paper, we shall assume that $B(x) = 1 - e^{-\mu x}$, $x \geq 0$.

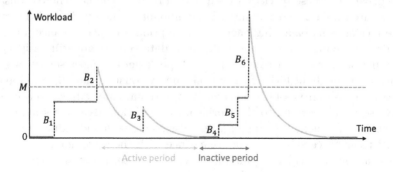

Fig. 1. The workload process.

The case without vacations has been the subject of several studies (see, e.g., [1] and its references). The stability condition for that case is that (cf. [4,5,10]),

$$\text{limsup}_{x \to \infty} \frac{\lambda \mathbb{E}(B)}{r(x)} < 1. \tag{1}$$

Clearly, the same condition should hold in case the server takes a vacation until workload level M is reached. From now on we assume that (1) holds. Below we focus on the steady-state workload distribution $V(\cdot)$ and its density $v(\cdot)$. We also need to take into account the steady-state workload distribution $V_I(\cdot)$ and its density $v_I(\cdot)$ during inactive (vacation) periods of the server; by $p_I := V_I(\infty)$ we denote the probability that the server is inactive.

Define

$$R(x, z) := \int_z^x \frac{1}{r(y)} dy, \quad 0 \leq z < x < \infty; \tag{2}$$

$R(x, z)$ represents the time required to move from level x down to level z in the absence of any arrivals. In particular, $R(x) := R(x, 0)$ denotes the time required to empty the system when starting at level x, in the absence of any arrivals. We assume in the sequel that $R(x) < \infty$ for $x < \infty$; notice that this excludes the choice $r(x) = rx$, which is sometimes termed the shotnoise case [12].

1.3 Related Literature

Our model is related to several topics in the queueing literature. First of all, it is a special example of a queue with vacations: the server takes a vacation when the system becomes empty, and resumes service when the workload reaches or exceeds a certain level. In the classical $M/G/1$ setting, such a D-policy has been extensively studied. We refer to [7] for references and, in particular, for the proof of optimality. For the case of switching costs and running costs, and with a holding cost per time unit which is a non-negative decreasing right-continuous function of the current workload, Feinberg and Kella [7] prove that D-policies are optimal for the average-cost-per-time-unit criterion. This means that there is an optimal policy that either runs the server all the time or switches the server off when the system becomes empty and switches it on when the workload reaches or exceeds some threshold D.

Secondly, our model touches upon the topic of speed scaling. We refer to [18] for an insightful discussion of speed scaling. Recent papers which consider single server queues with speed scaling where the speed of the server is adjusted according to the number of jobs in the system are, e.g., [15,19]. Multiserver queues with ON-OFF control have been extensively studied [8,9,13,14]. In the models in those papers, each server is turned off once it has no jobs to process and is turned on again when jobs are waiting.

Thirdly, there is an extensive literature on queues and dams with a level-dependent outflow rate. We mention the pioneering papers [10,11] and refer to [1] for some more recent results and further references.

1.4   The Structure of the Paper

Section 2 is devoted to a study of the steady-state workload distribution. A cost minimization problem is considered in Sect. 3, where also various numerical results are shown. Section 4 contains some suggestions for further research.

2   The Workload

In this section we first present integral equations for the steady-state workload density $v(\cdot)$ (Subsect. 2.1), while already deriving the workload density during inactive periods; then we formally solve those integral equations (Subsect. 2.2), and finally we present a detailed solution for two special cases: exponentially distributed service requirements (Subsect. 2.3) and generally distributed service requirements with $r(x) = r_1 x + r_0$ (Subsect. 2.4).

2.1   Integral Equations for the Workload Density

We use the level crossing technique (LCT), cf. [2,3,5], which is based on the principle that, in steady state, each level x is crossed just as often from above and from below. We need to distinguish between $x < M$ and $x \geq M$. When $x \geq M$, we have, with $V(0) = \mathbb{P}(V = 0)$ (see also Fig. 1):

$$r(x)v(x) = \lambda \int_{y=0}^{x} \mathbb{P}(B > x - y)v(y)\mathrm{d}y + \lambda\mathbb{P}(B > x)V(0), \qquad (3)$$

and when $x < M$ then

$$r(x)(v(x) - v_I(x)) = \lambda \int_{y=0}^{x} \mathbb{P}(B > x - y)v(y)\mathrm{d}y + \lambda\mathbb{P}(B > x)V(0). \qquad (4)$$

In both cases, the righthand side represents the upcrossing rate, which seems self-explanatory (see also Sections II.4.5 and III.5.10 of [6] for a similar integral equation for, respectively, the ordinary $M/G/1$ queue and the $M/G/1$ queue with service speed $r(x)$). The lefthand side gives the downcrossing rate. Here one has to realize that for $x \in (0, M)$ there can only be a downcrossing when the server is active; hence the term $v(x) - v_I(x)$ for $x \in (0, M)$. Let us now first determine $v_I(x)$ for $x \in (0, M)$.

The density $v_I(x)$.
One can write

$$v_I(x) = v(x|\text{server inactive})\mathbb{P}(\text{server inactive}), \quad 0 < x < M,$$
$$= 0, \quad x \geq M. \tag{5}$$

The probability p_I that the server is inactive equals the fraction of time that the server is inactive; hence, with m_0 and m_1, the means of inactive and active periods, we have

$$p_I = \frac{m_0}{m_0 + m_1}. \tag{6}$$

It is easy to determine m_0. Obviously,

$$m_0 = 1/\lambda \times (1 + m(M)), \tag{7}$$

where $m(x)$ is the renewal function, defined as $m(x) := \mathbb{E}N(x)$, with $\{N(x) := \sup\{n : B_1 + \cdots + B_n \leq x\}$ (cf. Chapter 3 of [16]). The conditional workload density given that the server is inactive also follows from renewal theory, and turns out to be closely related to the renewal function. Introducing the density

$$y(x) := v(x|\text{server inactive}), \tag{8}$$

with distribution $Y(\cdot)$, we shall prove the following.

Theorem 1.

$$Y(x) = \frac{1 + m(x)}{1 + m(M)}, \quad 0 \leq x \leq M. \tag{9}$$

Proof. Remove all the active periods, to obtain a sequence of successive inactive periods. Applying LCT to the thus obtained process, equating the rates of workload downcrossings and upcrossings of any level $x \in [0, M]$ we obtain:

$$\lambda \int_0^x \mathbb{P}(B > x - u)\mathrm{d}Y(u) = \frac{1}{m_0}, \quad 0 \leq x \leq M. \tag{10}$$

The righthand side of this equation reflects the event in which level M is *upcrossed*, which instantaneously (because we have omitted the active periods) is followed by a jump from above M to level 0 – and hence a *downcrossing* of each level $x \in [0, M]$. This happens once per inactive period; hence the term $\frac{1}{m_0}$. Divide both sides of (10) by λ and observe that (e.g., using (10) for $x = 0$)

$$Y(0) = \mathbb{P}(V = 0|\text{server inactive}) = \frac{1}{\lambda m_0}. \tag{11}$$

Rewrite (10) into

$$Y(x) - Y(0) = \int_0^x \mathbb{P}(B < x - u)dY(u)$$

$$= \mathbb{P}(B < x)Y(0) + \int_0^x \mathbb{P}(B < x - u)y(u)du, \quad 0 \le x \le M, \quad (12)$$

and subsequently into

$$\frac{Y(x) - Y(0)}{Y(0)} = \mathbb{P}(B < x) + \int_0^x \mathbb{P}(B < x - u)d\frac{Y(u) - Y(0)}{Y(0)}, \quad 0 \le x \le M. \quad (13)$$

Comparison with the well-known renewal equation (cf. Chapter 3 of [16])

$$m(x) = \mathbb{P}(B < x) + \int_0^x \mathbb{P}(B < x - u)dm(u), \quad (14)$$

shows that $\frac{Y(x) - Y(0)}{Y(0)} = m(x)$ and hence $Y(x) = Y(0)(1 + m(x))$. Finally use the fact that $Y(M) = 1$.

Remark 2.1. In the special case in which $B \sim \exp(\mu)$, one has $m(x) = \mu x$, and hence $y(x) = \frac{\mu}{1 + \mu M}$; the workload during an inactive period, when positive, is uniformly distributed on $(0, M)$.

Remark 2.2. For future use we observe that $v(\cdot)$ has a discontinuity in $x = M$, as revealed by (3) and (4):

$$v(M) - v(M-) = -v_I(M-). \quad (15)$$

Remark 2.3. We close this subsection by pointing out that, in all model variants to be studied in this paper, we have the following relation:

$$\frac{1}{\lambda V(0)} = m_0 + m_1. \quad (16)$$

Indeed, $\lambda V(0)$ is the rate of a customer arriving in an empty system, and hence $\frac{1}{\lambda V(0)}$ is the mean cycle time, viz., the sum of the means of an inactive period and an active period. Since m_0 is known, Formula (16) constitutes a relation between two important quantities: the probability $V(0)$ of an empty system, and the mean m_1 of an active period. These quantities will appear in most of the key workload formulas to be discussed in the sequel. Notice in particular, combining (5), (6), (7), (9) and (16), that

$$v_I(x) = V(0)m'(x), \quad 0 \le x < M. \quad (17)$$

2.2   Solution of the Integral Equations

In this subsection we present a formal solution of the integral equations (3) and (4). First rewrite these two equations into one integral equation:

$$v(x) = \int_{y=0}^{x} K(x,y)v(y)\mathrm{d}y + L(x), \tag{18}$$

where

$$K(x,y) := \frac{\lambda \mathbb{P}(B > x - y)}{r(x)}, \quad 0 \le y < x, \tag{19}$$

and (using (16) to express the unknown constant m_1 into $V(0)$):

$$L(x) := V(0)K(x,0), \quad x \ge M, \tag{20}$$
$$L(x) := V(0)K(x,0) + v_I(x) = V(0)[K(x,0) + m_0\lambda y(x)], \quad x < M,$$

where the last equality follows from (7), (9) and (17). Integral equation (18) is a Volterra integral equation of the second kind. The classical Picard iteration ([17], Chapter I) results in the following formal solution in terms of an infinite series of convolutions. Define recursively

$$K_n(x,y) := \int_{y}^{x} K(x,z)K_{n-1}(z,y)\mathrm{d}z, \quad 0 < y < x, \quad n = 2,3,\ldots,$$

where $K_1(x,y) := K(x,y)$. Then the Picard iteration applied to (18) yields

$$v(x) = L(x) + \int_{y=0}^{x} K(x,y)[L(y) + \int_{z=0}^{y} K(y,z)v(z)\mathrm{d}z]\mathrm{d}y$$

$$= \ldots = L(x) + \sum_{n=1}^{\infty} \int_{0}^{x} K_n(x,y)L(y)\mathrm{d}y. \tag{21}$$

One can follow the approach of [11] and use the bound $K(x,y) \le \frac{\lambda}{r(x)}$ to show inductively that $K_{n+1}(x,y) \le \frac{(\int_{y}^{x} \frac{\lambda}{r(u)}\mathrm{d}u)^n}{n!} \frac{\lambda}{r(x)}$. That implies the convergence of the infinite sum in (21).

What remains to be done is to find the unknown constant $V(0)$. This can be done by using the normalizing condition $\int_0^{\infty} v(x)\mathrm{d}x + V(0) = 1$.

Although one thus in principle obtains an expression for $v(\cdot)$, this solution is a rather formal one, expressed in terms of an infinite sum of non-explicit convolutions. Therefore we restrict ourselves in the next subsections to two special cases, for which we aim to derive more explicit expressions for $v(\cdot)$, viz., (i) the case of exponentially distributed service times, and (ii) the case of $r(x) = r_1 x + r_0$.

2.3   Solution of the Integral Equations in the Case of Exponentially Distributed Service Requirements

In this subsection we assume that $B(x) = 1 - \mathrm{e}^{-\mu x}$. After multiplication by $\mathrm{e}^{\mu x}$, Formula (3) reduces to

$$r(x)\mathrm{e}^{\mu x}v(x) = \lambda \int_{y=0}^{x} \mathrm{e}^{\mu y}v(y)\mathrm{d}y + \lambda V(0), \quad x \ge M, \tag{22}$$

which after differentiation and straightforward calculations yields:

$$v'(x) = \frac{\lambda - \mu r(x) - r'(x)}{r(x)} v(x), \quad x \geq M. \tag{23}$$

Hence, remembering that $R(x) = \int_0^x \frac{1}{r(y)} dy$, and introducing the yet unknown constant C:

$$v(x) = C \frac{e^{\lambda R(x) - \mu x}}{r(x)}, \quad x \geq M. \tag{24}$$

We now turn to (4). In the case of exponentially distributed service requirements, we already observed in Subsect. 2.1 that $v(x|\text{server inactive})$ is constant. Hence also $v_I(x)$ is constant: $v_I(x) = v_I(0)$, $0 \leq x < M$. After multiplication by $e^{\mu x}$, Formula (4) reduces to

$$r(x)e^{\mu x}v(x) = \lambda \int_{y=0}^x e^{\mu y}v(y)dy + \lambda V(0) + r(x)e^{\mu x}v_I(0), \quad x < M, \tag{25}$$

which after differentiation and straightforward calculations yields:

$$v'(x) = \frac{\lambda - \mu r(x) - r'(x)}{r(x)} v(x) + v_I(0)\left(\frac{r'(x)}{r(x)} + \mu\right), \quad x < M. \tag{26}$$

Using variation of constants to solve this inhomogeneous first-order differential equation, we obtain for $x < M$:

$$
\begin{aligned}
v(x) &= C^* \frac{e^{\lambda R(x) - \mu x}}{r(x)} + v_I(0) \int_{y=0}^x \left(\frac{r'(y)}{r(y)} + \mu\right) e^{\lambda R(x,y) - \mu(x-y)} \frac{r(y)}{r(x)} dy \\
&= \frac{e^{\lambda R(x) - \mu x}}{r(x)} [C^* + v_I(0) \int_{y=0}^x (r'(y) + \mu r(y)) e^{-\lambda R(y) + \mu y} dy].
\end{aligned}
\tag{27}
$$

We still need to determine several unknown constants: $V(0)$, $v_I(0)$ and the two constants C and C^*. For this, we have the following equations:

(i) The normalizing condition: $\int_0^\infty v(x)dx + V(0) = 1$.
(ii) Formula (17) for $x = 0$ yields $v_I(0) = \mu V(0)$.
(iii) It follows from (4) for $x = 0$ that $r(0)[v(0) - v_I(0)] = \lambda V(0)$, while (27) implies that $r(0)v(0) = C^*$; hence

$$C^* = \lambda V(0) + r(0)v_I(0) = V(0)[\lambda + \mu r(0)]. \tag{28}$$

(iv) Finally we use the discontinuity of $v(\cdot)$ in M, as described in Remark 2.2. After a lengthy calculation, C follows from (15), (24) and (27):

$$C = V(0)[\lambda + \lambda\mu \int_0^M e^{-\lambda R(y) + \mu y} dy]. \tag{29}$$

The fact that $v(x)$ is both for $x < M$ and $x > M$ linearly expressed in $V(0)$ makes it relatively straightforward to determine that remaining unknown $V(0)$ from the normalizing condition.

The following theorem summarizes our results of this subsection. The expression for $v(x)$, $x < M$ was obtained by using (27) and (28) and performing a partial integration.

Theorem 2.

$$v(x) = \mu V(0) + V(0)\frac{e^{\lambda R(x)-\mu x}}{r(x)}\lambda(1 + \mu\int_{y=0}^{x} e^{-\lambda R(y)+\mu y}dy), \quad x < M, \quad (30)$$

$$v(x) = V(0)\frac{e^{\lambda R(x)-\mu x}}{r(x)}\lambda(1 + \mu\int_{y=0}^{M} e^{-\lambda R(y)+\mu y}dy), \quad x \geq M, \quad (31)$$

with

$$V(0)^{-1} = 1 + \mu M + \int_{x=0}^{M}\frac{e^{\lambda R(x)-\mu x}}{r(x)}\lambda(1 + \mu\int_{y=0}^{x} e^{-\lambda R(y)+\mu y}dy)dx$$

$$+ \int_{x=M}^{\infty}\frac{e^{\lambda R(x)-\mu x}}{r(x)}\lambda(1 + \mu\int_{y=0}^{M} e^{-\lambda R(y)+\mu y}dy)dx. \quad (32)$$

2.4 Solution of the Integral Equations in the Case of Linear Service Speed

In this subsection we allow the service requirements to be generally distributed, but we assume the service speed to be linear: $r(x) = r_1 x + r_0$, where $r_0, r_1 > 0$. Notice that the stability condition (1) is always fulfilled, and that the condition that $R(x) < \infty$ for all finite x is also fulfilled. We apply Laplace transformation to (3) and (4), introducing

$$\phi(s) := \int_{x=0}^{\infty} e^{-sx}v(x)dx \quad (33)$$

$$= \int_{x=0}^{M-} e^{-sx}v(x)dx + \int_{x=M}^{\infty} e^{-sx}v(x)dx$$

for Re $s \geq 0$. It follows from (3) and (4) that

$$-r_1\frac{d}{ds}\phi(s) + r_0\phi(s) = \lambda\frac{1-\beta(s)}{s}\phi(s) + \lambda\frac{1-\beta(s)}{s}V(0) + \gamma(s), \quad (34)$$

where we introduce $\gamma(s) := \int_{x=0}^{M} e^{-sx}(r_1 x + r_0)v_I(x)dx$. According to (17) we have $v_I(x) = V(0)m'(x)$. Hence $\gamma(s)$ is known up to the yet unknown $V(0)$:

$$\gamma(s) = V(0)\int_{x=0}^{M} e^{-sx}(r_1 x + r_0)m'(x)dx =: V(0)\delta(s). \quad (35)$$

Solving the inhomogeneous first-order differential equation (34) yields, with D a yet unknown constant:

$$\phi(s) = e^{\frac{r_0}{r_1}s - \frac{\lambda}{r_1}\int_0^s \frac{1-\beta(u)}{u}du}[D - V(0)\int_{v=0}^s [\frac{\lambda}{r_1}\frac{1-\beta(v)}{v} + \frac{1}{r_1}\delta(v)]$$
$$\times e^{-\frac{r_0}{r_1}v + \frac{\lambda}{r_1}\int_0^v \frac{1-\beta(u)}{u}du}dv].$$
(36)

We still need to determine two unknown constants: $V(0)$ and D. Noticing that $\lim_{s\to\infty}\phi(s) = 0$ gives

$$D = V(0)\int_{v=0}^\infty [\frac{\lambda}{r_1}\frac{1-\beta(v)}{v} + \frac{1}{r_1}\delta(v)]e^{-\frac{r_0}{r_1}v + \frac{\lambda}{r_1}\int_0^v \frac{1-\beta(u)}{u}du}dv.$$
(37)

Indeed, it is easy to see that the exponential in (36), $e^{\frac{r_0}{r_1}s - \frac{\lambda}{r_1}\int_0^s \frac{1-\beta(u)}{u}du}$, tends to ∞ for $s \to \infty$, because the $\frac{r_0}{r_1}s$ term dominates for large s:

$$|\int_0^s \frac{1-\beta(u)}{u}du| \le \int_0^1 |\frac{\beta(u)-\beta(0)}{u}|du + \int_1^s \frac{1}{u}du \le \mathbb{E}B + \ln(s).$$

The normalizing condition states that $\phi(0) + V(0) = 1$, and hence

$$D = 1 - V(0).$$
(38)

We thus obtain one linear equation in the remaining unknown $V(0)$. The following theorem summarizes our results of this subsection.

Theorem 3.

$$\phi(s) = e^{\frac{r_0}{r_1}s - \frac{\lambda}{r_1}\int_0^s \frac{1-\beta(u)}{u}du}V(0)\int_{v=s}^\infty [\frac{\lambda}{r_1}\frac{1-\beta(v)}{v} + \frac{1}{r_1}\delta(v)]e^{-\frac{r_0}{r_1}v + \frac{\lambda}{r_1}\int_0^v \frac{1-\beta(u)}{u}du}dv,$$
(39)

with

$$V(0)^{-1} = 1 + \int_{v=0}^\infty [\frac{\lambda}{r_1}\frac{1-\beta(v)}{v} + \frac{1}{r_1}\delta(v)]e^{-\frac{r_0}{r_1}v + \frac{\lambda}{r_1}\int_0^v \frac{1-\beta(u)}{u}du}dv].$$
(40)

Remark 2.4. If $r_1 = 0$, our system reduces to an ordinary $M/G/1$ queue with a server which is switched off when the system becomes empty and gets activated again when the workload reaches a certain threshold (D-policy, cf. [7]). It readily follows from (34) (where the first term disappears for $r_1 = 0$) that $\phi(s)$ now becomes the product of the workload LST in an ordinary $M/G/1$ queue and an additional term that relates to the off-periods; such decomposition results are well-known in the literature of queues with vacations.

Remark 2.5. A tedious but straightforward calculation verifies that the results of Theorems 2 and 3 agree when $r(x) = r_1 x + r_0$ and $B(x) = 1 - e^{-\mu x}$. One first takes Laplace transforms in (30) and (31), obtaining

$$
\frac{\phi(s)}{V(0)} = \int_{x=0}^{M} \mu e^{-sx} dx + \lambda \int_{x=0}^{\infty} e^{-(s+\mu)x} \frac{(\frac{r_1 x + r_0}{r_0})^{\frac{\lambda}{r_1}}}{r_1 x + r_0} dx
$$

$$
+ \lambda \mu \int_{y=0}^{M} e^{\mu y} (\frac{r_1 y + r_0}{r_0})^{-\frac{\lambda}{r_1}} dy \int_{x=y}^{\infty} e^{-(s+\mu)x} \frac{(\frac{r_1 x + r_0}{r_0})^{\frac{\lambda}{r_1}}}{r_1 x + r_0} dx. \quad (41)
$$

One partial integration in the last integral of (41) gives a cancellation against the first term in the righthand side. Subsequently the transformation $\frac{r_1 x + r_0}{r_1 y + r_0} = \frac{v + \mu}{s + \mu}$ leads to the expression in (39).

Remark 2.6. From (36), using that $\mathbb{E}V = -\phi'(s)|_{s=0}$, it follows that

$$
\mathbb{E}V = -\frac{r_0}{r_1}(1 - V(0)) + \frac{\lambda \mathbb{E}B}{r_1}(1 - V(0)) + \frac{\lambda \mathbb{E}B}{r_1} V(0) + \frac{1}{r_1} \int_{x=0}^{M} (r_1 x + r_0) v_I(x) dx
$$

$$
= \frac{\lambda \mathbb{E}B - r_0}{r_1} + V(0)[\frac{r_0}{r_1} + \frac{1}{r_1} \int_{x=0}^{M} (r_1 x + r_0) m'(x) dx]. \quad (42)
$$

In the special case of $\exp(\mu)$ service times, $m'(x) = \mu$ and we have:

$$
\mathbb{E}V = \frac{\lambda - \mu r_0}{\mu r_1} + V(0)[\frac{r_0}{r_1}(1 + \mu M) + \frac{1}{2}\mu M^2]. \quad (43)
$$

3   Cost Optimization

Suppose that two types of costs are involved in the operation of the system: holding costs c_h per time unit for each unit of work present in the system, and switching costs c_s for each time the server is switched on. We are interested in choosing M such that the system costs are minimized. Hence we consider the following minimization problem (cf. (16)):

$$
\text{Minimize}_M \quad c_h \mathbb{E}V + c_s \frac{1}{m_0 + m_1} = c_h \mathbb{E}V + c_s \lambda V(0). \quad (44)
$$

In addition, the system might receive profits from each amount of work that is being served. However, we can ignore that profit, as it does not depend on the choice of M.

We focus on the case, studied in Subsect. 2.4, in which $r(x) = r_1 x + r_0$. It follows from (43) that our optimization problem becomes:

$$
\text{Min}_M \quad c_h \frac{\lambda \mathbb{E}B - r_0}{r_1} + c_h V(0)[\frac{r_0}{r_1} + \frac{1}{r_1} \int_{x=0}^{M} (r_1 x + r_0) m'(x) dx] + c_s \lambda V(0), \quad (45)
$$

which amounts to minimizing, w.r.t. M, the function

$$f(M) := V(0)[\frac{c_h r_0}{r_1} + \frac{c_h}{r_1} \int_{x=0}^{M} (r_1 x + r_0) m'(x) dx + c_s \lambda]; \tag{46}$$

here $V(0)$ depends on M, and is given by (40).

The derivative of $f(M)$ w.r.t. M should be zero, and hence M should satisfy

$$\frac{c_h}{r_1}(r_1 M + r_0) m'(M) = V(0)[c_s \lambda + \frac{c_h r_0}{r_1} + \frac{c_h}{r_1} \int_0^M (r_1 x + r_0) m'(x) dx]$$
$$\times \int_0^\infty \frac{1}{r_1} e^{-vM} (r_1 M + r_0) m'(M) e^{-\frac{r_0}{r_1} v + \frac{\lambda}{r_1} \int_0^v \frac{1-\beta(u)}{u} du} dv. \tag{47}$$

Let us now restrict ourselves to the case of $\exp(\mu)$ service times. Then (47) reduces to

$$\frac{c_h \mu}{r_1}(r_1 M + r_0) = V(0)[c_s \lambda + c_h \frac{r_0}{r_1} + \frac{c_h \mu}{r_1}(\frac{r_1}{2} M^2 + r_0 M)]$$
$$\times (M + \frac{r_0}{r_1}) \mu \int_0^\infty e^{-vM} e^{-\frac{r_0}{r_1} v} (\frac{\mu + v}{\mu})^{\frac{\lambda}{r_1}} dv. \tag{48}$$

Here $1/V(0)$ simplifies to

$$\frac{1}{V(0)} = 1 + \int_0^\infty \left(\frac{\lambda}{r_1} \frac{1}{\mu + v} + \frac{\mu}{r_1} \int_0^M e^{-vx} (r_1 x + r_0) dx \right) e^{-\frac{r_0}{r_1} v} (\frac{\mu + v}{\mu})^{\frac{\lambda}{r_1}} dv. \tag{49}$$

Remark 4.1. Matters simplify further if we assume that

$$\lambda = r_1. \tag{50}$$

By interchanging the two integrals in (49), we then obtain an explicit expression for $V(0)$:

$$\frac{1}{V(0)} = 1 + \frac{r_1}{\mu r_0} + \mu M + \ln \frac{\frac{r_0}{r_1} + M}{\frac{r_0}{r_1}}. \tag{51}$$

Remark 4.2. It should be observed that, if we take $r_0 = 0$, then the first integral in the right-hand side of (49) diverges, giving $V(0) = 0$. The explanation is that, when the service speed is $r_1 x$, the system never becomes zero.

3.1   Numerical Examples

We plot some graphs to show the behavior of the cost function as a function of the threshold M.

In our numerical experiments, we fix the arrival rate: $\lambda = 1$ and show the effect of other parameters on the cost function. Intuitively, on the one hand, a large threshold M leads to a larger workload in the system since the inactive period is longer. On the other hand, a large threshold may prevent frequent switching and thus may reduce the switching cost. Thus, it is expected to have an optimal M which balances the two types of costs. In all our numerical experiments, the cost function was convex, and we found a unique optimal M. However, we have not yet been able to analytically show convexity of the cost function in M.

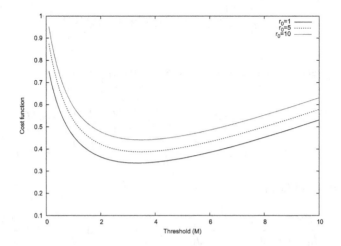

Fig. 2. Cost function for Case 1: $c_h = 0.1, c_s = 1$; various r_0.

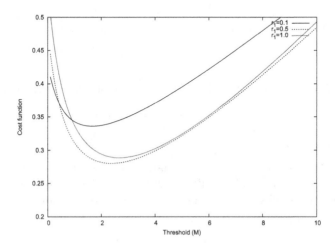

Fig. 3. Cost function for Case 2: $c_h = 0.1, c_s = 1$; various r_1.

Case 1: Cost function vs. M for various r_0
Figure 2 displays the cost function against the threshold M for several values of r_0; $r_0 = 1, 5, 10$. Other parameters are as follows: $c_h = 0.1, c_s = 1, \mu = 1, r_1 = 10$. Notice the above-mentioned convexity of the curves, guaranteeing that there is an optimal M that minimizes the cost function. We also observe that the optimal M is almost insensitive to r_0 in this case. A close inspection shows that the optimal M slightly increases with r_0.

Case 2: Cost function vs. M for various r_1
Figure 3 displays the cost function against the threshold M for several values of r_1; $r_1 = 0.1, 0.5, 1$. Parameters are fixed as follows: $c_h = 0.1, c_s = 1, \mu = 1, r_0 = 1$. The optimal value of M for a larger r_1 is seen to be bigger than that for a smaller r_1.

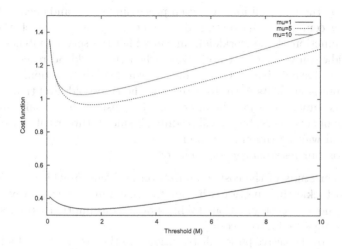

Fig. 4. Cost function for Case 3: $c_h = 0.1, c_s = 1, r_1 = 0.1$; various μ.

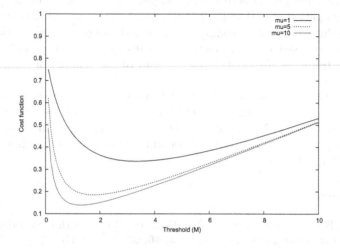

Fig. 5. Cost function for Case 3: $c_h = 0.1, c_s = 1, r_1 = 10$; various μ.

Case 3: Cost function vs. M for various μ

In this case, we display the cost function against the threshold M for various values of μ. Fixed parameters are as follows: $c_h = 0.1, c_s = 1, r_0 = 1$. Figure 4 is for the case $r_1 = 0.1$ while Fig. 5 is for the case $r_1 = 10$. We observe from Figs. 4 and 5 that the optimal value of M increases with μ.

Numerical Insights. Extensive numerical results suggest that the cost function is a convex function of M, so that there exists an optimal value of the threshold M. Furthermore, not surprisingly, the optimal threshold increases with r_0, r_1 and μ. A rigorous proof of the convexity of the cost function in the threshold M is left for future work.

4   Conclusion and Suggestions for Further Research

Motivated by the trade-off issue between processing speed and energy consumption in data centers, we have studied a queueing system in which the service speed is a function of the workload, and in which the server switches off when becoming idle, only to be activated again when the workload reaches a certain threshold. We have derived the (LST of the) workload distribution, and we have used an expression for its mean to determine the threshold level that minimizes a certain cost function. It may be interesting to consider the actual waiting time distribution of customers, but it will be difficult since future events have an effect on the actual waiting time in our model.

Topics on our research agenda include:

(i) A further study of the cost minimization problem, in which we also would like to tackle the question whether the cost function is convex. We wish to extend our cost function, taking power consumption as a function of processing speed into account.

(ii) A study of the active period distribution and the distribution of a full cycle, consisting of an inactive and subsequent active period. It should be observed that the length of an active period depends on the length of the preceding inactive period, but that the length of an inactive period does not depend on the length of the preceding active period; hence the distribution of the sum of the lengths of an inactive and subsequent active period in general differs from the distribution of the sum of the lengths of an active and subsequent inactive period.

(iii) We are presently analyzing the model variant in which the processing speed $r(x)$ is piecewise constant ($r(x) = r_i$ when the workload is lying in an interval J_i, $i = 1, 2, \ldots$), and in which the service requirement distribution $B(\cdot)$ is phase-type. This is a case for which it seems to be possible to obtain quite explicit results.

References

1. Bekker, R., Borst, S.C., Boxma, O.J., Kella, O.: Queues with workload-dependent arrival and service rates. Queueing Syst. **46**, 537–556 (2004)

2. Brill, P.H.: Level Crossing Methods in Stochastic Models. Springer International Publishing (2017)
3. Brill, P.H., Posner, M.J.M.: Level crossing in point processes applied to queues: single server case. Oper. Res. **25**, 662–674 (1977)
4. Browne, S., Sigman, K.: Workload-modulated queues with application to storage processes. J. Appl. Probab. **29**, 699–712 (1992)
5. Cohen, J.W.: On up- and downcrossings. J. Appl. Probab. **14**, 405–410 (1977)
6. Cohen, J.W.: The Single Server Queue, 2nd edn. North-Holland Publishing Company, Amsterdam (1982)
7. Feinberg, E.A., Kella, O.: Optimality of D-policies for an $M/G/1$ queue with a removable server. Queueing Syst. **42**, 355–376 (2002)
8. Gandhi, A., Harchol-Balter, M., Adan, I.: Server farms with setup costs. Perform. Eval. **67**, 1123–1138 (2010)
9. Gandhi, A., Doroudi, S., Harchol-Balter, M., Scheller-Wolf, A.: Exact analysis of the M/M/k/setup class of Markov chains via recursive renewal reward. ACM SIGMETRICS Perform. Eval. Rev. **41**, 153–166 (2013)
10. Gaver, D.P., Miller, R.G.: Limiting distributions for some storage problems. In: Arrow, K.J., Karlin, S., Scarf, H. (eds.) Studies in Applied Probability and Management Science, pp. 110–126. Stanford University Press, Stanford (1962)
11. Harrison, J.M., Resnick, S.I.: The stationary distribution and first exit probabilities of a storage process with general release rule. Math. Oper. Res. **1**, 347–358 (1976)
12. Koops, D., Boxma, O.J., Mandjes, M.R.H.: Networks of $\cdot/G/\infty$ queues with shot-noise-driven arrival intensities. Queueing Syst. **86**, 301–325 (2017)
13. Maccio, V.J., Down, D.G.: Structural properties and exact analysis of energy-aware multiserver queueing systems with setup times. Perform. Eval. **121**, 48–66 (2018)
14. Phung-Duc, T.: Exact solutions for M/M/c/setup queues. Telecommun. Syst. **64**, 309–324 (2017)
15. Phung-Duc, T., Rogiest, W., Wittevrongel, S.: Single server retrial queues with speed scaling: analysis and performance evaluation. J. Ind. Manage. Optim. **13**, 1927–1943 (2017)
16. Ross, S.M.: Stochastic Processes. Wiley, New York (1983)
17. Tricomi, F.G.: Integral Equations. Interscience Publishers, New York (1957)
18. Wierman, A., Andrew, L.L.H., Lin, M.: Speed scaling: an algorithmic perspective. In: Handbook of Energy-Aware and Green Computing. Chapman & Hall/CRC Computing and Information Science Series (2011)
19. Yajima, M., Phung-Duc, T.: Batch arrival single-server queue with variable service speed and setup time. Queueing Syst. **86**, 241–260 (2017)

Delay Analysis of a Two-Server Discrete-Time Queue Where One Server Is Only Intermittently Available

Freek Verdonck$^{(\boxtimes)}$, Herwig Bruneel , and Sabine Wittevrongel

Department of Telecommunications and Information Processing (TELIN),
SMACS Research Group, Ghent University (UGent),
Sint-Pietersnieuwstraat 41, 9000 Ghent, Belgium
{freek.verdonck,herwig.bruneel,sabine.wittevrongel}@ugent.be

Abstract. In this work we look at the delay analysis of a customer in a discrete-time queueing system with one permanent server and one occasional extra server. The arrival process is assumed to be general independent, the buffer size infinite and the service times deterministically equal to one slot. The system resides in one of two different states defined by the number of available servers. In the UP-state 2 servers are available and in the DOWN-state 1 server is available. State changes can only occur at slot boundaries. When the extra server becomes available, an UP-period starts (DOWN-period ends) and when the extra server becomes unavailable a DOWN-period starts (UP-period ends). The lengths of these periods, expressed in their number of slots, are assumed to follow a geometric distribution, with different parameter for UP-periods and DOWN-periods. Also, the extension is made to DOWN-periods according to a mixture of M geometric distributions. Using the technique of the dominant singularity, we provide a method to evaluate the tail characteristics of the delay of an arbitrary customer. The method is illustrated with a numerical example.

Keywords: Queueing theory · Discrete-time · Multiserver ·
Server interruptions · Delay · Tail

1 Introduction

This paper focusses on a discrete-time queueing system with two servers, where one server is permanently available and one server is subject to random interruptions. The buffer size is assumed to be infinite. The time horizon is divided into slots of equal length and the service times are deterministic and equal to one slot. The interruption process divides the system into two states: UP-states with two servers available and DOWN-states with one server available. State changes can only occur at slot boundaries, and these mark the beginnings and ends of UP-periods and DOWN-periods. The lengths of the periods are according

© Springer Nature Switzerland AG 2019
T. Phung-Duc et al. (Eds.): QTNA 2019, LNCS 11688, pp. 128–146, 2019.
https://doi.org/10.1007/978-3-030-27181-7_9

to a geometric distribution (with different parameter). This introduces correlation in the number of servers available from slot to slot. In a second stage we also consider a mixture of geometrics for the distribution of the lengths of the DOWN-periods.

In the earlier paper [3] the queue content of a system as such was described. The current paper is an extension to that work, we focus now on the delay of an arbitrary customer. We obtain the tail characteristics of the delay using the theory of the dominant singularity. This means that we obtain an approximation for the probability that the delay of an arbitrary customer is equal to, or larger than a certain value k for large k. In a numerical example the comparison is made between the proposed method and simulation. We obtain high accuracy even for not so large k while being much less demanding on computational resources.

Much research has been done on the delay characteristics of a customer in queueing systems. An important result for the delay characteristics of a multi-server model is [2], where in every slot c servers are available. In recent research, queueing systems with server interruptions gathered attention, see [4] for a comprehensive survey on both discrete-time and continuous-time models. When dealing with service interruptions, the majority of authors limit themselves to the analysis of the queue content and consider the delay only through Little's Law. Exceptions are [6, 7], where the distribution of the waiting time for a continuous-time queueing model with a single-server subject to interruptions is treated. In [10] a continuous-time model is proposed with a single server of which the service speed depends on the number of customers in the queue. The main contribution of the paper is the Laplace-Stieltjes transform of the sojourn time distribution. Furthermore, in [8] a discrete-time multiserver queueing system is analyzed where all m servers are subject to independent interruptions according to a Bernoulli process with the same parameter. A relationship is obtained between the pgf of the system content and the pgf of the delay. The analysis of the current paper stands out in the sense that it handles the delay in a multiserver queueing system with correlated server interruptions.

The study of this type of queueing systems is motivated by the many applications of queueing theory where the number of available servers is not constant over time. Examples are the airport check-in process [12] or production environments [11].

The outline of the paper is as follows. In the next section we describe the mathematical model under study. The queue content of this model has been analyzed in a recent paper [3] and in Sect. 3 we summarize some key results that are necessary for the delay analysis of this paper. Section 4 then presents the delay analysis for the case of geometric DOWN-periods, while Sect. 5 considers the extension where DOWN-periods are distributed according to a mixture of M geometric distributions. A numerical example is discussed in Sect. 6 and we conclude the paper in Sect. 7.

2   Queueing Model Under Study

The model under study is a discrete-time queueing system with infinite buffer size and service times deterministically equal to one slot. The system resides in one of two different states, based on the number of servers available. During a DOWN-slot, only one server is available, while during an UP-slot two servers are available. State changes can only occur at slot boundaries, and these mark the beginnings and ends of DOWN-periods and UP-periods. In this paper the lengths of the UP-periods are according to a geometric distribution with parameter α:

$$\text{Prob[UP-period has } n \text{ slots]} \triangleq (1 - \alpha)\alpha^{n-1}, \quad n > 0. \tag{1}$$

In general, the lengths of the DOWN-periods can be described by

$$\text{Prob[DOWN-period has } n \text{ slots]} \triangleq r(n), \quad n > 0, \tag{2}$$

$$R(z) \triangleq \sum_{n=1}^{\infty} r(n)z^n, \tag{3}$$

where we have introduced $R(z)$ as the pgf of the distribution of the DOWN-periods. The mean length of an UP-period is $\frac{1}{1-\alpha}$ and the mean length of a DOWN-period is given by $\bar{r}$, with

$$\bar{r} \triangleq R'(1). \tag{4}$$

If we denote with r_k the length of the kth DOWN-period, then the series $\{r_k\}$ is a set of independent and identically distributed (i.i.d.) random variables. The probability that an arbitrary slot is an UP-slot is given by σ and the probability that an arbitrary slot is a DOWN-slot is given by $(1 - \sigma)$, with

$$\sigma = \frac{1}{1 + (1 - \alpha)\bar{r}}. \tag{5}$$

The numbers of arrivals in a slot are i.i.d. and according to a general distribution:

$$\text{Prob[}n \text{ arrivals during a slot]} \triangleq c(n), \quad n \geq 0; \tag{6}$$

$$C(z) \triangleq \sum_{n=0}^{\infty} c(n)z^n; \tag{7}$$

$$\lambda \triangleq \sum_{n=0}^{\infty} nc(n). \tag{8}$$

The service of a customer can only start at the next slot boundary, even if a server is idle at the moment of arrival. The system is stable if the average arrival intensity is strictly smaller than the average number of servers available. The stability condition thus reads:

$$\lambda < 1 + \sigma = \frac{2 + (1 - \alpha)\bar{r}}{1 + (1 - \alpha)\bar{r}}. \tag{9}$$

For the delay analysis we consider the First In First Out (FIFO) policy. As the definition of delay we take the total system time of a customer. This includes the waiting time and service time, but not the remainder of the slot during which the customer arrives. This definition is illustrated in Fig. 1. This way, the total system time is an integer number of slots and it is not necessary to specify the exact arrival moment of a customer within its slot of arrival.

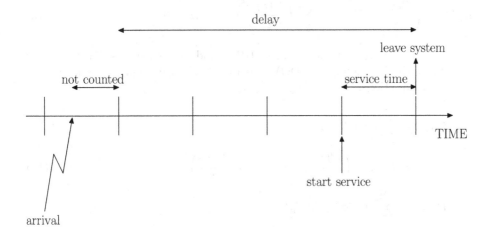

Fig. 1. Illustration of the delay of a customer

The system as described is also referred to as a Late Arrival System with Delayed Access (LAS-DA).

3   Queue Content

The delay analysis in this paper heavily relies on the distributions of the queue content obtained in [3]. In this section we briefly repeat some key results.

Let us denote with the stochastic variable g_n $(n \geq 0)$ the queue content at the beginning of the $(n+1)$st slot of an UP-period, with corresponding pgf $G_n(z)$. Analogously, we denote with the stochastic variable h_n $(n \geq 0)$ the queue content at the beginning of the $(n+1)$st slot of a DOWN-period, with corresponding pgf $H_n(z)$. The following recursive equations can be obtained for $n > 0$:

$$G_n(z) = \frac{C(z)}{z^2} \left[G_{n-1}(z) + G_{n-1}(0)\,(z^2 - 1) + G'_{n-1}(0)\,(z^2 - z) \right] ; \qquad (10)$$

$$H_n(z) = \frac{C(z)}{z} \left[H_{n-1}(z) + H_{n-1}(0)\,(z - 1) \right] . \qquad (11)$$

After recursive application, $G_n(z)$ and $H_n(z)$ can be expressed in terms of $G_0(z)$ and $H_0(z)$ and the unknowns $G_k(0)$, $G'_k(0)$ and $H_k(0)$ for $0 \le k < n$:

$$G_n(z) = \left(\frac{C(z)}{z^2}\right)^n G_0(z) + \sum_{i=1}^{n} \left(\frac{C(z)}{z^2}\right)^i \left[(z^2-1)G_{n-i}(0) + (z^2-z)G'_{n-i}(0)\right] ;$$

$$ \tag{12}$$

$$H_n(z) = \left(\frac{C(z)}{z}\right)^n H_0(z) + \sum_{i=1}^{n} \left(\frac{C(z)}{z}\right)^i (z-1)H_{n-i}(0) . \tag{13}$$

The queue content at the end of an UP-period can be expressed as the queue content at the beginning of a DOWN-period (and vice versa). This leads to a set of two equations:

$$G_0(z) = \sum_{n=1}^{\infty} r(n)H_n(z)$$

$$= H_0(z) R\left(\frac{C(z)}{z}\right) + (z-1)Q\left(\frac{C(z)}{z}\right) ; \tag{14}$$

$$H_0(z) = \sum_{n=1}^{\infty}(1-\alpha)\alpha^{n-1}G_n(z)$$

$$= \frac{C(z)}{z^2 - \alpha C(z)} \left[(1-\alpha)G_0(z) + (z-1)p(0) + (z^2-z)p(1)\right] , \tag{15}$$

with $p(0)$, $p(1)$ and $Q(z)$ still unknown. The latter is of the form

$$Q(z) \triangleq \sum_{i=1}^{\infty} q(i)z^i \quad ; \quad q(i) \triangleq \sum_{j=0}^{\infty} H_j(0) r(i+j). \tag{16}$$

The set of Eqs. (14) and (15) can be solved and leads to:

$$G_0(z) = (z-1)\frac{[p(0)+p(1)z]C(z)R\left(\frac{C(z)}{z}\right) + [z^2-\alpha C(z)]Q\left(\frac{C(z)}{z}\right)}{z^2 - C(z)\left[\alpha + (1-\alpha)R\left(\frac{C(z)}{z}\right)\right]} ; \tag{17}$$

$$H_0(z) = (z-1)C(z)\frac{p(0)+p(1)z+(1-\alpha)Q\left(\frac{C(z)}{z}\right)}{z^2 - C(z)\left[\alpha + (1-\alpha)R\left(\frac{C(z)}{z}\right)\right]} . \tag{18}$$

In [3] it is proven that if $R(z)$ is a rational function also $Q(z)$ is rational and has the same denominator as $R(z)$. The (finite number of) remaining unknowns can be determined by relying on the properties of pgfs, namely that they are normalized and analytical within the complex unit disk. The authors of [3] also found an expression for the queue content right after an arbitrary UP-slot or DOWN-slot. We adapt their results to have an expression for the queue content right **before** an arbitrary UP-slot or DOWN-slot as it is more useful for the delay-analysis. We call these queue contents g and h respectively with corresponding pgfs $G(z)$ and $H(z)$. Therefore we introduce the random variables K_{up} and K_{down} as the ordinate of an arbitrary slot within its period, with corresponding pgf $K_{up}(z)$ and $K_{down}(z)$:

$$\text{Prob}[K_{up} = k] = \frac{\sum_{n=k}^{\infty} r(n)}{\overline{r}} \quad ; \quad K_{up}(z) = \frac{z\,[R(z) - 1]}{(z-1)\overline{r}} \; ; \tag{19}$$

$$\text{Prob}[K_{down} = k] = (1-\alpha)\sum_{n=k}^{\infty}(1-\alpha)\alpha^{n-1} = (1-\alpha)\alpha^{k-1}. \tag{20}$$

These expressions are well known in probability theory, see for example [1]. Now we introduce the definitions of $G(z)$ and $H(z)$ and work them out.

$$G(z) \triangleq \sum_{k=1}^{\infty} \text{Prob}[K_{up} = k]\, G_{k-1}(z)$$

$$= \sum_{k=1}^{\infty}(1-\alpha)\alpha^{k-1} G_{k-1}(z)$$

$$= (1-\alpha)G_0(z) + \alpha H_0(z)\,, \tag{21}$$

and

$$H(z) \triangleq \sum_{k=1}^{\infty} \text{Prob}[K_{down} = k]\, H_{k-1}(z)$$

$$= \frac{1}{\overline{r}}H_0(z) + \sum_{k=2}^{\infty} \frac{\sum_{n=k}^{\infty} r(n)}{\overline{r}} \left[\left(\frac{C(z)}{z}\right)^{k-1} H_0(z)\right.$$

$$\left. + \sum_{i=1}^{k-1}\left(\frac{C(z)}{z}\right)^{i}(z-1)H_{k-1-i}(0)\right]$$

$$= z\frac{R\left(\frac{C(z)}{z}\right) - 1}{[C(z) - z]\overline{r}}H_0(z) + \frac{(z-1)}{\overline{r}}\sum_{k=2}^{\infty}\sum_{j=0}^{\infty}H_j(0)\,r(k+j)\sum_{i=1}^{k-1}\left(\frac{C(z)}{z}\right)^{i}$$

$$= z\frac{R\left(\frac{C(z)}{z}\right) - 1}{[C(z) - z]\overline{r}}H_0(z) + \frac{(z-1)}{\overline{r}}\sum_{k=2}^{\infty}q(k)\frac{\left(\frac{C(z)}{z}\right)^{k} - \frac{C(z)}{z}}{\frac{C(z)}{z} - 1}$$

$$= z\frac{R\left(\frac{C(z)}{z}\right) - 1}{[C(z) - z]\overline{r}}H_0(z) + (z-1)\frac{zQ\left(\frac{C(z)}{z}\right) - Q(1)\,C(z)}{[C(z) - z]\overline{r}}. \tag{22}$$

In the following sections we focus on the delay a customer experiences in a system as described in Sect. 2, with queue content distribution as above.

4   Delay Analysis for DOWN-Periods According to a Geometric Distribution

Let us first assume that the DOWN-periods are following a geometric distribution:

$$r(n) = (1 - \beta)\beta^{n-1}, \quad n > 0;$$
(23)

$$R(z) = \frac{(1 - \beta)z}{1 - \beta z}.$$
(24)

In this case $Q(z)$ is of the following form:

$$Q(z) = \frac{p(2)z}{1 - \beta z},$$
(25)

with $p(2)$ unknown. Substitution of (24) and (25) into (22) leads to

$$
\begin{aligned}
H(z) &= \frac{(1 - \beta)z}{z - \beta C(z)} H_0(z) + \frac{\beta(z-1)p(2)C(z)}{z - \beta C(z)} \\
&= (1 - \beta)H_0(z) + \frac{\beta(1 - \beta)C(z)}{z - \beta C(z)} H_0(z) + \frac{\beta(z-1)p(2)C(z)}{z - \beta C(z)} \\
&= (1 - \beta)H_0(z) + \beta R\left(\frac{C(z)}{z}\right) H_0(z) + \beta(z-1)Q\left(\frac{C(z)}{z}\right) \\
&= (1 - \beta)H_0(z) + \beta G_0(z),
\end{aligned}
$$
(26)

where in the last step we have made use of (14). When considering the memoryless property of the geometric distribution we can also directly obtain (26).

4.1   Delay of a Packet with k Customers Ahead

We now introduce the following notations: the stochastic variable u_k denotes the total system time of a customer entering the system during an UP-slot and with k customers ahead of it in the queue, thus excluding customers in service at the moment of arrival. The corresponding pgf is $U_k(z)$. Analogously, d_k denotes the total system time of a customer entering the system during a DOWN-slot and with k customers ahead of it in the queue. The corresponding pgf is $D_k(z)$. We can compose the following expressions:

$$U_0(z) = z;$$
(27)

$$U_1(z) = \alpha z + (1 - \alpha)z^2;$$
(28)

$$U_k(z) = \alpha z U_{k-2}(z) + (1 - \alpha)z D_{k-1}(z), \quad k \geq 2;$$
(29)

$$D_0(z) = z;$$
(30)

$$D_1(z) = \beta z^2 + (1 - \beta)z;$$
(31)

$$D_k(z) = \beta z D_{k-1}(z) + (1 - \beta)z U_{k-2}(z), \quad k \geq 2.$$
(32)

Let us now define the bivariate functions $U(x, z)$ and $D(x, z)$:

$$U(x, z) \triangleq \sum_{k=0}^{\infty} x^k U_k(z) ; \tag{33}$$

$$D(x, z) \triangleq \sum_{k=0}^{\infty} x^k D_k(z) . \tag{34}$$

Working out these definitions leads to

$$U(x, z) = U_0(z) + xU_1(z) + \sum_{k=0}^{\infty} \left\{ \alpha z U_k(z) x^{k+2} + (1 - \alpha)z D_{k+1}(z) x^{k+2} \right\}$$

$$= z + \alpha z x + \alpha z x^2 U(x, z) + (1 - \alpha)xz D(x, z) ; \tag{35}$$

$$D(x, z) = z + (1 - \beta)zx + \beta z x D(x, z) + (1 - \beta)zx^2 U(x, z) . \tag{36}$$

From the above two expressions we can get an explicit formula for both $U(x, z)$ and $D(x, z)$:

$$U(x, z) = \frac{f_1(x, z)}{g(x, z)} ; \tag{37}$$

$$D(x, z) = \frac{f_2(x, z)}{g(x, z)} , \tag{38}$$

with

$$f_1(x, z) = z \left\{ 1 + [\alpha + z(1 - \alpha - \beta)]x + (1 - \alpha - \beta)zx^2 \right\} ; \tag{39}$$

$$f_2(x, z) = z \left[1 + (1 - \beta)x + (1 - \alpha - \beta)zx^2 \right] ; \tag{40}$$

$$g(x, z) = 1 - \beta zx - \alpha zx^2 - (1 - \alpha - \beta)z^2 x^3 . \tag{41}$$

We expand (37) and (38) in partial fractions based on their poles in x which we denote as x_ϕ. These x_ϕ are roots of a third degree polynomial and are a function of z. For notational simplicity their argument is omitted. Note that there are only 2 x_ϕ when $\alpha + \beta = 1$. When that relation holds, there is a constant chance of α that the system is in an UP-slot and a constant chance of $\beta = 1 - \alpha$ that the system is in a down slot. In this simpler case without correlation on the number of servers available in a slot, the delay analysis can be performed in full as described in [5] so we leave this case out of consideration in the current analysis. Furthermore, we assume all x_ϕ to be distinct, i.e. all x_ϕ to have multiplicity 1. We then get:

$$U(x, z) = \sum_{\phi=1}^{3} \frac{f_1(x_\phi, z)}{g_x(x_\phi, z)(x - x_\phi)} ; \tag{42}$$

$$D(x, z) = \sum_{\phi=1}^{3} \frac{f_2(x_\phi, z)}{g_x(x_\phi, z)(x - x_\phi)} , \tag{43}$$

with

$$g_x(x, z) \triangleq \frac{\partial}{\partial x} g(x, z) = -\beta z - 2\alpha z x - 3(1 - \alpha - \beta)z^2 x^2. \tag{44}$$

We can then obtain an expression for $U_k(z)$ and $D_k(z)$:

$$U_k(z) = \frac{1}{k!} \frac{\partial^k}{\partial x^k} U(x, z) \Big|_{x=0} = \sum_{\phi=1}^{3} \frac{-f_1(x_\phi, z)}{x_\phi^{k+1} g_x(x_\phi, z)} \; ; \tag{45}$$

$$D_k(z) = \frac{1}{k!} \frac{\partial^k}{\partial x^k} D(x, z) \Big|_{x=0} = \sum_{\phi=1}^{3} \frac{-f_2(x_\phi, z)}{x_\phi^{k+1} g_x(x_\phi, z)}. \tag{46}$$

4.2   Delay of an Arbitrary Packet

Let us now look at an arbitrary packet P, arriving during slot S. There is a probability σ that S is an UP-slot and a probability $(1 - \sigma)$ that it is a DOWN-slot. Let us denote with the stochastic variable t_{up} the number of customers ahead of P at the end of S if S is an UP-slot, with corresponding pgf $T_{\text{up}}(z)$ and with the stochastic variable t_{down} the number of customers ahead of P at the end of S if S is a DOWN-slot, with corresponding pgf $T_{\text{down}}(z)$. The number of customers in the queue ahead of P are the customers that were present at the beginning of S, minus the customers that entered service during S and plus the customers that arrived during S but before P. The pgf $F(z)$ of this last amount is known, see for example [1]:

$$F(z) = \frac{C(z) - 1}{\lambda(z - 1)}. \tag{47}$$

We can write the following for $T_{\text{up}}(z)$ and $T_{\text{down}}(z)$:

$$T_{\text{up}}(z) = \left[G(z) + (z^2 - 1)G(0) + (z^2 - z)G'(0) \right] \frac{F(z)}{z^2} \; ; \tag{48}$$

$$T_{\text{down}}(z) = \left[H(z) + (z - 1)H(0) \right] \frac{F(z)}{z}. \tag{49}$$

We can now compose an expression for $W(z)$, the pgf of the delay of an arbitrary packet P:

$$W(z) = \sigma \sum_{k=0}^{\infty} \text{Prob}[t_{\text{up}} = k] U_k(z) + (1 - \sigma) \sum_{k=0}^{\infty} \text{Prob}[t_{\text{down}} = k] D_k(z)$$

$$= \sigma \sum_{\phi=1}^{3} \frac{-f_1(x_\phi, z)}{x_\phi g_x(x_\phi, z)} T_{\text{up}}\left(\frac{1}{x_\phi}\right) + (1 - \sigma) \sum_{\phi=1}^{3} \frac{-f_2(x_\phi, z)}{x_\phi g_x(x_\phi, z)} T_{\text{down}}\left(\frac{1}{x_\phi}\right). \tag{50}$$

Due to its complexity (remember that the x_ϕ are functions of z), the above expression cannot be easily inverted to give the full delay analysis of an arbitrary packet P entering the system. We can however aim to find the tail distribution. For sufficiently large k we have that:

$$\text{Prob}[\text{Delay} = k \text{ slots}] \approx -\frac{w_0}{z_0} z_0^{-k} ; \tag{51}$$

$$\text{Prob}[\text{Delay} > k \text{ slots}] \approx -\frac{w_0}{z_0(z_0 - 1)} z_0^{-k} , \tag{52}$$

with z_0 the pole of $W(z)$ with the smallest modulus and with

$$w_0 = \lim_{z \to z_0} [W(z)(z - z_0)] . \tag{53}$$

Note that z_0 is real-valued and larger than 1. We denote the corresponding x-value as x_ξ. The technique of the dominant singularity to derive tail distributions from a pgf is not new, see for example [2] and [9].

Let us take a closer look at the shape of $W(z)$ in (50) to determine where we can find this z_0. The functions $f_1(x, z)$ and $f_2(x, z)$ are polynomials in x and z and thus contain no poles. Furthermore, $x = 0$ is not a possible pole as $g(x, z)$ does not have a root in z in this case. As all x_ϕ are assumed distinct we get that z_0 must be found as a pole of $T_{\text{up}}(\frac{1}{x})$ or $T_{\text{down}}(\frac{1}{x})$. As $T_{\text{up}}(z)$ and $T_{\text{down}}(z)$ are pgfs, they cannot have poles within the complex unit disk; which means that $|x_\xi| < 1$. Further examination of (48) and (49) leads to the conclusion that z_0 must be found as a pole of $C(\frac{1}{x})$ or as a zero of:

$$\frac{1}{x^2} - C\left(\frac{1}{x}\right) \left[\alpha + (1 - \alpha)R\left(C\left(\frac{1}{x}\right)x\right)\right] . \tag{54}$$

Let us now first look at the pgfs $T_{\text{up}}(z)$ and $T_{\text{down}}(z)$ and their tail distributions. They describe the queue content as an arbitrary customer experiences on arrival. We can calculate a tail approximation of these distributions, based on their smallest pole in z, we call these z_{up} and z_{down} respectively. Lets take a closer look at (48) and (49) and fill in (17), (18), (21), (22) and (47):

$$T_{\text{up}}(z) = \left\{ (1 - \alpha)(z - 1) \frac{[p(0) + p(1)z] C(z) R\left(\frac{C(z)}{z}\right) + [z^2 - \alpha C(z)] Q\left(\frac{C(z)}{z}\right)}{z^2 - C(z)\left[\alpha + (1 - \alpha)R\left(\frac{C(z)}{z}\right)\right]} \right.$$

$$+ \alpha(z - 1)C(z) \frac{p(0) + p(1)z + (1 - \alpha)Q\left(\frac{C(z)}{z}\right)}{z^2 - C(z)\left[\alpha + (1 - \alpha)R\left(\frac{C(z)}{z}\right)\right]}$$

$$\left. + (z^2 - 1)G(0) + (z^2 - z)G'(0) \right\} \frac{C(z) - 1}{\lambda z^2(z - 1)} , \tag{55}$$

and

$$T_{\text{down}}(z) = \left\{ \beta(z-1) \frac{[p(0) + p(1)z]\, C(z)\, R\!\left(\frac{C(z)}{z}\right) + [z^2 - \alpha C(z)]\, Q\!\left(\frac{C(z)}{z}\right)}{z^2 - C(z)\left[\alpha + (1-\alpha)R\!\left(\frac{C(z)}{z}\right)\right]} \right.$$

$$+ (1-\beta)(z-1)C(z) \frac{p(0) + p(1)z + (1-\alpha)Q\!\left(\frac{C(z)}{z}\right)}{z^2 - C(z)\left[\alpha + (1-\alpha)R\!\left(\frac{C(z)}{z}\right)\right]}$$

$$\left. + (z-1)H(0) \right\} \frac{C(z) - 1}{\lambda z(z-1)}. \tag{56}$$

These are very similar and obviously have the same poles, and thus $z_{\text{up}} = z_{\text{down}}$. (Keep in mind that z_{up} is real-valued and larger than 1.) Poles can be present in $C(z)$ and as zeros of $z^2 - C(z)\left[\alpha + (1-\alpha)R\!\left(\frac{C(z)}{z}\right)\right]$.

Proposition 1. *The pole z_{up} cannot be a pole of $C(z)$ and must be found as a zero of*

$$z^2 - C(z)\left[\alpha + (1-\alpha)R\!\left(\frac{C(z)}{z}\right)\right].$$

Proof. We prove this by contradiction. Let us denote with z_c the smallest pole of $C(z)$ for which $|z| > 1$ and state that $z_{\text{up}} = z_c$. Then it must be real-valued and positive. Let us now look at the function $f(z)$ on $\mathbb{R} \to \mathbb{R}$:

$$f(z) \triangleq z^2 - C(z)\left[\alpha + (1-\alpha)R\!\left(\frac{C(z)}{z}\right)\right]. \tag{57}$$

Let $z_p : 1 < z_p \le z_c$ be the smallest pole of $f(z)$. Either z_p is a pole of $R\!\left(\frac{C(z)}{z}\right)$ or $z_p = z_c$. We have that $f(z)$ is continuous on the open interval $[1, z_p[$. Because $C(z)$ and $R(z)$ are pgfs, we have the following left-handed limit:

$$\lim_{z \to z_p^-} f(z) = -\infty. \tag{58}$$

Let us evaluate the function $f(z)$ and its derivative at $z = 1$:

$$f(1) = 0; \tag{59}$$
$$f'(1) = 2 + (1-\alpha)(1-\lambda)\bar{r}. \tag{60}$$

The derivative $f'(1)$ must be strictly positive due to the stability condition given in (9). We thus conclude that $f(z)$ equals 0 at $z = 1$, increases after $z = 1$ (and thus reaches positive values) before (continuously) going to $-\infty$ while z approaches z_p. Then, there exists (at least one) $z* : 1 < z* < z_p \le z_c$ where $f(z)$ changes sign. This is in contradiction with $z_{\text{up}} = z_c$.   □

For this result we did not rely on the restriction that the DOWN-periods are according to a geometric distribution, it is valid for all rational choices of $R(z)$, as long as the stability condition is fulfilled. Note that this does not prove that z_{up} exists. Indeed if we choose $C(z) = z$ (every slot contains 1 arrival), then it is obvious that every arriving customer experiences an empty queue ($T_{\text{up}}(z) = T_{\text{down}}(z) = 1$) and there is no pole z_{up}. The theory of the dominant singularity then stipulates that the tail probabilities can be approximated by 0, which in this case is exact for $k > 0$.

We can prove in a similar (but more involved) way that if z_0 exists, it must be a solution of (54) and it cannot be a pole of $C(\frac{1}{x})$. In stead of rigorously proving this, we provide an intuitive explanation. When an arbitrary arriving customer P experiences k customers in front of him in the queue, its delay must be between $k+1$ and $\lfloor \frac{k}{2} \rfloor + 1$, where $\lfloor ... \rfloor$ denotes the floor function. The delay is thus strongly correlated to the observed queue content, which also applies to the respective tail characteristics. Therefore it is intuitive that the dominant poles of $T_{\text{up}}(z)$ and $W(z)$ are found in the same expression.

Let us now assume that z_0 indeed exists. (It is our belief that for any case that is not trivial (such as the choice of $C(z) = z$), z_0 exists and can be found with a reasonable amount of computational effort.) Then we get for w_0 after applying L'Hôpital's rule on (53):

$$w_0 = \frac{-\sigma f_1(x_\xi, z_0)}{x_\xi g_x(x_\xi, z_0)} T^*_{\text{up}}\left(\frac{1}{x_\xi}\right) - \frac{(1-\sigma) f_2(x_\xi, z_0)}{x_\xi g_x(x_\xi, z_0)} T^*_{\text{down}}\left(\frac{1}{x_\xi}\right) ; \qquad (61)$$

with

$$T^*_{\text{up}}\left(\frac{1}{x_\xi}\right) \triangleq \lim_{z \to z_0} T_{\text{up}}\left(\frac{1}{x}\right)(z - z_0)$$

$$= \lim_{z \to z_0} \left[(1-\alpha)G_0\left(\frac{1}{x}\right) + \alpha H_0\left(\frac{1}{x}\right) + \left(\frac{1}{x^2} - 1\right)G(0) \right.$$

$$\left. + \left(\frac{1}{x^2} - \frac{1}{x}\right)G'(0) \right] F\left(\frac{1}{x}\right) x^2(z - z_0)$$

$$= \lim_{z \to z_0} \left[(1-\alpha)G_0\left(\frac{1}{x}\right) + \alpha H_0\left(\frac{1}{x}\right) \right] F\left(\frac{1}{x}\right) x^2(z - z_0) ; \qquad (62)$$

$$T^*_{\text{down}}\left(\frac{1}{x_\xi}\right) \triangleq \lim_{z \to z_0} T_{\text{down}}\left(\frac{1}{x}\right)(z - z_0)$$

$$= \lim_{z \to z_0} \left[(1-\beta)H_0\left(\frac{1}{x}\right) + \beta G_0\left(\frac{1}{x}\right) \right] F\left(\frac{1}{x}\right) x(z - z_0) . \qquad (63)$$

Bringing the limits within the square brackets gives:

$$T^*_{\text{up}}\left(\frac{1}{x_\xi}\right) = \left[(1-\alpha)G^*_0\left(\frac{1}{x_\xi}\right) + \alpha H^*_0\left(\frac{1}{x_\xi}\right) \right] F\left(\frac{1}{x_\xi}\right) x_\xi^2 \qquad (64)$$

$$T^*_{\text{down}}\left(\frac{1}{x_\xi}\right) = \left[(1-\beta)H^*_0\left(\frac{1}{x_\xi}\right) + \beta G^*_0\left(\frac{1}{x_\xi}\right) \right] F\left(\frac{1}{x_\xi}\right) x_\xi , \qquad (65)$$

where $G_0^*\left(\frac{1}{x}\right)$ and $H_0^*\left(\frac{1}{x}\right)$ are obtained by dividing the numerator of $G_0\left(\frac{1}{x}\right)$ and $H_0\left(\frac{1}{x}\right)$ by the derivative (with respect to z) of their respective denominators. This is not so difficult but rather tedious and leads to:

$$G_0^*\left(\frac{1}{x}\right) = \frac{(1-x)\left\{\begin{array}{l} xC\left(\frac{1}{x}\right)\left[p(0)x + p(1)\right]R\left(C\left(\frac{1}{x}\right)x\right) \\ + \left[1 - \alpha x^2 C\left(\frac{1}{x}\right)\right]Q\left(C\left(\frac{1}{x}\right)x\right)\end{array}\right\}}{\left\{\begin{array}{l}(1-\alpha)xC\left(\frac{1}{x}\right)\left[C'\left(\frac{1}{x}\right) - xC\left(\frac{1}{x}\right)\right]R'\left(C\left(\frac{1}{x}\right)x\right) \\ + 2\left[\alpha - (1-\alpha)R\left(C\left(\frac{1}{x}\right)x\right)\right]\left[xC\left(\frac{1}{x}\right) - \frac{1}{2}C'\left(\frac{1}{x}\right)\right]\end{array}\right\}\frac{dx}{dz}} ; \quad (66)$$

$$H_0^*\left(\frac{1}{x}\right) = \frac{(1-x)C\left(\frac{1}{x}\right)\left[p(0)x + p(1) + (1-\alpha)xQ\left(C\left(\frac{1}{x}\right)x\right)\right]}{\left\{\begin{array}{l}(1-\alpha)xC\left(\frac{1}{x}\right)\left[C'\left(\frac{1}{x}\right) - xC\left(\frac{1}{x}\right)\right]R'\left(C\left(\frac{1}{x}\right)x\right) \\ + 2\left[\alpha - (1-\alpha)R\left(C\left(\frac{1}{x}\right)x\right)\right]\left[xC\left(\frac{1}{x}\right) - \frac{1}{2}C'\left(\frac{1}{x}\right)\right]\end{array}\right\}\frac{dx}{dz}} . \quad (67)$$

Finally, in order to evaluate $\frac{dx}{dz}$ we remember that x is defined in terms of z as a solution of:

$$1 - \alpha z x^2 - (1-\alpha)zx^2 R(zx) = 0, \quad (68)$$

deriving both sides of the equations with respect to z and working out for $\frac{dx}{dz}$ leads to:

$$\frac{dx}{dz} = \frac{-\alpha x - (1-\alpha)xR(xz) - (1-\alpha)zx^2 R'(xz)}{2\alpha z + 2(1-\alpha)zR(xz) + (1-\alpha)z^2 x R'(xz)} . \quad (69)$$

5 Delay Analysis for DOWN-Periods According to a Mixture of M Geometric Distributions

The methodology as described in the above section can be expanded to the case where the DOWN-periods follow a distribution that is a mixture of M geometric distributions:

$$r(n) = \sum_{j=1}^{M} \omega_j(1-\beta_j)\beta_j^{(n-1)}, \quad n > 0; \quad (70)$$

$$R(z) = \sum_{j=1}^{M} \omega_j \frac{(1-\beta_j)z}{1-\beta_j z}, \quad (71)$$

with

$$\sum_{j=1}^{M} \omega_j = 1; \quad (72)$$

$$0 < \omega_j, \beta_j < 1. \quad (73)$$

We can think of this situation as if there are M different DOWN-substates, each with geometrically distributed sojourn times; and when an UP-period ends, the system has a probability ω_j to enter the jth substate.

5.1 Delay of a Packet with k Customers Ahead

We introduce again stochastic variables for the delay of a customer depending on the type of its arrival-slot and the number of customers ahead of it in the queue upon arrival. We use u_k and $U_k(z)$ for the delay of a customer arriving during an UP-slot and with k customers ahead of it in the queue and we introduce the stochastic variables d_k^j as the delay of a customer arriving during a slot of the jth DOWN-substate and with k customers ahead of it in the queue. The corresponding pgf is $D_k^j(z)$. We can establish the following relations:

$$U_0(z) = z \, ; \tag{74}$$

$$U_1(z) = \alpha z + (1 - \alpha) z^2 \, ; \tag{75}$$

$$U_k(z) = \alpha z U_{k-2}(z) + (1 - \alpha) z \sum_{j=1}^{M} \omega_j D_{k-1}^j(z) \, , \quad k \geq 2 \, ; \tag{76}$$

$$D_0^j(z) = z \, ; \tag{77}$$

$$D_1^j(z) = \beta_j z^2 + (1 - \beta_j) z \, ; \tag{78}$$

$$D_k^j(z) = \beta_j z D_{k-1}^j(z) + (1 - \beta_j) z U_{k-2}(z) \, , \quad k \geq 2 \, . \tag{79}$$

We work out the bivariate function $U(x, z)$ according to its definition in (33):

$$U(x, z) = z + \alpha z x + \alpha z x^2 U(x, z) + (1 - \alpha) z x \sum_{j=1}^{M} \omega_j D^j(x, z) \, , \tag{80}$$

where we have already introduced $D^j(x, z)$, which we define as:

$$D^j(x, z) \triangleq \sum_{k=0}^{\infty} x^k D_k^j(z) \tag{81}$$

$$= z + (1 - \beta_j) z x + \beta_j z x D^j(x, z) + (1 - \beta_j) z x^2 U(x, z) \, . \tag{82}$$

Combining (80) and (82) leads to:

$$U(x, z) = \frac{f_1(x, z)}{g(x, z)} \, ; \tag{83}$$

$$D^j(x, z) = \frac{f_2^j(x, z)}{g(x, z)} \, , \tag{84}$$

with

$$f_1(x, z) = \left[z + \alpha z x + (1 - \alpha) z x \sum_{j=1}^{M} \omega_j \frac{z + (1 - \beta_j) z x}{1 - \beta_j z x} \right] \prod_{j=1}^{M} (1 - \beta_j z x) \, ; \tag{85}$$

$$g(x, z) = \left[1 - \alpha z x^2 - (1 - \alpha) z^2 x^3 \sum_{j=1}^{M} \omega_j \frac{1 - \beta_j}{1 - \beta_j z x} \right] \prod_{j=1}^{M} (1 - \beta_j z x) \, ; \tag{86}$$

$$f_2^j(x, z) = \frac{[z + (1 - \beta_j) z x] \, g(x, z) + (1 - \beta_j) z x^2 \, f_U(x, z)}{(1 - \beta_j z x)} \, . \tag{87}$$

It can be verified that $f_1(x,z)$ and $f_2^j(x,z)$ are polynomial functions in x of degree $M+1$ and that $g(x,z)$ is a polynomial function in x of degree $M+2$. As before we can perform a partial fraction expansion based on the poles x_ϕ of $U(x,z)$ in x. Note that as before the x_ϕ are functions of z, but for notational simplicity the argument is omitted. We then get that $U_k(z)$ and $D_k^j(z)$ can be expressed as:

$$U_k(z) = \sum_{\phi=1}^{M+2} \frac{-f_1(x_\phi, z)}{x_\phi^{k+1} g_x(x_\phi, z)} ; \tag{88}$$

$$D_k^j(z) = \sum_{\phi=1}^{M+2} \frac{-f_2^j(x_\phi, z)}{x_\phi^{k+1} g_x(x_\phi, z)} , \tag{89}$$

with

$$g_x(x,z) \triangleq \frac{\partial}{\partial x} g(x,z) . \tag{90}$$

5.2   Delay of an Arbitrary Packet

Let us now consider an arbitrary packet P arriving during slot S. The probability that S is an UP-slot is given by σ. The probability that S is a DOWN-slot of substate j is given by

$$\text{Prob}[S \text{ is a DOWN-slot of substate } j] = \frac{(1-\sigma)\omega_j}{\overline{r}(1-\beta_j)} , \tag{91}$$

We now introduce the stochastic variable h^j as the queue content at the beginning of an arbitrary DOWN-slot of the jth substate, it is easily evaluated that its pgf $H^j(z)$ is given by:

$$H^j(z) = (1-\beta_j)H_0(z) + \beta_j G_0(z) , \tag{92}$$

with $G_0(z)$ and $H_0(z)$ given by (17) and (18) respectively. For the pgf of the queue content at the beginning of an arbitrary UP-slot, Eq. (21) applies. The number of customers in front of P at the end of S can be described by the pgf $T_{\text{up}}(z)$, as given in (48) if S is an UP-slot; or by the pgf $T_{\text{down}}^j(z)$ if S is a DOWN-slot of substate j:

$$T_{\text{down}}^j(z) = \left[H^j(z) + (z-1)H^j(0) \right] \frac{F(z)}{z} . \tag{93}$$

We now have all the tools to develop an expression for $W(z)$, the pgf of the delay of an arbitrary customer:

$$W(z) = \sum_{\phi=1}^{M+2} \frac{-\sigma f_1(x_\phi, z)}{x_\phi g_x(x_\phi, z)} T_{\text{up}}\left(\frac{1}{x_\phi}\right)$$

$$- \sum_{j=1}^{M} \frac{(1-\sigma)\omega_j}{\overline{r}(1-\beta_j)} \sum_{\phi=1}^{M+2} \frac{f_2^j(x_\phi, z)}{x_\phi g_x(x_\phi, z)} T_{\text{down}}^j\left(\frac{1}{x_\phi}\right) . \tag{94}$$

From here we can calculate z_0 and w_0 to obtain the tail-probabilities of the delay. As before, z_0 and the corresponding x-value x_ξ must be found as a solution of (54). After applying L'Hôpital's rule on (53) we get for w_0:

$$w_0 = \frac{-\sigma f_1(x_\xi, z_0)}{x_\xi g_x(x_\xi, z_0)} T_{\text{up}}^* \left(\frac{1}{x_\xi}\right) - \sum_{j=1}^{M} \frac{(1-\sigma)\omega_j}{\overline{r}(1-\beta_j)} \frac{f_2^j(x_\xi, z_0)}{x_\xi g_x(x_\xi, z_0)} T_{\text{down}}^{j*} \left(\frac{1}{x_\xi}\right), \quad (95)$$

with $T_{\text{up}}^*\left(\frac{1}{x}\right)$ as given in (64) and with $T_{\text{down}}^{j*}\left(\frac{1}{x}\right)$ found in a similar way:

$$T_{\text{down}}^{j*}\left(\frac{1}{x}\right) \triangleq \lim_{z \to z_0} T_{\text{down}}^j\left(\frac{1}{x}\right)(z-z_0)$$

$$= \left[(1-\beta_j)H_0^*\left(\frac{1}{x}\right) + \beta_j G_0^*\left(\frac{1}{x}\right)\right] F\left(\frac{1}{x}\right) x, \quad (96)$$

with $G_0^*\left(\frac{1}{x}\right)$ and $H_0^*\left(\frac{1}{x}\right)$ as given in (66) and (67).

6   Numerical Example

In this section we work out a numerical example to illustrate the method. We choose a mixture of 2 geometric distributions for the DOWN-periods with parameters as given in Table 1.

Table 1. Parameters of $R_1(z)$

β_1	β_2	ω_1	$\omega_2 = 1 - \omega_1$
0.878	0.922	0.609	0.391

The average length of a DOWN-period is then 10 slots. We choose a Poisson arrival process:

$$C(z) = e^{\lambda(z-1)}. \quad (97)$$

We look at two different scenarios. Firstly we choose a large α which leads to long UP-periods and combine this with a high arrival intensity. Secondly, we choose a smaller α and compensate for the shorter UP-periods with a smaller arrival intensity. Table 2 gives an overview of the considered scenarios.

Table 2. Overview scenarios

	Scenario 1	Scenario 2
λ	1.756	1.207
α	0.989	0.767
σ	0.9	0.3

The parameters have been chosen in such a way that in both situations the average queue content equals 12.0 customers. However, the tail characteristics of the delay are different. In Fig. 2 we plot the probability that a customer has a certain delay based on the method developed in this paper and based on simulation. It should be noted that the results obtained by simulation require a lot more computation time. The figure validates the method developed in this paper and also shows the importance of this work. The tail characteristics of the delay are not equal even though the average queue content is. Furthermore, it shows that the approximations are accurate already for small values of k.

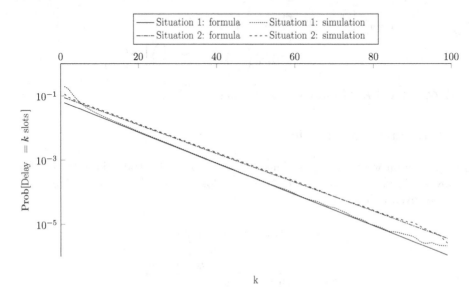

Fig. 2. Tail distribution of delay

6.1   Comments on Numerical Procedures

In this subsection we will briefly sketch how the numerical results can be obtained. Step 1 and 2 are direct application of the method developed in [3].

1. Numerically obtain the zeros z_i, $i = 1..M + 2$ of

$$z^2 - C(z)\left[\alpha + (1 - \alpha)R\left(\frac{C(z)}{z}\right)\right] = 0; \quad |z| \le 1, \tag{98}$$

with $z_1 = 1$ and with M the number of geometrics in $R(z)$.

2. Solve the set of linear equations

$$p(0) + p(1)z_i + (1 - \alpha)Q\left(\frac{C(z_i)}{z_i}\right) = 0; \quad i = 2..M + 2; \tag{99}$$

$$p(0) + p(1) + (1 - \alpha)Q(1) = \frac{1 + \sigma - \lambda}{\sigma}, \tag{100}$$

in order to obtain $p(0)$, $p(1)$ and the coefficients of the numerator of the rational function $Q(z)$.

3. Numerically obtain the zeros x_i of (54) for $|x| < 1$:

$$\frac{1}{x^2} - C\left(\frac{1}{x}\right)\left[\alpha + (1-\alpha)R\left(C\left(\frac{1}{x}\right)x\right)\right] = 0.$$

4. Use a polynomial rootsolver to obtain the zeros in z of $g(x_i, z) = 0$. Then z_0 is the smallest solution in z (real-valued and larger than 1) over all possible x_i. The corresponding x-value is then x_ξ.

After this, evaluation of w_0 is merely filling in all the values in (95). In all of the numerical examples we calculated, x_ξ is real-valued and the closest to 1 of all x_i obtained in Step 3.

7   Conclusions

In this paper we studied a discrete-time multiserver queueing model. Special about the considered model is that the number of available servers alternates over time (from 1 to 2 and back), this way correlation is introduced on the slot-to-slot server availability. Specifically in this paper we were interested in the delay characteristics of an arbitrary customer.

The queueing system is assumed to be in one of two different states: DOWN-state with 1 server available or UP-state with 2 servers available. The system resides in a given state for a stochastic number of slots before returning to the other state. This stochastic process is fully described by the distributions of the lengths of the state-periods and it introduces correlation on the number of servers available.

The paper is an extension to an earlier work that described the queue content distributions of a system as such.

The delay analysis is performed by linking the delay of a customer with k customers in front of it to the delay of a customer with $k-1$ and $k-2$ customers in front of it, using probability generating functions. The theory of the dominant singularity then provides a method to evaluate the tail probabilities of the delay of an arbitrary customer.

A numerical example validates the method and shows the importance of it. Different queueing systems with the same average queue content can show different delay characteristics. The numerical example also shows that the method which is an approximation for large delays, gives very accurate results even for small delays. It also uses a lot less computational resources as compared to simulation. Further research could include more general distributions for the DOWN-periods and the UP-periods.

References

1. Bruneel, H., Kim, B.G.: Discrete-Time Models for Communication Systems Including ATM. Kluwer Academic Publishers Group, Boston (1993)

2. Bruneel, H., Steyaert, B., Desmet, E., Petit, G.: Analytic derivation of tail probabilities for queue lengths and waiting times in atm multiserver queues. Eur. J. Oper. Res. **76**(3), 563–572 (1994)
3. Bruneel, H., Wittevrongel, S.: Analysis of a discrete-time single-server queue with an occasional extra server. Perform. Eval. **116**, 119–142 (2017)
4. Krishnamoorthy, A., Pramod, P., Chakravarthy, S.: Queues with interruptions: a survey. TOP **22**, 290–320 (2014)
5. Laevens, K., Bruneel, H.: Delay analysis for discrete-time queueing systems with multiple randomly interrupted servers. Eur. J. Oper. Res. **85**, 161–177 (1995)
6. Núñez-Queija, R.: Sojourn times in a processor sharing queue with service interruptions. Queueing Syst. **34**, 351–386 (2000)
7. Takine, T., Sengupta, B.: A single server queue with service interruptions. Queueing Syst. **26**(3–4), 285–300 (1997)
8. Vinck, B., Bruneel, H.: System delay versus system content for discrete-time queueing systems subject to server interruptions. Eur. J. Oper. Res. **175**(1), 362–375 (2006)
9. Woodside, C., Ho, E.: Engineering calculation of overflow probabilities in buffers with Markov-interrupted service. IEEE Trans. Commun. **35**(12), 1272–1277 (1987)
10. Yajima, M., Phung-Duc, T.: Batch arrival single server queue with variable service speed and setup time. Queueing Syst. **86**(3–4), 241–260 (2017)
11. Yue, D., Qin, Y.: A production inventory system with service time and production vacations. J. Syst. Sci. Syst. Eng. **28**, 168–180 (2019)
12. Zhou, Y., Anderson, R., Vakilzadian, H., Moeller, D., Deutschmann, A.: Developing a dynamic queueing model for the airport check-in process. In: 2018 IEEE International Conference on Electro/Information Technology (EIT), pp. 871–876 (2018)

Strategic Queues

A Closed Queueing Network
with Strategic Service Differentiation

Michal Benelli$^{(\boxtimes)}$ and Refael Hassin

Department of Statistics and Operations Research, Tel Aviv University,
Tel Aviv-Yafo, Israel
`michal.benelli@gmail.com`, `hassin@tauex.tau.ac.il`

Abstract. We consider a service system with a single server, exponentially distributed service time, and two types of service rates – high and low. A customer chooses to obtain a high rate or a low rate service, and then the customer is active for an exponentially distributed period of time with a given high or low rate, respectively, and returns to the queue to be served again. Customers strategically choose a service type in order to maximize their long-run activity time.

We investigate which strategies of the customers are socially optimal and explore conditions for Nash equilibria. We examine symmetric and asymmetric strategies, as well as behavioral strategies. We focus on the game with two customers.

We prove an equivalence of the conditions for the existence of pure and mixed equilibria to those in the behavioral model, though the value of the mixed equilibrium strategy differs from the value of the behavioral equilibrium strategy for the same parameters. We show that a pure asymmetric equilibrium does not exist, a pure asymmetric strategy cannot be socially optimal, and a pure symmetric equilibrium always exists.

Keywords: Strategic queueing · Networks · Nash equilibria ·
Social optimization

1 Introduction

We consider a system with a single server and N customers. The server supplies two service types - short or long. A served customer is active for a shorter or a longer period corresponding to the type of service he obtained, and when the activity period ends, returns to the queue to be served again. Consider, for example, a closed system of batteries with one charger. A battery can get a long or a short charging time, and will perform accordingly for a longer or a shorter time before recharging.

The motivation for our model is an application server shared by computer processes, described by Courcoubetis and Varaiya [3] and by Cheng and Kohler

This research was supported by the Israel Science Foundation (Grant No. 355/15).

© Springer Nature Switzerland AG 2019
T. Phung-Duc et al. (Eds.): QTNA 2019, LNCS 11688, pp. 149–165, 2019.
https://doi.org/10.1007/978-3-030-27181-7_10

[1]. Gallay and Hongler [4] consider a variation with electric vehicles with various charging facilities. The same model is described by Xu, Dai, Sykara and Lewis, [8] for a multi-robot control operator in a disaster area. The referred articles search for system optimization while we assume strategic customers and analyse equilibrium strategies.

We focus on the case with two customers and investigate equilibria and optimal solutions for maximum probability to be active. The main technical difficulty in the analysis is that in the closed queueing system there is a dependence between successive service and waiting times.

Our main results: The socially optimal strategy is the pure symmetric strategy with the smaller utilization factor (Theorem 1). When the low-rate service has the smaller utilization factor (activity rate/service rate), choosing the low-rate service is optimal and the unique equilibrium (Theorem 5). We conjecture that there always exists a pure symmetric equilibrium strategy (Conjecture 16). We show that a pure asymmetric strategy cannot be an equilibrium (Corollary 17). We prove that a set of parameters $(\rho_l, \rho_h, \lambda_l, \lambda_h)$ induces the same number and types of equilibria in the mixed-strategy model and in the behavioral-strategy model. We supply examples of the different cases.

Courcoubetis and Varaiya [3] describe two customers (processes) served by a single resource. The queueing network is similar to our model, but in their model the ratio between service time and activity time is fixed and they look for maximal utilization of the server.

Cheng and Kohler [1] deal with programs as customers too. They describe web-enabled application services. The customers are programs. A program sends a transaction to be processed by an ASP - Application Service Provider. Process time of a transaction is an exponential random variable, and so is the period of time between transactions. The paper compares the purchase of software for use in-house, with using the ASP's services, and analyzes the ASP's pricing scheme. This is a variation of the machine interference model described in [5]. Xu, Dai, Sykara and Lewis [8] describe a multi-robot control operator. There are N robots operated by a single server. The operator interacts with the robot for a period of time (IT - Interaction Time) raising its performance above an upper threshold, after which the robot is neglected for a period of time (NT - Neglect Time) until its performance deteriorates below a lower threshold, when the operator must again interact with the server. Both IT and NT are exponential random variables. In their paper the operator is free to choose between high quality or low quality interaction, or a mixed strategy, i.e. provide high quality interaction with a certain probability. They look for the probability p that maximizes the value of the utility function per cycle, whereas in our paper we look for the maximal utility per unit time.

Chu, Wan and Zhan [2] consider a ride-hailing platform where idle taxi drivers accept or ignore a rider's request depending on profitability considerations. There are two types of riders with a different profitability (price) but the same service rate. If instead of discriminating between riders by price, the more profitable rider would have a higher service rate, and the utility function would be the proportion of time the taxi is busy, the model will be similar to ours.

Strategic queueing is introduced and surveyed by Hassin and Haviv [7] and Hassin [6]. Hassin [6] §4.7 and §6.3.2 survey the literature on expert systems with duration-dependent service value. In our model customers return to service and are interested in maximizing their long-run utility. Customers act strategically choosing their service rate.

Section 2 formally defines the model and summarises our results. Section 3 analyses the model when each customer sticks to the same service type repeatedly, and Sect. 4 analyses the model when each customer draws a service type with the same probability in each cycle. Section 5 shows that the conditions for equilibria are equivalent in both models, and in Sect. 6 we show that there always exists a pure equilibrium in the first model, and use the equivalence to prove that there always exists a pure equilibrium in the second model. Section 7 elaborates on research continuation of the subject.

2   The Model

We study a closed queueing network with a single server and two customers. A customer who enters service chooses either a low-rate or a high-rate service. After the end of the service the customer leaves for a low-rate or a high-rate activity period, respectively, and then returns to the queue. See Fig. 1.

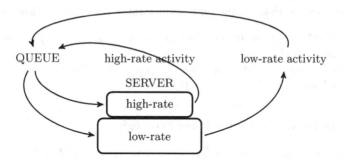

Fig. 1. The model.

Service time is exponentially distributed. The service rate is μ_l or μ_h, corresponding to low and high service rate, $\mu_h > \mu_l$. Activity time is exponentially distributed with rate λ_l or λ_h corresponding to low and high activity rate, $\lambda_h > \lambda_l$. The service discipline is FCFS. When both customers choose the same service rate the model behaves as the Machine Interference Model (e.g. Gross and Harris [5] Section 2.7).

The utility of a customer is equal to the steady-state probability to be active. The social utility is the average utility of the customers. We denote the utility factor $\rho_\theta = \frac{\lambda_\theta}{\mu_\theta}, \theta \in \{l, h\}$.

2.1   Model Variations

Let customer i draw low service rate with probability $p^i, i = 1, 2$. We analyse social optima and existence of equilibria using the following strategies:

- pure strategy - customer i draws once a service rate (low service rate with probability p^i, $p^i \in \{0, 1\}$) and sticks to it.
- mixed strategy - customer i draws once a service rate (low service rate with probability p^i, $p^i \in [0, 1]$) and sticks to it.
- behavioral strategy - customer i independently draws low service rate with probability p^i, $p^i \in (0, 1)$ each time he enters service. The customer sticks to the same probability throughout the game.

The utility function for a single customer is the expected fraction of time in activity, or, in other words, the probability to find the customer in the active period. $U(p^1, p^2)$ is the value of the utility function for customer 1, when each customer 2 draws low service rate with probability p^2. The social utility is the average utility of the customers.

We discuss pure symmetric strategies separately as they yield the same utilities for the mixed game and for the behavioral game.

3   Mixed Strategies

In the mixed strategy version customer i sticks to low service rate with probability p^i, otherwise he sticks to the high service rate. We consider three scenarios:

- The pure symmetric solution where $p^1 = p^2$, $p^i \in \{0, 1\}$ - the machine interference model.
- The pure asymmetric solution where $p^1 \neq p^2$, $p^i \in \{0, 1\}$.
- The mixed solution where $p^i \in [0, 1]$.

3.1   Steady-State Solution - Pure Strategies

Suppose the customers choose different service types. Let (k, m) define a state of the system, where $k, m \in \{a, s, w\}$ are the states of the first and the second customers, respectively. a - active, s - in service, w - waiting for service.

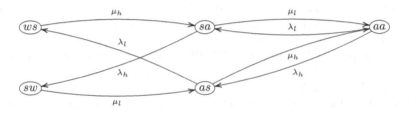

Fig. 2. Transition diagram for the asymmetric strategy, $p^1 = 1, p^2 = 0$.

Let $U(p^1, p^2)$ denote the utility of customer 1, under the strategy profile (p^1, p^2). Let π_{km} denote the steady-state probability for state (k, m) under the asymmetric pure strategy case. We compute $U(1, 0) = \pi_{aa} + \pi_{as}$ - the probability that customer 1 is active, and $U(0, 1) = \pi_{aa} + \pi_{sa}$ - the probability that customer 2 is active, using Fig. 2.

$$U(1, 0) = \left(1 + \rho_l + \frac{\lambda_l \rho_h^2 (\lambda_l + \lambda_h \rho_l + \lambda_l \rho_l)}{(\lambda_l + \lambda_h \rho_l)(\lambda_h + \lambda_h \rho_h + \lambda_l \rho_h)}\right)^{-1} \tag{1a}$$

$$U(0, 1) = \left(1 + \rho_h + \frac{\lambda_h \rho_l^2 (\lambda_h + \lambda_l \rho_h + \lambda_h \rho_h)}{(\lambda_h + \lambda_l \rho_h)(\lambda_l + \lambda_l \rho_l + \lambda_h \rho_l)}\right)^{-1} \tag{1b}$$

We compute $U(1, 1)$ and $U(0, 0)$ when both customers choose the same service rate, using (1a) and (1b) respectively: with $\rho_l = \rho_h, \lambda_l = \lambda_h$:

$$U(1, 1)) = \left(1 + \rho_l + \frac{\rho_l^2}{1 + \rho_l}\right)^{-1} \tag{2a}$$

$$U(0, 0) = \left(1 + \rho_h + \frac{\rho_h^2}{1 + \rho_h}\right)^{-1} \tag{2b}$$

3.2   Optimal Strategy

There are cases (see Subsect. 3.4 for examples), where the pure symmetric strategy with the larger ρ is an equilibrium, but we now show that the socially optimal strategy is always the pure symmetric strategy with the smaller utilization factor ρ.

Theorem 1. *The pure symmetric strategy with the smaller ρ is optimal.*

Proof. We first note that the function $\frac{1+\rho}{1+2\rho+2\rho^2}$ is monotone decreasing in ρ and therefore among pure symmetric strategies, choosing the service with the smaller ρ is optimal.

We now proceed to show that the optimal pure symmetric strategy is better than the asymmetric pure strategy. When the customers draw different pure strategies, the social utility is the average utility of the customers, i.e. $SU(0, 1) = \frac{U(0,1)+U(1,0)}{2}$. By (1):

$$SU(0, 1) = \frac{\left(1 + \rho_l + \frac{\lambda_l \rho_h^2 (\lambda_l + \lambda_h \rho_l + \lambda_l \rho_l)}{(\lambda_l + \lambda_h \rho_l)(\lambda_h + \lambda_h \rho_h + \lambda_l \rho_h)}\right)^{-1}}{2} + \frac{\left(1 + \rho_h + \frac{\lambda_h \rho_l^2 (\lambda_h + \lambda_l \rho_h + \lambda_h \rho_h)}{(\lambda_h + \lambda_l \rho_h)(\lambda_l + \lambda_l \rho_l + \lambda_h \rho_l)}\right)^{-1}}{2} \tag{3}$$

Lemma 2. *When $\rho_l = \rho_h$, the socially-optimal pure strategy is symmetric.*

Proof. Given ρ, the value $U(0, 0)$ of the pure symmetric equilibrium is independent of λ_h and λ_l. We show that for any λ_h and λ_l, $SU(0, 1) \leq U(0, 0)$. We first find the parameters that maximize $SU(0, 1)$ by computing $\frac{\partial SU(0,1)}{\partial \lambda_l}$ and $\frac{\partial SU(0,1)}{\partial \lambda_h}$, when $\rho_l = \rho_h$:

$$\frac{\partial SU(0,1)}{\partial \lambda_l} = \lambda_h(\lambda_h^2 - \lambda_l^2)D$$

$$\frac{\partial SU(0,1)}{\partial \lambda_h} = \lambda_l(\lambda_l^2 - \lambda_h^2)D$$

where $D = \frac{\rho_h^2(2\rho_h+1)}{2\left(\lambda_l^2\rho_h(\rho_h+1)^2+\lambda_l\lambda_h\left(2\rho_h\left(\rho_h^2+\rho_h+1\right)+1\right)+\lambda_h^2\rho_h(\rho_h+1)^2\right)^2} > 0.$

By equating both derivatives to zero we see that SU(0,1) is extreme when $\lambda_l = \lambda_h$. The second derivatives are negative when we assign $\lambda_l = \lambda_h$, hence $SU(0,1) \leq U(0,0)$ is maximized when $\lambda_l = \lambda_h$ in which case it is equal to $U(0,0)$. $\square$

Lemma 3. *A pure asymmetric strategy is never strictly better than the best pure symmetric strategy.*

Proof. For $\rho_l = \rho_h$ the claim follows from Lemma 2. Assume first that $\rho_l > \rho_h$. By (3):

$$\frac{\partial SU(0,1)}{\partial \rho_l} =$$

$$-\frac{(\rho_h(\lambda_l+\lambda_h)+\lambda_h)\left(\lambda_l^3\rho_h(\rho_h+1)+\lambda_l^2\lambda_h((\rho_l(\rho_l+4)+1)\rho_h+1)+\lambda_l\lambda_h^2\rho_l(2(\rho_l+1)\rho_h+\rho_l+4)+\lambda_h^3\rho_l^2(\rho_h+2)\right)}{2\left(\lambda_l\rho_h^2(\rho_l(\lambda_l+\lambda_h)+\lambda_l)+(\rho_l+1)\rho_h(\lambda_l+\lambda_h)(\lambda_l+\lambda_h\rho_l)+\lambda_h(\rho_l+1)(\lambda_l+\lambda_h\rho_l)\right)^2} < 0,$$

Given ρ_h, $SU(0,1)$ is decreasing in ρ_l and the maximum in the interval $\rho_l \in [\rho_h, \infty]$ is achieved when $\rho_l = \rho_h$.

Assume now $\rho_l < \rho_h$, By (3):

$$\frac{\partial SU(0,1)}{\partial \rho_h} =$$

$$-\frac{(\rho_l(\lambda_l+\lambda_h)+\lambda_l)\left(\lambda_l^3(\rho_l+2)\rho_h^2+\lambda_l^2\lambda_h\rho_h(2\rho_l(\rho_h+1)+\rho_h+4)+\lambda_l\lambda_h^2(\rho_l\rho_h(\rho_h+4)+\rho_l+1)+\lambda_h^3\rho_l(\rho_l+1)\right)}{2\left(\lambda_l\rho_h^2(\rho_l(\lambda_l+\lambda_h)+\lambda_l)+(\rho_l+1)\rho_h(\lambda_l+\lambda_h)(\lambda_l+\lambda_h\rho_l)+\lambda_h(\rho_l+1)(\lambda_l+\lambda_h\rho_l)\right)^2} < 0,$$

Given ρ_l, $SU(0,1)$ is decreasing in ρ_h and the maximum in the interval $\rho_h \in [\rho_l, \infty]$ is achieved when $\rho_l = \rho_h$.

We showed in Lemma 2 that when $\rho_l = \rho_h$ the maximum is achieved when $\lambda_l = \lambda_h$. $\square$

Lemma 4. *A mixed asymmetric strategy is never optimal*

Proof. A mixed strategy is a weighted average of the pure strategies $U(0,0), U(1,1), U(0,1), U(1,0)$, therefore it cannot be strictly better than all of them. $\square$

3.3 Equilibria

We use Fig. 3, where $\lambda_h = 1.5, \rho_h = 0.1, \lambda_l = 1$. We draw the utilities on the range $\rho_l \in [0.100, 0.110]$.

W.l.o.g. suppose λ_l, ρ_h and λ_h stay fixed. We start with $\rho_l = \rho_h$ (In the example in Fig. 3 $\rho_h = 0.100$) and analyse the change in equilibria as ρ_l increases.

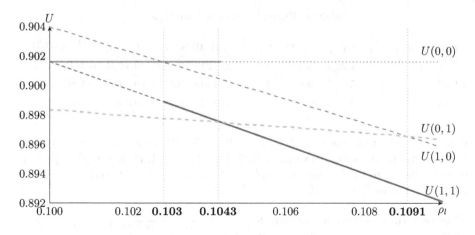

Fig. 3. $\lambda_h = 1.5, \rho_h = 0.1, \lambda_l = 1, \rho_l > \rho_h$

Theorem 5. *When $\rho_l \leq \rho_h$, $p^1 = p^2 = 1$ is the only pure-strategy equilibrium.*

Proof. We prove the claim by showing that $U(0,1) < U(1,1)$, and $U(0,0) < U(1,0)$.

By definition $\lambda_h > \lambda_l$. By (1b) and (2a), $U(0,1) < U(1,1)$ is equivalent to

$$1 + \rho_h + \frac{\lambda_h \rho_l^2 (\lambda_h + \lambda_l \rho_h + \lambda_h \rho_h)}{(\lambda_h + \lambda_l \rho_h)(\lambda_l + \lambda_l \rho_h + \lambda_h \rho_l)} > 1 + \rho_l + \frac{\rho_l^2}{1 + \rho_l} \Leftrightarrow$$

$$\lambda_h (1 + \rho_l)(\lambda_h + \lambda_l \rho_h + \lambda_h \rho_h) > (\lambda_h + \lambda_l \rho_h)(\lambda_l + \lambda_l \rho_h + \lambda_h \rho_l) \Leftrightarrow$$

$$\lambda_h^2 + \lambda_h^2 \rho_h + \lambda_h^2 \rho_l \rho_h > \lambda_h \lambda_l + \lambda_l^2 \rho_h + \lambda_l^2 \rho_h \rho_l$$

$\lambda_h^2 > \lambda_h \lambda_l$, $\lambda_h^2 \rho_h > \lambda_l^2 \rho_h$ and $\lambda_h^2 \rho_l \rho_h > \lambda_l^2 \rho_h \rho_l$ prove the claim.

In the same way, $U(1,0) > U(0,0)$ is equal to

$$1 + \rho_l + \frac{\lambda_l \rho_h^2 (\lambda_l + \lambda_h \rho_l + \lambda_l \rho_l)}{(\lambda_l + \lambda_h \rho_l)(\lambda_h + \lambda_h \rho_h + \lambda_l \rho_h)} < 1 + \rho_h + \frac{\rho_h^2}{1 + \rho_h} \Leftrightarrow$$

$$\frac{\lambda_l (\lambda_l + \lambda_h \rho_l + \lambda_l \rho_l)}{(\lambda_l + \lambda_h \rho_l)(\lambda_h + \lambda_h \rho_h + \lambda_l \rho_h)} < \frac{1}{1 + \rho_h} \Leftrightarrow$$

$$\lambda_l (\rho_h + 1)(\lambda_l \rho_l + \lambda_l + \lambda_h \rho_l) < \lambda_l (\rho_h + 1)(\lambda_l \rho_l + \lambda_l + \lambda_h \rho_l) \Leftrightarrow$$

$$\lambda_l^2 + \lambda_l^2 \rho_l + \lambda_l \lambda_h \rho_l + \lambda_l^2 \rho_h + \lambda_l^2 \rho_l \rho_h + \lambda_l \lambda_h \rho_l \rho_h <$$
$$\lambda_h^2 + \lambda_l \lambda_h \rho_l + \lambda_h^2 \rho_l + \lambda_l \lambda_h \rho_h + \lambda_l^2 \rho_l \rho_h + \lambda_l \lambda_h \rho_l \rho_h \Leftrightarrow$$
$$\lambda_l^2 + \lambda_l^2 \rho_l + \lambda_l^2 \rho_h < \lambda_h^2 + \lambda_h^2 \rho_l + \lambda_l \lambda_h \rho_h.$$

$\square$

3.4 Examples

In Table 1 we show examples of pure and mixed strategies. In all three examples $p^1 = p^2 = 0$ is socially optimal among pure symmetric and asymmetric strate-

Table 1. Examples of mixed strategies.

Name	λ_l	μ_l	λ_h	μ_h	ρ_l	ρ_h	$U(1,1)$	$U(0,0)$	$U(1,0)$	$U(0,1)$	$\frac{U(0,1)+U(1,0)}{2}$
Ex1	1	1	1.1	1.2	1	0.9167	0.4	0.4246	0.4149	0.4083	0.4116
Ex2	1	1	4	5	1	0.8	0.4	0.4639	0.4771	0.3435	.4103
Ex3	1	1	3	4	1	0.75	0.4	0.4828	0.4723	0.369	.4207

gies, but in the first example, Ex1, $U(1,0) < U(0,0)$, hence $p^1 = p^2 = 0$ is an equilibrium. In the second example, Ex2, $U(0,1) < U(1,1)$, hence $p^1 = p^2 = 1$ is an equilibrium. In the third example, Ex3, both $p^1 = p^2 = 0$ and $p^1 = p^2 = 1$ are equilibria, as $U(0,1) < U(1,1)$ and $U(1,0) < U(0,0)$.

Theorem 6. *A unique mixed equilibrium* $q = \frac{U(0,0)-U(1,0)}{U(0,0)-U(1,0)+U(1,1)-U(0,1)}$ *exists iff there exist two pure equilibria, either both symmetric or both asymmetric.*

Proof. Suppose both $p^1 = p^2 = 0$ and $p^1 = p^2 = 1$ are equilibria, then q is the probability that enforces, for a given customer, indifference between the strategies, when the other customer sticks to it. It is the solution of the equation $qU(1,1) + (1-q)U(1,0) = qU(0,1) + (1-q)U(0,0)$.

If both pure strategies are equilibria then $U(0,1) < U(1,1)$ and $U(1,0) < U(0,0)$ and therefore $0 < q < 1$.

Similarly, if the asymmetric strategies are equilibria then $U(0,1) > U(1,1)$ and $U(1,0) > U(0,0)$ so that again $0 < q < 1$.

If q is a valid mixed strategy, i.e. $0 < q < 1$, then the numerator and the denominator both have the same sign and either $U(0,0) > U(1,0)$ and $U(1,1) > U(1,0)$ and both pure symmetric strategies are equilibria, or $U(0,0) < U(1,0)$ and $U(1,1) < U(1,0)$ and then both pure *asymmetric* strategies are equilibria. $\square$

In example Ex3, $q = \frac{0.4828-0.4723}{0.4828-0.4723+0.4-0.369} = 0.2530$.

4   Behavioral Strategies

We now consider symmetric behavioral strategies, where all customers independently draw low service rate with probability p, *every time they enter service*.

4.1   Social Optimum of the Behavioral-Strategy Game

We are looking for p that maximizes the probability to be active when all customers follow the same strategy p.

Let (k,m) define a state of the system, where, $k, m \in \{a_l, a_h, s_l, s_h, w\}$ are the states of the customers without ordering (as customers are homogeneous). a_l, a_h - low-rate or high-rate activity, s_l, s_h - low-rate or high-rate service, w - waiting for service.

There are nine possible states. The vector π provides the stationary probabilities for each state:

$$\pi = \left(\pi_{w,s_l}, \pi_{w,s_h}, \pi_{s_l,a_l}, \pi_{s_h,a_l}, \pi_{s_l,a_h}, \pi_{s_h,a_h}, \pi_{a_l,a_l}, \pi_{a_l,a_h}, \pi_{a_h,a_h}\right).$$

We compute π from the transition-rate diagram of the queueing system (see Fig. 4), and derive the probabilities P_i, $i = 0, 1, 2$, for i active customers. $U(p,p)$ is the utility of a customer when both customers select low-service rate with probability p.

$$U(p,p) = \frac{P_1}{2} + P_2.$$
$$P_1 = \pi_{s_l,a_l} + \pi_{s_h,a_l} + \pi_{s_l,a_h} + \pi_{s_h,a_h}.$$
$$P_2 = \pi_{a_l,a_l} + \pi_{a_l,a_h} + \pi_{a_h,a_h}.$$

In the resulting formula of $U(p,p)$, p appears always as a divisor of $1 - p$, hence we simplify the presentation by searching for $r = \frac{1-p}{p}$ that maximizes $V(r) = U(p,p)$, and then $p = \frac{1}{r+1}$.

$$V(r) = \frac{a + br + cr^2}{\frac{1+2\rho_l+2\rho_l^2}{1+\rho_l}a + er + \frac{1+2\rho_h+2\rho_h^2}{1+\rho_h}cr^2} \tag{4}$$

where

$a = (\lambda_l + \lambda_l\rho_l + \lambda_h\rho_l)(\lambda_h + \rho_h)\lambda_h^2,$

$b = \lambda_l\lambda_h(\lambda_l^2\rho_h(2 + \rho_l) + \lambda_h^2\rho_l(2 + \rho_h) + \lambda_l\lambda_h(2 + \rho_l + \rho_h + 2\rho_l\rho_h)),$

$c = (\lambda_h\rho_l + \lambda_l)(\lambda_h + \lambda_l\rho_h + \lambda_h\rho_h),$

$d = 2\lambda_l\lambda_h((\lambda_h^2\rho_l + \lambda_l^2\rho_h)(\rho_l + 1)(\rho_h + 1) + \lambda_l\lambda_h(\rho_l(\rho_h(\rho_l + \rho_h + 2) + 1) + \rho_h + 1)).$

In particular, when $p^1 = p^2 = 1$, $U(1,1) = \frac{1+\rho_l}{1+2\rho_l+2\rho_l^2}$ which is equal to the utility computed in (2a). Similarly, for $p = 0$, using L'Hôpital's rule, we obtain $U(0,0) = \frac{1+\rho_h}{1+2\rho_h+2\rho_h^2}$, as in (2b).

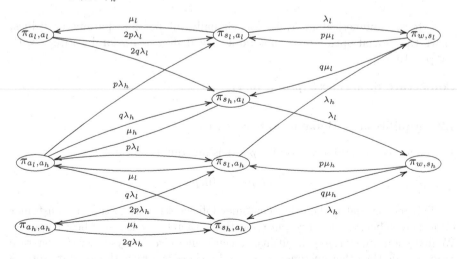

Fig. 4. Steady-state transition diagram for $U(p,p)$, $q = 1 - p$.

Theorem 7. *If $\rho_l = \rho_h$ then $U(p,p)$ has two maximum points, namely $p^* = 0$ and $p^* = 1$.*

Proof. By (4), for every $0 \le p \le 1$,
$$U(p,p) = V(r) = \frac{a+br+cr^2}{\frac{a}{U(1,1)}+dr+\frac{c}{U(1,1)}r^2} = U(1,1)\frac{a+br+cr^2}{a+U(1,1)dr+cr^2}.$$
$$U(1,1)d > b \quad \text{as} \quad \frac{1+\rho}{1+2\rho+2\rho^2}d - b = \frac{\lambda_l\lambda_h(\lambda_l-\lambda_h)^2\rho^2}{1+2\rho+2\rho^2} > 0,$$
hence $U(p,p) < U(1,1) = U(0,0)$.   □

We now deal with any ρ_l and ρ_h, not necessarily $\rho_l = \rho_h$. Figure 5 shows a graphical representation of $U(p,p)$ with two numeric examples. $U(p,p)$ is unimodal with one minimum point in $[0,1]$.

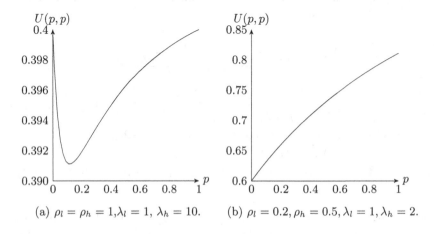

(a) $\rho_l = \rho_h = 1, \lambda_l = 1, \lambda_h = 10$.   (b) $\rho_l = 0.2, \rho_h = 0.5, \lambda_l = 1, \lambda_h = 2$.

Fig. 5. Examples of $U(p,p)$.

Extensive numerical analysis, for virtually all possible parameter values, shows that social optimum cannot be achieved by a behavioral strategy with $0 < p < 1$.

Conjecture 8. $\arg\max U(p,p) \in \{0,1\}$.

4.2   Equilibria of Behavioral Strategies

The mixed behavioral strategy $0 < p < 1$ is an equilibrium if

$$U(1,p) = U(0,p). \tag{5}$$

This means that if the other customer follows strategy p, the first customer is indifferent between the two pure selections, therefore p is also a best response. We first compute $U(1,p)$ - the utility of a customer that always takes the low-rate service while the other customer - denoted p chooser- draws the low-rate service with probability p. We build the steady-state transition diagram (see Fig. 6). Let (k, m) define a state of the system, where, $k, m \in \{a_l, a_h, s_l, s_h, w\}$ are the states of the first and the second customer, respectively. a_l, a_h - low-rate or high-rate activity, s_l, s_h - low-rate or high-rate service, w - waiting for service. The vector π gives the stationary probabilities for each state. The utility $U(1, p) = \pi_{a_l, s_l} + \pi_{a_l, a_l} + \pi_{a_l, s_h} + \pi_{a_l, a_h}$. In the same way we build the corresponding transition diagram for $U(0, p)$, and compute the utilities:

$$U(1, p) = (1 + \rho_l)\frac{ap + b}{c_1 p + d_1} \tag{6a}$$

$$U(0, p) = (1 + \rho_h)\frac{ap + b}{c_0 p + d_0} \tag{6b}$$

$$U(p, 1) = (1 + \rho_l)\frac{ap + b - k(1 - p)\lambda_l}{c_1 p + d_1} = U(1, p) - (1 + \rho_l)\lambda_l\frac{(1 - p)k}{c_1 p + d_1} \tag{6c}$$

$$U(p, 0) = (1 + \rho_h)\frac{ap + b + kp\lambda_h}{c_0 p + d_0} = U(0, p) + (1 + \rho_h)\lambda_h\frac{pk}{c_0 p + d_0} \tag{6d}$$

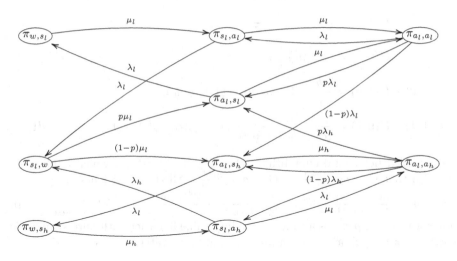

Fig. 6. Steady-state transition diagram for $U(1, p)$.

where

$$a = \lambda_h^3 \rho_l - \lambda_l^3 \rho_h + \lambda_l \lambda_h^2 - \lambda_l^2 \lambda_h$$

$$b = \lambda_l(\lambda_h \rho_l + \lambda_l)((\lambda_l + \lambda_h)\rho_h + \lambda_h)$$

$$c_1 = \lambda_h^3 \rho_l \left(2\rho_l^2 + 2\rho_l + 1\right) + \lambda_l \lambda_h^2 \left(\rho_l^3(\rho_h + 1) + 2\rho_l^2 + 2\rho_l + 1\right)$$
$$\quad + \lambda_l^2 \lambda_h(\rho_l + 1)\left(\rho_l^2 \rho_h - \rho_l\left(\rho_h^2 + 1\right) - 1\right) - \lambda_l^3(\rho_l + 1)^2 \rho_h(\rho_h + 1)$$

$$d_1 = \lambda_l(\rho_l + 1)(\lambda_h^2 \rho_l(\rho_l + 1)(\rho_h + 1) + \lambda_l \lambda_h(\rho_l(\rho_h(\rho_l + \rho_h + 2) + 1) + \rho_h + 1)$$
$$\quad + \lambda_l^2(\rho_l + 1)\rho_h(\rho_h + 1))$$

$$c_0 = \lambda_h^3 \rho_l (\rho_l + 1)(\rho_h + 1)^2 - \lambda_l \lambda_h^2 (\rho_h + 1)\left(-\left(\rho_l^2 + 1\right)\rho_h + \rho_l \rho_h^2 - 1\right)$$
$$\quad - \lambda_l^2 \lambda_h \left((\rho_l + 1)\rho_h^3 + 2\rho_h^2 + 2\rho_h + 1\right) - \lambda_l^3 \rho_h \left(2\rho_h^2 + 2\rho_h + 1\right)$$
$$d_0 = \lambda_l (2\rho_h(\rho_h + 1) + 1)(\lambda_h \rho_l + \lambda_l)(\lambda_h \rho_h + \lambda_h + \lambda_l \rho_h))$$
$$k = (\lambda_h^2 - \lambda_l^2)\rho_l \rho_h + \lambda_l \lambda_h (\rho_h - \rho_l)$$

Theorem 9. *A unique mixed behavioral equilibrium* $p = \frac{(1+\rho_h)d_1 - (1+\rho_l)d_0}{(1+\rho_l)c_0 - (1+\rho_h)c_1}$ *exists iff there exist two pure equilibria, either both symmetric or both asymmetric.*

Proof. According to Theorem 5 more than one pure equilibrium is possible only if $\rho_l > \rho_h$. Recall that we defined $\lambda_h > \lambda_l$, it is easy to see that in that case all the coefficients in the following equations are positive. By (6)

$$U(1,0) = (1 + \rho_l)\frac{b}{d_1}.$$

$$U(0,1) = (1 + \rho_h)\frac{a+b}{c_0 + d_0}.$$

$$U(1,1) = (1 + \rho_l)\frac{a+b}{c_1 + d_1}.$$

$$U(0,0) = (1 + \rho_h)\frac{b}{d_0}.$$

$$U(0,0) - U(1,0) = \frac{b}{d_0 d_1}((1 + \rho_h)d_1 - (1 + \rho_l)d_0).$$

$$U(1,1) - U(0,1) = \frac{a+b}{(c_0 + d_0)(c_1 + d_1)}(1 + \rho_l)(c_0 + d_0) - (1 + \rho_h)(c_1 + d1).$$

p is a mixed behavioral equilibrium when $U(1,p) = U(0,p)$, which implies, by (6a, 6b) that $(1 + \rho_l)\frac{ap+b}{c_1 p + d_1} = (1 + \rho_h)\frac{ap+b}{c_0 p + d_0}$. All the coefficients are positive hence $p = \frac{(1+\rho_h)d_1 - (1+\rho_l)d_0}{(1+\rho_l)c_0 - (1+\rho_h)c_1} = \frac{\frac{d_0 d_1}{b}(U(0,0) - U(1,0))}{\frac{d_0 d_1}{b}(U(0,0) - U(1,0)) + \frac{(c_0 + d_0)(c_1 + d_1)}{a+b}(U(1,1) - U(0,1))}$. In other words, $0 < p < 1$ is a valid probability value iff either both pure symmetric strategies are equilibria or both pure asymmetric strategies are equilibria. $\square$

For example, when $\lambda_l = 1, \rho_l = 1, \lambda_l = 2, \rho_l = 1, p^1 = p^2 = 1$ is the only behavioral equilibrium and $p < 0$. Suppose $\lambda_l = 1, \rho_l = 1, \lambda_l = 3, \rho_l = 0.75$, $p = \frac{41}{482}$. In this case, $p^1 = p^2 = 0$, $p^1 = p^2 = 1$ and $p^1 = p^2 = \frac{41}{482}$ are equilibria. In Fig. 7 we show the functions $U(1,p)$ and $U(0,p)$ intersecting at $p = \frac{41}{482}$. The best-response function BR is FTC - Follow the Crowd, i.e. BR = 0 for $p < \frac{41}{482}$, BR = 1 for $p > \frac{41}{482}$, and indifference exists when $p = \frac{41}{482}$.

Note that there is a mixed equilibrium (see Example Ex3 in Subsect. 3.1) for the same parameters, but the value is different - $p = 0.2530$. These are two different strategies but the conditions for the strategies to be equilibria are equal, as we prove in the next section.

5 Equilibrium Equivalence

In this section we prove that an equilibrium in the mixed strategy game exists iff there exists an equilibrium in the behavioral game for exactly the same parameters, though the value may be different.

Lemma 10. *In the behavioral model the functions* $U(p,1), U(p,0), U(1,p)$ *and* $U(0,p)$ *are monotone in the interval* $0 \le p \le 1$.

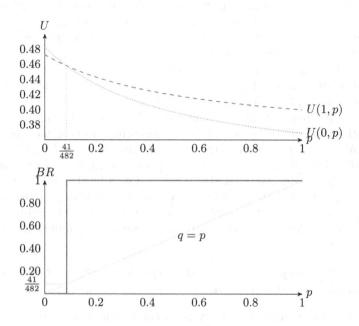

Fig. 7. Best response - follow the crowd.

Proof. From (6) all these functions have the form

$$U(p) = \frac{Ap + B}{Cp + D}$$

$$\frac{\partial U(p)}{\partial p} = \frac{AD - BC}{(Cp + D)^2}$$

where A, B, C, D are constants depending on the parameters $\rho_l, \rho_h, \lambda_l, \lambda_h$ alone. The numerator of the derivative does not contain p hence $U(p)$ is a monotone function of p. If $AD = BC$ then $U(p)$ is constant. Note that the denominator of $U(p)$ cannot be zero in the interval $[0,1]$ as $U(p)$ is bounded by 1, being a probability value. $\square$

Theorem 11. *A set of parameters* $(\rho_l, \rho_h, \lambda_l, \lambda_h)$ *induces the same number and types of equilibria in the mixed-strategy model and in the behavioral-strategy model.*

Proof. We prove for pure symmetric equilibria, for asymmetric pure equilibria and then for a mixed equilibrium.

A pure equilibrium in the behavioral-strategy model is a pure equilibrium in the mixed-strategy model as the condition for behavioral-strategy equilibrium is needed for a repeated draw in every cycle hence stronger than the condition needed for mixed-strategy equilibrium. So it is enough to prove that a pure mixed-strategy equilibrium is also an equilibrium in the behavioral-strategy model.

There are two pure symmetric strategies - $p^1 = p^2 = 0$ and $p^1 = p^2 = 1$. We prove for $p^1 = p^2 = 1$ and the proof for $p^1 = p^2 = 0$ is similar. $p^1 = p^2 = 1$ is an equilibrium in the mixed-strategy model if $U(1,1) \geq U(0,1)$. $p^1 = p^2 = 1$ is an equilibrium in the behavioral-strategy model if $U(1,1) \geq U(p,1)$ $\forall 0 \leq p \leq 1$. By Lemma 10 $U(p,1)$ is monotone. When $p^1 = p^2 = 1$ is an equilibrium in the mixed-strategy model $U(1,1) \geq U(0,1)$, hence $U(p,1)$ is monotone *non-decreasing* and $U(1,1) \geq U(p,1)$ $\forall 0 \leq p \leq 1$.

There are two pure asymmetric strategies - $p^1 = 1, p^2 = 0$ and $p^1 = 0, p^2 = 1$. Either both are equilibria or neither is an equilibrium, as customers are homogeneous. When both pure asymmetric strategies are equilibria, $U(0,1) > U(1,1)$ and $U(1,0) > U(0,0)$ and therefore neither pure symmetric strategy is an equilibrium in either model. We showed in Theorems 6 and 9 that in either model a mixed strategy exists iff there are two pure strategy equilibria, i.e. the formulae that compute q - the mixed equilibrium, and p - the behavioral equilibrium, will yield valid probability values for exactly the same set of parameters, although in general $q \neq p$. □

Corollary 12. *If no pure symmetric strategy is an equilibrium then both asymmetric pure strategies are equilibria.*

Proof. If no pure symmetric strategy is an equilibrium then $U(0,1) \geq U(1,1)$ and $U(1,0) \geq U(0,0)$ which defines both asymmetric pure equilibria.

6   Graphical Analysis of Equilibria

In Theorem 11 we showed the equivalence of equilibria in the mixed and the behavioral models, and the following discussion refers to both. We illustrate the region of each equilibrium type by a graph in the (ρ_l, ρ_h) plane, for $\lambda_l = 1$ and various λ_h. The points (ρ_l, ρ_h) on the boundary of the region where $p^1 = p^2 = 1$ is an equilibrium satisfy $U(0,1) = U(1,1))$, and the points (ρ_l, ρ_h) on the boundary of the region where $p^1 = p^2 = 0$ is an equilibrium satisfy $U(1,0) = U(0,0))$. By (6) we get the two equations that define the relation between (ρ_l, ρ_h) on the boundaries:

$$(1 + \rho_h)(c_1 + d_1) = (1 + \rho_l)(c_0 + d_0). \tag{7}$$

$$(1 + \rho_h)d_1 = (1 + \rho_l)d_0. \tag{8}$$

Suppose λ_h is fixed. For every given ρ_l we define

$$\underline{\rho}_h(\rho_l) = \min\{\rho_h | p_0 = p_1 = 1 \text{ is an equilibrium}\}.$$
$$\bar{\rho}_h(\rho_l) = \max\{\rho_h | p_0 = p_1 = 0 \text{ is an equilibrium}\}.$$

Figure 8 shows the region of each equilibrium for $(\rho_l, \rho_h) \in [0, 1.5]$. Subfigures 8a, b, c and d show that as λ_h increases, the region where both pure strategies are equilibria increases as well. We observe in these figures that $\bar{\rho}_h(\rho_l) > \underline{\rho}_h(\rho_l)$, implying the existence of a region with both $p^1 = p^2 = 1$ and $p^1 = p^2 = 0$ equilibria, and by Theorem 9 a mixed equilibrium exists as well, for every $\rho_l > 0$.

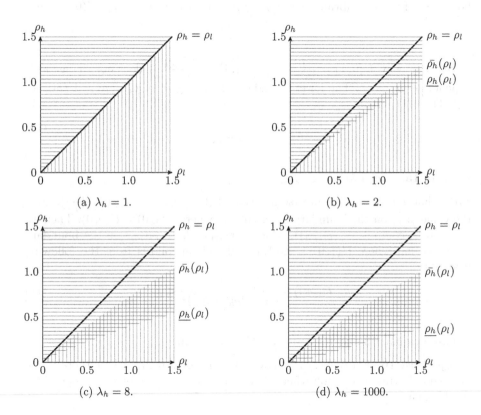

(a) $\lambda_h = 1$.

(b) $\lambda_h = 2$.

(c) $\lambda_h = 8$.

(d) $\lambda_h = 1000$.

Fig. 8. $p^1 = p^2 = 1$ is equilibrium in the region marked by vertical lines, $p^1 = p^2 = 0$ is equilibrium in the region marked by horizontal lines.

The next lemma proves the extreme cases. When $\lambda_h = \lambda_l = 1$ the region where both equilibria exist reduces to the line $\rho_h = \rho_l$, and when $\lambda_h \to \infty$ the region where both symmetric pure strategies are equilibria increases gradually with λ_h and the boundaries converge, as we show in Fig. 8.

Lemma 13. *Suppose $\lambda_h = \lambda_l = 1$. Then $p^1 = p^2 = 1$ is an equilibrium when $\rho_h \geq \rho_l$, and $p^1 = p^2 = 0$ is an equilibrium when $\rho_h \leq \rho_l$. (See Fig. 8a).*

Proof. When $\lambda_h = \lambda_l = 1$ choosing the larger μ (smaller ρ) is a dominant strategy as it achieves the same activity time for a smaller service time.

Lemma 14. *When $\lambda_h \to \infty$, $\bar{\rho}_h(\rho_l) = \frac{1}{4}(\rho_l - 1 + \sqrt{1 + 6\rho_l + \rho_l^2})$ and $\underline{\rho}_h(\rho_l) = \frac{\rho_l^2}{(1+\rho_l)^2}$ (See Figs. 8b, c and d).*

Proof. For specific ρ_l, ρ_h and $\lambda_h \to \infty$, let μ_h grow accordingly to keep ρ_h fixed and note that the service time of the customer with $p^2 = 0$ is negligible. $U(1,0) = \frac{1}{1+\rho_l}$ as there is no waiting time for the first customer with $p^1 = 1$. $U(0,0) = \frac{1+\rho_h}{1+2\rho_h+2\rho_h^2}$ does not change as it depends on ρ_h alone. The upper boundary of the equilibrium region $p_1 = p_0 = 0$ satisfies $U(1,0) = U(0,0)$, i.e.,

$$\bar{\rho}_h(\rho_l) = \frac{1}{4}\left(\rho_l - 1 + \sqrt{1 + 6\rho_l + \rho_l^2}\right) \tag{9}$$

In the same way, $U(0,1) = \frac{1}{1+\rho_h}\frac{1}{1+\rho_l}$. $U(1,1) = \frac{1+\rho_l}{1+2\rho_l+2\rho_l^2}$ does not change as it depends on ρ_l alone. The lower boundary of the region of $p_1 = p_0 = 1$ satisfies $U(0,1) = U(1,1)$ i.e.,

$$\underline{\rho}_h(\rho_l) = \frac{\rho_l^2}{(1+\rho_l)^2}. \tag{10}$$

$\square$

Lemma 15. *When $\lambda_h \to \infty$, $\bar{\rho}_h(\rho_l) > \underline{\rho}_h(\rho_l)$.*

Proof. For a given ρ_h, the inverse function of (9) - $\bar{\rho}_l(\rho_h) = \rho_h(1 + \frac{\rho_h}{1+\rho_h})$, gives the value of ρ_l on the boundary of the region where $U(1,0) = U(0,0)$. Then we use (10) to compute $\underline{\rho}_h(\bar{\rho}_l(\rho_h))$, which corresponds to $\bar{\rho}_l(\rho_h)$ on the boundary of the region where $U(1,0) = U(1,1)$. The claim follows from the following inequality:

$$\underline{\rho}_h(\bar{\rho}_l(\rho_h)) = \frac{(\rho_h + 2\rho_h^2)^2}{(1 + 2\rho_h + 2\rho_h^2)^2} < \rho_h. \tag{11}$$

$\square$

We use Lemmas 13 and 15 to suggest Conjecture 16.

Conjecture 16. *Every (ρ_l, ρ_h) is covered either by a region where there is one pure equilibrium or a region where there are two pure equilibria and one mixed equilibrium.*

We proved the conjecture for when $\lambda_h = \lambda_l$ in Lemma 13. Our numerical results indicate that when λ_h increases the equilibrium curves move downwards as shown in Fig. 8, and when $\lambda_h \to \infty$ we proved the conjecture in Lemma 15.

Corollary 17. *Assuming Conjecture 16 is correct, the pure asymmetric strategies $p_1 = 0, p_0 = 1$ and $p_1 = 1, p_0 = 0$ cannot be equilibria.*

Proof. In each point (ρ_l, ρ_h) there is at least one pure equilibrium. If each customer draws a different pure strategy than at least one of them can improve his result by changing to the equilibrium pure strategy. Hence there is no asymmetric equilibrium.

$\square$

7   Concluding Remarks

We propose future research in the following directions: Generalize the model with more service types and more customers. Complete conjectures in this paper.

In the current analysis the decision is taken at the entry, and at that point in time the other customer is always active. If the decision is taken at the entry to the **activity period** there are two possible states of the other customer - either in service or in activity. We want to determine the conditions for optima and equilibria, and the difference compared to the model discussed in this paper.

References

1. Cheng, H.K., Kohler, G.J.: Optimal pricing policies of web-enabled application services. Decis. Support Syst. **35**, 59–272 (2003)
2. Chu L. Y., Wan Z., Zhan D.: Harnessing the double-edged sword via routing: Information provision on ride-hailing platforms. Manuscript (2018)
3. Courcoubetis, C., Varaiya, P.: A game theoretic view of two processes using a single resource. IEEE Trans. Autom. Control AC **28**, 1059–1061 (1983)
4. Gallay, O., Hongler, M.O.: Circulation of autonomous agents in production and service networks. Int. J. Prod. Econ. **120**, 378–388 (2009)
5. Gross D., Harris C. M.: Fundamentals of Queueing, 2nd edn. Wiley, New York, Chapter 2.7, pp. 104–114 (1985)
6. Hassin R.: Rational Queueing. CRC Press, Chapter 6.3.1, p. 153, Chapter 4.7, p. 108 (2016)
7. Hassin R., Haviv M.: To Queue or Not to Queue: Equilibrium Behavior in Queueing Systems. Kluwer Academic Publishers, Chapter 8.6, p. 167 (2003)
8. Xu Y., Dai T., Sycara K., Lewis M.: Service level differentiation in multi-robots control. In: The 2010 IEEE/RSJ International Conference on Intelligent Robots and Systems, pp. 2224–2229 (2010)

On Rational Behavior in a Loss System with One Observable Queue and One Unobservable Queue

Refael Hassin and Jonathan H. P. Milo[✉]

Department of Statistics and Operations Research, School of Mathematical Sciences,
Tel Aviv University, Tel Aviv-Yafo, Israel
hassin@post.tau.ac.il, milojona@tauex.tau.ac.il

Abstract. We examine a system with two heterogeneous servers. An arriving customer first observes the queue length at the slower server and decides whether to join it or join the unobservable queue of the faster server. Customers arrive to the system and decide which queue to join according to the reward, waiting cost, and service rates. Once a customer chooses a queue, she cannot change her decision. We analyze a special case of this model where there is no waiting space except for the customer in service. The probability for entering the observable queue (if the server idle) is denoted by p, and this is the strategy of the customers. We analyze and characterize the Nash equilibria and the socially-optimal probabilities of the system, and the relation between the two as function of the model's parameters. We also examine throughput maximization.

Keywords: Strategic queueing · Nash equilibrium · Social optimization

1 Introduction

We often make decisions using partial or incomplete information, and some degree of uncertainty. Such situations are frequent in queueing systems, where the incomplete information can be characterized by unknown service rates, unknown arrival rates, unobservable queue length, etc. For example:

1. Gas stations - You arrive at a gas station. Your options are to wait in line for the pump, or to take a chance and proceed to another gas station down the road.
2. Emergency rooms - Some emergency rooms provide information regarding the length of the line for treatment or the expected waiting time for a new patient, while other emergency rooms provide no such service.
3. Parking spots - You wish to park your car. When arriving at a parking lot, you are faced with the option to park there or to try and find a closer lot further down the road.

This research was supported by the Israel Science Foundation (grant No. 355/15).

T. Phung-Duc et al. (Eds.): QTNA 2019, LNCS 11688, pp. 166–182, 2019.
https://doi.org/10.1007/978-3-030-27181-7_11

In these examples one of the queues is unobservable. The third example is different in the sense that there is no waiting line, the parking lot is either full or has a free spot.

Hassin's "two gas stations" model [2] describes a system of two identical queues, Q_1 and Q_2, where only the line of Q_1 is observable, and customers minimize expected waiting time. Hassin showed that the model leads to "avoid-the-crowd" (ATC) behavior, and thus a unique equilibrium exists. Further analysis of his model proved to be hard. Altman, Jimenez, Núñez-Queija, and Yechiali [1] considered a more general model, in which the two queues have different service rates. They left the question of whether this extension is also ATC open.

In this paper we look at the model of [1] with exponential service time and heterogeneous service rates, with the added assumption that customers cannot wait in line for a server, i.e. the queues are $\cdot/M/1/1$. We find and characterize the Nash equilibrium, and the socially-optimal strategies for joining Q_1. We also discuss throughput maximization (deleted "the") and show that the optimal strategy does not necessarily prescribe joining Q_1 with probability 1.

Hassin and Roet-Green describe a model [5] in which customers face parallel unobservable queues. Upon arrival to the system, every customer picks a queue randomly and observes it. After learning the length of the queue, the customer can either stay at that queue, or pay a fixed cost (not necessarily the same cost for every queue) to observe another queue. The customer then joins the shorter queue. Customers incur a cost for every unit of time spent waiting for service. The customers' objective is to minimize their expected cost. This model resembles our model in the sense that customers face one observable queue and one unobservable queue, and then decide what action to take according to the future potential costs. The difference between the two models is that in our model there is no cost of observing the queues, and the decision to skip the first queue and observe the second queue is irrevocable.

Singh, Delasay, and Scheller-Wolf [6] consider a system with two independent M/M/1 queues (service providers), a leader and a follower. Customers are time sensitive. The leader chooses whether to reveal the queue length. If the leader chooses to reveal this information, then the follower has the option to reveal as well. The service providers compete for market share. The authors find that, in equilibrium, the leader reveals the queue length if and only if its service rate is slower than the follower, while the follower's best response depends on the parameters of the model.

The subject of queueing systems with strategic players who act to maximize expected utility is extensively reviewed in [3] and [4].

2   Setting

The system consists of two queues, Q_1 and Q_2, with exponential service distributions with rates μ_1 and μ_2 respectively, $\mu_1 < \mu_2$. The arrival process to the system is Poisson with parameter λ. For simplicity we use the notation Q_i for both the ith queue and the ith server, $i = 1, 2$. Q_1 is M/M/1/1, while Q_2 is

$\cdot/M/1/1$ since the arrival rate to Q_2 is dependent on Q_1. Customers obtain a reward of R by getting served from either Q_1 or Q_2, and incur a cost of C per unit of time spent in service (this is a loss system and no time is spent waiting for service). To avoid trivial solutions, we also assume that $R - \frac{C}{\mu_1} > 0$, i.e. $\frac{R\mu_1}{C} > 1$.

We use normalized parameters: $\rho_i = \frac{\lambda}{\mu_i}$, and $\nu_i = \frac{R\mu_i}{C}$, for $i = 1, 2$, where ν_i is the normalized reward in units of waiting costs per expected service period, $\frac{C}{\mu_i}$. Notice that $\frac{\nu_1}{\nu_2} = \frac{\mu_1}{\mu_2} = \frac{\rho_2}{\rho_1}$, so there are only three independent parameters. Subsequently we refer only to ρ_1, ρ_2, ν_1, and disregard ν_2. Note that $\rho_1 > \rho_2$ since $\mu_1 < \mu_2$, and $\nu_1 > 1$ by assumption. For simplicity we denote $\theta_1 := \rho_1^{-1} = \frac{\mu_1}{\lambda}$, $\theta_2 := \rho_2^{-1} = \frac{\mu_2}{\lambda}$. Hence $0 < \theta_1 < \theta_2$.

3   The Model

When a customer arrives to the system, she sees Q_1. If the server in Q_1 is busy, she proceeds to Q_2. If the server in Q_1 is idle, she begins service there with probability p, followed by her departure from the system, and with probability $1 - p$ she moves to Q_2. On arrival to Q_2 she will be served if the server is idle, and if the server is busy she will leave the system unserved. This process is illustrated in Fig. 1.

Clearly, if $\mu_1 \geq \mu_2$ the customer has no reason to skip Q_1, when the first server is idle. This is why we focus on the case where $\mu_1 < \mu_2$.

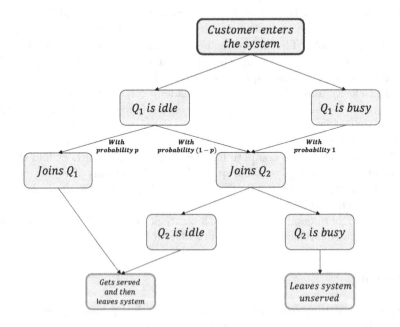

Fig. 1. Flowchart of the system.

3.1 Steady-State Probabilities

As seen in Fig. 2, the system can be in one of four states, each represented by an ordered pair of $0/1$ where the first position refers to Q_1 and the second to Q_2, 0 indicates that the server is free, and 1 indicates it is busy. We denote by $\pi_{(0,0)}, \pi_{(0,1)}, \pi_{(1,0)}, \pi_{(1,1)}$ the stationary probabilities of the four states.

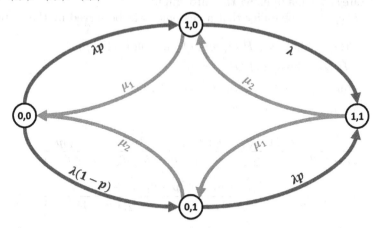

Fig. 2. Flow diagram of the system.

The steady-state probabilities are determined by the following equations:

1. $\lambda\pi_{(0,0)} = \mu_1\pi_{(1,0)} + \mu_2\pi_{(0,1)}$.
2. $(\mu_2 + p\lambda)\pi_{(0,1)} = (1-p)\lambda\pi_{(0,0)} + \mu_1\pi_{(1,1)}$.
3. $(\mu_1 + \lambda)\pi_{(1,0)} = p\lambda\pi_{(0,0)} + \mu_2\pi_{(1,1)}$.
4. $\pi_{(0,0)} + \pi_{(0,1)} + \pi_{(1,0)} + \pi_{(1,1)} = 1$.

Solving the linear system yields:

$$-\ \pi_{(0,0)} = \frac{\mu_1\mu_2(\lambda(1+p)+\mu_1+\mu_2)}{\mu_1\lambda^2(1+p-p^2)+\mu_2(\mu_1\lambda(p+2)+2p\lambda^2+\mu_1^2)+\mu_2^2(p\lambda+\mu_1)+p\lambda^3+\mu_1^2\lambda(1-p)}.$$

$$-\ \pi_{(0,1)} = \frac{\mu_1\lambda((\mu_1+\mu_2)(1-p)+\lambda)}{\mu_1\lambda^2(1+p-p^2)+\mu_2(\mu_1\lambda(p+2)+2p\lambda^2+\mu_1^2)+\mu_2^2(p\lambda+\mu_1)+p\lambda^3+\mu_1^2\lambda(1-p)}.$$

$$-\ \pi_{(1,0)} = \frac{\mu_2\lambda p(\mu_1+\mu_2+\lambda)}{\mu_1\lambda^2(1+p-p^2)+\mu_2(\mu_1\lambda(p+2)+2p\lambda^2+\mu_1^2)+\mu_2^2(p\lambda+\mu_1)+p\lambda^3+\mu_1^2\lambda(1-p)}.$$

$$-\ \pi_{(1,1)} = \frac{\lambda^2 p(\mu_1(1-p)+\mu_2+\lambda)}{\mu_1\lambda^2(1+p-p^2)+\mu_2(\mu_1\lambda(p+2)+2p\lambda^2+\mu_1^2)+\mu_2^2(p\lambda+\mu_1)+p\lambda^3+\mu_1^2\lambda(1-p)}.$$

Dividing the numerator and denominator by λ^3, and rearranging gives:

$$-\ \pi_{(0,0)} = \frac{\theta_1\theta_2(1+\theta_1+\theta_2)+(\theta_1\theta_2)p}{\theta_1(1+\theta_1+\theta_2)(\theta_2+1)+\big((1+\theta_1+\theta_2)(\theta_2+1)-\theta_1^2\big)p-\theta_1 p^2}.$$

$$-\ \pi_{(0,1)} = \frac{\theta_1(1+\theta_1+\theta_2)-(\theta_1(1+\theta_2))p}{\theta_1(1+\theta_1+\theta_2)(\theta_2+1)+\big((1+\theta_1+\theta_2)(\theta_2+1)-\theta_1^2\big)p-\theta_1 p^2}.$$

$$-\ \pi_{(1,0)} = \frac{(\theta_2(1+\theta_1+\theta_2))p}{\theta_1(1+\theta_1+\theta_2)(\theta_2+1)+\big((1+\theta_1+\theta_2)(\theta_2+1)-\theta_1^2\big)p-\theta_1 p^2}.$$

$$-\ \pi_{(1,1)} = \frac{(1+\theta_1+\theta_2)p-\theta_1 p^2}{\theta_1(1+\theta_1+\theta_2)(\theta_2+1)+\big((1+\theta_1+\theta_2)(\theta_2+1)-\theta_1^2\big)p-\theta_1 p^2}.$$

Recall that $\theta_i = \frac{1}{\rho_i}$, for $i = 1, 2$.

4   Throughput Maximization

In this section, our objective is to compute the value of p that maximizes the system's throughput, i.e., the average rate of customers served by the system. In particular, we wish to characterize the condition that guarantees $p = 1$ is the optimal strategy for maximizing the throughput.

Denote by $T(p)$ the probability that a customer will be served by the system.

$$
\begin{aligned}
T(p) &= P(Q_1 \text{ is idle}) \cdot p + P(Q_1 \text{ and } Q_2 \text{ are idle}) \cdot (1 - p) \\
&\quad + P(Q_1 \text{ is busy and } Q_2 \text{ is idle}) \\
&= (\pi_{(0,0)} + \pi_{(0,1)})p + \pi_{(0,0)}(1 - p) + \pi_{(1,0)} \\
&= \pi_{(0,0)} + \pi_{(1,0)} + \pi_{(0,1)}p
\end{aligned}
$$

$$
= \frac{\theta_1\theta_2 (1 + \theta_1 + \theta_2) + \theta_1\theta_2 p + \theta_2(1 + \theta_1 + \theta_2)p}{\theta_1 (1 + \theta_1 + \theta_2)(\theta_2 + 1) + ((1 + \theta_1 + \theta_2)(\theta_2 + 1) - \theta_1^2) p - \theta_1 p^2}
$$

$$
+ \frac{\theta_1(1 + \theta_1 + \theta_2)p - \theta_1 (\theta_1 + \theta_2) p^2}{\theta_1 (1 + \theta_1 + \theta_2)(\theta_2 + 1) + ((1 + \theta_1 + \theta_2)(\theta_2 + 1) - \theta_1^2) p - \theta_1 p^2}
$$

$$
= \frac{\theta_1\theta_2\kappa + (\theta_1\theta_2 + \kappa(\theta_1 + \theta_2)) p - \theta_1 (\theta_1 + \theta_2) p^2}{\theta_1\kappa (\theta_2 + 1) + (\kappa (\theta_2 + 1) - \theta_1^2) p - \theta_1 p^2}
$$

$$
= \frac{-p^2 (\theta_1 + \theta_2) + p \left(\kappa \left(\frac{\theta_1 + \theta_2}{\theta_1} \right) + \theta_2 \right) + \kappa\theta_2}{-p^2 + p \left(\kappa \left(\frac{\theta_2 + 1}{\theta_1} \right) - \theta_1 \right) + \kappa (1 + \theta_2)},
$$

where $\kappa = 1 + \theta_1 + \theta_2$.

Our objective is to maximize $T(p)$.

Lemma 1. $T(p)$ *is concave, for* $0 \le p \le 1$.

Proof. We need to show that $\frac{dT}{dp}$ is a monotonically decreasing function of p. In order to do so, we look at the second derivative, $\frac{d^2T}{dp^2}$.

$$
\begin{aligned}
\frac{d^2T}{dp^2} &= \frac{2\theta_1^2 \left[\left(\theta_2^3 + (2\theta_1 + 1)\theta_2^2 - \theta_1^3 \right) p^3 + 3\theta_1 \left(\theta_2^2 + \theta_1\theta_2 + \theta_1 \right) \kappa p^2 \right]}{((\theta_1 + p)(\theta_1 p - (\theta_2 + 1)\kappa))^3} \\
&\quad + \frac{2\theta_1^2 \left[-3\theta_1\kappa^2(2\theta_2 + 1)p + \kappa^3(1 + 3\theta_2 + \theta_2^2(\theta_1 + 2)) \right]}{((\theta_1 + p)(\theta_1 p - \theta_2 + 1)\kappa))^3}.
\end{aligned}
$$

The denominator is always negative:

- $\theta_1 + p > 0$.
- $\theta_1 p - (\theta_2 + 1)\kappa < \theta_1 p - \theta_2\kappa \le \theta_1 - \theta_2\kappa < \theta_1 - \theta_2 < 0$,   since $0 < \theta_1 < \theta_2 < \kappa$, $p \le 1, 1 < \kappa$.

The numerator is always positive:

- $-\left((\theta_2)^3 + (2\theta_1 + 1)(\theta_2)^2 - (\theta_1)^3\right)p^3 \geq 0$,   since $0 < \theta_1 < \theta_2$ and $0 \leq p$.

- $3\theta_1\left((\theta_2)^2 + \theta_1\theta_2 + \theta_1\right)\kappa p^2 \geq 0$,   since all elements are non-negative.

- $-3\theta_1\kappa^2(2\theta_2 + 1)p + \kappa^3(1 + 3\theta_2 + \theta_2^2(\theta_1 + 2)) = -\kappa^2(6\theta_1\theta_2 + 3\theta_1)p + \kappa^2(\kappa + 3\kappa\theta_2 + \kappa\theta_2^2(\theta_1 + 2)) > 0$,   since $0 \leq p \leq 1$, $0 < \theta_1 < \theta_2 < \kappa = (1 + \theta_1 + \theta_2)$, and $6\theta_1\theta_2 + 3\theta_1 < 3\kappa\theta_2 = 3\theta_2 + 3\theta_2^2 + 3\theta_1\theta_2$.

$\square$

Corollary 1. *The optimal value, i.e. the p that maximizes $T(p)$ is strictly positive, since $T'(0) = \frac{1+2\theta_2}{(1+\theta_2)^2} > 0$.*

Corollary 2. $p = 1$ *is the optimal strategy for maximizing throughput iff $\frac{dT}{dp}(1) \geq 0$.*

$$T'(1) = \frac{-\theta_1\left(\theta_2^3 - (\theta_1 - 1)\theta_2^2 - 2\theta_1(\theta_1 + 2)\theta_2 - \theta_1\right)}{(\theta_1 + 1)^2(\theta_2(\theta_1 + \theta_2 + 2) + 1)^2} \geq 0$$
$$\iff \theta_2^3 - \theta_2^2(\theta_1 - 1) - 2\theta_1(\theta_1 + 2)\theta_2 - \theta_1 \leq 0,$$

since $0 < \theta_1$.

For any given θ_2, we can write this expression as a second-degree polynomial of θ_1:

$T'(1) \geq 0 \iff (2\theta_2)\theta_1^2 + (1 + 4\theta_2 + \theta_2^2)\theta_1 - (\theta_2^2 + \theta_2^3) \geq 0$.

We notice that this polynomial has two roots, one of which is always positive and the other one is always negative. We also notice that the second derivative of the polynomial is strictly positive. Therefore (Fig. 3),

Proposition 1. $T'(1) \geq 0 \iff \theta_1 \geq \frac{-(1+4\theta_2+\theta_2^2)+\sqrt{(1+4\theta_2+\theta_2^2)^2+8(\theta_2^3+\theta_2^4)}}{4\theta_2} > 0$.

The border between the region where $p = 1$ is optimal and the region where $p < 1$ is optimal, is represented by the function

$$\theta_1 = F(\theta_2) := \frac{-(1 + 4\theta_2 + \theta_2^2) + \sqrt{(1 + 4\theta_2 + \theta_2^2)^2 + 8(\theta_2^3 + \theta_2^4)}}{4\theta_2}.$$

By differentiating $F(\theta_2)$ we can show that F is monotonically increasing in θ_2, and its derivative is monotonically approaching $\frac{1}{2}$.

Following Corollary 1 and Corollary 2, we now find the explicit expression for p that maximizes $T(p)$:

$$\frac{dT}{dp} = \frac{p^2\left((\theta_1 + \theta_2)\left(\theta_1 - \kappa\frac{\theta_2}{\theta_1}\right) + \theta_2\right) + p\left(2\kappa(-\theta_1 - \theta_2(\theta_1 + \theta_2)) + \kappa^2(1 + 2\theta_2)\right)}{\left(p^2 - p\left(\kappa\left(\frac{\theta_2+1}{\theta_1}\right) - \theta_1\right) - \kappa(1 + \theta_2)\right)^2}.$$

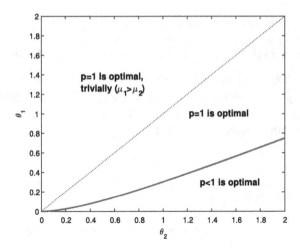

Fig. 3. Regions where $p < 1$ and $p = 1$ are the optimal strategies.

To find the roots of this derivative, we only look at the roots of the numerator, since the denominator is strictly positive. The numerator is a quadratic expression, so its roots are:

$$p = \frac{\kappa(\theta_1 + \theta_2(\theta_1 + \theta_2))}{(\theta_1 + \theta_2)\left(\theta_1 - \kappa\frac{\theta_2}{\theta_1}\right) + \theta_2}$$

$$\pm \frac{\sqrt{(\kappa(\theta_1 + \theta_2(\theta_1 + \theta_2)))^2 - \left((\theta_1 + \theta_2)\left(\theta_1 - \kappa\frac{\theta_2}{\theta_1}\right) + \theta_2\right)\kappa^2(1 + 2\theta_2)}}{(\theta_1 + \theta_2)\left(\theta_1 - \kappa\frac{\theta_2}{\theta_1}\right) + \theta_2}.$$

Notice that $(\theta_1 + \theta_2)\left(\theta_1 - \kappa\frac{\theta_2}{\theta_1}\right) + \theta_2 < 0$, since $\kappa = 1 + \theta_1 + \theta_2$. All other elements are positive, thus the expression under the square root is greater than $\kappa(\theta_1 + \theta_2(\theta_1 + \theta_2))$. Therefore, in order to keep p positive we are left with only one root:

$$p = \frac{\kappa(\theta_1 + \theta_2(\theta_1 + \theta_2))}{(\theta_1 + \theta_2)\left(\theta_1 - \kappa\frac{\theta_2}{\theta_1}\right) + \theta_2}$$

$$- \frac{\sqrt{(\kappa(\theta_1 + \theta_2(\theta_1 + \theta_2)))^2 - \left((\theta_1 + \theta_2)\left(\theta_1 - \kappa\frac{\theta_2}{\theta_1}\right) + \theta_2\right)\kappa^2(1 + 2\theta_2)}}{(\theta_1 + \theta_2)\left(\theta_1 - \kappa\frac{\theta_2}{\theta_1}\right) + \theta_2}.$$

$$\tag{1}$$

To conclude, from Corollary 1 we know that the p that maximizes the throughput is strictly positive, from Corollary 2 we get the condition for $p = 1$, and any other $p < 1$ is described by the expression in Eq. 1.

5 Equilibrium

Decisions made by previous customers affect the state probabilities of the system, therefore the strategy of a new customer is dependent on strategies of others. Since customers are homogeneous, we consider symmetric equilibrium strategies.

We have three possibilities:

1. The strategy of all customers is $p = 1$.
2. The strategy of all customers is $p = 0$.
3. The strategy of all customers is $0 < p < 1$, and customers are indifferent between joining Q_1 when the server is idle, or proceeding to Q_2.

We now consider a customer who observes that Q_1 is idle.

The probability that Q_2 is idle, conditioned on Q_1 being idle, is:

$$\widehat{P} := P(Q_2 \text{ is idle} \mid Q_1 \text{ is idle}) = \frac{\pi_{(0,0)}}{\pi_{(0,0)} + \pi_{(0,1)}}$$

$$= \frac{\theta_1\theta_2(1 + \theta_1 + \theta_2) + \theta_1\theta_2 p}{[\theta_1\theta_2(1 + \theta_1 + \theta_2) + \theta_1\theta_2 p] + [\theta_1(1 + \theta_1 + \theta_2) - (\theta_1 + \theta_2)p]}$$

$$= \frac{\theta_2(1 + \theta_1 + \theta_2) + \theta_2 p}{(1 + \theta_2)(1 + \theta_1 + \theta_2) - \theta_1 p} = \frac{\theta_2\kappa + \theta_2 p}{(1 + \theta_2)\kappa - \theta_1 p}.$$

We use this probability to characterize the equilibria of customer behavior by comparing the expected utility of joining Q_1 when it is idle, i.e. $R - \frac{C}{\mu_1}$, with the expected utility of not joining Q_1 when it is idle, i.e. $\widehat{P}(R - \frac{C}{\mu_2}) + (1 - \widehat{P}) \cdot 0 = \widehat{P}(R - \frac{C}{\mu_2})$.

Theorem 1. *There exists a unique symmetric equilibrium strategy, p_e.*

$$p_e = \begin{cases} 0 & \text{if } 1 < \nu_1 \leq A \\ \frac{\kappa(\nu_1 - A)}{\nu_1(\theta_1 + \theta_2) - 2\theta_1} & \text{if } A \leq \nu_1 \leq B \\ 1 & \text{if } \nu_1 \geq B \end{cases} \tag{2}$$

where $\kappa = 1 + \theta_1 + \theta_2$, $A = \kappa - 2\theta_1$, and $B = \kappa A - 2\theta_1$ (Fig. 4).

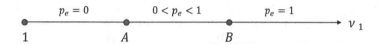

Fig. 4. Graphical representation of the regions of each equilibrium type.

Proof. In order to prove the theorem, we will show that $p = 0$ is an equilibrium iff $\nu_1 \in (1, A]$, $p = 1$ is an equilibrium iff $\nu_1 > B$, and when $\nu_1 \in (A, B]$ there exists a unique equilibrium $0 < p < 1$.

- $p = 0$ is an equilibrium iff when all join Q_2 with probability 1 then it is still preferable to join Q_2. In other words, when we plug $p = 0$ in $\widehat{P}$, we get

  $R - \frac{C}{\mu_1} \leq \widehat{P}(R - \frac{C}{\mu_2})$, or equivalently $\frac{R}{C} \leq \frac{\frac{1}{\mu_1} - \frac{\widehat{P}}{\mu_2}}{1 - \widehat{P}}$. Substituting $p = 0$ in $\widehat{P}$

  gives $\frac{\frac{1}{\mu_1} - \frac{\widehat{P}}{\mu_2}}{1 - \widehat{P}} = \frac{\frac{1}{\mu_1} - \frac{\theta_2}{(\theta_2+1)\mu_2}}{\frac{1}{\theta_2+1}} = \frac{\theta_2+1}{\mu_1} - \frac{\theta_2}{\mu_2}$.

  Multiplying by μ_1 gives us the condition: $\frac{R\mu_1}{C} \leq \theta_2 + 1 - \theta_2\frac{\mu_1}{\mu_2} = 1 + \theta_2 - \theta_1 = \kappa - 2\theta_1 = A$. Notice that $A > 1$, since $\theta_2 > \theta_1$.

  We conclude that $p = 0$ is an equilibrium iff $\nu_1 \leq A$.

- $p = 1$ is an equilibrium iff when all join Q_1 with probability 1 then it is still preferable to join Q_1. In other words, when we plug $p = 1$ in $\widehat{P}$, we get

  $R - \frac{C}{\mu_1} \geq \widehat{P}(R - \frac{C}{\mu_2})$, or equivalently $\frac{R}{C} \geq \frac{\frac{1}{\mu_1} - \frac{\widehat{P}}{\mu_2}}{1 - \widehat{P}}$. Substituting $p = 0$ in $\widehat{P}$

  gives $\frac{\frac{1}{\mu_1} - \frac{\widehat{P}}{\mu_2}}{1 - \widehat{P}} = \frac{\frac{1}{\mu_1} - \frac{\theta_2(1+\kappa)}{(\theta_2(1+\kappa)+1)\mu_2}}{\frac{1}{\theta_2(1+\kappa)+1}} = \frac{\theta_2(1+\kappa)+1}{\mu_1} - \frac{\theta_2(1+\kappa)}{\mu_2}$.

  Multiplying by μ_1 gives us the condition: $\frac{R\mu_1}{C} \geq \theta_2(1+\kappa) + 1 - \frac{\mu_1}{\mu_2}\theta_2(1+\kappa) = \theta_2(1+\kappa) + 1 - \theta_1(1+\kappa) = \kappa(1 + \theta_2 - \theta_1) - 2\theta_1 = \kappa A - 2\theta_1 = B$.

  We conclude that $p = 1$ is an equilibrium iff $\nu_1 \geq B$.

  Notice that $B > A$, since $\kappa > \theta_1$, and $A > 1$.

- $0 < p < 1$ is an equilibrium iff a customer who sees Q_1 idle is indifferent between the two options, i.e.

$$R - \frac{C}{\mu_1} = \widehat{P}\left(R - \frac{C}{\mu_2}\right),$$

or

$$\frac{R - \frac{C}{\mu_1}}{R - \frac{C}{\mu_2}} = \widehat{P} = \frac{\theta_2\kappa + \theta_2 p}{(1 + \theta_2)\kappa - \theta_1 p}.$$

After rearranging the elements and solving for p, we get: $p = \frac{\kappa(\nu_1 - A)}{\nu_1(\theta_1+\theta_2) - 2\theta_1}$.

Notice that the denominator is always positive:

$$\nu_1(\theta_1 + \theta_2) - 2\theta_1 > 0 \Leftrightarrow \nu_1 > \frac{2\theta_1}{\theta_1 + \theta_2}.$$

This follows since $\nu_1 > 1$ and $\theta_1 < \theta_2$.

We now show that $0 < \frac{\kappa(\nu_1 - A)}{\nu_1(\theta_1+\theta_2) - 2\theta_1} < 1$ iff $A < \nu_1 < B$:

$\frac{\kappa(\nu_1 - A)}{\nu_1(\theta_1+\theta_2) - 2\theta_1} < 1 \Longleftrightarrow \kappa(\nu_1 - A) < \nu_1(\theta_1 + \theta_2) - 2\theta_1$. This is equivalent to $\nu_1(\kappa - \theta_1 - \theta_2) < \kappa A - 2\theta_1$, which after substituting $\kappa = 1 + \theta_1 + \theta_2$ gives $\nu_1 < \kappa A - 2\theta_1 = B$, So $\frac{\kappa(\nu_1 - A)}{\nu_1(\theta_1+\theta_2) - 2\theta_1} < 1 \Longleftrightarrow \nu_1 < B$.

It is clear that $0 < \frac{\kappa(\nu_1 - A)}{\nu_1(\theta_1+\theta_2) - 2\theta_1}$ iff $A < \nu_1$.

We conclude that $0 < p < 1$ is an equilibrium iff $A < \nu_1 < B$.

Finally, we notice that $\frac{\kappa(\nu_1 - A)}{\nu_1(\theta_1+\theta_2) - 2\theta_1} = 1$ when $\nu_1 = B$, and $\frac{\kappa(\nu_1 - A)}{\nu_1(\theta_1+\theta_2) - 2\theta_1} = 0$ when $\nu_1 = A$,

thus completing the proof. $\square$

6   Social Optimality

In order to define the social welfare function we define the following probabilities:

- $\widetilde{P}$ is the probability that Q_1 is idle. Therefore, $\widetilde{P} = \pi_{(0,0)} + \pi_{(0,1)}$.
  We can calculate $\pi_{(0,0)} + \pi_{(0,1)}$ directly: Since Q_2 does not affect $\pi_{(0,0)} + \pi_{(0,1)}$, we can look at Q_1 as a single $M/M/1/1$ queue with arrival rate of $p\lambda$ and service time of μ_1.
  Hence $\widetilde{P} = \dfrac{\frac{1}{\lambda p}}{\frac{1}{\lambda p} + \frac{1}{\mu_1}} = \dfrac{\frac{1}{\lambda p}}{\frac{\lambda p + \mu_1}{\lambda p \mu_1}} = \dfrac{1}{\frac{\lambda p}{\mu_1} + 1} = \dfrac{1}{\rho_1 p + 1} = \dfrac{\theta_1}{p + \theta_1}.$

- $\check{P}$ is the probability Q_2 is idle given that Q_1 is busy. Therefore,
  $\check{P} = \dfrac{\pi_{(1,0)}}{\pi_{(1,0)} + \pi_{(1,1)}} = \dfrac{\theta_2 \kappa p}{\theta_2 \kappa p + \kappa p - \theta_1 p^2} = \dfrac{\theta_2 \kappa}{(1+\theta_2)\kappa - \theta_1 p}.$

Recall that $\widehat{P}$ is the probability Q_2 is idle given that Q_1 is also idle, i.e. $\widehat{P} = \dfrac{\theta_2 \kappa + \theta_2 p}{(1+\theta_2)\kappa - \theta_1 p}.$

The social welfare function is then:

$$U(p) = \widetilde{P}\left[p\left(R - \frac{C}{\mu_1}\right) + (1-p)\,\widehat{P}\left(R - \frac{C}{\mu_2}\right)\right] + \left(1 - \widetilde{P}\right)\check{P}\left(R - \frac{C}{\mu_2}\right).$$

Normalizing $U(p)$ by R gives:

$$\frac{U(p)}{R} = \widetilde{P}\left[p\left(1 - \frac{1}{\nu_1}\right) + (1-p)\,\widehat{P}\left(1 - \frac{1}{\nu_2}\right)\right] + \left(1 - \widetilde{P}\right)\check{P}\left(1 - \frac{1}{\nu_2}\right)$$

$$\overset{(*)}{=} \widetilde{P}\left[p\left(1 - \frac{1}{\nu_1}\right) + (1-p)\,\widehat{P}\left(1 - \frac{\theta_1}{\theta_2 \nu_1}\right)\right] + \left(1 - \widetilde{P}\right)\check{P}\left(1 - \frac{\theta_1}{\theta_2 \nu_1}\right)$$

$$= \widetilde{P}\left[p\left(1 - \frac{1}{\nu_1}\right) + \left(1 - \frac{\theta_1}{\theta_2 \nu_1}\right)\left((1-p)\,\widehat{P} + \left(\frac{1 - \widetilde{P}}{\widetilde{P}}\right)\check{P}\right)\right]$$

$$\overset{(**)}{=} \frac{\theta_1}{p + \theta_1}\left[p\left(1 - \frac{1}{\nu_1}\right)\right]$$

$$+ \frac{\theta_1}{p + \theta_1}\left[\left(1 - \frac{\theta_1}{\theta_2 \nu_1}\right)\left((1-p)\frac{\theta_2 \kappa + \theta_2 p}{(1+\theta_2)\kappa - \theta_1 p} + \frac{p}{\theta_1}\frac{\theta_2 \kappa}{(1+\theta_2)\kappa - \theta_1 p}\right)\right]$$

$$= \frac{\theta_1}{p + \theta_1}\left[p\left(1 - \frac{1}{\nu_1}\right) + \left(\theta_2 - \frac{\theta_1}{\nu_1}\right)\left(\frac{\kappa + p\left(1 - \kappa + \frac{\kappa}{\theta_1} - p\right)}{(1+\theta_2)\kappa - \theta_1 p}\right)\right]$$

$$= \frac{1}{p + \theta_1}\left[p\left(\theta_1 - \frac{\theta_1}{\nu_1}\right) + \left(\theta_2 - \frac{\theta_1}{\nu_1}\right)\left(\frac{\kappa + p\left(1 - \kappa + \frac{\kappa}{\theta_1} - p\right)}{\left(\frac{1+\theta_2}{\theta_1}\right)\kappa - p}\right)\right]$$

$$= \frac{-p^2\left(\theta_1 + \theta_2 - \frac{2\theta_1}{\nu_1}\right) + p\left(\theta_2 - \frac{\theta_1}{\nu_1} + \frac{\kappa}{\theta_1}\left(\theta_1 + \theta_2 + \frac{\theta_1}{\nu_1}\left(\theta_1 - \theta_2 - 2\right)\right)\right)}{-p^2 + p\left(\kappa\left(\frac{\theta_2 + 1}{\theta_1}\right) - \theta_1\right) + \kappa\left(1 + \theta_2\right)}$$

$$+\frac{\kappa\left(\theta_2 - \frac{\theta_1}{\nu_1}\right)}{-p^2 + p\left(\kappa\left(\frac{\theta_2+1}{\theta_1}\right) - \theta_1\right) + \kappa(1+\theta_2)}.$$

We get equality (*) by substituting $\nu_2 = \nu_1\frac{\mu_2}{\mu_1} = \nu_1\frac{\theta_2}{\theta_1}$, and equality (**) by plugging $\widetilde{P}$, $\widehat{P}$ and $\check{P}$. The socially optimal probability $0 \le p^* \le 1$ maximizes $U(p)$, i.e. $p^* = \arg\max_{0\le p\le 1}\{U(p)\} = \arg\max_{0\le p\le 1}\left\{\frac{U(p)}{R}\right\}$.

7   Comparison of Equilibrium and Optimal Joining Strategies

We look at p^* and p_e as functions of ν_1, while we fix θ_1 and θ_2. We define an intersection point as the point where the sign of $(p^* - p_e)$ changes from positive to negative.

Lemma 2. $p^* = 0 \implies p_e = 0$. *If an intersection exists, it occurs for $\nu_1 > 1 + \theta_2 - \theta_1$.*

Proof. p^* is defined as argmax of the function $U(p)$, therefore if $p^* = 0$ it follows that $\frac{\partial U(p)}{\partial p}(0) < 0$, i.e. $\frac{(2\theta_2+1)\nu_1 - \theta_2(\theta_2-\theta_1+2)-1}{(\theta_2+1)^2\nu_1} < 0$. After rearranging we get $\nu_1 < \frac{\theta_2(\theta_2-\theta_1+2)+1}{2\theta_2+1}$.

From Theorem 1, $\nu_1 < 1 + \theta_2 - \theta_1(= A) \implies p_e = 0$.

To prove the lemma, we show that $\frac{\theta_2(\theta_2-\theta_1+1)+1}{2\theta_2+1} < 1 + \theta_2 - \theta_1$:

$\frac{\theta_2(\theta_2-\theta_1+1)+1}{2\theta_2+1} < 1 + \theta_2 - \theta_1 \iff$

$(\theta_2)^2 - \theta_1\theta_2 + 2\theta_2 + 1 < 2\theta_2 + 1 + 2(\theta_2)^2 + \theta_2 - 2\theta_1\theta_2 - \theta_1 \iff$

$0 < (\theta_2)^2 + \theta_2 - \theta_1\theta_2 - \theta_1 = \theta_2(\theta_2 - \theta_1) + (\theta_2 - \theta_1)$. The last inequality holds since $\theta_2 > \theta_1$. $\square$

Theorem 2. *p_e and p^* intersect at most once in the domain where $0 < p_e$, $p^* < 1$.*

Such an intersection exists iff $\theta_1 < \sqrt{1 + \theta_2 + \theta_2^2} - 1$.

Proof. According to Lemma 2 $p^* > 0$ if $p_e > 0$, and according to Theorem 1 $p_e > 0$ iff $\nu_1 > 1 + \theta_2 - \theta_1$. Therefore we limit our search or an intersection point to $\nu_1 > 1 + \theta_2 - \theta_1$. We can also limit the search for an intersection by discarding the region where at least one of the functions is 1. In the domain where $0 < p_e, p^* < 1$ the functions are:

- $p_e(\nu_1) = \frac{\kappa(\nu_1+\theta_1-\theta_2-1)}{\nu_1(\theta_1+\theta_2)-2\theta_1}$, from expression (2).
- $p^*(\nu_1) = \frac{-\nu_1\kappa\theta_1(\theta_2^2+\theta_1\theta_2+\theta_1)+\kappa\theta_1(2\theta_1\theta_2+\theta_1)}{\nu_1\left((\kappa+\theta_1)\theta_2^2-\theta_1^3\right) - \left((\kappa+\theta_1)\theta_1\theta_2-\theta_1^3\right)}$

$\quad + \frac{\sqrt{\theta_1\theta_2\kappa^3\left[\nu_1^2\theta_2(\tau+\theta_1\theta_2) - \nu_1\left(\theta_2(\tau+\theta_2^2)+\theta_1(\tau+\theta_1\theta_2)\right)+\theta_1(\tau+\theta_2^2)\right]}}{\nu_1\left((\kappa+\theta_1)\theta_2^2-\theta_1^3\right) - \left((\kappa+\theta_1)\theta_1\theta_2-\theta_1^3\right)}$,

where $\tau = 1 + \theta_1 + 2\theta_2 + \theta_1^2$.

Explanation: $\left\{ p \mid \dfrac{dU(p)}{dp} = 0 \right\} = \dfrac{-\nu_1\kappa\theta_1\left(\theta_2^2+\theta_1\theta_2+\theta_1\right)+\kappa\theta_1(2\theta_1\theta_2+\theta_1)}{\nu_1\left((\kappa+\theta_1)\theta_2^2-\theta_1^3\right)-\left((\kappa+\theta_1)\theta_1\theta_2-\theta_1^3\right)}$

$\pm\dfrac{\sqrt{\theta_1\theta_2\kappa^3\left[\nu_1^2\theta_2(\tau+\theta_1\theta_2)-\nu_1\left(\theta_2(\tau+\theta_2^2)+\theta_1(\tau+\theta_1\theta_2)\right)+\theta_1(\tau+\theta_2^2)\right]}}{\nu_1\left((\kappa+\theta_1)\theta_2^2-\theta_1^3\right)-\left((\kappa+\theta_1)\theta_1\theta_2-\theta_1^3\right)}.$

Notice that the root corresponding to the minus sign is always negative, since the denominator is always positive and the numerator is always negative. Thus the only feasible option is the root corresponding to the plus sign.

It should be mentioned that the expression above for $p^*(\nu_1)$ is not always valid.

We need to make sure that the discriminant is non-negative:

$$\theta_1\theta_2\kappa^3\left[\nu_1^2\theta_2\left(\tau+\theta_1\theta_2\right)-\nu_1\left(\theta_2\left(\tau+\theta_2^2\right)+\theta_1\left(\tau+\theta_1\theta_2\right)\right)+\theta_1\left(\tau+\theta_2^2\right)\right]\geq 0.$$

We solve this inequality by finding the roots of the quadratic equation. The solutions are $\nu_1 \leq \frac{\theta_1}{\theta_2}$ and $\nu_1 \geq \frac{\tau+\theta_2^2}{\tau+\theta_1\theta_2}$. Since ν_1 must be greater than 1, and $\frac{\theta_1}{\theta_2} < 1$, the only feasibly solution is

$$\nu_1 \geq \frac{\tau+\theta_2^2}{\tau+\theta_1\theta_2} = \frac{1+\theta_1+2\theta_2+\theta_1^2+\theta_2^2}{1+\theta_1+2\theta_2+\theta_1^2+\theta_1\theta_2}.$$

In addition to this condition on $\nu_1, \theta_1, \theta_2$, we must demand that the expression for $p^*(\nu_1)$ satisfies the following inequalities (Otherwise, we get $p^*(\nu_1)=0$ or $p^*(\nu_1)=1$):

1. $0 \leq p^*(\nu_1) \leq 1$.
2. $U(p^*(\nu_1)) \geq U(0)$ and $U(p^*(\nu_1)) \geq U(1)$.

There are three roots to the expression $\{p_e(\nu_1)-p^*(\nu_1)\}$:

1. $\nu_{1,1} = \frac{\theta_1}{\theta_2}$.

2. $\nu_{1,2} = \dfrac{(\theta_1+2)\theta_2^2+(2-\theta_1)\theta_2-\theta_1^3-\theta_1^2-\left[(\theta_2-\theta_1)\sqrt{\theta_1\kappa((\theta_1+4)\theta_2+\theta_1(\theta_1+1))}\right]}{2\theta_2}$.

3. $\nu_{1,3} = \dfrac{(\theta_1+2)\theta_2^2+(2-\theta_1)\theta_2-\theta_1^3-\theta_1^2+\left[(\theta_2-\theta_1)\sqrt{\theta_1\kappa((\theta_1+4)\theta_2+\theta_1(\theta_1+1))}\right]}{2\theta_2}$.

Notice that $\nu_{1,1}$, $\nu_{1,2}$, $\nu_{1,3}$ are solutions of the equation $(\theta_1 - \nu_1\theta_2)(a\nu_1^2 + b\nu_1 + c) = 0$, where $a = \theta_2$, $b = \theta_1^2(\theta_1+1) - \theta_2(\theta_1^2 + 3\theta_1 + 2)$,

and $c = \frac{(\theta_1^2(\theta_1+1)-\theta_2(\theta_1^2+3\theta_1+2))^2 - (\theta_1-\theta_2)^2\theta_1(\theta_1+\theta_2+1)(\theta_2^2+\kappa\theta_1)}{4\theta_2}$.

$\nu_{1,1}$ and $\nu_{1,2}$ are smaller than $1+\theta_2-\theta_1$, so we ignore them. $\nu_{1,3}$ is larger than $1+\theta_2-\theta_1$. Thus $\nu_{1,3}$ is the only candidate for a point of intersection. In order for p_e and p^* to intersect at $\nu_{1,3}$, we need to make sure that $0 < p_e(\nu_{1,3}) = p^*(\nu_{1,3}) < 1$.
We examine this with $p_e(\nu_{1,3})$:

$$p_e(\nu_{1,3}) = \dfrac{\kappa\left(\sqrt{\theta_1\kappa\left((\theta_1+4)\theta_2+\theta_1\left(\theta_1+1\right)\right)}-\theta_1\left(\kappa-2\right)\right)}{2\left((\theta_1+\theta_2)^2+\theta_2\right)}.$$

The inequality $p_e(\nu_{1,3}) > 0$ always holds, since we assume that $0 < \theta_1 < \theta_2$.
The inequality $p_e(\nu_{1,3}) < 1$ holds iff $0 < \theta_1 < \sqrt{1+\theta_2+\theta_2^2}-1$. $\square$

We now show that if the inequality from Theorem 2 holds then the model's assumption ($\theta_1 < \theta_2$) also holds:

$\sqrt{1 + \theta_2 + \theta_2^2} - 1 < \theta_2 \Longleftrightarrow$

$1 + \theta_2 + \theta_2^2 < (\theta_2 + 1)^2 \Longleftrightarrow$

$1 + \theta_2 + \theta_2^2 < 1 + 2\theta_2 + \theta_2^2$, and this is always true, since $\theta_1 > 0$.

We conclude the the model's assumption does in fact hold.

We notice that when $\nu_1 \to \infty$, social optimality is equivalent to throughput maximization.

Recall from Sect. 4,

$$F(\theta_2) := \frac{-(1 + 4\theta_2 + \theta_2^2) + \sqrt{(1 + 4\theta_2 + \theta_2^2)^2 + 8(\theta_2^3 + \theta_2^4)}}{4\theta_2},$$

and define

$$G(\theta_2) := \sqrt{1 + \theta_2 + \theta_2^2} - 1.$$

See Fig. 5.

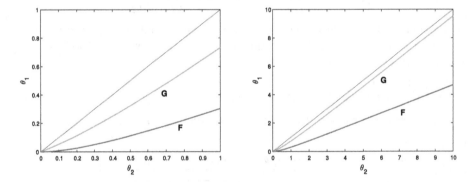

Fig. 5. The functions F and G.

Therefore from Corollary 2 we directly obtain:

- $\lim_{\nu_1 \to \infty} p^*(\nu_1) = 1 \Longleftrightarrow \theta_1 \geq F(\theta_2)$.
- $\lim_{\nu_1 \to \infty} p^*(\nu_1) < 1 \Longleftrightarrow \theta_1 < F(\theta_2)$.

Returning to the condition for intersection in Theorem 2 gives:

- $p^*(\nu_1)$ and $p_e(\nu_1)$ intersect $\Longleftrightarrow 0 < \theta_1 < G(\theta_2)$.
- $p^*(\nu_1)$ and $p_e(\nu_1)$ do not intersect $\Longleftrightarrow G(\theta_2) < \theta_1$.

The following properties hold:

1. $F(\theta_2) < G(\theta_2)$, $\forall \theta_2 > 0$. Therefore, if $\lim_{\nu_1 \to \infty} p^*(\nu_1) < 1$, then $p^*(\nu_1)$ and $p_e(\nu_1)$ intersect.
2. $\lim_{\theta_2 \to \infty} F'(\theta_2) = \frac{1}{2}$.

3. $\lim_{\theta_2 \to 0} F'(\theta_2) = 0$.
4. $\lim_{\theta_2 \to \infty} G'(\theta_2) = 1$.
5. $\lim_{\theta_2 \to 0} G'(\theta_2) = \frac{1}{2}$.

We now consider three possible relations between θ_1 and θ_2:

1. $\theta_1 < F(\theta_2)$, i.e. $\lim_{\nu_1 \to \infty} p^*(\nu_1) < 1$ and $p_e(\nu_1)$ intersects with $p^*(\nu_1)$.
2. $F(\theta_2) \le \theta_1 < G(\theta_2)$, i.e. $\lim_{\nu_1 \to \infty} p^*(\nu_1) = 1$ and $p_e(\nu_1)$ intersects with $p^*(\nu_1)$.
3. $\theta_1 \ge G(\theta_2)$, i.e. $\lim_{\nu_1 \to \infty} p^*(\nu_1) = 1$ and $p_e(\nu_1)$ does not intersect with $p^*(\nu_1)$.

It should be mentioned that in the second case, if $\theta_1 = F(\theta_2)$ then $p^*(\nu_1)$ will asymptotically approach 1 as ν_1 goes to infinity, while if $\theta_1 > F(\theta_2)$ then $p^*(\nu_1)$ will reach 1 at a finite value of ν_1.

Examples of these three cases:

1. In Fig. 6, $\theta_1 = 0.1$, $\theta_2 = 2$.
$$0.1 < \frac{-(1+4\cdot 2+2^2)+\sqrt{(1+4\cdot 2+2^2)^2+8(2^3+2^4)}}{4\cdot 2} \approx 0.75.$$
2. In Fig. 7, $\theta_1 = 0.13$, $\theta_2 = 0.5$.
$$0.11 \approx \frac{-(1+4\cdot 0.5+0.5^2)+\sqrt{(1+4\cdot 0.5+0.5^2)^2+8(0.5^3+0.5^4)}}{4\cdot 0.5} \quad < \quad 0.13 \quad <$$
$$\sqrt{1+0.5+0.5^2}-1 \approx 0.32.$$
3. In Fig. 8, $\theta_1 = 1$, $\theta_2 = 1.1$.
$$1 > \sqrt{1+1.1+1.1^2}-1 \approx 0.82.$$

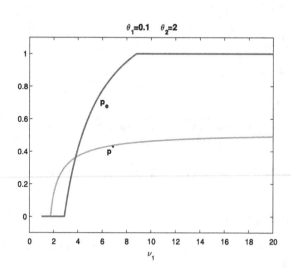

Fig. 6. $p^* < p_e$ for $\nu_1 > \nu_{1,3}$, and $\lim_{\nu_1 \to \infty} p^*(\nu_1) < 1$.

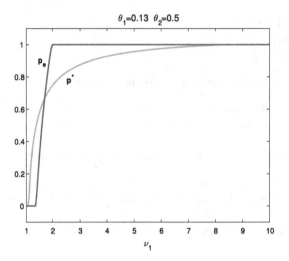

Fig. 7. $p^* \leq p_e$ for $\nu_1 > \nu_{1,3}$, and $\lim_{\nu_1 \to \infty} p^*(\nu_1) = 1$.

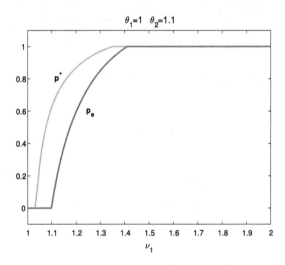

Fig. 8. $p^* \geq p_e$ for all $\nu_1 \geq 1$, and $\lim_{\nu_1 \to \infty} p^*(\nu_1) = 1$.

7.1   Economic Explanation

Increasing the probability one enters Q_1 influences the system in both positive and negative ways:

– Positive: It increases the likelihood that others will be served by the faster server.
  This has greater effect for large values of C (relative to R).

– Negative: It increases the overall congestion of the system, since one is more likely to be served by the slower server, and thus others are more likely to leave the system unserved.
  This has greater effect for large values of R (relative to C).

When $\frac{R}{C}$ is small the positive externality is larger than the negative externality. Customers prefer that others will choose Q_1 with greater probability (a large value of p), in order to have Q_2 free for themselves. This is because in this case a short service time is more important to customers than the guarantee of service. However, for high values of R (or small values of C) the opposite is also possible.

8 Avoid the Crowd Vs. Follow the Crowd

In order to examine the question whether customers' best response is to avoid or follow the crowd we observe $\widehat{P}$, i.e. the probability that Q_2 is idle given that Q_1 is idle.

If $\widehat{P}$ increases in p, then customers would tend to avoid the crowd. When $\widehat{P}$ is increasing, it becomes more likely that Q_2 is idle, so the customer would avoid the crowd and use lower values of p.

If $\widehat{P}$ decreases in p, then customers would tend to follow the crowd. When $\widehat{P}$ is decreasing, it becomes more likely that Q_2 is busy, so the customer would follow the crowd and use higher values of p. Recall:

$$\widehat{P} = \frac{\theta_2 \kappa + \theta_2 p}{(1 + \theta_2)\kappa - \theta_1 p}.$$

The derivative of $\widehat{P}$ is:

$$\frac{d\widehat{P}}{dp} = \frac{\theta_2 \kappa^2}{\left((1 + \theta_2)\kappa - \theta_1 p\right)^2}.$$

This expression is positive since both the numerator and denominator are positive.

Hence we conclude that $\widehat{P}$ is monotonically increasing, and customers avoid the crowd. This explains the uniqueness of the equilibrium probability p_e.

9 Concluding Remarks

We managed to describe and analyze the Nash equilibrium and socially-optimal strategies of the system, finding closed formulas, and showing that it is beneficial for customers to avoid the crowd, and therefore the equilibrium is unique.

When describing these strategies as a function of the normalized reward, ν_1, we find that the two functions sometimes intersect, depending on the rates of service. Another property is that in certain cases, the socially-optimal strategy asymptotically converges to a value strictly smaller than 1. We demonstrate that the throughput-maximizing strategy does not necessarily dictate entering the observable queue with probability of 1.

For further research, we propose the following generalizations of the model:

- Adding more queues to the system:
  In this case, the queues need to be in a sequence $Q_1, Q_2, ..., Q_n$. Upon arrival to the system, customers can observe only Q_1. When customers arrive to queue Q_i $1 \leq i \leq n-1$, they choose their action according to the history they observed, i.e. the status in queues $Q_1, ..., Q_{i-1}$ (if $i = 1$, the history is empty). Of course, if Q_i is busy, the customer will automatically move to Q_{i+1}, with probability 1. If customers arrive to the last queue, Q_n, there is no decision for them to make, since they will always prefer getting service rather than leaving the system unserved.
  Notice that here the service rates do not have to be monotonically increasing.
- Adding waiting spots, finite or infinite, making the queues $\cdot/M/1/s$, or $\cdot/M/1$.

References

1. Altman, E., Jiménez, T., Núñez-Queija, R., Yechiali, U.: Optimal routing among $\cdot/M/1$ queues with partial information. Stoch. Models **20**(2), 149–171 (2004)
2. Hassin, R.: On the advantage of being the first server. Manage. Sci. **42**(4), 618–623 (1996)
3. Hassin, R.: Rational Queueing. Chapman & Hall, CRC Press, Boca Raton (2006)
4. Hassin, R., Haviv, M.: To Queue or Not to Queue: Equilibrium Behavior in Queueing Systems. Kluwer Academic Publishers, Norwell (2003)
5. Hassin, R., Roet-Green, R.: Cascade equilibrium strategies in a two-server queueing system with inspection cost. Eur. J. Oper. Res. **267**, 1014–1026 (2018)
6. Singh, S.P., Delasay, M., Scheller-Wolf, A.: Evaluating the First-Mover's Advantage in Announcing Real-Time Delay Information. Available at SSRN-id3081583 (2019)

Strategic Joining in a Single-Server Retrial Queue with Batch Service

Ke Sun[1], Jinting Wang[1(✉)], and George Zhang[2,3]

[1] Department of Mathematics, Beijing Jiaotong University, Beijing 100044, China
{12271044,jtwang}@bjtu.edu.cn
[2] Department of Decision Sciences, Western Washington University,
Bellingham, WA 98225, USA
[3] Beedie School of Business, Simon Fraser University, Burnaby, BC V5A 1S6, Canada

Abstract. We consider a single-server retrial queue with batch service where potential customers arrive according to a Poisson process. The service is divided into two periods: busy period and admission period, and they are corresponding to whether the server is in service or not respectively. These two periods constitute an alternate renewal process. When arrivals find busy period, they make join-or-balk decisions and the joining ones will stay at the orbit and try to get into server at constant rate; when arrivals find admission period, they get into the server directly. At the end of each admission period, all customers in the server will be served together regardless of the size of the batch. Therefore, we give the assumption of service reward that vary with the service size. Furthermore, customers in the orbit fail to get into the server before the end of each service cycle were forced to leave the system. We identify the (Nash) equilibrium joining strategies and the social- and profit- maximization problems of arrivals by assuming that they are informed about the service period upon arrivals. Finally, the optimal joining strategies are showed by numerical examples.

Keywords: Retrial queue · Batch service · Strategic customers · Quality-capacity tradeoff · Busy period · Admission period

1 Introduction

In most of queueing literature, it is usually assumed that customers observe unavailable server upon arrivals will abandon the system or wait in line for service. However, in practical applications, blocked customers may leave the service area temporarily but repeat their service requests after a random period of time. These so-called retrial queues have been widely studied in recent decades to model various problems, such as telecommunication networks, cognitive radio networks and telephone switching systems, see Falin and Templeton (1997) and Artalejo and Gómez-Corral (2008). In the majority of these studies, the joining customers stay at a virtual space (orbit) and retry to occupy the server at a

© Springer Nature Switzerland AG 2019
T. Phung-Duc et al. (Eds.): QTNA 2019, LNCS 11688, pp. 183–198, 2019.
https://doi.org/10.1007/978-3-030-27181-7_12

random rate continuously. In recent years, there are many researchers focus on queueing systems with retrial policy. Interested readers may refer to Artalejo (1995), Artalejo and Gómez-Corral (1997) and Falin (2008) for more details. Unlike the classical queueing models with FCFS (first-come first-served) discipline, customers' join-or-balk decisions in retrial queues depend not only on the current state but also the behavior of future arrivals. Recently, Wang and Zhang (2011), Wang and Zhang (2013) have studied customers' strategic behavior in retrial queueing systems.

The service capacity has always been the focus of queueing studies, while the vast majority of them, however, take the single service discipline. In fact, batch service occur quite frequently in practice. For example, in public transportation system like shuttle, it begins only when a certain number of passengers or a period of time have been accumulated. In cloud computing service center, the server handles multiple data at the same time. Studies on batch service include the monograph of Kleinrock (1976), in which the quantitative batch service queue ($M/M^r/1$) was explored by solving the corresponding performance measures. Early researches of batch service queues focused more on performance evaluation and optimal control, see, for example, Bailey (1954), Chaudhry and Templeton (1983), Downton (1955), Deb and Serfozo (1973), Medhi (1975), among others. In addition, performance analysis in an $MAP/PH/1$ queue with flexible group service were carried out in Brugno (2017), Brugno et al. (2017, 2018) recently.

Initiated by Naor (1969) with a single-server observable queue scenario, game-theoretic analysis of queueing systems has received great attention during the past decades. There has been a growing number of studies from the viewpoint of economics, see Edelson and Hilderbrand (1975), Economou and Kanta (2011), Guo and Hassin (2011), and among others. The monographs Hassin and Haviv (2003) and Hassin (2016) summarized the main models and methodologies in a variety of queueing systems by using game analysis. However, there exists only a few game-theoretic research of batch service queues. Bountali and Economou (2017a, 2017b) explored the strategic behavior of customers in a single-server queue with a fixed service scale K in different information cases.

In retrial queues, server remains idle in a random period of time after each service completion until a customer in the orbit retries or a primary customer arrives. Hence, the server always be wasted unreasonably. Therefore, Manou et al. (2014) and Dudin et al. (2015) proposed that server can provide service in groups for improving throughput. Besides, they considered the admission strategy, and the service process is divided into busy period and admission period. At the end of busy period, the admission period starts and the external arrivals and customers in the orbit who make attempts can be served together at the end of this period. While in Manou et al. (2014), the authors modeled a transportation station based on batch service and admission strategy in a retrial queue. Besides, the station is left empty after each departure of facility, that is, customers in the orbit failed get into server will be cleared when the admission period ends. It was also obvious that, the shortcoming of admission strategy is that customers do not receive instant service upon arrivals.

Inspired by the aforementioned studies, we consider a Markovian single-server retrial queue with infinite service capacity and admission strategy. By assuming that each busy and admission period with exponential duration time, we find that the service process constitutes an alternate renewal process. Customers make join-or-balk decisions upon arrivals based on linear reward-cost structure from an economic viewpoint. We give a specific example of the service process of shuttle in the starting station, where the admission and busy periods are corresponding to whether the facility stay in the station or not, respectively. Moreover, passengers find the busy period would like to stroll at the nearby shopping mall and return after a random period of time. The shuttle departs with all customers in the server at the end of admission period and a new busy period start at the same time.

In our model, we adopt a nonconstant service reward scenario, related studies see Anand et al. (2011), Li et al. (2016), Li et al. (2017), Xu et al. (2015), Xu et al. (2016) and the reward is decreasing with service congestion in our paper here. Therefore, the service provider faces a quality-capacity tradeoff: reducing the congestion in a service cycle increases individual's service reward, but decreases the customers' possibility of getting into server. To summarize, this paper is the first work that carries out the game-theoretic analysis of a batch service retrial queue with two-stage service process. Besides, the nonconstant service reward function is adopted, which is inversely proportional to the congestion level of each service. Sensitivity analysis of optimal joining strategies is presented from different perspectives.

The remainder of this paper is organized as follows. Section 2 gives our assumptions and preliminary results. In Sect. 3, we analyse the steady-state performance measures of the system. The corresponding equilibrium, social- and profit- maximizing joining strategies are derived in Sect. 4. In Sect. 5, we consider linear-reward function as a special case. Section 6 carries out sensitively analysis of the customers' behavior and social- and profit- maximization strategies through different numerical scenarios, then it followed by the conclusions in Sect. 7.

2   Model Description

We consider a single-server retrial queue with batch service, where potential customers arrive in a Poisson process with rate λ. When the server is busy, the service time is exponentially distributed with parameter μ, regardless of the size of a batch. There are two periods in each service process, busy period and admission period. In the busy period, each arrival decides whether to join the orbit or not based on a reward-cost structure. It consists of the service reward minus waiting cost. If he enters the orbit, then he retries to get into server with constant rate θ, and he will success when meet admission period. At the end of the busy period, the admission period starts and it is exponentially distributed with parameter γ. During this period, all external arrivals and customers in the orbit who retry can get into the server. At the end of the admission period, all customers in the server can be served together and customers in the orbit who fail to get into server were forced to leave the system at the same time. Finally, we assume that the inter-arrival times, service times, admission times and retrial times are mutually independent.

Each customer receives a nonconstant reward after completing service, but receives no reward if he balks or failed to retry. We denote the reward function as $R(N)$ (>0), where N is the average number of customers be served in each service cycle. Also, customers stay at the orbit accumulate waiting cost at rate C per time unit and there is no waiting cost for these who get into the server. Besides, every customer is risk neutral and aims to maximize his own expected benefit, and their join-or-balk decisions are irrevocable, i.e., reneging after joining and re-enter after balking are forbidden.

We define the service period by $I(t)$ at time t, where $I(t) = 0$ and 1 (0: busy period, 1: admission period). Note that the two periods are exponentially distributed with parameter μ and γ, respectively. Denote B_k and A_k as the busy period and admission period of the kth service cycle T_k respectively, where B_k and A_k ($k = 1, 2, \cdots$) constitute an alternate renewal process. Moreover, we denote $N_{or}(t)$ and $N_{se}(t)$ as the number of customers in the orbit and in the server at time t, respectively. It is clear that the stochastic process $\{I(t), N_{or}(t), N_{se}(t), t \geq 0\}$ is a three-dimensional continuous time Markov chain with state space $\Omega = \{1\} \times \{0, 1, \cdots\} \times \{0, 1, \cdots\} \cup \{0\} \times \{0, 1, \cdots\} \times \{0\}$.

Assumption 1. The service reward $R(N)$ is a continuous function with N, and it is decreasing and concave in the average number of customers be served in a service cycle N, i.e., $\frac{dR(N)}{dN} < 0$ and $\frac{dR^2(N)}{dN^2} \leq 0$.

This assumption satisfies the characteristics of the quality-capacity service. The inequality $\frac{dR(N)}{dN} < 0$ implies that the service reward decreasing with the service congestion. Regarding $\frac{dR^2(N)}{dN^2} \leq 0$, the reward function $R(N)$ is concave in the number of customers be served in a service cycle. This makes sense: when N is small, the quality of service is less affected by increasing one unit, however, when the number of customers of a batch is relatively large, due to the unlimited size of a batch in our paper, the marginal effect of adding one customer shall be increasing. Therefore, the marginal utility due to the increase of a customer becomes more evident. In the literature, Li et al. (2016) adopted a linear reward function which satisfies $R'(\mu_i) < 0$ and $R''(\mu_i) = 0$, where they focus on the quality-speed service. As an extension, a decreasing and concave reward function $R(\mu_i)$ ($R''(\mu_i) < 0$) was considered in Li et al. (2017).

In what follows, we assume that all arrivals can not observe the number of customers in the system but the service period. Recall that external arrival definitely join the system due to zero waiting cost and a positive reward. Thus we concentrate on the these who find busy period upon arrivals.

Remark 1. *As we have known, the two periods constitute an alternate renewal process in a service cycle, and they are exponentially distributed with independent parameters respectively. According to the knowledge in Stochastic Process, the* *stationary probabilities of the two periods in each service cycle only depend on parameters μ and γ, i.e.,*

$$p_0 = \frac{\gamma}{\mu + \gamma}, \quad p_1 = \frac{\mu}{\mu + \gamma}. \tag{1}$$

3   Steady-State Analysis

In this Section, we study steady-state results of the system and assume that arrivals find busy period join the system with probability q. Let $p_{k,i,j}$ represents the stationary probability of state (k, i, j). Besides, we define the corresponding partial generating functions as

$$p_1(x, y) = \sum_{i=0}^{\infty} \sum_{j=0}^{\infty} p_{1,i,j} x^i y^j, \quad p_0(x) = \sum_{i=0}^{\infty} p_{0,i,0} x^i. \tag{2}$$

Then we have the following results.

Lemma 1. *In the single-server retrial queue with batch service, arrivals find busy period join with probability q, then the average number of customers in the orbit in the two periods and the mean number of customers in the server at the admission period are given as*

$$N_{0,or} = \frac{\lambda q}{\mu}, \tag{3}$$

$$N_{1,or} = \frac{\lambda q \gamma}{\mu(\theta + \gamma)}, \tag{4}$$

$$N_{se} = \frac{\lambda \mu(\theta + \gamma) + \lambda q \gamma \theta}{\mu \gamma(\theta + \gamma)}. \tag{5}$$

Proof. The balance equations for the stationary distribution of the Markov chain $\{I(t), N_{or}(t), N_{se}(t), t \geq 0\}$ are given as follows:

$$(\lambda q + \mu) p_{0,0,0} = \gamma \sum_{i=0}^{\infty} \sum_{j=0}^{\infty} p_{1,i,j}, \tag{6}$$

$$(\lambda q + \mu) p_{0,i,0} = \lambda q p_{0,i-1,0}, \quad i = 1, 2, 3, \cdots, \tag{7}$$

$$(\lambda + \gamma + i\theta) p_{1,i,0} = \mu p_{0,i,0}, \quad i = 0, 1, 2, \cdots, \tag{8}$$

$$(\lambda + \gamma + i\theta) p_{1,i,j} = \lambda p_{1,i,j-1} + (i+1)\theta p_{1,i+1,j-1}, \quad i = 0, 1, 2, \cdots, j = 1, 2, 3, \cdots. \tag{9}$$

The stationary probability of the two periods can be defined as:

$$p_0 = \sum_{i=0}^{\infty} p_{0,i,0}, \quad p_1 = \sum_{i=0}^{\infty} \sum_{j=0}^{\infty} p_{1,i,j}. \tag{10}$$

Multiplying both sides of (8) by x^i and summing over all i, respectively. We obtain

$$\sum_{i=0}^{\infty}(\lambda + \gamma + i\theta)p_{1,i,0}x^i = \mu p_0(x). \tag{11}$$

Multiplying both sides of (9) by $x^i y^j$ and summing over all i, j, respectively. With the help of (11), we can easily have the following after some simple algebraic operations

$$(\lambda(1-y)+\gamma)\,p_1(x,y) + \theta(x-y)\frac{\partial p_1(x,y)}{\partial x} = \mu p_0(x). \tag{12}$$

By using (6), (7) and (1), we have

$$p_{0,i,0} = \frac{\gamma p_1}{\lambda q + \mu}\left(\frac{\lambda q}{\lambda q + \mu}\right)^i = \frac{\mu \gamma}{(\mu+\gamma)(\lambda q + \mu)}\left(\frac{\lambda q}{\lambda q + \mu}\right)^i, \quad i = 0,1,2,\cdots, \tag{13}$$

then, in the busy period, the mean number of customers in the orbit is

$$N_{0,or} = \sum_{i=1}^{\infty}\frac{i p_{0,i,0}}{p_0} = \sum_{i=1}^{\infty}i\left(\frac{\lambda q}{\lambda q + \mu}\right)^i\frac{\mu}{\lambda q + \mu} = \frac{\lambda q}{\mu}. \tag{14}$$

In the admission period, there are customers both in the orbit and the server. We denote the corresponding average number of customers as $N_{1,or}$ and N_{se}, and take the derivative of the Eq. (12) with respect to x, y, respectively. Then, by setting $x = 1, y = 1$, we have the following equations:

$$(\theta + \gamma)p_1 N_{1,or}^{ob} = \mu p_0 N_{0,or}, \tag{15}$$

$$\gamma p_1 N_{se} = \lambda p_1 + \theta p_1 N_{1,or}, \tag{16}$$

from which we obtain

$$N_{1,or} = \frac{\lambda q \gamma}{\mu(\theta + \gamma)}, \tag{17}$$

$$N_{se} = \frac{1}{\gamma}(\lambda p_1 + \theta N_{1,or}) = \frac{\lambda \mu(\theta + \gamma) + \lambda q \gamma \theta}{\mu \gamma(\theta + \gamma)}. \tag{18}$$

This completes the proof. □

Remark 2. *Obviously, in the steady state, the expected number of customers be served in a service cycle is exactly the the average number of customers in the server at the admission period, so we immediately have $N = N_{se} = \frac{\lambda \mu(\theta+\gamma)+\lambda q\gamma\theta}{\mu\gamma(\theta+\gamma)}$. This varies with customers' joining probability q and we denote it as $N(q)$.*

Remark 3. *Since $\frac{dN(q)}{dq} = \frac{\lambda\theta}{\mu(\theta+\gamma)} > 0$ is a constant here, combining with Assumption 1, the variation tendency of the reward function $R(N)$ on q is completely the same as it on N. That is to say, $\frac{dR(N(q))}{dq} < 0$ and $\frac{dR^2(N(q))}{dq^2} \leq 0$ are also established.*

We then consider the mean sojourn time in the orbit and the probability that customers in the orbit getting into server in a service cycle, and we denote them as T and p_s, respectively.

Proposition 1. *In the single-server retrial queue with batch service, the mean sojourn time before getting into server for arrivals in the orbit and the probability they get service in a service cycle are as follows:*

$$T = \frac{\mu + \theta + \gamma}{\mu(\theta + \gamma)}, \tag{19}$$

$$p_s = \frac{\theta}{\theta + \gamma}. \tag{20}$$

Proof. By considering arrivals find busy period, their sojourn time in the orbit T can be divided into two parts in a service cycle: T_0 and T_1, which represent the time they stay at the busy period and admission period, respectively. Where $T_0 = \frac{1}{\mu}$ due to the memoryless property of exponential distribution on the busy period. Besides, customers in the admission period leave the orbit either the admission period is over or they have retrials in this period. For the admission period is over, they have to wait for an exponentially distributed time with parameter γ; for having retrials, they have to wait for an exponentially distributed time with parameter θ. Therefore, T_1 is the time until the occurrence of one of the two events, that is to say, $T_1 = \frac{1}{\theta + \gamma}$, then we can get (19) easily. In addition, the probability that customers in the orbit get into the server is $P(X < Y)$, where X follows an exponential distribution with parameter θ and Y is an exponentially distributed random variable with rate γ. Accordingly, we obtain that $p_s = \frac{\theta}{\theta + \gamma}$.   □

4   Optimal Joining Strategies

We then explore the equilibrium joining strategies of arrivals who find busy period and the socially optimal strategy from the viewpoint of social planner. Besides, the revenue-maximizing problem by imposed an admission fee is considered. The results are summarized in the following.

Theorem 1. *In the single-server retrial queue with batch service, a unique Nash equilibrium joining strategy 'join the orbit with probability q_e whenever finding the busy period' exists, and is given as*

$$q_e = \begin{cases} 0 & \text{if } \frac{CT}{p_s} > R(N(0)), \\ \frac{\mu\gamma(\theta+\gamma)R^{-1}\left(\frac{CT}{p_s}\right) - \lambda\mu(\theta+\gamma)}{\lambda\theta\gamma} & \text{if } R(N(1)) \le \frac{CT}{p_s} \le R(N(0)), \\ 1 & \text{if } \frac{CT}{p_s} < R(N(1)), \end{cases} \tag{21}$$

where $R^{-1}(\cdot)$ is the inverse of function $R(N)$.

Proof. Considering a tagged customer who join the orbit when find busy period and the other blocked customers follow the given joining strategy q. So the expected net benefit of him is

$$S(q) = p_s R(N(q)) - CT. \tag{22}$$

Since we have $\frac{dR(N(q))}{dq} < 0$, then the benefit $S(q)$ is strictly deceasing with q in $[0, 1]$ and has a unique maximum

$$S(0) = p_s R(N(0)) - CT, \tag{23}$$

and a unique minimum

$$S(1) = p_s R(N(1)) - CT. \tag{24}$$

Then we have to consider the following three cases:

- When $\frac{CT}{p_s} > R(N(0))$, $S(q)$ is negative for every q, then the best response is balking and the unique Nash equilibrium strategy is $q_e = 0$;
- when $R(N(1)) \leq \frac{CT}{p_s} \leq R(N(0))$, there exists a unique solution of the equation $S(q) = 0$ which lies in the interval $[0, 1]$, and it is exactly the optimal strategy;
- when $\frac{CT}{p_s} < R(N(1))$, $S(q)$ is positive for every q, in other words, the best response is joining and $q_e = 1$.

The results of the three cases are corresponding to the three parts in (21). $\square$

Since $S(q)$ is a decreasing function of q, whenever the joining probability $q < q_e$, the expected benefit of a joining customer at the busy period is positive, thus the optimal response is to join the system. Similarly, when $q > q_e$, the unique best response is balking. Moreover, every strategy is the best response when $q = q_e$. Therefore, we can find that the optimal strategy of a customer is a decreasing function of the strategy adopted by the other customers. This correspond to the "avoid the crowd" (ATC) situation. What's more, $R(N(0)) \geq 0$ is required for ruling out the situation that no one will join the system.

We can now proceed to the problem of social optimization which aim to maximize the sum of the benefit of all customers per time unit. The results can be summarized in the following theorem.

Theorem 2. *In the single-server retrial queue with batch service, a unique mixed strategy 'join the orbit with probability q_{soc} whenever finding the busy period' that maximizes the social benefits per time unit exists and is given as*

$$q_{soc} = \begin{cases} 0 & if \quad q_{max} < 0, \\ q_{max} & if \quad 0 \leq q_{max} \leq 1, \\ 1 & if \quad q_{max} > 1. \end{cases} \tag{25}$$

where q_{max} is the unique solution of the equation

$$p_0 p_s R(N(q)) + (p_0 p_s q + p_1) \times \frac{dR(N(q))}{dq} - p_0 CT = 0. \tag{26}$$

Proof. For a given joining strategy q of arrivals who find busy period, and the operation process in steady-state constitute an alternate renewal process as described in Sect. 2. Hence, the social benefits per time unit is

$$S_{soc}(q) = \lambda p_0 q(p_s R(N(q)) - CT) + \lambda p_1 R(N(q))$$
$$= \lambda(q p_0 p_s + p_1) R(N(q)) - \lambda p_0 q CT. \tag{27}$$

The second-order derivative of $S_{soc}(q)$ with respect to q is given as

$$\frac{d^2 S_{soc}(q)}{dq^2} = \lambda \left(2 p_0 p_s \times \frac{dR(N(q))}{dq} + (p_0 p_s q + p_1) \times \frac{d^2 R(N(q))}{dq^2} \right). \tag{28}$$

Also, with the help of Remark 2 in Sect. 3, we have $\frac{d^2 S_{soc}(q)}{dq^2} < 0$. Then, there exists a unique q_{max} that maximizes $S_{soc}(q)$, where q_{max} is the unique solution of the equation $\frac{dS_{soc}(q)}{dq} = 0$. We then consider the location of q_{max} and interval $[0, 1]$ in the following three cases:

- When $q_{max} < 0$, $S_{soc}(q)$ is decreasing in q, thus $S_{soc}(q) \leq S_{soc}(0)$ in the domain of $[0, 1]$, the best response is to balk, then $q_{soc} = 0$;
- when $0 \leq q_{max} \leq 1$, $S_{soc}(q)$ is concave with q in $[0, 1]$, and q_{max} is exactly the maximum point, then $q_{soc} = q_{max}$;
- when $q_{max} > 1$, $S_{soc}(q)$ is increasing with q, so $S_{soc}(q) \leq S_{soc}(1)$ in the domain of $[0, 1]$, the best response is to join the system, then $q_{soc} = 1$.

The results of the three cases are corresponding to the three parts of (25). $\square$

Finally, we focus our attention on the problem of profit maximization, that is, the administrator aims to maximize his profit by imposing an admission fee p on the customers who get into the server. By imposing this fee, the reward of the customer changes from $R(N(q))$ to $R(N(q)) - p$. Therefore, we consider a new equilibrium joining strategy $q_{prof}(p)$. In addition, the nonnegativity of $R(N(q)) - p$ is required since arrivals find the admission period prefer to join the system. That is to say, the price $p \leq R(N(0))$ is necessary. We are interested in studying the strategy q_{prof} in this way, then we have the following results.

Theorem 3. *In the single-server retrial queue with batch service, a unique mixed strategy 'join the orbit with probability q_{prof} whenever finding the busy period' that maximizes the administrator's net profit per time unit exists and is given as*

$$q_{prof} = \begin{cases} 0 & \text{if } \tilde{q}_{max} < 0 \quad \text{or} \quad \tilde{q}_{max} \geq 0 \text{ \& } \Pi_2(0) \geq max\{\Pi_1(\tilde{q}_{max}), \Pi_1(1)\}, \\ \tilde{q}_{max} & \text{if } 0 \leq \tilde{q}_{max} \leq 1 \text{ \& } \Pi_1(\tilde{q}_{max}) > \Pi_2(0), \\ 1 & \text{if } \tilde{q}_{max} > 1 \text{ \& } \Pi_1(1) > \Pi_2(0). \end{cases} \tag{29}$$

where $\tilde{q}_{max} = q_{max}$ and $\Pi_1(q), \Pi_2(q)$ are constructed in (34).

Proof. Suppose that all arrivals adopt the joining strategy q when observe the busy period and join directly when finding admission period. We define $\Pi(q)$ to be the net profit per time unit of the administrator, when blocked customers follow the strategy q. Let $p(q)$ be the admission fee that induce the strategy q. Then we have

$$\Pi(q) = \lambda(qp_0p_s + p_1)p(q). \tag{30}$$

To determine $p(q)$ in terms of q, a reasonable range is $p(q) \leq R(N(0))$, if not, no customer will join the system in each service cycle. First, we consider an arrival who find busy period when an admission fee $p(q)$ is imposed. Then his optimal response is to replace $R(N(q))$ in Theorem 2 with $R(N(q)) - p(q)$.

with $R(N(q))$ replaced by $R(N(q)) - p(q)$

This customer responses optimally according to Theorem 2 with $R(N(q))$ replaced by $R(N(q)) - p(q)$. Therefore we have to solve the equation

$$p_s\left(R(N(q)) - p_1(q)\right) = CT, \tag{31}$$

from which we obtain

$$p_1(q) = R(N^{ob}(q)) - \frac{CT}{p_s}, \quad q \in [0,1]. \tag{32}$$

Subsequently, we pay attention to the arrival who find admission period, similarly, he responses optimally according to the following:

$$R(N(q)) - p_2(q) = 0. \tag{33}$$

Note that arrival who find busy period has a negative benefit at this time if he joins the system, thus $q = 0$, that is $p_2(q) = R(N(0))$.

Then the profit of the administrator assumes the form

$$\Pi(q) = \begin{cases} \Pi_1(q) = \lambda(p_0p_sq + p_1)(R(N(q)) - \frac{CT}{p_s}), & \text{if } p = p_1(q), \\ \Pi_2(q) = \lambda p_1 R(N(0)), & \text{if } p = p_2(q). \end{cases} \tag{34}$$

The second-order derivative of $\Pi_1(q)$ is

$$\frac{d^2\Pi_1(q)}{dq^2} = \lambda\left(2p_0p_s \times \frac{dR(N(q))}{dq} + (p_0p_sq + p_1) \times \frac{d^2R(N(q))}{dq^2}\right) < 0. \tag{35}$$

Thus $\Pi_1(q)$ is concave with q and there exists a unique point $\tilde{q}_{max}$ that maximizes $\Pi_1(q)$ which was solved by $\frac{d\Pi_1(q)}{dq} = 0$. It is remarkable that the value of $\Pi_2(q)$ is not depend on q. We then consider the following three cases:

- When $\tilde{q}_{max} < 0$, $\Pi_1(q)$ is decreasing in q in $[0,1]$, thus $\Pi_1(q) \leq \Pi_1(0) < \Pi_2(0)$ in the domain of q, the best response is balking, then $q_{prof} = 0$, the optimal price $p(q) = R(N(0))$;

- when $0 \leq \tilde{q}_{max} \leq 1$, $\Pi_1^{ob}(q)$ is concave with q in $[0,1]$, the maximum of $\Pi_1(q)$ is achieved at $\tilde{q}_{max}$. If $\Pi_1(\tilde{q}_{max}) > \Pi_2(0)$, we have $q_{prof} = \tilde{q}_{max}$, the corresponding optimal price $p(q) = R(N(\tilde{q}_{max})) - \frac{CT}{p_s}$, otherwise, $q_{prof} = 0$, and $p(q) = R(N(0))$;
- when $\tilde{q}_{max}^{ob} > 1$, $\Pi_1(q)$ is increasing in q in $[0,1]$, thus $\Pi_1(q) \leq \Pi_1(1)$ in the domain of q. If $\Pi_1(1) > \Pi_2(0)$, we have $q_{prof} = 1$, the corresponding optimal price $p(q) = R(N(1))$, otherwise, $q_{prof} = 0$, and $p(q) = R(N(0))$.

The results of the three cases are summarized in (29).   □

5   Linear Reward Case

Based on our previous assumption, the service reward $R(N)$ is concave in the average number of customers be served in a service cycle N. A simple but special case $\frac{dR^2(N)}{dN^2} = 0$ is worth studying, and we denote it as $R(N) = R - kN$, where R is a nonnegative constant, and k is the unit reward loss with the increase of one customer in a service cycle ($k \geq 0$).

We aim to explore the optimal joining strategies of blocked customers.

Theorem 4. *In the single-server retrial queue with batch service and linear service reward, the individual, social and profit optimal joining strategies* q_e, q_{soc}, q_{prof} *are given as*

$$q_e = \begin{cases} 0 & if \ \ k > \frac{p_s R - CT}{p_s N(0)}, \\ \frac{p_s R - CT - p_s kN(0)}{p_s k(N(1) - N(0))} & if \ \ \frac{p_s R - CT}{p_s N(1)} \leq k \leq \frac{p_s R - CT}{p_s N(0)}, \\ 1 & if \ \ k < \frac{p_s R - CT}{p_s N(1)}, \end{cases} \quad (36)$$

$$q_{soc} = \begin{cases} 0 & if \ \ k > \frac{p_s R - CT}{2p_s N(0)}, \\ \frac{p_s R - CT - 2p_s kN(0)}{2p_s k(N(1) - N(0))} & if \ \ \frac{p_s R - CT}{2p_s N(1)} \leq k \leq \frac{p_s R - CT}{2p_s N(0)}, \\ 1 & if \ \ k < \frac{p_s R - CT}{2p_s N(1)}, \end{cases} \quad (37)$$

$$q_{prof} = \begin{cases} 0 & if \ \ k > \frac{p_s R - CT}{2p_s N(0)} \ \ or \ k \leq \frac{p_s R - CT}{2p_s N(0)} \\ & \& \ \ \Pi_2(0) \geq max\{\Pi_1(\tilde{q}_{max}), \Pi_1(1)\}, \\ \frac{p_s R - CT - 2p_s kN(0)}{2p_s k(N(1) - N(0))} & if \ \ \frac{p_s R - CT}{2p_s N(1)} \leq k \leq \frac{p_s R - CT}{2p_s N(0)} \\ & \& \ \ \Pi_1(\tilde{q}_{max}) > \Pi_2(0), \\ 1 & if \ \ k < \frac{p_s R - CT}{2p_s N(1)} \ \ \& \ \ \Pi_1(1) > \Pi_2(0), \end{cases}$$

$$\quad (38)$$

where $\tilde{q}_{max} = \frac{p_s R - CT - 2p_s kN(0)}{2p_s k(N(1) - N(0))}$, $\Pi_1(q)$ *and* $\Pi_2(q)$ *are given by* (45).

Proof. Based on the reward-cost structure, the expected benefit of a joining customer in the busy period when others follow the strategy q is given as

$$S(q) = p_s(R - kN(q)) - CT. \quad (39)$$

We can easily find that $S(q)$ is deceasing in q. Thus the unique maximum and minimum are $S(0)$ and $S(1)$, respectively. As a result, we consider the following three cases:

- When $p_s((R - kN(0)) - CT < 0$, $S(q)$ is negative for every $q \in [0, 1]$, then $q_e = 0$;
- when $p_s((R - kN(1)) - CT \leq 0 \leq p_s((R - kN(0)) - CT$, there exists a unique solution of the equation $S(q) = 0$ which lies in the interval $[0, 1]$, and it is exactly the optimal strategy;
- when $p_s((R - kN(1)) - CT > 0$, $S(q)$ is positive for every $q \in [0, 1]$, then $q_e = 1$.

The social benefits per time unit when all blocked customers follow the strategy q in the busy period is

$$S_{soc}(q) = \lambda(p_0 p_s q + p_1)(R - kN(q)) - \lambda q p_0 CT, \tag{40}$$

and it is concave with q.

Thus there exists a unique point q_{max} that maximizes $S_{soc}(q)$. Where q_{max} is the unique solution of the equation

$$p_0 p_s(R - kN(q)) - (p_0 p_s q + p_1)k \times \frac{dN(q)}{dq} - p_0 CT = 0. \tag{41}$$

Then we obtain $q_{max} = \frac{p_s R - CT - 2p_s kN(0)}{2p_s k(N(1) - N(0))}$ and we discuss it in the following three cases:

- When $q_{max} < 0$, $S_{soc}(q)$ is decreasing in q in $[0, 1]$, the maximum is obtained at 0, then $q_{soc} = 0$;
- when $0 \leq q_{max} \leq 1$, $S_{soc}(q)$ is concave with q in $[0, 1]$, then $q_{soc} = q_{max}$;
- when $q_{max} > 1$, $S_{soc}(q)$ is increasing in q in $[0, 1]$, then $q_{soc} = 1$.

As we have described before, when arrivals can observe the service period of the system, they join directly at the admission period. In the profit maximizing problem, a reasonable admission fee $p(q) \leq R - kN(0)$ is considered. When customers adopt the strategy q in the busy period, the profit per time unit is

$$\Pi(q) = \lambda(p_0 p_s q + p_1)p(q). \tag{42}$$

By the analysis in Theorem 3, the two possible price which base on the customers' optimum in two periods respectively are

$$p_1(q) = R - kN(q) - \frac{CT}{p_s}, \tag{43}$$

$$p_2(q) = R - kN(0). \tag{44}$$

The corresponding revenue function is

$$\Pi(q) = \begin{cases} \Pi_1(q) = \lambda(p_0 p_s q + p_1)(R - kN(q) - \frac{CT}{p_s}), & \text{if } p = p_1(q), \\ \Pi_2(0) = \lambda p_1(R - kN(0)), & \text{if } p = p_2(q). \end{cases} \quad (45)$$

Since $\Pi_1(q)$ is concave with q, thus there exists a unique point $\widetilde{q}_{max}$ that maximizes $\Pi_1(q)$. And it is the unique solution of the equation

$$p_0 p_s(R - kN(q)) - (p_0 p_s q + p_1)k \times \frac{dN(q)}{dq} - p_0 CT = 0. \quad (46)$$

The value of $\widetilde{q}_{max}$ are discussed in the three cases:

- When $\widetilde{q}_{max} < 0$, $\Pi_1(q)$ is decreasing in q in $[0, 1]$, thus $\Pi_1(q) \leq \Pi_1(0) < \Pi_2(0)$ in the domain of q, the best response is balking, then $q_{prof} = 0$ and the optimal admission fee is $R - kN(0)$;
- when $0 \leq \widetilde{q}_{max} \leq 1$, $\Pi_1(q)$ is concave with $q \in [0, 1]$, the maximum of $\Pi_1(q)$ is achieved at $\widetilde{q}_{max}$. If $\Pi_1(\widetilde{q}_{max}) > \Pi_2(0)$, we have $q_{prof} = \widetilde{q}_{max}$, otherwise, $q_{prof} = 0$;
- when $\widetilde{q}_{max} > 1$, $\Pi_1(q)$ is increasing in q in $[0, 1]$, thus $\Pi_1(q) \leq \Pi_1(1)$ in the domain of q. If $\Pi_1(1) > \Pi_2(0)$, we have $q_{prof} = 1$, otherwise, $q_{prof} = 0$.

The above results are summarized in Eqs. (36), (37) and (38).   □

Remark 4. *Based on the Theorem above, by comparing the Eqs. (36) and (37) concretely, the order of the individual joining probability and the socially optimal one is $q_{soc} \leq q_e$. However, the order between q_{prof} and q_{soc}, q_e is not clear, and we shall explore it by numerical examples.*

6   Numerical Examples

In this section, we conduct numerical analysis to gain more insights on the optimal joining probabilities. In addition, customers' joining probabilities are all increasing with R, and decreasing with the unit waiting cost C intuitively. Thus in all applications, we assume $R = 4, C = 1$, and aim to obtain a direct observation of the joining probabilities with respect to the parameters λ, μ and γ, respectively. Throughout the examples, we use the linear function of reward as a main research object, that is $R(N) = R - kN$.

More concretely, in Fig. 1, it shows the impact of arrival rate on the three joining probabilities. As we can see, q_e, q_{soc}, q_{prof} are non-increasing functions of λ. It can be explained by the *ATC* situation we have analysed before, that is a higher congestion level decrease the service reward $R(N)$ as well as customers's joining willing. Figure 2 shows that a customer arrive at the busy period find a higher service rate μ prefers to join the system, since this means a shorter waiting time in the orbit and less customers be served in a service cycle. In Fig. 3, it shows the impact of admission rate γ on the joining strategies. As we

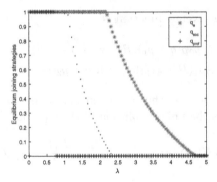

Fig. 1. The three joining probabilities with respect to λ, $\mu = 1, \gamma = 2, k = 1.5, \theta = 5$.

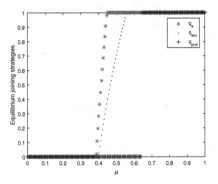

Fig. 2. The three joining probabilities with respect to μ, $\lambda = 0.7, \gamma = 2, k = 1.5, \theta = 5$.

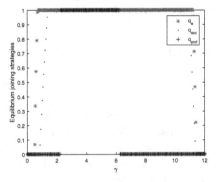

Fig. 3. The three joining probabilities with respect to γ, $\lambda = 0.8, \gamma = 2, k = 2, \theta = 5$.

have seen, the three joining probabilities are nondecreasing with a small γ or a long admission period. With the increase of γ, the admission period is shorter and it decreases the average number of customers in the server N and increases the service reward $R(N)$ at this moment. However, for a large enough γ, the increase of it has a great negative impact on the final probability p_s of getting into the server. Therefore, γ has a dual influence on joining probabilities. In addition, the numerical scenarios in Figs. 1, 2 and 3 all indicate the order $q_{prof} \leq q_{soc} \leq q_e$. It is a result valid in many other queueing systems due to the negative externality of individuals.

7 Conclusions

In this paper, we analyzed the strategic customers' behavior in a Markovian retrial queueing system with batch service where the number of customers in the system is not revealed. The system's operation process is divided into two periods: busy period and admission period, and it can be regard as a renewal process. The equilibrium, social- and profit- maximizing joining strategies were derived and compared numerically. The general reward function scenario as well as the linear reward function were explored. It was observed that the equilibrium social benefit is better off when the service period is revealed to customers. In addition, the order $q_{prof} \leq q_{soc} \leq q_e$ is obtained numerically. For the future work, one may extend the system have a constant admission period, or wait for a given number of customers before service start, or combine the two by setting a maximum period length and load simultaneously.

Acknowledgments. This work was supported in part by the National Natural Science Foundation of China under grant nos. 71871008 and 71571014. It was also supported by the Fundamental Research Funds for the Central Universities under grant 2019YJS196.

References

Anand, K.S., Paç, M.F., Veeraraghavan, S.: Quality-speed conundrum: trade-offs in customer-intensive services. Manage. Sci. **57**, 40–56 (2011)

Artalejo, J.R.: A queueing system with returning customers and waiting line. Oper. Res. Lett. **17**, 191–199 (1995)

Artalejo, J.R., Gómez-Corral, A.: Steady state solution of a single-server queue with linear repeated requests. J. Appl. Probab. **34**, 223–233 (1997)

Artalejo, J.R., Gómez-Corral, A.: Retrial Queueing Systems: A Computational Approach. Springer, Heidelberg (2008). https://doi.org/10.1007/978-3-540-78725-9

Bailey, N.T.: On queueing processes with bulk service. J. Royal Stat. Soc. Ser. B (Methodol.) **16**, 80–87 (1954)

Bountali, O., Economou, A.: Equilibrium joining strategies in batch service queueing systems. Eur. J. Oper. Res. **260**, 1142–1151 (2017a)

Bountali, O., Economou, A.: Equilibrium threshold joining strategies in partially observable batch service queueing systems. Ann. Oper. Res. 1–23 (2017b)

Brugno, A.: MAP/PH/1 systems with group service: performance analysis under different admission strategies (2017)

Brugno, A., D'Apice, C., Dudin, A., Manzo, R.: Analysis of an MAP/PH/1 queue with flexible group service. Int. J. Appl. Math. Comput. Sci. **27**, 119–131 (2017)

Brugno, A., Dudin, A.N., Manzo, R.: Analysis of a strategy of adaptive group admission of customers to single server retrial system. J. Ambient Intell. Humanized Comput. **9**, 123–135 (2018)

Chaudhry, M.L., Templeton, J.G.C.: A First Course in Bulk Queues. Wiley, New York (1983)

Downton, F.: Waiting time in bulk service queues. J. Roy. Stat. Soc. B Ser. B **17**, 256–261 (1955)

Deb, R.K., Serfozo, R.F.: Optimal control of batch service queues. Adv. Appl. Probab. **5**, 340–361 (1973)

Dudin, A.N., Manzo, R., Piscopo, R.: Single server retrial queue with group admission of customers. Comput. Oper. Res. **61**, 89–99 (2015)

Economou, A., Kanta, S.: Equilibrium customer strategies and social-profit maximization in the single-server constant retrial queue. Naval Res. Logistics **58**, 107–122 (2011)

Edelson, N.M., Hilderbrand, D.K.: Congestion tolls for Poisson queuing processes. Econometrica **43**, 81–92 (1975)

Falin, G.I.: The M/M/1 retrial queue with retrials due to server failures. Queueing Syst. **58**, 155–160 (2008)

Falin, G., Templeton, J.G.: Retrial Queues. Chapman & Hall, London (1997)

Guo, P., Hassin, R.: Strategic behavior and social optimization in Markovian vacation queues. Oper. Res. **59**, 986–997 (2011)

Hassin, R., Haviv, M.: To Queue or Not to Queue: Equilibrium Behavior in Queueing Systems. Kluwer Academic Publishers, Boston (2003)

Hassin, R.: Rational Queueing. CRC Press, Boca Raton (2016)

Kleinrock, L.: Queueing Systems, Volume 2: Computer Applications, vol. 66. Wiley, New York (1976)

Li, X., Guo, P., Lian, Z.: Quality-speed competition in customer-intensive services with boundedly rational customers. Prod. Oper. Manage. **25**, 1885–1901 (2016)

Li, X., Li, Q., Guo, P., Lian, Z.: On the uniqueness and stability of equilibrium in quality-speed competition with boundedly-rational customers: the case with general reward function and multiple servers. Int. J. Prod. Econ. **193**, 726–736 (2017)

Manou, A., Economou, A., Karaesmen, F.: Strategic customers in a transportation station: when is it optimal to wait? Oper. Res. **62**, 910–925 (2014)

Medhi, J.: Waiting time distribution in a Poisson queue with a general bulk service rule. Manage. Sci. **21**, 777–782 (1975)

Naor, P.: The regulation of queue size by levying tolls. Econometrica **37**, 15–24 (1969)

Wang, J., Zhang, F.: Equilibrium analysis of the observable queues with balking and delayed repairs. Appl. Math. Comput. **218**, 2716–2729 (2011)

Wang, J., Zhang, F.: Strategic joining in M/M/1 retrial queues. Eur. J. Oper. Res. **230**, 76–87 (2013)

Xu, Y., Scheller-Wolf, A., Sycara, K.: The benefit of introducing variability in single-server queues with application to quality-based service domains. Oper. Res. **63**, 233–246 (2015)

Xu, X., Lian, Z., Li, X., Guo, P.: A hotelling queue model with probabilistic service. Oper. Res. Lett. **44**, 592–597 (2016)

Queueing Networks

A Linear Programming Approach
to Markov Reward Error Bounds
for Queueing Networks

Xinwei Bai[(✉)] and Jasper Goseling

Department of Applied Mathematics, University of Twente, P.O. Box 217,
7500 AE Enschede, The Netherlands
{x.bai,j.goseling}@utwente.nl

Abstract. In this paper, we present a numerical framework for constructing bounds on stationary performance measures of random walks in the positive orthant using the Markov reward approach. These bounds are established in terms of stationary performance measures of a perturbed random walk whose stationary distribution is known explicitly. We consider random walks in an arbitrary number of dimensions and with a transition probability structure that is defined on an arbitrary partition of the positive orthant. Within each component of this partition the transition probabilities are homogeneous. This enables us to model queueing networks with, for instance, break-downs and finite buffers. The main contribution of this paper is that we generalize the linear programming approach of [10] to this class of models.

Keywords: Multi-dimensional random walk ·
Stationary performance measures · Error bound ·
Markov reward approach · Linear programming

1 Introduction

We present a framework for establishing bounds on stationary performance measures of a class of discrete-time random walks in the M-dimensional positive orthant, i.e., with state space $S = \{0, 1, \dots\}^M$. This class of random walks enables us to model queueing networks with nodes of finite or infinite capacity, and with transition rates that depend on the number of jobs in the nodes. The latter allows us to consider, for instance, queues with break-downs or networks with overflow. The stationary performance measures that can be considered in our framework include average number of jobs in a queue, throughput and blocking probabilities.

More precisely, for a random walk R we assume that a unique stationary probability distribution $\pi : S \to [0, 1]$ for which the balance equations hold exists, *i.e.*, there exists π that satisfies

$$\pi(n) \sum_{n' \in S} P(n, n') = \sum_{n' \in S} \pi(n') P(n', n), \qquad \forall n \in S, \qquad (1)$$

© Springer Nature Switzerland AG 2019
T. Phung-Duc et al. (Eds.): QTNA 2019, LNCS 11688, pp. 201–220, 2019.
https://doi.org/10.1007/978-3-030-27181-7_13

where $P(n, n')$ denotes the transition probability from n to n'. For a non-negative function $F : S \rightarrow [0, \infty)$, we are interested in the stationary performance measure,

$$\mathcal{F} = \sum_{n=(n_1,\ldots,n_M) \in S} \pi(n) F(n). \tag{2}$$

For example, if $F(n) = n_1$, then $\mathcal{F}$ represents the average number of jobs in the first node.

If π is known explicitly, $\mathcal{F}$ can be derived directly. However, in general it is difficult to obtain an explicit expression for the stationary probability distribution of a random walk. In this paper, we do not focus on obtaining the stationary probability distribution. Instead, our interest is in providing a general numerical framework to obtain upper and lower bounds on $\mathcal{F}$ for general random walks. In line with this goal, we do not establish existence of $\mathcal{F}$ a priori. Instead we will see that if our method successfully finds an upper and lower bound, then $\mathcal{F}$ exists.

Consider a perturbed random walk $\bar{R}$, of which the stationary probability distribution $\bar{\pi}$ is known explicitly. Moreover, we consider an $\bar{F} : S \rightarrow [0, \infty)$ for $\bar{R}$, which can be different from F. The bounds on $\mathcal{F}$ are established in terms of

$$\bar{\mathcal{F}} = \sum_{n \in S} \bar{\pi}(n) \bar{F}(n). \tag{3}$$

We use the Markov reward approach, as introduced in [4], to build up these bounds. The method has been applied to various queueing networks in [3,5,7,8] and an overview of this approach has been given in [6]. In the works mentioned above, error bounds have been manually verified for each specific model. The verification can be quite complicated. Thus, a linear programming approach has been presented in [10] that provides bounds on $\mathcal{F}$ for random walks in the quarter plane ($M = 2$). In particular, in [10] the quarter plane is partitioned into four components, namely the interior, the horizontal axis, the vertical axis and the origin. Homogeneous random walks with respect to this partition, *i.e.*, transition probabilities are the same everywhere within a component, are considered there.

In this paper, we extend the linear programming approach in [10]. The contribution of this paper is two-fold. First, we build up a numerical program that can be applied to general models. In [10], an R in the quarter plane with a specific partition is considered. The numerical program used in [10] cannot be easily implemented for general partitions or multi-dimensional cases. In this paper, we are able to consider an R in an arbitrary dimensional state space. Moreover, we allow for general transition probability structures. For example, we can consider models such as a two-node queue with one finite buffer and one infinite buffer. We can also consider models in which the transition probabilities are dependent on the number of jobs in a node. Secondly, in the linear programming approach in [10], one important step is that in the optimization problem established for obtaining the performance bounds we first assign values to a set of variables using their interpretation such that all the constraints hold. Next we see these variables as parameters in the problem. In this paper we formulate a linear program to obtain values for this set of variables while in [10] the values are manually chosen and then verified. We show that this linear program is always feasible.

The problem of obtaining the stationary probability distribution has been considered in various works. For instance, methods have been developed to find π through its probability generating function in [2,9,19]. It is shown that for random walks in the quarter plane a boundary value problem can be formulated for the probability generating function. However, the boundary value problem has an explicit solution only in special cases (for example in [19]). If the probability generating function is obtained, the algorithm developed in [1] can provide a numerical inversion of the probability generating function. In addition, the matrix geometric method has been discussed in [16,18] for Quasi-birth-and-death (QBD) processes with finite phases, which provides an algorithmic approach to obtain the stationary probability distribution numerically. In [17] perturbation analysis has been applied to various QBD processes in the quarter plane. Under certain drift conditions, explicit expressions are derived for the error bound. One advantage of the approach in [10] is that its approach can be applied when the drift conditions are not satisfied. In the works mentioned above, only random walks in the two-dimensional orthant have been considered. As is mentioned above, our main contribution is to be able to establish performance bounds for random walks in the multi-dimensional positive orthant.

The remainder of this paper is structured as follows. In Sect. 2, we define the model and notation. Then, in Sect. 3 we review the results of the Markov reward approach. In Sect. 4, we formulate optimization problems for the upper and lower bounds, which are non-convex and have countably infinite number of variables and constraints. Next, in Sect. 5 we apply the linear programming approach and establish linear programs for the bounds. Finally, in Sect. 6 we present some numerical examples.

2   Model and Notation

Let R be a discrete-time random walk in $S = \{0, 1, \dots\}^M$. Denote by $P : S \times S \to [0, 1]$ the transition probability matrix of R. In this paper, only transitions between the nearest neighbors are allowed, $i.e.$, $P(n, n') > 0$ only if $n' - n \in N(n)$, where $N(n)$ denotes the set of possible transitions from n, $i.e.$,

$$N(n) = \left\{ u \in \{-1, 0, 1\}^M \mid n + u \in S \right\}. \tag{4}$$

For a finite index set K, we define a partition of S as follows.

Definition 1. $C = \{C_k\}_{k \in K}$ is called a partition of S if

1. $S = \cup_{k \in K} C_k$.
2. For all $j, k \in K$ and $j \neq k$, $C_j \cap C_k = \emptyset$.
3. For any $k \in K$, $N(n) = N(n')$, $\forall n, n' \in C_k$.

The third condition, which is non-standard for a partition, ensures that all the states in a component have the same set of possible transitions. With this condition, we are able to define homogeneous transition probabilities within a component, meaning that the transition probabilities are the same everywhere in a component. Denote by $c(n)$ the index of the component of partition C that n is located in. We call $c : S \rightarrow K$ the index indicating function of partition C. Throughout the paper, various partitions will be used. We will use capital letters to denote partitions and the corresponding small letters to denote their index indicating functions.

We restrict our attention to an R that is homogeneous with respect to a partition C of the state space, i.e., $P(n, n + u)$ depends on n only through the component index $c(n)$. Therefore, we denote by $N_{c(n)}$ and $p_{c(n),u}$ the set of possible transitions from n and transition probability $P(n, n + u)$, respectively. To illustrate the notation, we present the following example.

Example 1. Consider $S = \{0, 1, \dots\}^2$. Suppose that C consists of

$$C_1 = \{0\} \times \{0\}, \quad C_2 = \{1, 2, 3, 4\} \times \{0\}, \quad C_3 = \{5, 6, \dots\} \times \{0\},$$
$$C_4 = \{0\} \times \{1, 2, \dots\}, C_5 = \{1, 2, 3, 4\} \times \{1, 2, \dots\},$$
$$C_6 = \{5, 6, \dots\} \times \{1, 2, \dots\}.$$

The components and their sets of possible transitions are shown in Fig. 1.

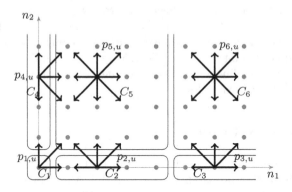

Fig. 1. A finite partition of $S = \{0, 1, \dots\}^2$ and the sets of possible transitions for its components

Based on a partition, we now define a component-wise linear function.

Definition 2. *Let C be a partition of S. A function $H : S \rightarrow [0, \infty)$ is called C-linear if there exists $h_{k,0}, \dots, h_{k,M} \in \mathbb{R}$ such that*

$$H(n) = \sum_{k \in K} \mathbf{1}\left(n \in C_k\right) \left(h_{k,0} + \sum_{i=1}^{M} h_{k,i} n_i \right). \tag{5}$$

In this paper, we often consider transformations of H of the form $G(n) = H(n + u)$, $u \in N(n)$. It is of interest to consider a partition Z of S such that G is Z-linear when H is C-linear.

Definition 3. *Given a finite partition C, $Z = \{Z_j\}_{j \in J}$ is called a refinement of C if*

1. *Z is a finite partition of S.*
2. *For any $j \in J$, any $n \in Z_j$ and any $u \in N_j$, $c(n + u)$ depends only on j and u, i.e.,*

$$c(n + u) = c(n' + u), \quad \forall n, n' \in Z_j. \tag{6}$$

Remark that a refinement of C is not unique. To give more intuition, in the following example we give a refinement of C that is given in Example 1.

Example 2. In this example, consider the partition C given in Example 1. A refinement of C is shown in Fig. 2.

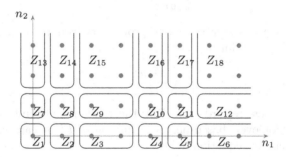

Fig. 2. A refinement of C defined in Example 1

Since R is homogeneous with respect to partition C, it is homogeneous with respect to partition Z as well. Next, we present the result that $H(n + u)$ is Z-linear if H is C-linear. The proof of the lemma is straightforward and is hence omitted.

Lemma 1. *Let $H : S \to [0, \infty)$ be a C-linear function. Moreover, let Z be a refinement of C. For any $u \in \{-1, 0, 1\}^M$, define $G : S \to [0, \infty)$ as $G(n) = 1(n + u \in S)H(n + u)$. Then, G is Z-linear.*

Let us consider an example to demonstrate the intuition behind this lemma. Consider the partition C and partition Z given in Example 1 and 2 respectively. Moreover, let

$$H(n) = \begin{cases} 1, & \text{if } n \in C_1, C_2, C_4, C_5, \\ 0, & \text{if } n \in C_3, C_6. \end{cases} \tag{7}$$

and let $u = (1,0)$. It is easy to check that $H(n)$ is C-linear. Now consider $n_0 = (4,0)$ and $n'_0 = (5,0)$, which are both in C_2. We can verify that $G(n_0) = H(n_0 + u) = 0$ and $G(n'_0) = 1$. Therefore, $G(n)$ is not C-linear. In partition Z, n_0 and n'_0 are located in two different components. Then it can be checked that $G(n)$ is indeed Z-linear.

3   Preliminaries: Markov Reward Approach

Suppose that we have obtained an $\bar{R}$ for which $\bar{\pi}$ is known explicitly. Then, we build up upper and lower bounds on $\mathcal{F}$ using the Markov reward approach, an introduction to which is given in [6]. In this section, we give a review of this approach including its main result.

In the Markov reward approach, $F(n)$ is considered as a reward if R stays in n for one time step. Let $F^t(n)$ be the expected cumulative reward up to time t if R starts from n at time 0, $i.e.$,

$$F^t(n) = \sum_{k=0}^{t-1} \sum_{m \in S} P^k(n,m) F(m), \tag{8}$$

where $P^k(n,m)$ is the k-step transition probability from n to m. Then, since R is ergodic and $\mathcal{F}$ exists, for any $n \in S$,

$$\lim_{t \to \infty} \frac{F^t(n)}{t} = \mathcal{F}, \tag{9}$$

$i.e.$, $\mathcal{F}$ is the average reward gained by the random walk independent of the starting state. Moreover, based on the definition of F^t, it can be verified that the following recursive equation holds,

$$F^{t+1}(n) = F(n) + \sum_{n' \in S} P(n,n') F^t(n') \tag{10}$$

and $F^0(n) = 0$. Next, we define the bias terms as follows.

Definition 4. *For any* $t = 0, 1, \ldots,$ *the bias terms* $D^t : S \times S \to \mathbb{R},$ *are defined as*

$$D^t(n,n') = F^t(n') - F^t(n). \tag{11}$$

We present the main result of the Markov reward approach below.

Theorem 1 (Result 9.3.5 in [6]). *Suppose that* $\bar{F} : S \to [0,\infty)$ *and* $G : S \to [0,\infty)$ *satisfy*

$$\left| \bar{F}(n) - F(n) + \sum_{n' \in S} \left(\bar{P}(n,n') - P(n,n') \right) D^t(n,n') \right| \leq G(n), \tag{12}$$

for all $n \in S, t \geq 0$. *Then* $|\bar{\mathcal{F}} - \mathcal{F}| \leq \sum_{n \in S} \bar{\pi}(n) G(n)$.

In this paper, we obtain bounds on $\mathcal{F}$ by finding $\bar{F}$ and G for which (12) holds.

We do not need R and $\bar{R}$ to be irreducible. More generally, it is sufficient that there is a single absorbing communicating class (which can be different for R and $\bar{R}$). This implies that we allow for transient states. Even though we are only interested in the steady-state behavior of our processes, it will be important for the application of the Markov reward approach to explicitly model these transient states. Existing proofs of the Markov reward approach considers only irreducible processes. It is clear from the proof of Result 9.3.2 in [6] that this result can be straightforwardly generalized to processes with transient states. We will use this extended result in a numerical example in Sect. 6. In addition to the bound on $|\bar{\mathcal{F}} - \mathcal{F}|$, the following comparison result, which sometimes gives a better upper bound, is given in [6].

Theorem 2 (Result 9.3.2 in [6]). *Suppose that* $\bar{F} : S \to [0, \infty)$ *satisfies*

$$\bar{F}(n) - F(n) + \sum_{n' \in S} \left(\bar{P}(n, n') - P(n, n') \right) D^t(n, n') \geq 0, \tag{13}$$

for all $n \in S$, $t \geq 0$. *Then* $\mathcal{F} \leq \bar{\mathcal{F}}$.

Similarly, if the LHS of (13) is non-positive, then $\mathcal{F} \geq \bar{\mathcal{F}}$.

4   Problem Formulation

Recall that $P(n, n')$ and $\bar{P}(n, n')$ denote the transition probability of R and $\bar{R}$, respectively. Let $\Delta(n, n') = \bar{P}(n, n') - P(n, n')$. From the result of Theorem 1, the following optimization problem comes up naturally to provide an upper bound on $\mathcal{F}$.

Problem 1 (Upper bound).

$$\min \sum_{n \in S} \left[\bar{F}(n) + G(n) \right] \bar{\pi}(n),$$

$$\text{s.t.} \left| \bar{F}(n) - F(n) + \sum_{n' \in S} \Delta(n, n') D^t(n, n') \right| \leq G(n), \quad \forall n \in S, t \geq 0, \tag{14}$$

$$\bar{F}(n) \geq 0, G(n) \geq 0, \quad \forall n \in S.$$

In this problem, $\bar{F}(n)$, $G(n)$ and $D_u^t(n)$ are variables and $\bar{\pi}(n)$, $\Delta(n, n')$ are parameters. Similarly, using max $\sum_{n \in S} \left[\bar{F}(n) - G(n) \right] \bar{\pi}(n)$ as the objective function, we obtain a problem that returns a lower bound on $\mathcal{F}$.

In addition, the following problems provide a direct upper or lower bound on $\mathcal{F}$, which follows from the comparison result introduced in Sect. 3.

Problem 2 (Comparison upper bound).

$$\min \sum_{n \in S} \bar{F}(n)\bar{\pi}(n),$$

$$\text{s.t.}\ \bar{F}(n) - F(n) + \sum_{n' \in S} \Delta(n, n')D^t(n, n') \geq 0, \quad \forall n \in S, t \geq 0, \quad (15)$$

$$\bar{F}(n) \geq 0, \quad \forall n \in S.$$

Problem 3 (Comparison lower bound).

$$\max \sum_{n \in S} \bar{F}(n)\bar{\pi}(n),$$

$$\text{s.t.}\ \bar{F}(n) - F(n) + \sum_{n' \in S} \Delta(n, n')D^t(n, n') \leq 0, \quad \forall n \in S, t \geq 0, \quad (16)$$

$$\bar{F}(n) \geq 0, \quad \forall n \in S.$$

Note that if Problem 2 is feasible then Problem 3 will be unbounded or infeasible and vice versa. It will be seen from numerical results that in some cases the comparison result can provide a better upper or lower bound than that obtained from Problem 1. In the remainder of this paper, we only consider Problem 1, since the other problems can be solved in the same fashion. There are countably infinite variables and constraints in Problem 1. In the next two sections, we will reduce Problem 1 to a linear program with a finite number of variables and constraints.

5   Linear Programming Approach to Error Bounds

In this section we first present the theory of the linear programming approach and formulate a linear problem for obtaining the lower bound. Then we reduce the linear problem with infinite number of variables and constraints to one with finite variables and constraints by restricting our consideration to C-linear functions. For the linear programming approach, we use the idea from [10] that we consider bounding functions on $D^t(n, n')$ which are independent of t. Replacing $D^t(n, n')$ with these bounding functions in (14), we get rid of t in the constraints and obtain sufficient conditions for (14). Simultaneously, we add several extra constraints to ensure that these newly introduced functions are indeed upper and lower bounds on $D^t(n, n')$.

In (12), since only transitions between the nearest neighbors are allowed, we have $\Delta(n, n + u) = 0$ for $u \notin N_{c(n)}$. Then, $\Delta(n, n')D^t(n, n')$ vanishes from (12) for all $n' - n \notin N_{c(n)}$. Thus, it is sufficient to only consider the bias terms between nearest neighbors, *i.e.*, $D^t(n, n + u)$ for $u \in N_{c(n)}$.

More precisely, consider functions $A : S \times S \to [0, \infty)$ and $B : S \times S \to [0, \infty)$, for which

$$-A(n, n + u) \leq D^t(n, n + u) \leq B(n, n + u), \quad (17)$$

for all $t \geq 0$. Then, in Problem 1, replacing $D^t(n, n')$ with the bounding functions, we get rid of the time-dependent terms and obtain the following constraints that guarantee (14),

$$\bar{F}(n) - F(n) + \sum_{u \in N_{c(n)}} \max \{\Delta(n, n+u)B(n, n+u), -\Delta(n, n+u)A(n, n+u)\}$$

$$\leq G(n), \tag{18}$$

$$F(n) - \bar{F}(n) + \sum_{u \in N_{c(n)}} \max \{\Delta(n, n+u)A(n, n+u), -\Delta(n, n+u)B(n, n+u)\}$$

$$\leq G(n). \tag{19}$$

Besides the constraints given above, additional constraints are necessary to guarantee that (17) holds. In the next part, we establish these additional constraints.

Recall that $D^t(n, n+u) = F^t(n+u) - F^t(n)$. We will show in the next section that $D^{t+1}(n, n+u)$ can be expressed as a linear combination of $D^t(m, m+v)$ where $v \in N_{c(m)}, m \in S$. More precisely, there exists $\phi(n, u, m, v) \geq 0$ for which the following equation holds,

$$D^{t+1}(n, n+u) = F(n+u) - F(n) + \sum_{m \in S} \sum_{v \in N_{c(m)}} \phi(n, u, m, v)D^t(m, m+v),$$

$$\tag{20}$$

for $t \geq 0$. We will reduce the sum in the equation above to a sum over a finite number of states. Therefore, the convergence of the sum is not an issue. Then, the following inequalities are sufficient conditions for $-A(n, n')$ and $B(n, n')$ to be a lower and upper bound on $D^t(n, n+u)$, respectively,

$$F(n+u) - F(n) + \sum_{m \in S} \sum_{v \in N_{c(m)}} \phi(n, u, m, v)B(m, m+v) \leq B(n, n+u), \quad (21)$$

$$F(n+u) - F(n) - \sum_{m \in S} \sum_{v \in N_{c(m)}} \phi(n, u, m, v)A(m, m+v) \geq -A(n, n+u).$$

$$\tag{22}$$

Summarizing the discussion above, the following problem gives an upper bound on $\mathcal{F}$.

Problem 4.

$$\min \sum_{n \in S} [\bar{F}(n) + G(n)] \, \bar{\pi}(n),$$

$$\text{s.t. } \bar{F}(n) - F(n) + \sum_{u \in N_{c(n)}} \max \{\Delta(n, n+u)B(n, n+u), -\Delta(n, n+u)A(n, n+u)\}$$

$$\leq G(n), \tag{23}$$

$$F(n) - \bar{F}(n) + \sum_{u \in N_{c(n)}} \max \{\Delta(n, n+u)A(n, n+u), -\Delta(n, n+u)B(n, n+u)\}$$

$$\leq G(n), \tag{24}$$

$$D^{t+1}(n, n+u) = F(n+u) - F(n) + \sum_{m \in S} \sum_{v \in N_{c(m)}} \phi(n, u, m, v) D^t(m, m+v),$$
(25)

$$F(n+u) - F(n) + \sum_{m \in S} \sum_{v \in N_{c(m)}} \phi(n, u, m, v) B(m, m+v) \leq B(n, n+u),$$
(26)

$$F(n) - F(n+u) + \sum_{m \in S} \sum_{v \in N_{c(m)}} \phi(n, u, m, v) A(m, m+v) \leq A(n, n+u),$$
(27)

$$\phi(n, u, m, v) \geq 0, \quad \text{for } n, m \in S, u \in N_{c(n)}, v \in N_{c(m)}$$
$$A(n, n+u) \geq 0, B(n, n+u) \geq 0, \bar{F}(n) \geq 0, G(n) \geq 0, \quad \text{for } n, n' \in S.$$

In this problem the variables are $\phi(n, u, m, v)$, $A(n, n+u)$, $B(n, n+u)$, $D^t(n, n+u)$, $\bar{F}(n)$, $G(n)$ and the parameters are $\bar{\pi}(n)$, $F(n)$, $\Delta(n, n+u)$. Problem 4 is non-linear since there are product terms such as $\phi(n, u, m, v) A(n, n')$ and $\phi(n, u, m, v) B(n, n')$. Therefore, we apply the idea proposed in [10]. More precisely, first we obtain values of a set of $\phi(n, u, m, v)$, for which (25) holds. Then, we plug the obtained $\phi(n, u, m, v)$ into Problem 4 as parameters and remove (25) from the problem. As a consequence, Problem 4 becomes linear. In [10], the set of $\phi(n, u, m, v)$ is obtained by manual derivation. In the following part, we formulate a linear program where the variables are $\phi(n, u, m, v)$ and they are interpreted as flows among states.

5.1   Linear Program for Finding $\phi(n, u, m, v)$

In this section, we formulate a linear program to obtain $\phi(n, u, m, v)$ for which (25) holds. For the bias terms, using (10), we get

$$D^{t+1}(n, n+u) = F^{t+1}(n+u) - F^{t+1}(n)$$
$$= F(n+u) - F(n) + \sum_{m \in S}[P(n+u, m) - P(n, m)]F^t(m). \quad (28)$$

Thus, (25) holds if and only if

$$\sum_{m \in S} \sum_{v \in N_{c(m)}} \phi(n, u, m, v) D^t(m, m+v) = \sum_{m \in S}[P(n+u, m) - P(n, m)]F^t(m).$$
(29)

Rewriting the LHS of (29), we have

$$\sum_{m \in S} \sum_{v \in N_{c(m)}} \phi(n, u, m, m+v) D^t(m, m+v)$$

$$= \sum_{m \in S} \sum_{v \in N_{c(m)}} \phi(n, u, m, m+v)[F^t(m+v) - F^t(m)]$$

$$= \sum_{m \in S} \left\{ \sum_{v \in N_{c(m)}} [\phi(n, u, m+v, -v) - \phi(n, u, m, v)] \right\} F^t(m). \quad (30)$$

In comparison with the RHS of (29), we obtain the following constraint that is sufficient for (29) as well as for (25),

$$\sum_{v \in N_{c(m)}} [\phi(n, u, m+v, -v) - \phi(n, u, m, v)] = P(n+u, m) - P(n, m), \quad (31)$$

for all $n, m \in S$, $u \in N_{c(n)}$. Intuitively, for any fixed $n \in S$ and any fixed $u \in N_{c(n)}$, $\phi(n, u, m, v)$ can be interpreted as a flow from state m to state $m+v$, and $P(n+u, m) - P(n, m)$ can be seen as the demand at state m. Then, intuitively (31) means that the demand at every state m is equal to the difference between the inflow and outflow of m.

Next we formulate a linear program with a finite number of constraints and variables. Moreover, we show that based on the solution of this linear program we can obtain $\phi(n, u, m, v) \geq 0$ that satisfies (31) and hence satisfies (25). The objective of this linear program is to minimize the sum of all $\phi(n, u, m, v)$. We remark that in this paper we do not optimize with respect to the overall objective, which is to find the best error bound. In the discussion section, we provide an outlook on alternative objective functions that may be used.

We need a final piece of notation. Let $Z = \{Z_j\}_{j \in J}$ be a refinement of partition C defined in Definition 3. Then, for any $n \in Z_j$ and $u \in N_j$, let $c(j, u)$ be the index of the component of partition C that $n + u$ is located in. For $j \in J$ and $u \in N_j$, let

$$N_{j,u} = N_j \cup \left(u + N_{c(j,u)}\right). \quad (32)$$

Now, we consider the following problem and present Theorem 3.

Problem 5.

$$\min \quad \sum_{j \in J} \sum_{u \in N_j} \sum_{d \in N_{j,u}} \sum_{v \in N_{c(j,d)}} \varphi_{j,u,d,v},$$

$$\text{s.t.} \quad \sum_{v \in N_{c(j,d)}} \mathbf{1}(d + v \in N_{j,u}) [\varphi_{j,u,d+v,-v} - \varphi_{j,u,d,v}] = p_{c(j,u),d-u} - p_{j,d},$$

$$\forall j \in J, u \in N_j, d \in N_{j,u}, \quad (33)$$

$$\varphi_{j,u,d,v} \geq 0, \quad \forall j \in J, u \in N_j, d \in N_{j,u}, v \in N_{c(j,d)}.$$

Theorem 3. *Problem 5 is feasible and has a finite number of variables and constraints. Suppose that $\varphi_{j,u,d,v}$ is the optimal solution of Problem 5. Then,*

$$\phi(n, u, m, v) = \begin{cases} \varphi_{z(n),u,m-n,v}, & \text{if } m \in n + N_{z(n),u} \text{ and } m + v \in n + N_{z(n),u}, \\ 0, & \text{otherwise,} \end{cases}$$

$$(34)$$

satisfies (31).

Proof. We present the proof in two steps. First, for every $j \in J$ and $u \in N_j$, we consider a specific state in Z_j and its neighbors. Using its neighborhood structure we show that a feasible solution $\varphi_{j,u,d,v}$ of Problem 5 exists. Next we show that based on the optimal solution and using the homogeneous property, we can assign values to the flows between the states of S, i.e., $\phi(n, u, m, v)$. After this assignment, we can obtain flows $\phi(n, u, m, v)$ for which (31) holds.

Consider some fixed $j \in J$ and $u \in N_j$. Let n_0 be some state in Z_j. Consider an undirected graph $\mathcal{G} = (\mathcal{V}, \mathcal{E})$, where $\mathcal{V}$ contains all the nearest neighbors of n_0 and of $n_0 + u$. Moreover, $e \in \mathcal{E}$ if and only if e connects two nearest neighbors. It is easy to see that $\mathcal{G}$ is connected. From the discussion after (31), we see that (31) intuitively means to find flows on $e \in \mathcal{E}$ such that the demand at every node $m \in \mathcal{V}$ is equal to the difference between the inflow and outflow of m.

This is a classical flow problem in graph theory and combinatorial optimization. In our case, the graph is connected. Moreover, there is no capacity for the flows and all the demands sum up to 0. Thus, there exists a feasible non-negative flow on $\mathcal{G}$ (see, for instance, Exercise 5 in Chap. 8 in [15]). In other words, there exists $\phi_0(n_0, u, m, v) \geq 0$, where $m, m + v \in \mathcal{V}$, such that for all $m \in \mathcal{V}$,

$$\sum_{v \in N_{c(m)}} \mathbf{1}\,(m + v \in \mathcal{V})\,[\phi_0(n_0, u, m + v, -v) - \phi_0(n_0, u, m, v)]$$

$$= P(n_0 + u, m) - P(n_0, m). \tag{35}$$

From (32), we see that $m \in \mathcal{V}$ if and only if $m = n + d$ for some $d \in N_{j,u}$. Take $\varphi_{j,u,d,v} = \phi_0(n_0, u, n + d, v)$. Since R is homogeneous with respect to partition C as well as partition Z, $P(n_0 + u, m) = p_{c(j,u),d-u}$ and $P(n_0, m) = p_{j,d}$. Therefore, we can verify that (35) is equivalent to (33) hence Problem 5 is feasible.

Suppose that $\varphi_{j,u,d,v}$ is the optimal solution of Problem 5. Then consider $\phi(n, u, m, v)$ where $n, m \in S$, $u \in N_{c(n)}$ and $v \in N_{c(m)}$. If $m \in n + N_{z(n),u}$ and $m + v \in n + N_{z(n),u}$, then $\varphi_{z(n),u,m-n,v}$ is well defined and satisfies (33). Thus, using $\phi(n, u, m, v) = \varphi_{z(n),u,m-n,v}$ we can verify that (31) holds. Otherwise if $m \notin n + N_{z(n),u}$ or $m + v \notin n + N_{z(n),u}$, (31) holds since $\phi(n, u, m, v) = 0$ and for its RHS, $P(n + u, m) - P(n, m) = 0$.

Finally we argue that Problem 5 has a finite number of variables and constraints. Since there are $|J|$ components in partition Z and at most 3^M possible transitions for every component, the number of the variables in Problem 5 is bounded by $2\,|J| \cdot 27^M$ from above. Moreover, the number of the constraints is bounded from above by $2\,|J| \cdot 9^M$.

5.2   Implementation of Problem 4

Suppose that we have obtained a set of coefficients $\varphi_{j,u,d,v}$ from Problem 5. In this section, we show that by restricting $F(n)$, $A(n, n')$, $B(n, n')$ to be C-linear and using the partition structure of S described in Sect. 2, Problem 4 can be reduced to a linear program with a finite number of variables and constraints.

Since we only consider the bias terms between the nearest neighbors, we rewrite the bounding functions as $A_u(n)$ and $B_u(n)$ for $n \in S$ and $u \in N_{c(n)}$.

Then, using the result of Theorem 3, plugging $\phi(n, u, m, v)$ as parameters into Problem 4 and removing (25), Problem 4 is equivalent to the following problem.

Problem 6.

$$\min \sum_{n \in S} \left[\bar{F}(n) + G(n) \right] \bar{\pi}(n),$$

$$\text{s.t. } \bar{F}(n) - F(n) + \sum_{u \in N_{c(n)}} \max \left\{ \Delta_{c(n),u} B_u(n), -\Delta_{c(n),u} A_u(n) \right\} - G(n) \le 0,$$
$$(36)$$

$$F(n) - \bar{F}(n) + \sum_{u \in N_{c(n)}} \max \left\{ \Delta_{c(n),u} A_u(n), -\Delta_{c(n),u} B_u(n) \right\} - G(n) \le 0,$$
$$(37)$$

$$F(n+u) - F(n) + \sum_{d \in N_{z(n),u}} \sum_{v \in N_{c(z(n),d)}} \varphi_{z(n),u,d,v} B_v(n+d) - B_u(n) \le 0,$$
$$(38)$$

$$F(n) - F(n+u) + \sum_{d \in N_{z(n),u}} \sum_{v \in N_{c(z(n),d)}} \varphi_{z(n),u,d,v} A_v(n+d) - A_u(n) \le 0,$$

$$A_u(n) \ge 0, B_u(n) \ge 0, \bar{F}(n) \ge 0, G(n) \ge 0, \quad \text{for } n \in S, u \in N_{c(n)}. \quad (39)$$

In the problem the variables are $A_u(n)$, $B_u(n)$, $\bar{F}(n)$ and $G(n)$. Here Problem 6 is already linear in its variables. It remains to reduce it to a problem with finite number of constraints and variables.

Next, we give the reduction by restricting $\bar{F}, G, A_u$ and B_u to be C-linear. From Lemma 1, we know that $A_v(n+d)$ and $B_v(n+d)$ are Z-linear. Thus, it is easy to check that all the constraints in Problem 6 have the form,

$$H(n) \le 0,$$

where $H(n)$ is Z-linear.

For any Z_j and $i \in \{1, \ldots, M\}$, define $L_{j,i}$ and $U_{j,i}$ as

$$L_{j,i} = \min_{n \in Z_j} n_i, \qquad U_{j,i} = \sup_{n \in Z_j} n_i. \quad (40)$$

Notice that Z_j can be unbounded in dimension i, in which case $U_{j,i} = \infty$. Moreover, let $I(Z_j)$ be the set containing all the unbounded dimensions of Z_j and ∂Z_j be the corners of Z_j, i.e.,

$$I(Z_j) = \{ i \in \{1, 2, \ldots, M\} \mid U_{j,i} = \infty \}, \quad (41)$$
$$\partial Z_j = \{ n \in Z_j \mid n_i = L_{j,i}, \ \forall i \in I(Z_j), \quad n_k \in \{L_{j,k}, U_{j,k}\}, \ \forall k \notin I(Z_j) \}. \quad (42)$$

For example, for the Z partition in Example 2, $I(Z_3) = \emptyset, \partial Z_3 = \{(2,0), (3,0)\}$ and $I(Z_6) = \{1\}, \partial Z_6 = \{(6,0)\}$. Then, for the constraint $H(n) \le 0$ for $n \in Z_j$, sufficient and necessary conditions can be obtained in terms of the coefficients $h_{j,i}$. We give the following lemma to specify these conditions. The proof for this lemma is straightforward and hence is omitted.

Lemma 2. *Suppose that $H(n)$ is Z-linear. Then, $H(n) \leq 0$ for all $n \in n \in Z_j$ if and only if*

$$H(n) \leq 0, \ \forall \, n \in \partial Z_j, \qquad h_{j,i} \leq 0, \ \forall i \in J(Z_j). \tag{43}$$

For any $n \in \partial Z_j$, clearly $H(n) = h_{j,0} + \sum_{i=1}^{M} h_{j,i} n_i$ is linear in the coefficients $h_{j,i}$. For each bounded dimension, there are at two corners of Z_j. Thus, (43) contains at most 2^M linear constraints in $h_{j,i}$.

Next, consider the objective function of Problem 6. In the next lemma, we show that it can be written as a linear combination of the coefficients $\bar{f}_{k,i}$ and $g_{k,i}$. The proof for the lemma is straightforward and hence is omitted.

Lemma 3. *Suppose that $\bar{F} : S \to [0,\infty)$ and $G : S \to [0,\infty)$ are C-linear. Then,*

$$\sum_{n \in S} \left[\bar{F}(n) + G(n) \right] \bar{\pi}(n) = \sum_{k \in K} \left(\bar{f}_{k,0} + g_{k,0} \right) \sum_{n \in C_k} \bar{\pi}(n)$$

$$+ \sum_{k \in K} \sum_{i=1}^{M} \left(\bar{f}_{k,i} + g_{k,i} \right) \sum_{n \in C_k} n_i \bar{\pi}(n). \tag{44}$$

Therefore, based on the two lemmas above, we give the main result of this section in the following theorem.

Theorem 4. *Suppose that $\bar{F}$, G, A_u and B_u are C-linear. Then, Problem 6 can be reduced to a linear program with a finite number of variables and constraints.*

Proof. From Lemmas 2 and 3, we see that Problem 6 can be reduced to a linear program where the coefficients of the functions are variables. Next, we will show that there is a finite number of variables and constraints in the reduced problem.

There are at most $|K|$ components and at most 3^M transitions from each state. Since $\bar{F}$, G, A_u and B_u are C-linear, the total number of coefficients is at most $2 |K| (3^M + 1)(M + 1)$. Hence, the number of variables in Problem 6 is finite. Moreover, for each component Z_j, there are at most 2^M corners and at most M unbounded dimensions. Hence, each constraint in Problem 6 can be reduced to at most $|J| (M + 2^M)$ constraints. Then, the number of constraints is finite.

6   Numerical Experiments

In this section, we consider some numerical examples for various queueing networks and establish upper and lower bounds on various performance measures. We have used Pyomo [13], a Python-based, open-source optimization modeling language package, to implement the optimization problems. The Gurobi solver [12] has been used to obtain solutions to these problems.

6.1   Finite Two-Node Tandem System

Consider a tandem system containing two nodes. Every job arrives at Node 1 according to a Poisson process and then goes to Node 2 after receiving its service at Node 1. Each node has a capacity for jobs that can be allowed. Let N_1 and N_2 denote the capacity of Node 1 and Node 2, respectively. An arriving job is rejected and lost if Node 1 is saturated. When Node 2 is saturated, a job remains at Node 1 upon completion. Let λ be the arrival rate. For Node 1, we consider a threshold $T \leq N_1$. The service rate is μ_1 if the number of jobs in Node 1 is no more than T and μ_1^* otherwise. The service rate of Node 2 is always μ_2. Assume that $\lambda < \mu_1$, $\lambda < \mu_1^*$ and $\lambda < \mu_2$. This system does not have a product-form stationary probability according to [4].

The Original Random Walk. Let $n = (n_1, n_2)$ represent the number of jobs in the system. Then the state space is $S = \{0, 1, \dots\}^2$. Note that the tandem system is a continuous-time system. We apply the uniformization technique introduced in [11] to transform the system into a discrete-time random walk R. Without loss of generality, we assume that $\lambda + \max\{\mu_1, \mu_1^*\} + \mu_2 \leq 1$ and take uniformization constant 1. First we describe the resulting transition probabilities for $n \in \{0, 1, \dots, N_1\} \times \{0, 1, \dots, N_2\}$

$$P(n, n + e_1) = \lambda \mathbf{1}(n_1 < N_1), \quad P(n, n - e_2) = \mu_2 \mathbf{1}(n_2 > 0), \qquad (45)$$

$$P(n, n + d_1) = \begin{cases} \mu_1 \mathbf{1}(n_1 > 0, n_2 < N_2), & n_1 \leq T, \\ \mu_1^* \mathbf{1}(n_2 < N_2), & n_1 > T, \end{cases} \qquad (46)$$

$$P(n, n) = 1 - \sum_{u \in \{e_1, d_1, -e_2\}} P(n, n + u), \qquad (47)$$

where $e_1 = (1, 0)$, $d_1 = (-1, 1)$ and $e_2 = (0, 1)$.

We see that $\{0, 1, \dots, N_1\} \times \{0, 1, \dots, N_2\}$ forms a communicating class. Next, we define transition probabilities for the states outside $\{0, 1, \dots, N_1\} \times \{0, 1, \dots, N_2\}$ in such a way that these states are transient. The remaining transition probabilities are

$$P(n, n + e_1) = \lambda, \qquad (48)$$

$$P(n, n - e_2) = \mu_2 \mathbf{1}(n_2 > 0), \qquad (49)$$

$$P(n, n + d_1) = \begin{cases} \mu_1 \mathbf{1}(n_1 > 0), & n_1 \leq T, \\ \mu_1^*, & n_1 > T, \end{cases} \qquad (50)$$

$$P(n, n) = 1 - \sum_{u \in \{e_1, d_1, -e_2\}} P(n, n + u), \qquad (51)$$

The transition probabilities of R are shown in Fig. 3.

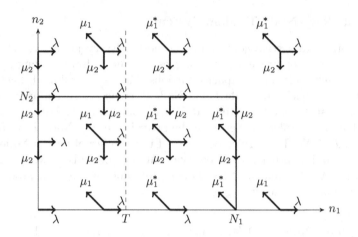

Fig. 3. Transition probabilities of R

The Perturbed Random Walk. For the perturbed random walk, consider an $\bar{R}$ in S with the transition probabilities

$$\bar{P}(n, n + e_1) = \lambda \mathbf{1}(n + e_1 \in S), \quad \bar{P}(n, n - e_2) = \mu_2 \mathbf{1}(n - e_2 \in S), \tag{52}$$

$$\bar{P}(n, n + d_1) = \begin{cases} \mu_1 \mathbf{1}(n + d_1 \in S), & n_1 \leq T, \\ \mu_1^* \mathbf{1}(n + d_1 \in S), & n_1 > T, \end{cases} \tag{53}$$

$$\bar{P}(n, n) = 1 - \sum_{u \in \{e_1, d_1, -e_2\}} P(n, n + u). \tag{54}$$

The transition probabilities of $\bar{R}$ are shown in Fig. 4. We can verify that the stationary probability distribution of $\bar{R}$ is

$$\bar{\pi}(n) = \begin{cases} C \cdot \rho_1^{n_1} \sigma^{n_2}, & n_1 \leq T, \\ C \cdot \rho_1^T \rho_2^{n_1 - T} \sigma^{n_2}, & n_1 > T, \end{cases} \tag{55}$$

where $\rho_1 = \lambda/\mu_1$, $\rho_2 = \lambda/\mu_1^*$, $\sigma = \lambda/\mu_2$ and C is the normalization constant, i.e., $C^{-1} = (1 - \rho_1)^{-1}(1 - \rho_1^{T+1})(1 - \sigma)^{-1} + \rho_1^T \rho_2 (1 - \rho_2)^{-1}(1 - \sigma)^{-1}$.

We consider two performance measures, namely the probability that an arriving job is rejected and the average number of jobs in the system. For the first performance measure $F(n) = \mathbf{1}(n_1 = N_1)$ and for the second $F(n) = n_1 + n_2$. Consider a symmetric scenario, i.e., $N_1 = N_2$. In the numerical example, take for example $T = 4$, $\lambda/\mu_1 = 1/2$, $\lambda/\mu_1^* = 1/3$ and $\lambda/\mu_2 = 1/3$. In Fig. 5, we plot bounds on $\mathcal{F}$ for various N_1. In addition, we plot the upper bound given by the comparison result in Problem 2. The upper and lower bounds are denoted by $\mathcal{F}_u$ and $\mathcal{F}_l$ respectively, and the upper bound given by comparison result is denoted by $\mathcal{F}_u^{(c)}$. Note that in Fig. 5(a) the y-axis is in logarithm scale. Moreover, in Fig. 5(a), although it is subtle $\mathcal{F}_u^{(c)}$ still provides a slightly better upper bound than $\mathcal{F}_u$.

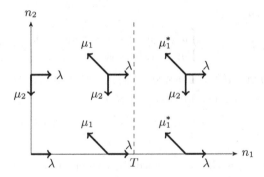

Fig. 4. State space and transition rates of $\bar{R}$

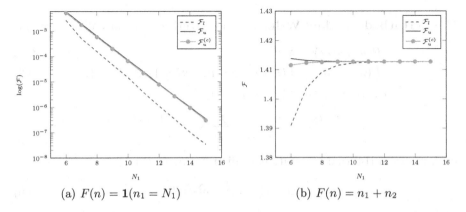

(a) $F(n) = \mathbf{1}(n_1 = N_1)$ (b) $F(n) = n_1 + n_2$

Fig. 5. Bounds on $\mathcal{F}$ for various N_1: $N_1 = N_2$, $T = 4$, $\lambda/\mu_1 = 1/2$, $\lambda/\mu_1^* = 1/3$, $\lambda/\mu_2 = 1/3$

6.2 Tandem System with Boundary Speed-Up or Slow-Down

Consider a tandem system containing M nodes. Every job arrives at node 1 and goes through all the nodes to receive its service from each node. In the end, the job leaves the system through node M. Let λ be the arrival rate. Moreover, we assume that each server has the service rate μ when there are jobs in the queue. For server 1, the service rate changes to μ^*, if all the other queues become empty. Let $\mu^* = \eta \cdot \mu$. For the stability of the system, assume that $\lambda/\mu < 1$.

The Original Random Walk. In this example, we have $S = \{0, 1, \ldots\}^M$. Notice that the tandem system described above is a continuous-time system. Therefore, we use the uniformization method to transform the continuous-time tandem system into a discrete-time R. Without loss of generality, assume that $\lambda + \max\{\mu, \mu^*\} + 2\mu \leq 1$. Hence, we take the uniformization constant 1. Then, the non-zero transition probabilities of the discrete-time R are given below.

$$P(n, n + e_1) = \lambda, \tag{56}$$

$$P(n, n + d_1) = \begin{cases} \mu^*, & \text{if } n_2 = \cdots = n_M = 0, \\ \mu, & \text{otherwise,} \end{cases} \tag{57}$$

$$P(n, n + d_i) = \mathbf{1}\,(n + d_i \in S)\,\mu, \quad \forall i = 2, \ldots, M - 1, \tag{58}$$

$$P(n, n - e_M) = \mathbf{1}\,(n - e_M \in S)\,\mu, \tag{59}$$

$$P(n, n) = 1 - \sum_{u \in \{e_1, d_1, \ldots, d_{M-1}, e_M\}} P(n, n + u), \tag{60}$$

for all $n \in S$, where e_i is the vector with the i-th entry being 1 and all the other entries being 0 and d_i is the vector with the i-th entry being -1, the $i + 1$-th entry being 1 and all the others being 0.

The Perturbed Random Walk. For the perturbed random walks $\bar{R}$, we take

$$\bar{P}(n, n + e_1) = \lambda, \tag{61}$$

$$\bar{P}(n, n + d_i) = \mathbf{1}\,(n + d_i \in S)\,\mu, \quad \forall i = 1, \ldots, M - 1, \tag{62}$$

$$\bar{P}(n, n - e_M) = \mathbf{1}\,(n - e_M \in S)\,\mu, \tag{63}$$

$$\bar{P}(n, n) = 1 - \sum_{u \in \{e_1, d_1, \ldots, d_{M-1}, e_M\}} \bar{P}(n, n + u). \tag{64}$$

We know from [14] that the stationary distribution of $\bar{R}$ is,

$$\bar{\pi}(n) = (1 - \rho)^M \cdot \rho^{\sum_{i=1}^{M} n_i}, \tag{65}$$

where $\rho = \lambda/\mu$.

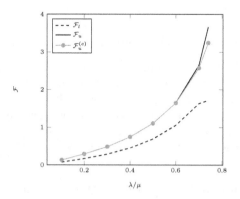

Fig. 6. Bounds on $\mathcal{F}$ for various λ/μ: $F(n) = n_1$, $\mu^* = 1.5\mu$

As the performance measure, we consider the average number of jobs in the first queue (including the job in service), *i.e.*, $F(n) = n_1$. First we consider

Table 1. Bounds on $\mathcal{F}$ for various M: $F(n) = n_1$, $\lambda/\mu = 0.5$, $\mu^* = 0.75\mu$

M	$\mathcal{F}_l$	$\mathcal{F}_u$	$\mathcal{F}_u^{(c)}$
2	1.0	1.4667	1.4667
3	1.0	1.26	1.26
4	1.0	1.0843	1.0843
5	1.0	1.0717	1.0717

the case that $M = 3$, *i.e.*, a three-node tandem system. We take $\eta = 1.5$ and consider various values of λ/μ. The bounds and comparison result are given in Fig. 6. When the load is larger than 0.75, the problems for both upper and lower bound are infeasible. Hence, the results for these cases are not included.

Next, we fix $\lambda/\mu = 0.5$, $\eta = 0.75$ and consider various values for M. The bounds on $\mathcal{F}$ and the comparison result are given in Table 1. As we see from Table 1, our numerical program can be applied to higher-dimensional space.

7   Conclusions and Discussion

In this paper, we have considered random walks in M-dimensional positive orthant. Given a non-negative C-linear function, we have formulated optimization problems that provide upper and lower bounds on the stationary performance measure. Moreover, we have shown that these optimization problems can be reduced to linear programs with a finite number of variables and constraints.

Through numerical experiments, we see that the linear programs for upper and lower bounds are not always feasible. In particular, for some models, once the load exceeds some threshold the problems become infeasible. It will be of interest in future work to gain additional insight and to generalize our methodology to also work with these models. Another interesting direction is to explore how to choose the objective function of Problem 5 such that it improves the error bound. Finally, as the dimension increases, the number of variables and constraints in Problems 5 and 6 increase exponentially. It is of interest to develop methods to reduce the number of variables and constraints, especially for models where M is large.

Acknowledgments. Xinwei Bai acknowledges support by a CSC scholarship [No. 201407720012].

References

1. Abate, J., Whitt, W.: Numerical inversion of probability generating functions. Oper. Res. Lett. **12**(4), 245–251 (1992)
2. Cohen, J.W., Boxma, O.J.: Boundary Value Problems in Queueing System Analysis. North Holland (1983)

3. van Dijk, N.M.: Perturbation theory for unbounded Markov reward processes with applications to queueing. Adv. Appl. Probab. **20**(1), 99–111 (1988)
4. van Dijk, N.M.: Simple bounds for queueing systems with breakdowns. Perform. Eval. **8**(2), 117–128 (1988)
5. van Dijk, N.M.: Bounds and error bounds for queueing networks. Ann. Oper. Res. **79**, 295–319 (1998)
6. van Dijk, N.M.: Error bounds and comparison results: the Markov reward approach for queueing networks. In: Boucherie, R.J., van Dijk, N.M. (eds.) Queueing Networks: A Fundamental Approach, International Series in Operations Research & Management Science, vol. 154. Springer, Boston (2011). https://doi.org/10.1007/978-1-4419-6472-4_9
7. van Dijk, N.M., Lamond, B.F.: Simple bounds for finite single-server exponential tandem queues. Oper. Res. **36**, 470–477 (1988)
8. van Dijk, N.M., Miyazawa, M.: Error bounds for perturbing nonexponential queues. Math. Oper. Res. **29**(3), 525–558 (2004)
9. Fayolle, G., Iasnogorodski, R., Malyshev, V.A.: Random Walks in the Quarter-Plane: Algebraic Methods, Boundary Value Problems and Applications, vol. 40. Springer, Heidelberg (1999). https://doi.org/10.1007/978-3-642-60001-2
10. Goseling, J., Boucherie, R.J., van Ommeren, J.C.W.: A linear programming approach to error bounds for random walks in the quarter-plane. Kybernetika **52**(5), 757–784 (2016)
11. Grassmann, W.K.: Transient solutions in Markovian queueing systems. Comput. Oper. Res. **4**(1), 47–53 (1977)
12. Gurobi Optimization, LLC: Gurobi Optimizer Reference Manual (2018). http://www.gurobi.com
13. Hart, W.E., et al.: Pyomo–Optimization Modeling in Python, vol. 67, 2nd edn. Springer, Cham (2017). https://doi.org/10.1007/978-3-319-58821-6
14. Jackson, J.R.: Networks of waiting lines. Oper. Res. **5**(4), 518–521 (1957)
15. Korte, B., Vygen, J.: Combinatorial Optimization, vol. 12. Springer, Heidelberg (2012). https://doi.org/10.1007/978-3-642-24488-9
16. Latouche, G., Ramaswami, V.: Introduction to Matrix Analytic Methods in Stochastic Modeling, vol. 5. SIAM (1999)
17. Masuyama, H., Sakuma, Y., Kobayashi, M.: Simple error bounds for the QBD approximation of a special class of two dimensional reflecting random walks. In: Proceedings of the 11th International Conference on Queueing Theory and Network Applications, p. 23. ACM (2016)
18. Neuts, M.F.: Matrix-Geometric Solutions in Stochastic Models: An Algorithmic Approach. Dover Publications (1981)
19. Resing, J., Örmeci, L.: A tandem queueing model with coupled processors. Oper. Res. Lett. **31**(5), 383–389 (2003)

Class Aggregation for Multi-class Queueing Networks with FCFS Multi-server Stations

Pasquale Legato[(⊠)] and Rina Mary Mazza

University of Calabria, 87036 Rende, CS, Italy
{legato,rmazza}@dimes.unical.it

Abstract. The analytical solution of a queueing network is an appreciable first option in performance studies of service systems. Under the so-called product-form conditions, Mean Value Analysis (MVA) is the standard algorithm still adopted, but the user has to face the exponential computational complexity in the number of customer classes. In the last three decades, some (pseudo) polynomial approximated variants to MVA have been proposed in literature. These approximations are based on the transformation of the recursive MVA equations into a system of nonlinear equations to be solved iteratively. They are consolidated only with reference to (fixed-rate) single-server stations and are used in practice even though theoretical convergence remains an open problem. In this paper we exploit the possibility of aggregating customer classes in order to replace the exact multi-class MVA by new approximated procedures where MVA has to be run under at the most two customer classes. The resulting procedures are developed around a nested fixed-point iteration schema and are especially suitable for solving large size multi-class networks with multi-server stations under a first-come-first-served discipline. Convergence and accuracy of our procedures are numerically assessed through a very large set of experiments against the exact solution by the multi-class MVA.

Keywords: Queueing networks · Mean Value Analysis · Fixed-point approximation

1 Introduction

Queueing networks with multiple customer classes and multiple server stations are a consolidated mathematical tool for the performance evaluation of both computing and communication systems since a long time. Nowadays, the increasing diffusion of web-based service systems [9] and cloud computing architectures [12], as well as wireless sensor networks [8] calls for the use of analytical tools in service quality evaluation and capacity planning. Despite the celebrated "bigger is better" principle introduced by Kleinrock [7], availability requirements under server failures and other causes of service interruption suggest to partition the given capacity of one big server into a modular set of identical smaller servers.

This paper deals with the approximate analytical solution of multi-class queueing network models of the BCMP type [3]. Service stations can be equipped with multiple identical servers (mutually independent) each having a common fixed-service rate (i.e.

The original version of this chapter was revised: a few formulas were corrected. The correction to this chapter is available at https://doi.org/10.1007/978-3-030-27181-7_24

T. Phung-Duc et al. (Eds.): QTNA 2019, LNCS 11688, pp. 221–239, 2019.
https://doi.org/10.1007/978-3-030-27181-7_14

non load-dependent). To apply the BCMP modeling framework under the FCFS discipline, one has to assume that the average service time is the same for all the job classes and it follows a (negative) exponential probability distribution function. One such application could occur in evaluating query-based wireless sensor networks [6] whenever one has to adopt a closed queueing network model.

The standard way to get the exact solution of a multi-class queueing network of the closed type is by resorting to the classic MVA algorithm [13]. MVA is easy to understand and implement and also amenable to heuristic modifications to cover specific models at hand that do not fall within the BCMP class of queueing networks. Unfortunately, its computational cost becomes prohibitive for models with more than four or five classes and not restricted to a few customers per class. This occurs because the recursive MVA algorithm is characterized by a dependency between some variables being updated at the current iteration and some others that have been calculated at the previous iteration and whose number rapidly increases with the number of classes. Hence, in a multi-class queueing network the computational complexity of MVA is exponential in both the number of iterations and the number of variables.

The contribution of this paper is the proposal and numerical evaluation of a new approximate but (pseudo) polynomial MVA-based algorithm especially aimed at covering the multi-server case under FCFS. For this type of networks, due to the same average service time per class, by running the exact MVA one may verify that just a slight variability of the waiting time per class of customers is usually observed at any given station. Hence, the core idea of our approximation is the aggregation of all the customer classes into one or two classes, followed by the solution of the resulting network by a single or two-class MVA, respectively. The paper is organized as follows. Background and previous work are provided in Sect. 2. The new approximate algorithms based on class-aggregation are presented in Sect. 3. Numerical experiments for accuracy validation are given in Sect. 4. Conclusions are in Sect. 5.

2   Background and Previous Work

To make the paper self-contained, the MVA algorithm of our concern is first resumed.

2.1   MVA Algorithm for Multi-class Multi-Server Networks

The following notation is used from now on.

M = number of stations.

j = $1, \ldots, M$ as station index.

m_j = number of servers at station j.

c = $1, \ldots, C$ as class index.

N_c = number of class c customers in network population.

n_c = number of class c customers in network population at current MVA iteration, ranging from 0 to N_c.

N   = network population vector $\mathbf{N} = (N_1, \ldots, N_c, \ldots, N_C)$.

V_{jc} = expected number of visits to station j by a class c customer.

R_{jc} = expected service duration for a class c customer at station j.

$\mathbf{n}$ = network population vector $\mathbf{n} = (n_1, \ldots, n_c, \ldots, n_C)$ at current MVA iteration.

$\mathbf{n} - 1_c$ = network population vector minus one class c customer $\mathbf{n} - 1_c = (n_1, \ldots, n_c - 1, \ldots, n_C)$ at current MVA iteration.

$D_{jc}(\mathbf{n})$ = expected sojourn time per visit at station j by a class c customer, under $\mathbf{n}$.

$Q_{jc}(\mathbf{n})$ = expected number of class c customers at station j, under $\mathbf{n}$.

$T_c(\mathbf{n})$ = expected network throughput for class c customers, under $\mathbf{n}$.

$T_{jc}(\mathbf{n})$ = expected station j throughput for class c customers, under $\mathbf{n}$.

$P_j(l|\mathbf{n})$ = steady-state marginal probability that $l = \sum_{c=1}^{C} n_c$ customers are present at station j, under $\mathbf{n}$.

According to the above notation, the multi-server MVA algorithm under FCFS discipline and the common expected service duration over all classes ($R_{jc} = R_j$, $c = 1, \ldots, C$) is given below.

MVA Algorithm

For $j = 1$ *to* M **Do**

  $P_j(0|0_1, \ldots, 0_C) = 1$

  For $c = 1$ to C **Do**

    $Q_{jc}(0_1, \ldots, 0_C) = 0$

    For $l = 1$ to m_{j-1} **Do**

      $P_j(l|0_1, \ldots, 0_C) = 0$

For $\mathbf{n} = (0_1, \ldots, 0_C)$ to $(N_1, \ldots, N_C)$ **Do**

  For $c = 1$ to C **Do**

    For $j = 1$ to M **Do**

$$D_{jc}(\mathbf{n}) = \frac{R_j}{m_j}\left[1 + \sum_{d=1}^{C} Q_{jd}(\mathbf{n} - 1_c) + \sum_{l=1}^{m_j - 1}(m_j - l)P_j(l - 1|\mathbf{n} - 1_c)\right] \quad (1)$$

  For $c = 1$ to C **Do**

$$T_c(\mathbf{n}) = n_c / \sum_{j=1}^{M} V_{jc}D_{jc}(\mathbf{n}) \quad (2)$$

  For $j = 1$ to M **Do**

    For $c = 1$ to C **Do**

$$Q_{jc}(\mathbf{n}) = V_{jc}T_c(\mathbf{n})D_{jc}(\mathbf{n}) \quad (3)$$

    For $l = 1$ to m_{j-1} **Do**

$$P_j(l|\mathbf{n}) = (1/l)\left[\sum_{c=1}^{C} R_j V_{jc}T_c(\mathbf{n})P_j(l - 1|\mathbf{n} - 1_c)\right] \quad (4)$$

$$P_j(0|\mathbf{n}) = 1 - 1/m_j\left[\sum_{c=1}^{C} R_j V_{jc}T_c(\mathbf{n}) + \sum_{l=1}^{m_j-1}(m_j - l)P_j(l|\mathbf{n})\right] \quad (5)$$

It is easy to recognize that both the sojourn time Eq. (1) and the marginal probability Eq. (4) are responsible for the exponential computational complexity of MVA. In fact, due to the recursive dependence of both $Q_{jc}(\mathbf{n})$ and $P_j(l|\mathbf{n})$ from the corresponding measures related to the population $\mathbf{n}$ with one class c customer removed, the computational time complexity of the algorithm is:

$$O\left(\prod_{j=1}^{M} m_j \cdot C \cdot \prod_{c=1}^{C} (N_c + 1)\right) \tag{6}$$

Observe that the marginal probabilities are needed for computing the expected values of the customer sojourn time at each station only in the multi-server case. In the particular case of just one server per station, the sojourn time (fundamental) Eq. (1) of the MVA algorithm reduces to the following:

$$D_{jc}(\mathbf{n}) = R_j \left[1 + Q_{jc}(\mathbf{n} - 1_c) + \sum_{d=1, d \neq c}^{C} Q_{jd}(\mathbf{n} - 1_c) \right] \tag{7}$$

This applies whenever the user is not interested in computing marginal probabilities, but expected performance measures only (e.g. sojourn times, queue lengths and throughputs for each station).

2.2 Heuristic Relationships

The previous work that appeared in the literature immediately after the publication of the MVA algorithm [2, 5, 10, 14] is guided by the idea of circumventing the recursive nature of the MVA algorithm. This is accomplished by introducing an approximate relationship between $Q_{jc}(\mathbf{n} - 1_c)$ and $Q_{jc}(\mathbf{n})$ and exploiting the possibility of setting $P_j(l - 1|\mathbf{n} - 1_c)$ equal to $P_j(l - 1|\mathbf{n})$.

The first goal is pursued by introducing a measure (δ) of the change in the fraction of the total number of customers found in station j resulting from the removal of one class c customer out of the (current) network population $\mathbf{n}$:

$$\delta_{jc}(\mathbf{n}) = Q_{jc}(\mathbf{n} - 1_c)/(n_c - 1) - Q_{jc}(\mathbf{n})/n_c \tag{8}$$

Then a relationship amenable to heuristic particularizations of the δ measure is the following:

$$Q_{jc}(\mathbf{n} - 1_c) = (n_c - 1)\left[Q_{jc}(\mathbf{n})/n_c + \delta_{jc}(\mathbf{n})\right] \tag{9}$$

Clearly, the first option consists in setting the δ factor to zero, thus obtaining a proportionality assumption ($(n_c - 1)/n_c$ as proportionality factor) on the relationship between $Q_{jc}(\mathbf{n} - 1_c)$ and $Q_{jc}(\mathbf{n})$. This is known as the Bard-Schweitzer heuristic relationship [2, 14]. According to these authors, one has to further assume that removing one class c customer does not affect the expected queue length of customers belonging to different classes:

$$Q_{jd}(\mathbf{n} - 1_c) = Q_{jd}(\mathbf{n}), \ d = 1, \ldots, C; d \neq c \tag{10}$$

Hence, the complete approximation to Eq. (7) of the exact MVA algorithm for (fixed-rate) single-server stations is obtained:

$$D_{jc}(\mathbf{n}) = R_j \left[1 + \frac{n_c - 1}{n_c} Q_{jc}(\mathbf{n}) + \sum_{d=1, d \neq c}^{C} Q_{jd}(\mathbf{n}) \right] \tag{11}$$

Provided that marginal probabilities are not required but only expected performance measures are required, formula (11) is all the user needs to define a (fixed-point) iterative heuristic algorithm (known as the proportional estimation (PE) algorithm) for queueing networks with fixed-rate single-server stations only. This because the expected system throughput Eq. (2) can be inserted in the formula for the expected station queue length:

$$Q_{jc}(\mathbf{n}) = V_{jc}T_c(\mathbf{n})D_{jc}(\mathbf{n}) \tag{12}$$

thus obtaining the following equation:

$$Q_{jc}(\mathbf{n})/n_c = V_{jc}D_{jc}(\mathbf{n}) / \sum_{i=1}^{M} V_{ic}D_{ic}(\mathbf{n}) \tag{13}$$

Equations (11) and (13) are at the basis of an approximated fixed-point iteration algorithm.

The computational complexity of PE is $O(MC)$ in both space and time (per iteration) requirements. An extensive theoretical study on the existence and uniqueness of the solution of PE, as well as on convergence, has been carried out by Pattipati et al. [11]. In particular, the existence of the solution is established for monotonic, but single-class networks, i.e. networks where the service rates are monotonically non decreasing functions of the number of customers at the stations. Uniqueness and convergence results have been obtained only under the limiting condition that the number of customers of each class grows to infinity.

To cover the case of fixed-rate services in multi-server stations, under the PE assumption on marginal queue lengths, Pattipati et al. [11] mention the following assumption

$$P_j(l-1|\mathbf{n}-1_c) = P_j(l-1|\mathbf{n}) \tag{14}$$

as the simplest choice to be used within Eq. (4). This corresponds to using Poisson arrivals to each station under the further assumption that each station is constrained to accept at the most $N = N_1 + N_2 + \ldots + N_C$ customers; therefore, it is viewed as an $M/M/m_j/N$ station, and then solved as a birth-death process. To employ the formula for marginal probabilities returned by the $M/M/m_j/N$ queueing model for the open station, the average arrival rates at any given station, cumulated over all customer classes $(\lambda_j, j = 1, \ldots, M)$, are set equal to the average station throughput under the further approximation:

$$T_j(\mathbf{n}-1_c) = T_j(\mathbf{n}), T_j(\mathbf{n}) = \sum_{c=1}^{C} V_{jc}T_c(\mathbf{n}) \triangleq \lambda_j, \quad j = 1, \ldots, M \tag{15}$$

More recently, the QD-MVA algorithm [4] provides an extension of the Bard-Schweitzer approximation to the load-dependency function admitted in the BCMP class of closed queueing networks with queue-dependent service requirements. It has been proposed to avoid the computation of state probabilities and operates on mean values only. Numerical results have only been provided for a pure-delay station which supplies one-out-of-two multi-server stations in tandem, under a processor-sharing discipline.

Research efforts aimed to improve the PE approximation have been based on the assumption that the change in the fraction of class r customers present at station j resulting from the removal of one class c customer is constant around the current population vector $\mathbf{n}$. This corresponds to replacing the null setting of the (δ) factor by the constant setting:

$$\delta_{jc}(\mathbf{n}) = \delta_{jc}(\mathbf{n} - 1_c), \quad \forall \mathbf{n} \tag{16}$$

Hence, at the price of introducing a nested fixed-point iteration within the PE (to estimate the constant δ measure), the so-called "Linearizer" heuristics, first proposed by Chandy and Neuse [5], has been obtained. The new assumption on the $\delta_{jc}(\mathbf{n})$ yields an improvement over the PE algorithm at the expense of an acceptable increase of the space complexity $O(MC^2)$ and the time (per iteration) complexity $O(MC^3)$. Some years later, Zahorjan et al. [18] propose to work with cumulative (i.e. over all classes rather than per class) queue lengths in formula (9), namely:

$$Q_j(\mathbf{n} - 1_c) = (n-1)\left[Q_j(\mathbf{n})/n + \delta_{jc}(\mathbf{n})\right] \tag{17}$$

where $\delta_{jc}(\mathbf{n}) = \left[Q_j(\mathbf{n} - 1_c)/(n-1)\right] - \left[Q_j(\mathbf{n})/n\right]$.

Their resulting Aggregate Queue Length (AQL) algorithm allows reducing the space complexity of Linearizer from $O(MC^2)$ to $O(MC)$ and time/iteration complexity from $O(MC^3)$ to $O(MC^2)$, without significantly affecting the accuracy of the algorithm. They have not established formal convergence results for the original Linearizer algorithm, nor for their own variant. Since then, some other variants to both PE and Linearizer and more sophisticated implementations have been proposed [16, 17], but always restricted to fixed-rate services in single-server stations. In our numerical experiments we adopt both the consolidated PE and AQL algorithms as (alternative) inner procedures.

To cope with the case of service stations with load-dependent service rates (hence the multi-server case of interest in this paper), Neuse and Chandy [10] have introduced the SCAT algorithm for the approximate solution of multi-class queueing networks with load-dependent service rate stations. SCAT is based on the idea of reconstructing the profile of the distribution of the marginal probability of having one customer at station j from the estimate of the mean queue length at the same station. To this purpose, Chandy and Neuse assign the whole marginal probability "mass" to the first two integer values neighboring the estimated (generally fractional) value of the mean queue length. This idea is later refined by Akyildiz and Bolch [1], who propose to scatter the assignment of probability mass to a wider range of neighboring integer values. Their scattering is carried out according to a (pseudo) normal distribution function. In this way, they are able to achieve significant improvements upon the original SCAT in some numerical instances. Unfortunately, other numerical evidence for non heavy-loaded networks states that the profile of marginal queue length probability strongly departs from a normal-like one. More recently Suri et al. [15] have pursued the idea of multiplying $Q_{jc}(\mathbf{n} - 1_c)$ by a correction factor within formula (11), in order to cope with the multi-server case. Their factor is aimed to capture the reduction of the expected queue length resulting from the presence of many servers instead of one at the same station. The above factor is defined as an empirical function of both the number of servers and their utilization factor for each station at hand. Unfortunately, this is not appealing for the applications of generic queueing networks in real practice.

3   A New Approximation for Class Invariant Service Times

Our approach for reducing the computational complexity of the multi-class multi-server MVA from exponential to (pseudo) polynomial when computing both queue lengths and marginal probability distributions in multi-server networks under the FCFS discipline is presented in this section. It aims at providing an alternative practical approach to the open station approximation to each multi-server station (the one mentioned by Pattipati et al. [11]). In our opinion, the open station approximation could be poor, especially for service stations supplied by lightly-loaded stations. On the other hand, our approximation method is inspired by the observation that just a slight variability of the customer sojourn time per class occurs under a FCFS discipline. The different average number of visits to the same station by customers belonging to different classes, besides the different population per class, is responsible for the (slight) variability of the waiting time per class. We argue that the different average number of visits per class produces different sampling patterns according to which a longer or shorter queue at that station is observed by arriving customers and then suffered in terms of longer or shorter waiting times.

3.1   One-Phase Class Aggregation

Let us introduce $P_j(l|N)$, $l = 1, \ldots, N$ as the probability that l customers are present at station j under a network population size of $N = N_1 + N_2 + \ldots + N_C$ customers. Instead of $P_j(l - 1|\mathbf{N} - 1_c)$ $l = 1, \ldots, N$, we propose using $P_j(l - 1|N - 1)$ obtained by aggregating all customer classes into a unique representative class of size N. Then, to get an approximate sojourn time equation for our approximate MVA algorithm, we adopt either the PE or AQL approximation for queue lengths.

The concept of representative class is now introduced. We associate a single-class network to a multi-class queueing network characterized by a population vector $\mathbf{N} = (N_1, \ldots, N_c)$, a matrix of visits $\mathbf{V} = (V_{jc}, j = 1, \ldots, M; c = 1, \ldots, C)$ and a vector of service requests $\mathbf{R} = (R_j, j = 1, \ldots, M)$. The single class, here called "representative class", is defined by the following formulas:

$$N = \sum_{c=1}^{C} N_c \tag{18}$$

$$V_j = \sum_{c=1}^{C} V_{jc} T_c(\mathbf{N}) / \sum_{c=1}^{C} T_c(\mathbf{N}), \quad j = 1, \ldots, M \tag{19}$$

It is worth observing that, by simply replacing V_j (V_{jc}) with R_j (R_{jc}) in (19) one could further define the average service time for the representative class whenever one has to deal with different average service times per class, under processor-sharing or other no-waiting service disciplines covered by the celebrated BCMP theorem [3]. In this case, as it will be made clear in the following, one could aggregate all the classes except for the one under evaluation and then repeat the evaluation for the remaining classes one at a time.

For the time being, our reasoning is restricted to the task of aggregating all classes of customers to be serviced under the FCFS discipline. So, the expected throughput (TH) values per class, $T_c(\mathbf{N})$, required by the (aggregation) formulas (18) and (19) to define the representative class are computed iteratively by a fixed-point procedure based either on the PE approximation or the AQL approximation for average queue lengths. Whatever be the approximation adopted, the marginal queue length probabilities referred to the representative class have to be embedded. For sake of clarity, in the following we give the simpler procedure obtained by using the PE approximation (aka *TH_PE*):

***TH_PE* Procedure**
Read $P_j(l - 1|N - 1), l = 1, ..., m_j - 1; j = 1, ..., M$
Initialize $D_{jc}(\mathbf{N}) = R_j, c = 1, ..., C; j = 1, ..., M$
Repeat
  For $c = 1$ to C **Do**
    $$T_c(\mathbf{N}) = N_c / \sum_{i=1}^{M} V_{ic} D_{ic}(\mathbf{N}) \tag{20}$$
    For $j = 1$ to M **Do**
    $$Q_{jc}(\mathbf{N}) = V_{jc} T_c(\mathbf{N}) D_{jc}(\mathbf{N}) \tag{21}$$
    $$D_{jc}(\mathbf{N}) = \frac{R_j}{m_j} \left[\begin{array}{l} 1 + \frac{N_c - 1}{N_c} Q_{jc}(\mathbf{N}) + \sum_{\substack{d=1 \\ d \neq c}}^{C} Q_{jd}(\mathbf{n}) \\ + \sum_{l=1}^{m_j - 1} (m_j - 1) P_j(l - 1|N - 1) \end{array} \right] \tag{22}$$
Until convergence upon $D_{jc}(\mathbf{N}), c = 1, ..., C; j = 1, ..., M$

The computational time complexity of *TH_PE* is:

$$O\left(\sum_{j=1}^{M} m_j \cdot C \right) \text{ per iteration} \tag{23}$$

Observe that replacing the PE approximation with the AQL approximation requires using also $P_j(l - 1|N - 2), j = 1, ..., M, l = 1, ..., m_j - 1$ because a nested fixed-point iteration occurs in order to calculate δ_{jc} according to (17). The resulting *TH_AQL* procedure is a bit more elaborated than *TH_PE* but straightforward from the algorithmic presentation in [18] and, therefore, it is omitted here.

Whatever be the per class throughput approximation (by PE or AQL) used for computing the average number of visits per station (19) related to the representative class, the resulting procedure (*TH_PE* or *TH_AQL*) has to be nested into an outer fixed point-iterative algorithm. This algorithm will be referred as *1P_MP_PE* or *1P_MP_AQL* to specify that, under a one-phase aggregation (1P) of customer classes, marginal probabilities (MP) are computed by the per class throughputs returned from *TH_PE* or *TH_AQL*, respectively.

Both *1P_MP_PE* and *1P_MP_AQL* refine the definition of the representative class and evaluate the marginal queue length probabilities at each station for the representative class on the basis of the following reasoning. If we had the exact throughput values $T_c(\mathbf{N})$ $c = 1, \ldots, C$, then the expected number of visits to define the representative class would be determined from (19) once and for all. Instead, since this is not the case and moreover the above throughputs depend in turn from the marginal probabilities of the representative class, we have to resort to a nested fixed-point algorithm where both the expected throughputs and the marginal probabilities are iteratively refined. Thus, *1P_MP_PE* and *1P_MP_AQL* require an initial estimate of the marginal probabilities of the representative class at each station and then they iteratively updates the initial estimate until convergence is achieved. Both *TH _PE* and *TH_AQL* procedures operate as a nested "repeat until" statement to produce an accurate approximation of the per class network throughputs. Given the current estimate of the marginal probabilities, *TH_PE* or *TH_AQL* computes the corresponding expected throughput values per class needed to update the visits parameters (19) for the representative class.

For sake of clarity, the *1P_MP_PE* procedure is summarized here:

1P_MP_PE Procedure
Initialize $P_j(l - 1|N - 1)$, $l = 1, \ldots, m_j - 1; j = 1, \ldots, M$
Repeat

    1. Compute class throughput by *TH_PE*
    2. Define representative class
    3. Solve multi-server network under representative class by single-class MVA
    4. Update $P_j(l - 1|N - 1)$, $l = 1, \ldots, m_j - 1; j = 1, \ldots, M$

Until convergence upon $P_j(l - 1|N - 1)$, $l = 1, \ldots, m_j - 1; j = 1, \ldots, M$
Return expected performance measures per class and station

Returned performance measures for each (original) class of customers, e.g. expected throughputs and queue lengths at each station, are computed by the following formulas:

$$T_{jc}(\mathbf{N}) = V_{jc}N_c / \sum_{i=1}^{M} V_{ic}D_{ic}(\mathbf{N}) \tag{24}$$

$$Q_{jc}(\mathbf{N}) = T_{jc}(\mathbf{N})D_{jc}(\mathbf{N}) \tag{25}$$

where $D_{jc}(\mathbf{N})$ values, for $= 1, \ldots, M$ and $c = 1, \ldots, C$, are those returned by the inner *TH_PE* procedure once that the *1P_MP_PE* outer procedure is no longer able to update the marginal queue length probabilities $P_j(l - 1|N - 1)$, $l = 0, 1, \ldots, m_j - 1$ which are used in the Eq. (22).

The computational time complexity of $1P\_MP\_PE$ is:

$$O\left(\sum_{j=1}^{M} m_j \cdot (C+N)\right) \text{ per iteration} \tag{26}$$

For later numerical comparisons it is worth presenting the procedure of reference called $M/M/m_j/N\_AQL$ [11] where the AQL approximation is used in conjunction with the open station approximation under Poisson arrivals and at the most N customers in that station.

For sake of clarity, the procedure is resumed here:

$M/M/m_j/N\_$ AQL Procedure
Initialize $P_j(l-1|N-1), P_j(l-1|N-2), l = 1,\ldots,m_j-1; j = 1,\ldots,M$
Repeat

  1. Compute network throughput for each class by TH_AQL
  2. Compute Poisson arrival rates for each station by (15)
  3. Compute the marginal probabilities from the basic model $M/M/m_j/N$
  4. Update $P_j(l-1|N-1), P_j(l-1|N-2)\ l = 1,\ldots,m_j-1; j = 1,\ldots,M$

Until convergence upon $P_j(l-1|N-1),\ P_j(l-1|N-2)\ \ l = 1,\ldots,m_j-1;\ j = 1,\ldots,M$
Return expected performance measures per class and station by (24) and (25)

Observe that the set of probability values $P_j(l-1|N-2)\ l = 1,\ldots,m_j-1$ are required by the need of estimating the δ factor in (17) within the AQL approximation at the basis of the TH_AQL procedure time.

3.2 Two-Phase Class Aggregation

Whatever be the inner approximation (PE or AQL) used in the one-phase aggregation ($1P\_MP\_*$) algorithm to estimate the expected marginal queue length at each multi-server station, $Q_j(\mathbf{N}-1_c)$, it is clear that it plays a significant role in the quality of the overall approximation algorithm. In this section, we present an improved procedure which is based on the idea of using class aggregation two times. The first time (i.e. phase_1) it is used to obtain an estimate of $Q_j(\mathbf{N}-1_c)$; the second time (i.e. phase_2) it is used to obtain $P_j(l-1|N-1)$. To estimate $Q_j(\mathbf{N}-1_c)$ we aggregate all the customer classes except for class c (when class c is under evaluation) and then we use a two-class MVA algorithm where all the other classes are aggregated into a unique representative class, under an estimate of $P_j(l-1|N-2)$ which will be refined at the end of the second phase. The second phase uses $Q_j(\mathbf{N}-1_c)\ c = 1,\ldots,C$ as returned from phase_1 and executes just one step of the multi-class MVA algorithm to estimate the per class network throughputs ($T_c(\mathbf{N})$) under the network population vector $\mathbf{N}$. This allows to use class aggregation again, but now for all the customer classes. The network under the new representative class is then solved by a single-class MVA in order to update the estimates of the marginal queue lengths probabilities until convergence. In particular, $P_j(l-1|N-2)$ is used in phase_1 to estimate $Q_j(\mathbf{N}-1_c)$ while

$P_j(l - 1|N - 1)$ is subjected to an iterative refinement in phase_2, by running the single-class MVA algorithm under the representative class.

The two-phase (2P) class aggregation procedures based on the PE or AQL approximations will be referred as $2P\_MP\_PE$ or $2P\_MP\_AQL$, respectively. For completeness, the $2P\_MP\_PE$ is resumed in the following.

$2P\_MP\_PE$ Procedure
Phase 1: aggregate all classes except for class c and compute $Q_j(N - 1_c)$
For $c = 1$ to C Do

1. Compute $T_d(N - 1_c)$, $d = 1, \ldots, C$ and $d \neq c$ by TH_PE using $P_j(l - 1|N - 2)$, $l = 1, \ldots, m_j - 1$
2. Define the representative class for all $C - 1$ classes (except for c)
3. Execute the two-class MVA and return $Q_j(N - 1_c)$, $j = 1, \ldots, M$

Phase 2: aggregate all classes and compute $P_j(l - 1|N - 2)$ and $P_j(l - 1|N - 1)$
Initialize $P_j(l - 1|N - 1)$, $l = 1, \ldots, m_j - 1$; $j = 1, \ldots, M$
Repeat

1. Execute one step of MVA with all C classes using $Q_j(N - 1_c)$, $j = 1, \ldots, M$, $c = 1, \ldots, C$
2. Define the representative class for all classes
3. Solve multi-server network under unique representative class by MVA
4. Update $P_j(l - 1|N - 1)$, $l = 1, \ldots, m_j - 1$; $j = 1, \ldots, M$

Until convergence upon $P_j(l - 1|N - 2)$ and $P_j(l - 1|N - 1)$, $l = 1, \ldots, m_j - 1$; $j = 1, \ldots, M$

The two-phase class aggregation procedure results in an acceptable increase of the computational time complexity with respect to the simpler one-phase aggregation procedure:

$$O\left(\sum_{j=1}^{M} m_j \cdot C^2 \cdot N^2\right) \tag{27}$$

Providing a formal analysis of convergence for any of our aggregation based fixed-point iteration procedures is very unlikely, as experienced since the proposal of the (inner) PE procedure and related successive efforts [11]. Rather, we assess the mechanism of class aggregation through extensive numerical experiments. It provides an accuracy level that is deemed adequate for practical purposes.

4   Numerical Validation

To assess both the accuracy and the convergence of our procedures, an ad-hoc suite has been implemented for the random generation and automated solution of hundreds of networks all at one time. The input data setting defining each network instance may be changed by using a text editor. Clearly, a set of solutions may be identified and, upon occurrence, rerun by simply repeating the initial seed which identifies the specific experimental run. Here we give a report on our extensive numerical experience with all the procedures based on the idea of class aggregation in comparison with the $M/M/m_j/N\_AQL$ consolidated procedure of reference from the literature [11]. Numerical results from all the approximate procedures are compared against the exact solution returned by the multi-class multi-server MVA algorithm. Using the one- and two-phase class aggregations, we have tested both the PE and AQL variants, thus obtaining, respectively, *1P_MP_PE, 1P_MP_AQL, 2P_MP_PE* and *2P_MP_AQL*. These procedures have been evaluated by solving thousands of network instances bearing a classical central server topology. This topology does not restrict the generality of our results since our aggregation procedures combined with the single class MVA-based approximation do not depend on network topology. Our approximations are obviously sensitive to the different average number of visits per station attributed to each class of customers to be aggregated. On the other hand, the open station decomposition approximation underlying the $M/M/m_j/N\_AQL$ procedure is expected to result more adequate when the central station acts as a bottleneck for the queueing network. In this case, the output flow of customers is more likely to be approximated by an exponential renewal (service) process supplying all the peripheral stations.

This stated, it is our belief that the generation of network instances at the basis of a validation study has to be random but not entirely blind, i.e. the generation has to be driven and controlled to make the set of solved instances significant for practice and, therefore, free from outliers, i.e. cases where the server utilization factor at any given station is too close to 1 or too close to 0 and from here on referred to as "anomalous". To avoid outliers, special attention during the generation of random instances has been paid by controlling the resulting "loading factor" per station that is defined as the product of "the average number of visits x the average service time/the number of servers". The loading factor per station determines the related "server utilization factor" (SUF) as the major output measure upon which we have focused our effort in order to solve random, but significant queueing networks. A significant network instance, for validation purposes, occurs when high SUF values occur at one or more (bottleneck) stations, i.e. SUF $\geq$ 0.90. High SUFs have been pursued in our experiments in order to get an effective evaluation of the numerical accuracy by which the procedures under testing are able to return good estimations of the marginal queue length probabilities involved in the fundamental MVA Eq. (1). As a matter of fact, at low values of SUF the weight of the above probabilities becomes negligible in (1) since the servers are idle most of the time. So, given that we aim to solve a random set of instances that share a pursued configuration of SUF values among the network stations, we have generated network instances where the population size per class and the station visits per class are differentiated enough. This ensures carrying out an effective evaluation of the procedures based on class aggregation.

In all our experiments we have considered from three to five customer classes, while the population size per class has been uniformly generated within ranges that keep computational time tolerable even when solving a thousand instances. The ranges are: [9–15] for 3-class instances, [6–10] for 4-class instances and [3–6] for 5-class instances. These ranges follow the idea of generating both instances with "many" classes and a "few" customers per class, as well as instances with a "few" classes and "many" customers per class. Before considering the key parameter for evaluating the effect of class aggregation, i.e. the average number of visits per class at the same service station, we have fixed four service stations: a central station with six (fast) servers with an average service time fixed to 0.01 t.u. and three peripheral stations with six (slow) servers with an average service time ranging between 0.02 and 0.04 t.u. The latter service setting at the peripheral stations allows us to push the bottleneck status from the central station to one or a couple of the peripheral stations as the average service values are changed from 0.02 to 0.04. To complete service settings, observe that the average service times have to be considered in conjunction with the randomly generated number of visits. In the experiments presented in this section the average number of visits per class at each station has been uniformly generated within the interval [1–30]. The resulting loading factor per station (product of the service time per customer visit to a station multiplied by the average number of visits) is wide enough to produce one or multiple bottlenecks in unbalanced network instances that seem appropriate for our numerical tests. A typical sample of SUF values for non-anomalous instances are shown in Table 1. They share the common feature of belonging to unbalanced networks with:

(i)   a bottleneck occurring either at the central station (as in model n°1) or at one of the three peripheral stations (as in model n°2);

(ii) a strong bottleneck occurring either at the central station (as in model n°3) or at one of three peripheral stations (as in model n°4);

(iii) two bottlenecks occurring at the central station and at one of the peripheral stations (as in models n°5 and n°6);

(iv) two bottlenecks occurring at two of the three peripheral stations (as in models n° 7 and 8);

(v)   no bottlenecks occurring in moderate-high loaded networks (as in model n°9 and n°10).

As usual with the numerical evaluation of MVA-based approximation algorithms, our tests focus on the expected values for the waiting time, queue length and throughput per class computed at each station to evaluate the accuracy of our procedures based on both the one-phase and two-phase aggregation of customer classes.

Table 1. Sample of SUF values for numerical tests.

Model	Server Utilization Factor (SUF)			
	Station 1	Station 2	Station 3	Station 4
n°1	0.9513	0.7735	0.6999	0.8313
n°2	0.8171	0.9765	0.8538	0.5265
n°3	0.9867	0.7889	0.7995	0.7967
n°4	0.7213	0.9895	0.6667	0.7817
n°5	0.9623	0.6565	0.6994	0.9693
n°6	0.9606	0.5265	0.8567	0.9416
n°7	0.7751	0.9051	0.9785	0.3359
n°8	0.7219	0.4974	0.9536	0.9788
n°9	0.6394	0.7507	0.6525	0.5542
n°10	0.8436	0.7667	0.7903	0.8765

Clearly, due to Little's law, the accuracy on the expected queue lengths depends on the combination of the errors separately reported on waiting times and throughputs.

For each of the above performance indices (I), both the average (*AVG_ER*) and maximum relative error (*MAX_ER*) have been computed by using their exact evaluation returned by the multi-class multi-server MVA. The former is averaged over all the evaluations of the same performance index in every network generated and solved, while the latter is observed throughout the evaluations returned at each station of every network belonging to the set of solved instances (say K instances and $k = 1, \ldots K$), i.e.:

$$MAX\_ER = \max_{k,j,c} \frac{\left| I_{cj}^k(\mathbf{N})^{APPROX} - I_{cj}^k(\mathbf{N})^{EXACT} \right|}{I_{cj}^k(\mathbf{N})^{EXACT}}, \tag{28}$$

$$AVG\_ER = \frac{\sum_{k=1}^{K} \sum_{c=1}^{C_k} \sum_{j=1}^{M} \frac{\left| I_{cj}^k(\mathbf{N})^{APPROX} - I_{cj}^k(\mathbf{N})^{EXACT} \right|}{I_{cj}^k(\mathbf{N})^{EXACT}}}{K \cdot M \cdot \sum_k C_k} \tag{29}$$

where I_{cj}^k refers to the waiting time, queue length and throughput values per class and per station.

Numerical validation of our one-phase and two-phase aggregation-based procedures against results from the multi-class exact MVA algorithm is executed once that convergence is achieved by means of the method of successive substitutions in both the inner and outer procedures implemented within our fixed-point iteration algorithms. Usually, i.e. for non-anomalous instances similar to those shown in Table 1, the (outer) convergence test (on marginal queue length probabilities) is successful within 10 iterations under a level of tolerance that has been fixed to 10^{-3} when computing the relative error, element by element, on the entire updated vector of marginal queue length probabilities. The same tolerance level has been kept for convergence upon the average performance measures (queue lengths, waiting times and throughputs) returned by the PE or AQL procedures at the inner level of our aggregation procedures. The maximum number of allowed iterations has been fixed to 200.

From our extensive numerical experience obtained through thousands of instances, in Table 2 we report on an example of 1000 generated instances of which 856 instances under $0.10 \leq SUF \leq 0.99$ have been selected for solution and the remaining ones (144) have been skipped because anomalous. Observe that the analytical evaluation of the average performance measures under SUF values ranging from 0.99 on is neither necessary in practice, nor reliable from a numerical point when the standard MVA algorithm is implemented (as in our numerical investigations). Under quite unrealistic SUF values greater than 0.99 at any given station, the probability of finding 0, 1 or a few customers at that station (equal to finding 0, 1 or a few busy servers) becomes very close to zero. This may cause numerical instability and large errors in updating probability values from one iteration to another, thus, inflating the relative error upon which the tolerance level for convergence test is performed.

From Table 2, one may immediately appreciate the improvements obtained with the aggregation-based procedures. Whether they be one- or two-phased, AQL-based

Table 2. Results for 856 networks under SUF $\leq$ 0.99 at each station.

Error on expected performance measure (%) per class & station	Algorithm for network solution				
	1P_MP_PE	1P_MP_AQL	2P_MP_PE	2P_MP_AQL	M/M/m$_j$/ N_AQL
Waiting time AVG_ER	4.10	1.39	0.78	0.76	1.83
Waiting time MAX_ER	14.25	12.26	7.44	7.32	16.34
Queue length AVG_ER	4.40	1.45	0.75	0.73	1.86
Queue length MAX_ER	15.48	13.51	9.14	9.02	17.60
Throughput AVG_ER	1.02	0.36	0.21	0.22	0.99
Throughput MAX_ER	6.15	7.95	3.59	3.63	12.32

aggregation procedures always outperform the $M/M/m_j/N\_AQL$ procedure. The PE-based one-phase procedure outperforms the $M/M/m_j/N\_AQL$ procedure in terms of maximum errors for all the performance measures, but it underperforms in the estimation of their average values. Most importantly, the results in Table 2 show the significant improvement of accuracy achieved by the two-phase aggregation procedures: the average relative error is kept well below 1% for both waiting times and queue lengths and 0.5% for throughput values. Taking a deeper look at the numerical differences between $2P\_MP\_PE$ and $2P\_MP\_AQL$, one may recognize that there is no reason to implement the more sophisticated AQL approximation in place of the PE variant. For the same instances, it is even more important to observe that the maximum relative error registered is below 10% for both the waiting times and queue lengths and 5% for the throughput estimates.

To further assess the accuracy of our aggregation-based procedures under a couple of lower-range SUF values (i.e. 0.95 and 0.90 instead of 0.99) that are more likely to occur in practical usage of analytical solution of queueing networks, we now show in Tables 3 and 4 the results of the numerical comparisons for the two subsets of instances sharing the new constraints on the maximum SUF value per station.

Table 3. Results for 391 networks under SUF $\leq$ 0.95 at each station.

Error on expected performance measure (%) per class & station	Algorithm for network solution				
	1P_MP_PE	1P_MP_AQL	2P_MP_PE	2P_MP_AQL	M/M/m$_j$/ N_AQL
Waiting time AVG_ER	0.61	0.81	0.37	0.38	1.04
Waiting time MAX_ER	12.52	12.26	6.50	6.56	16.34
Queue length AVG_ER	1.72	0.85	0.36	0.36	1.08
Queue length MAX_ER	11.93	13.51	6.40	6.39	17.60
Throughput AVG_ER	0.45	0.23	0.12	0.13	0.78
Throughput MAX_ER	5.67	7.95	3.59	3.63	12.32

Table 4. Results for 169 networks under SUF $\leq$ 0.90 at each station.

Error on expected performance measure (%) per class & station	Algorithm for network solution				
	1P_MP_PE	1P_MP_AQL	2P_MP_PE	2P_MP_AQL	M/M/m$_j$/ N_AQL
Waiting time AVG_ER	0.60	0.39	0.17	0.18	0.52
Waiting time MAX_ER	10.12	11.14	6.16	6.20	15.18
Queue length AVG_ER	0.63	0.42	0.17	0.17	0.53
Queue length MAX_ER	10.36	12.22	6.40	6.39	14.83
Throughput AVG_ER	0.17	0.12	0.06	0.06	0.50
Throughput MAX_ER	4.40	6.35	3.59	3.63	11.02

Both the average relative error and the maximum error on the performance measures decrease as the maximum SUF decreases. In particular, we believe that the *2P_MP_PE* procedure may be recommended in practice: the maximum error on the most critical performance measures, i.e. the expected waiting time and the expected queue length, decreases from 7.44% to 6.16% and from 9.14% to 6.40%, respectively, as the maximum SUF of the solved models decreases from 0.990 to 0.900. Moreover, it is worth remarking that numerical performance of *2P_MP_PE* with respect to throughput evaluation is confirmed as even better than the one achieved by *2P_MP_AQL*. This evidence is not new among researchers who have experienced the use of PE and AQL approximations.

Once that we have successfully assessed the very good accuracy of our two-phase aggregation procedures, some final examples are given in the following to summarize our extensive investigations on the possibility of convergence failure, due to the standard MVA implementation, under very high SUF values in unbalanced networks. To this purpose, we have tested our two-phase aggregation procedures on an additional set of 248 non-anomalous instances out of 1000 that have been generated, all with an SUF ranging from 0.95 to 0.999. The rationale of choosing similar SUF values is twofold: on one hand, we investigate the effect of numerical instability on the convergence of the marginal queue length probabilities; on the other, we validate the superiority of our two-phase procedures also for those type of instances where the $M/M/m_j/N\_AQL$ procedure should be likely to perform better due to the fact that the underlying renewal exponential service assumption on the output flow of each station is more adequate under the above range of SUF values.

We have registered 2 cases of convergence failure for our two-phase aggregation procedures. Our numerical experience suggests that these failures may occur due to stations bearing SUF value very close to 0.999. These failures have been recovered by simply increasing the tolerance level from 0.001 to 0.01 without experiencing an appreciable loss of accuracy, as illustrated in Table 5.

The adequacy of fixing the tolerance level at 0.01 for the above border-line instances when evaluating the relative error on single components of the entire probability vector is confirmed by a further larger sample of 722 non-anomalous instances solved out of 1000 generated. The growth from 248 to 722 is achieved by randomly "slowing down" some servers in order for the corresponding SUF values to rise above

Table 5. Results for 248 networks under $0.95 < \text{SUF} < 0.999$ at each station and TL 0.001/0.01.

Error on expected performance measure (%) per class & station	Algorithm for network solution					
	2P_MP_PE		2P_MP_AQL		M/M/m$_j$/N_AQL	
	Tolerance value for convergence					
	0.001	0.01	0.001	0.01	0.001	0.01
Waiting time AVG_ER	0.24	0.24	0.24	0.24	0.43	0.44
Waiting time MAX_ER	6.48	6.50	7.40	7.42	8.32	8.34
Queue length AVG_ER	0.23	0.23	0.23	0.23	0.44	0.45
Queue length MAX_ER	7.76	7.78	8.97	8.99	7.49	7.56
Throughput AVG_ER	0.04	0.04	0.04	0.04	0.10	0.10
Throughput MAX_ER	1.22	1.21	1.62	1.61	3.00	3.04

0.95. Companion results are reported in Table 6, after that 10 occurrences of convergence failure in our two-phase aggregation procedures have been excluded. Numerical instabilities and negative probabilities observed in the standard implementation of the MVA algorithm under SUF values higher than 0.995 are responsible for these failures.

Table 6. Results for 722 networks under $0.95 < \text{SUF} < 0.999$ at each station.

Error on expected performance measure (%) per class & station	Algorithm for network solution					
	2P_MP_PE		2P_MP_AQL		M/M/m$_j$/N_AQL	
	Tolerance value for convergence					
	0.001	0.01	0.001	0.01	0.001	0.01
Waiting time AVG_ER	0.53	0.53	0.50	0.50	1.31	1.33
Waiting time MAX_ER	5.48	6.32	5.19	5.51	9.12	9.13
Queue length AVG_ER	0.52	0.52	0.47	0.50	1.37	1.40
Queue length MAX_ER	7.99	8.04	6.94	6.98	7.56	7.55
Throughput AVG_ER	0.11	0.11	0.11	0.11	0.26	0.27
Throughput MAX_ER	1.69	1.69	1.38	1.39	3.22	3.23

Overall, we may conclude that final results in Tables 5 and 6 confirm the effectiveness of our two-phase aggregation procedures, in particular the *2P_MP_PE*, whatever be the position of the bottleneck within the network (central of peripheral), if any, as long as it remains limited to non-anomalous SUF values (below 0.99) when using the standard implementation of the MVA algorithm.

5   Conclusions

Out of all of the approximation procedures investigated, the two-phase fixed-point iteration schema may be recommended for solving large multi-class queueing networks with multi-server stations. Polynomial complexity per iteration is obtained by the aggregation of all customer classes except for one in the first phase and then by aggregating all the classes into a unique representative in the second phase. Besides the average performance measures per class at each service station, the marginal queue length probabilities per class are returned upon convergence. A formal proof of convergence seems difficult and even unlikely, considering the past efforts put into this type of iterative schemas. Nevertheless, a very safe behavior in convergence, along with the capability of keeping the maximum relative error upon standard performance measures below 10%, is assessed after thousands of numerical experiments. Unless very high levels of the utilization factor of the generic server (SUF) at each station are overcome, our two-phase aggregation procedure, under both the so-called proportional estimation or the aggregate queue length approximation used in conjunction with the standard implementation of the MVA algorithm, may be useful in real practice. No convergence problems have to be taken into account under a server utilization factor less than 0.99, when solving queueing networks with one or more bottleneck stations. Whenever this safeguard is respected, just a few iterations (less than 10 outer iterations) are usually needed to achieve convergence on the marginal probabilities for the representative class at each station.

References

1. Akyildiz, I.F., Bolch, G.: Mean value analysis approximation for multiple server queueing networks. Perform. Eval. **8**, 77–91 (1988)
2. Bard, J.: Some extensions to multi-class queueing network analysis. In: Arato, M., Butrimenko, A., Gelenbe, E. (eds.) Performance of Computer Systems, pp. 51–62. North Holland, Amsterdam (1979)
3. Baskett, F., Chandy, K.M., Muntz, R.R., Palacios, F.G.: Open, closed, and mixed networks of queues with different classes of customers. J. ACM **22**(2), 248–260 (1975)
4. Casale, G., Pérez, J.F., Wang, W.: QD-AMVA: evaluating systems with queue-dependent service requirements. Perform. Eval. **91**, 80–98 (2015)
5. Chandy, K.M., Neuse, D.: Linearizer: a heuristic algorithm for queueing network models of computing systems. Commun. ACM **25**(2), 126–134 (1982)
6. Degirmenci, G., Kharoufeh, J.P., Baldwin, R.O.: On the performance evaluation of query-based wireless sensor networks. Perform. Eval. **70**, 124–147 (2013)
7. Kleinrock, L.: Queueing Systems: Computer Applications, vol. 2, 1st edn. Wiley, New York (1976)
8. Lenin, R.B., Ramaswamy, S.: Performance analysis of wireless sensor networks using queueing networks. Ann. Oper. Res. **233**, 237–261 (2015)
9. Menascé, D.A., Almeida, V.A.: Capacity Planning for Web Services: Metrics, Models, and Methods. Prentice Hall, Upper Saddle River (2002)
10. Neuse, D., Chandy, K.M.: SCAT: a heuristic algorithm for queueing network models of computing systems. ACM SIGMETRICS Perform. Eval. Rev. **10**(3), 59–79 (1981)

11. Pattipati, K.R., Kostreva, M.M., Teele, J.L.: Approximate mean value analysis algorithms for queueing networks: existence, uniqueness, and convergence results. J. ACM **37**, 643–673 (1990)
12. Raei, H., Yazdani, N., Shojaee, R.: Modeling and performance analysis of cloudlet in mobile cloud computing. Perform. Eval. **107**, 34–53 (2017)
13. Reiser, M., Lavenberg, S.S.: Mean-value analysis of closed multichain queueing networks. J. ACM **27**(2), 312–322 (1980)
14. Schweitzer, P.J.: Approximate analysis of multi-class closed networks of queues. In: Arato, M., Butrimenko, A., Gelenbe, E. (eds.) Proceedings of the International Conference on Stochastic Control and Optimization, pp. 25–29, Amsterdam, Netherlands (1979)
15. Suri, R., Sahu, S., Vernon, M.: Approximate mean value analysis for closed queueing networks with multiple-server stations. In: Bayraksan, G., Lin, W., Son, Y., Wysk, R. (eds.) Proceedings of the 2007 Industrial Engineering Research Conference, pp. 1–6. Tennessee, USA (2007)
16. Wang, H., Sevcik, K.C.: Experiments with improved approximate mean value analysis algorithms. Perform. Eval. **39**, 189–206 (2000)
17. Wang, H., Sevcik, K.C., Serazzi, G., Wang, S.: The general form linearizer algorithms: a new family of approximate mean value analysis algorithms. Perform. Eval. **65**, 129–151 (2008)
18. Zahorjan, J., Eager, J.D., Sweillam, H.M.: Accuracy, speed and convergence of approximate mean value analysis. Perform. Eval. **8**, 255–270 (1988)

Stationary Analysis of a Tandem Queue with Coupled Processors Subject to Global Breakdowns

Ioannis Dimitriou$^{(\boxtimes)}$ [ID]

Department of Mathematics, University of Patras, 26500 Patras, Greece
`idimit@math.upatras.gr`
`https://thalis.math.upatras.gr/~idimit/`

Abstract. We consider a tandem queue with coupled processors, which is subject to global breakdowns. When the network is in the operating mode and both queues are non empty, the total service capacity is shared among the stations according to fixed proportions. When one of the stations becomes empty, the total service capacity is assigned to the non-empty station. Moreover, arrival rates depend on the state of the network. The system is described by a Markov modulated random walk in the quarter plane representing the number of jobs in the two stations, and the state of the network. We first apply the power series approximation method to obtain power series expansions of the generating function of the stationary joint queue length distribution for both network states. We also provide a way to derive the generating function of the stationary joint queue length distribution for both network states in terms of the solution of a Riemann-Hilbert boundary value problem. Numerical results are obtained to show insights in the system performance.

Keywords: Tandem queues · Coupled processors ·
Power-series approximation · Boundary value problems

1 Introduction

Queueing networks with service interruptions are known to be adequate models to handle realistic problems in manufacturing, telecommunications etc. Despite their great importance, there has been done very few works involving more than one queue, since even under favourable assumptions, the existence of service interruptions destroys the separability [5] of the appropriate multidimensional Markov process and renders its solution intractable.

The problem becomes even more challenging when we further assume that the nodes are interacting with each other. With the term "interaction", we mean that the service rate at a node depends on the state of the other nodes, i.e., a networks with coupled processors [20,26,32]. For such systems, which do not possess the "product form" solution, analytic methods have been developed in [10,21].

© Springer Nature Switzerland AG 2019
T. Phung-Duc et al. (Eds.): QTNA 2019, LNCS 11688, pp. 240–259, 2019.
https://doi.org/10.1007/978-3-030-27181-7_15

In this work, we go one step further and consider a two-node tandem network with coupled processors, which is subject to global (i.e., network) breakdowns. It is assumed that jobs arrive at the first station according to a Poisson process depended on the state of the network, and require service at both stations before leaving the network. The amounts of work that a job requires at each of the stations are independent, exponentially distributed random variables. When the network is in the operating mode, and both stations are non-empty, the total service capacity is shared between the stations according to fixed proportions.

When one of the stations becomes empty, the total service capacity is given to the non-empty station. When the network is in the set-up mode after a failure occurrence, both stations stop working[1], for an exponentially distributed time period. During set-up period, jobs continue to arrive, but now at a decreased rate in order to avoid further congestion[2]. Such a system is fully described by a random walk in the quarter plane (representing the number of jobs in each node), which is modulated by a two state Markov chain (representing the state of the network).

For such a network we provide two different approaches to investigate its stationary behaviour, named the Power Series Approximation (PSA) method, and the theory of Riemann-Hilbert boundary value problems (BVP).

1.1   Related Work

Most of the existing studies involving breakdowns have concentrated on models with a single job queue served by one or more processors, e.g. [28, 34]. Other related results where jobs from a single source are directed to one of several parallel queues, and where breakdowns result in the loss of jobs, or the direct transfer to other queues are given in [27, 35]. A two-node network subject to breakdowns and repairs was also analysed in [29]. Approximate solutions to obtain performance measures, based on replacing an interruptable server with an uninterruptable but slower one, choosing the new service rate without affecting the overall service capacity was given in [31, 39].

Queues with coupled processors were initially studied in [20]. To gain quantitative insights about the queueing process, the probability generating function (pgf) of the stationary joint queue-length distribution is derived by using the theory of Riemann-Hilbert boundary value problems. Later, in [10] a systematic and detailed study of the technique of studying two dimensional random walks to a boundary value problem was presented, while several numerical issues were also discussed. Important generalizations were given in [4, 9, 13–17, 19, 24, 26, 32] (not exhaustive list) where various two-dimensional queueing models with the aid of the theory of Riemann (-Hilbert) boundary value problems. A tandem queue with two coupled processors was analyzed in [32], while later, computational issues as well as asymptotic results were discussed and presented in [23, 26], respectively.

---

[1] This is natural when we are dealing with queues in series.

[2] Such an operation can be performed by a central scheduler, responsible for the congestion management.

Other approaches to analyze two-dimensional queueing models have been developed in [1] (compensation method), and [6, 25] (power series algorithm). In the latter one, power series expansions of steady-state probabilities as functions of a certain parameter of the system (usually the load) were derived. Recently, the authors in [41] studied generalized processor sharing queues and introduced an alternative method to provide approximated stationary metrics in slotted time generalized processor sharing queues, in which at the beginning of a slot, if both queues are nonempty, a type 1 (resp. type 2) customer is served with probability β (resp. $1 - \beta$); see also [18, 36–38]. There, the authors focus in constructing power series expansions for the pgf of the joint stationary queue length distribution directly from the functional equation, while the obtained power series expansions are in β.

Applications of coupled processor models arise in systems where limited resources are dynamically shared among processors, e.g., in data transfer in bidirectional cable and data networks [26], in bandwidth sharing of data flows [24], in the performance modeling of device-to-device communication [40], to model the complex interdependence among transmitters due to interference [7, 15, 17, 19], as well as in assembly lines in manufacturing [2].

1.2   Our Contribution

In this work we focus on the stationary analysis of two-node tandem queue with coupled processors and network breakdowns.

Applications. Potential application of our system are found in systems with limited capacity, which must be shared in multiple operations. For example, in the modeling of virus attacks or other malfunctions in cable access networks regulated by a request-grant mechanism.

Another application of the model can be found in manufacturing [2], and in particular in an assembly line. There, two operations on each job must be performed using a limited service capacity. To increase the network throughput, we couple the service rates at each of the operations, and thus we use the service capacity of an operation for which no jobs are waiting for the other operation. Such a system is heavily affected by the presence of failures during the job processing, which in turn will definitely deteriorate the system throughput.

Fundamental Contribution. Based on the generating function approach, we apply two different methods to investigate the stationary behaviour of the underline Markov modulated random walk in the quarter plane. First, we apply the power series approximation method, initially introduced in [41] (see also [11, 18, 36, 38]), for two-parallel generalized processor sharing queues, described by a typical random walk in the quarter plane (RWQP). In this work we show that this method is still valid for related *Markov modulated* RWQP (e.g., [4, 13, 14, 30, 33]), and thus, extend the class of models that can be applied. Under such a method we obtain power series expansions of the pgf of the joint stationary distribution for either state of the network. A recursive technique to derive their coefficients is also presented.

Secondly, we also derive the pgfs of the joint stationary queue-length distribution for either state of the network with the aid of the theory of Riemann-Hilbert boundary value problems [10, 20, 21]. It is seen that applying the theory of boundary value problems some further technical difficulties are also arise. More precisely, by applying the generating function approach we first come up with a system of functional equations, which is then reduced to a single fundamental equation.

The rest of the paper is summarized as follows. In Sect. 2 we present the mathematical model in detail and obtain the functional equations along with some preliminary results. Section 3 is devoted to the analysis of the two extreme cases where the total capacity is allocated to one of the two stations, even if both stations are nonempty. In Sect. 4 we apply the power series approximation method for the case where the service capacity is shared by the two stations, where in Sect. 5, we provide a complete analysis on how to obtain the pgfs of the stationary joint queue length distribution in terms of a solution of a Riemann-Hilbert boundary value problem. Numerical validations of the performance metrics obtained by using the PSA and the BVP methods, as well as some observations about how the system parameters affect the system performance for a near priority system are given in Sect. 6.

2   The Model and the Functional Equations

Consider a two-stage tandem queue, where jobs arrive at queue 1 according to a Poisson process with rate depending on the state of the network. In particular, the network is subject to breakdowns which occur according to a Poisson process with rate γ. When a breakdown occurs, both stations stop working for an exponentially distributed time period with rate τ. Thus, the network alternates between the *operating mode* and the *set-up mode*. Denote by $C(t)$ the state of the network at time t, with $C(t) = 0$ (resp. 1), when network is in operating (resp. set-up). When $C(t) = i$, jobs arrive in station 1 according to a Poisson process with rate λ_i, $i = 0, 1$.

Each job demands service at both queues before departing from the network. More precisely, at station j, a job requires an exponentially distributed amount of service with parameter ν_j, $j = 1, 2$. The total service capacity of the tandem network equals one unit of work per time unit. In particular, when both stations are non-empty, station j is served at a rate ϕ_j, $j = 1, 2$, and at a rate 1 when it is the only non empty. Without loss of generality we assume hereon that $\phi_1 = p$ and $\phi_2 = 1 - p$, where $0 \leq p \leq 1$.

Let $Q_j(t)$, $j = 1, 2$, be the number of customers at queue j at time t. Under usual assumptions the stochastic process $X(t) = \{(C(t), Q_1(t), Q_2(t)); t \geq 0\}$ is an irreducible and aperiodic continuous time Markov chain with state space $E = \{0, 1\} \times \mathbb{Z}^+ \times \mathbb{Z}^+$. Denote by $\pi_i(n, k)$ the stationary probability of having n and k customers at stations 1 and 2, respectively, when the network is in state i. The balance equations are given by

$$(\lambda_0 + \gamma)\pi_0(0,0) = \nu_2\pi_0(0,1) + \tau\pi_1(0,0),$$
$$(\lambda_0 + \gamma + \nu_2)\pi_0(0,1) = \nu_1\pi_0(1,0) + \nu_2\pi_0(0,2) + \tau\pi_1(0,1), \quad (1)$$
$$(\lambda_0 + \gamma + \nu_2)\pi_0(0,k) = p\nu_1\pi_0(1,k-1) + \nu_2\pi_0(0,k+1) + \tau\pi_1(0,k), \, k \geq 2$$

$$(\lambda_0 + \gamma + \nu_1)\pi_0(n,0) = \lambda_0\pi_0(n-1,0) + (1-p)\nu_2\pi_0(n,1) + \tau\pi_1(n,0),$$

$$(\lambda_0 + \gamma + p\nu_1 + (1-p)\nu_2)\pi_0(n,1) = \lambda_0\pi_0(n-1,1) + \nu_1\pi_0(n+1,0)$$
$$+ (1-p)\nu_2\pi_0(n,2) + \tau\pi_1(n,1), \, n \geq 1, \quad (2)$$

$$(\lambda_0 + \gamma + p\nu_1 + (1-p)\nu_2)\pi_0(n,k) = \lambda_0\pi_0(n-1,k) + p\nu_1\pi_0(n+1,k-1)$$
$$+ (1-p)\nu_2\pi_0(n,k+1) + \tau\pi_1(n,k), \, n \geq 1, k \geq 2$$

$$(\lambda_1 + \tau)\pi_1(n,k) = \gamma\pi_0(n,k) + \lambda_1\pi_1(n-1,k). \quad (3)$$

Define the probability generating functions of the joint stationary queue length distribution

$$\Pi_i(x,y) = \sum_{n=0}^{\infty}\sum_{k=0}^{\infty} \pi_i(n,k)x^n y^k, \, i = 0, 1, \, |x| \leq 1, |y| \leq 1.$$

Using the balance equations we obtain after some algebra the following system of functional equations

$$R(x,y)\Pi_0(x,y) = A(x,y)\Pi_0(x,0) + B(x,y)\Pi_0(0,y)$$
$$+ C(x,y)\Pi_0(0,0) + \tau xy\Pi_1(x,y), \quad (4)$$

$$\Pi_1(x,y) = \tfrac{\gamma}{D(x)}\Pi_0(x,y),$$

where, $D(x) = \lambda_1(1-x) + \tau$ and

$$R(x,y) = xy(\lambda_0(1-x) + \gamma) + \nu_1 py(x-y) + \nu_2(1-p)x(y-1),$$
$$A(x,y) = (1-p)[\nu_2 x(y-1) + \nu_1 y(y-x)],$$
$$B(x,y) = -\tfrac{p}{1-p}A(x,y), \quad (5)$$
$$C(x,y) = \nu_1(1-p)y(x-y) + \nu_2 px(y-1).$$

Our aim is to solve the system of functional Eq. (4). Substituting the second in (4) to the first one, we obtain the following fundamental functional equation

$$\Pi_0(x,y)[D(x)R(x,y) - \tau\gamma xy] = D(x)\{A(x,y)\Pi_0(x,0) + B(x,y)\Pi_0(0,y) \atop + C(x,y)\Pi_0(0,0)\}. \quad (6)$$

Clearly, $\Pi_0(1,1) + \Pi_1(1,1) = 1$, while by using the second in (4) we obtain $\Pi_1(1,1) = \tfrac{\tau}{\gamma}\Pi_0(1,1)$. Thus, the probabilities of the network state are easily given by

$$\Pi_0(1,1) = \frac{\tau}{\tau + \gamma}, \, \Pi_1(1,1) = \frac{\gamma}{\tau + \gamma}.$$

Let $x = s(y) := \frac{\nu_1 y^2}{\nu_1 + \nu_2(1-y)}$. Note that $A(s(y),y) = B(s(y),y) = 0$. Then, using (6) we obtain

$$\Pi_0(s(y),y) = \frac{D(s(y))C(s(y),y)}{D(s(y))R(s(y),y) - \tau\gamma s(y)y}\Pi_0(0,0). \quad (7)$$

Letting $y \to 1$ in (7) we obtain

$$\Pi_0(0,0) = \frac{\tau}{\tau + \gamma} - \left(\lambda_0 \frac{\tau}{\tau + \gamma} + \lambda_1 \frac{\gamma}{\tau + \gamma}\right)\left(\frac{1}{\nu_1} + \frac{1}{\nu_2}\right). \qquad (8)$$

Note that (8) implies that our network is stable when

$$\lambda_0\left(\frac{1}{\nu_1} + \frac{1}{\nu_2}\right)\frac{\tau}{\tau + \gamma} + \lambda_1\left(\frac{1}{\nu_1} + \frac{1}{\nu_2}\right)\frac{\gamma}{\tau + \gamma} < \frac{\tau}{\tau + \gamma}, \qquad (9)$$

which can be explained by realizing that the left hand side of (9) equals the amount of work brought into the system per time unit, and in order the system to be stable, should be less than the amount of work departing the system per time unit.

Let $\rho_{kj} := \frac{\lambda_k}{\nu_j}$, $k = 0, 1$, $j = 1, 2$, and $\rho_k = \rho_{k1} + \rho_{k2}$, $k = 1, 2$. Then, (8) is rewritten as

$$\Pi_0(0,0) = \frac{\tau}{\tau + \gamma}(1 - \frac{\rho_0 \tau + \rho_1 \gamma}{\tau}).$$

Remark 1. Note that $\lambda_0\left(\frac{1}{\nu_1} + \frac{1}{\nu_2}\right)\frac{\tau}{\tau+\gamma}$ (resp. $\lambda_1\left(\frac{1}{\nu_1} + \frac{1}{\nu_2}\right)\frac{\gamma}{\tau+\gamma}$) refers to the amount of work that arrive at the system per time unit when the network is in the operating mode (resp. in the set-up mode), while a job can depart from the network only when it is in the operating mode, and this is happening with probability $\tau/(\tau + \gamma)$.

3 The Cases $p = 0$ and 1

When $p = 0$ (resp. $p = 1$), the model can be seen as a tandem queues served by a single server, in which preemptive priority is given to station 2 (resp. station 1). It is easily seen that in such cases, the functional Eq. (4) can be easily solved since either the coefficient of $\Pi_0(0, y)$ (when $p = 0$), or the one of $\Pi_0(x, 0)$ (when $p = 1$) is equal to zero.

In case $p = 0$ (i.e., $B(x, y) = 0$), upon a service completion in station 1, the server continues serving the customer in station 2, since station 2 has priority. Thus, in such a case our system reduces to an unreliable queueing system, in which the service time consists of two exponential phases with parameters ν_1 and ν_2, respectively. Note that for $p = 0$,

$$y := \xi(x) = \frac{\nu_2 D(x)}{D(x)(\nu_2 + \lambda_0(1 - x)) + \lambda_1 \gamma(1 - x)},$$

vanishes the left-hand side of (4), and yields

$$\Pi_0(x, 0) = \left(\frac{\tau}{\tau + \gamma}\right)\frac{(1 - \frac{\rho_0 \tau + \rho_1 \gamma}{\tau})\nu_1 \xi(x)(\xi(x) - x)}{\nu_2 x(\xi(x) - 1) + \nu_1 \xi(x)(\xi(x) - x)}.$$

Substituting back in (4) yields

$$\Pi_0(x,y) = \frac{(\frac{\tau}{\tau+\gamma})(1-\frac{\rho_0\tau+\rho_1\gamma}{\tau})D(x)\nu_1\nu_2 x}{D(x)[\lambda_0 y(1-x)+\nu_2(y-1)]+\lambda_1\gamma y(1-x)}$$

$$\times \left\{ \frac{\xi(x)(y-1)(\xi(x)-x)+y(x-y)(\xi(x)-1)}{\nu_2 x(\xi(x)-1)+\nu_1\xi(x)(x-\xi(x))} \right\}, \tag{10}$$

$$\Pi_1(x,y) = \frac{\gamma}{D(x)}\Pi_0(x,y).$$

The case $p=1$ is even more interesting and corresponds to an unreliable tandem queue attended by a single server and preemptive priority for the first station. That is, if upon a customer arrival the server is at the second station, it switches immediately to the first one. Moreover, upon a set-up completion, after a global breakdown, the server will start serving at the first station if there are customers waiting. And this is the case even if a breakdown occurs when was serving a customer at the second station. To the author's best knowledge, that case has never considered before. For $p=1$, (4) reduces to

$$[D(x)(\lambda_0 x(1-x)+\nu_1(x-y))+\lambda_1\gamma x(1-x)]y\Pi_0(x,y)$$
$$= D(x)\{\Pi_0(0,y)[\nu_2 x(1-y)+\nu_1 y(x-y)]+\Pi_0(0,0)\nu_2 x(y-1)\}. \tag{11}$$

Let $x:=u(y)$ the unique root of $D(x)(\lambda_0 x(1-x)+\nu_1(x-y))+\lambda_1\gamma x(1-x)=0$ inside the unit circle. Then, the right-hand side should also vanish and thus,

$$\Pi_0(0,y) = \Pi(0,0)\frac{\nu_2 u(y)(1-y)}{\nu_2 u(y)(y-1)+\nu_1 y(y-u(y))}.$$

Substituting back in (11) yields

$$\Pi_0(x,y) = \frac{D(x)\nu_1 y(x-u(y))}{u(y)[D(x)(\lambda_0 x(1-x)+\nu_1(x-y))+\lambda_1\gamma x(1-x)]}\Pi_0(0,y),$$

$$\Pi_1(x,y) = \frac{\gamma}{D(x)}\Pi_0(x,y). \tag{12}$$

4   The Case $0 < p < 1$: Power Series Approximation in p

In the following, we are going to construct a power series expansion of the pgf $\Pi_0(x,y)$ in p starting by (6). Then, having that result we are able to construct power series expansions of $\Pi_1(x,y)$ in p using the second in (4). With that in mind, let

$$\Pi_j(x,y) = \sum_{m=0}^{\infty} V_m^{(j)}(x,y)p^m, \quad j=0,1. \tag{13}$$

Our aim in the following, is to obtain $V_m^{(j)}(x,y)$, $m \geq 0$, $j=0,1$, by employing an approach similar to the one developed in [11,18,41][3]. Equation (6) is rewritten as

$$G(x,y)\Pi_0(x,y) - G_{10}(x,y)\Pi_0(x,0) - G_{00}(x,y)\Pi_0(0,0)$$
$$= pG_{10}(x,y)[\Pi_0(x,y) - \Pi_0(x,0) - \Pi_0(0,y) + \Pi_0(0,0)], \tag{14}$$

---

[3] See Appendix C for the analyticity of $\Pi_j(x,y)$ close to $p=0$.

where,

$$G(x, y) = D(x)[\lambda_0 y(1 - x) + \nu_2(y - 1)] + \lambda_1 \gamma y(1 - x),$$
$$G_{10}(x, y) = D(x)[\nu_2(y - 1) - \nu_1 y(1 - yx^{-1})],$$
$$G_{00}(x, y) = D(x)\nu_1 y(1 - yx^{-1}).$$

The major difficulty in solving (6) corresponds to the presence of the two unknown boundary functions $\Pi_0(x, 0)$, $\Pi_0(0, y)$. Having in mind that in the left-hand side of (14) there is only one boundary function, we are able to follow the approach in [11, 18, 41]. The next Theorem summarizes the basic result of this section.

Theorem 1. *Under stability condition (9),*

$$V_0^{(0)}(x, y) = \frac{D(x)\nu_1\nu_2(\tilde{Y}(x) - x)(\tilde{Y}(x)(y - 1) + x - y)}{G(x, y)[\nu_2 x(\tilde{Y}(x) - 1) - \nu_1 \tilde{Y}(x)(x - \tilde{Y}(x))]} V^{(0)}(0, 0),$$

$$V_m^{(0)}(x, y) = \frac{G_{10}(x, y)}{G(x, y)} Q_{m-1}(x, y), \; m > 0, \tag{15}$$

$$V_m^{(1)}(x, y) = \frac{\gamma}{D(x)} V_m^{(0)}(x, y), \; m \geq 0,$$

where

$$\tilde{Y}(x) = \frac{\nu_2 D(x)}{D(x)(\nu_2 + \lambda_0(1 - x)) + \lambda_1 \gamma(1 - x)},$$
$$Q_m(x, y) = V_m^{(0)}(x, y) - V_m^{(0)}(x, \tilde{Y}(x)) - V_m^{(0)}(0, y) + V_m^{(0)}(0, \tilde{Y}(x)), \; m \geq 0,$$

and $Q_{-1}(x, y) := 0$.

Proof. The proof follows the lines in [18, 41]. Note that $\Pi_0(x, y)$ is analytic function of p in a neighbourhood of 0. We start by expressing $\Pi_0(x, y)$ in power series expansion of p by using (14), and equating the corresponding powers of p at both sides. This yields

$$V_m^{(0)}(x, y)G(x, y) = G_{10}(x, y)[V_m^{(0)}(x, 0) + P_{m-1}(x, y)] + G_{00}(x, y)V_m^{(0)}(0, 0), \; m \geq 0, \tag{16}$$

where

$$P_m(x, y) = V_m^{(0)}(x, y) - V_m^{(0)}(x, 0) - V_m^{(0)}(0, y) + V_m^{(0)}(0, 0),$$

with $P_{-1}(x, y) = 0$. Note that $G(x, y) = 0$ has a unique zero $y = \tilde{Y}(x)$ such that

$$\tilde{Y}(x) = \frac{\nu_2 D(x)}{D(x)(\nu_2 + \lambda_0(1 - x)) + \lambda_1 \gamma(1 - x)}^4.$$

---

[4] Note that $\tilde{Y}(x) = \xi(x)$, where $\xi(x)$ was defined in Sect. 3 for the case $p = 0$.

It is easy to realize that $|\tilde{Y}(x)| < 1$, for $|x| = 1$. Substituting in (16) we eliminate its left-hand side yielding

$$V_m^{(0)}(x,0) = -\frac{G_{00}(x,\tilde{Y}(x))}{G_{10}(x,\tilde{Y}(x))}V_m^{(0)}(0,0) - P_{m-1}(x,y). \tag{17}$$

Substituting (17) back in (16) yields the first in (15). Then, using the second equation in (4) we derive the coefficients $V_m^{(1)}(x,y)$ in terms of $V_m^{(0)}(x,y)$ as given in the second in (15). From (8) it is readily seen that

$$V_0^{(0)}(0,0) = \frac{\tau}{\tau+\gamma} - \left(\lambda_0\frac{\tau}{\tau+\gamma} + \lambda_1\frac{\gamma}{\tau+\gamma}\right)\left(\frac{1}{\nu_1} + \frac{1}{\nu_2}\right),$$
$$V_m^{(0)}(0,0) = 0, \ m > 0.$$

4.1 Performance Metrics

We focus on the mean queue lengths given by

$$
\begin{aligned}
E(Q_1) &= \sum_{m=0}^{\infty}p^m\frac{\partial}{\partial x}[V_m^{(0)}(x,1) + V_m^{(1)}(x,1)]|_{x=1} \\
&= \frac{\lambda_1\gamma}{\tau(\tau+\gamma)} + (1+\frac{\gamma}{\tau})\sum_{m=0}^{\infty}p^m\frac{\partial}{\partial x}V_m^{(0)}(x,1)|_{x=1}, \\
E(Q_2) &= \sum_{m=0}^{\infty}p^m\frac{\partial}{\partial y}[V_m^{(0)}(1,y) + V_m^{(1)}(1,y)]|_{y=1} \\
&= (1+\frac{\gamma}{\tau})\sum_{m=0}^{\infty}p^m\frac{\partial}{\partial y}V_m^{(0)}(1,y)|_{y=1},
\end{aligned}
\tag{18}
$$

Let

$$v_{m,1} = \frac{\partial}{\partial x}V_m^{(0)}(x,1)|_{x=1}, \ v_{m,2} = \frac{\partial}{\partial y}V_m^{(0)}(1,y)|_{y=1}.$$

Truncation of the power series in (18) yields,

$$
\begin{aligned}
E(Q_1) &= \frac{\lambda_1\gamma}{\tau(\tau+\gamma)} + (1+\frac{\gamma}{\tau})\sum_{m=0}^{M}p^m v_{m,1} + O(p^{M+1}), \\
E(Q_2) &= (1+\frac{\gamma}{\tau})\sum_{m=0}^{\infty}p^m v_{m,2} + O(p^{M+1}).
\end{aligned}
\tag{19}
$$

Truncation yields accurate approximations for p close to 0. However, we have to note the actual calculation of the expressions in (18) requires the computation of the first derivatives of $V_m^{(0)}(x,y)$ for $m \geq 0$. Although we provided an algorithm (see Theorem 1) to calculate these coefficients, the calculation of their first derivatives is far from straightforward due to the extensive use of L'Hopital's rule.

5   The Case $0 < p < 1$: A Riemann-Hilbert Boundary Value Problem

In the following, we proceed with the determination of $\Pi_j(x,y)$, $j = 0,1$, $|x| \leq 1$, $|y| \leq 1$ with the aid of the theory of boundary value problems (BVP). In particular, we first obtain $\Pi_0(x, y)$ in terms of the solution of a Riemann-Hilbert boundary value problem by using (6), and then, we use the second in (4), to finally derive $\Pi_1(x, y)$.

A key step to analyze the functional Eq. (6) is the careful examination of the algebraic curve defined by the kernel equation,

$$H(x, y) := D(x)R(x, y) - \tau\gamma xy = 0. \tag{20}$$

It is easily seen that $H(x, y)$ is a polynomial of third degree in x, and of second degree in y. The study of $H(x, y) = 0$ (see Appendix A) allows to continue the unknown functions $\Pi_0(x, 0)$, $\Pi_0(0, y)$ analytically outside the unit disk, and to reduce their determination to a Dirichlet boundary value problem.

Let $\mathcal{C}_x = \{x \in \mathbb{C} : |x| = 1\}$, $\mathcal{C}_y = \{y \in \mathbb{C} : |y| = 1\}$, $\mathcal{D}_x = \{x \in \mathbb{C} : |x| \leq 1\}$, $\mathcal{D}_y = \{y \in \mathbb{C} : |y| \leq 1\}$, and denote by $\mathcal{U}^+$ (resp. $\mathcal{U}^-$) the interior (resp. the exterior) domain bounded by the contour $\mathcal{U}$. Then the following lemma provides information about the location of the zeros of the kernel $H(x, y)$.

Lemma 1. *If $y \in \mathcal{C}_y$ (resp. $x \in \mathcal{C}_x$), $H(x, y) = 0$ has a unique root, say $X_0(y) \in \mathcal{D}_x$ (resp. $Y_0(x) \in \mathcal{D}_y$).*

Proof. See Appendix A

By the implicit function theorem, we see that the algebraic function $Y(x)$ (respectively $X(y)$) defined by $H(x, Y(x)) = 0$ (resp. $H(X(y), y) = 0$) is analytic except at branch points. Denote $X_1(y)$, $X_2(y)$ the other two in x, with $|X_1(1)| < |X_2(1)|$, by $Y_1(x)$ the other one in y.

Lemma 2. *The algebraic function $Y(x)$, defined by $H(x, Y(x)) = 0$, has six real positive branch points, and two of them, say x_1, x_2, are such that $0 = x_1 < x_2 < 1$.*

Proof. See Appendix A.

Define the image contour, $\mathcal{L} = Y_0(\overrightarrow{[0, x_2]})$, where $[\overrightarrow{u, v}]$ stands for the contour traversed from u to v along the upper edge of the slit $[u, v]$ and then back to u along the lower edge of the slit. The following lemma shows that the mapping $Y(x)$, $x \in [0, x_2]$, gives rise to the smooth and closed contour $\mathcal{L}$.

Remark 2. The study of $H(x, y) = 0$ with respect to x is slightly more difficult, but allows to also show that the algebraic function $X(y)$ has also two real and non-negative branch points inside $\mathcal{D}_y$, say $0 \leq y_1 < y_2 < 1$. Similarly, for $y \in [y_1, y_2]$, $X(y)$ lies on a closed contour $\mathcal{M}$. Further details are omitted due to space limitations, and mainly due to the fact that our main contribution relies on the use of PSA method.

Lemma 3. *For $x \in [0, x_2]$, the algebraic function $Y(x)$ lies on a closed contour $\mathcal{L}$, which is symmetric with respect to the real line and such that*

$$|y|^2 = \frac{(1-p)\nu_2}{p\nu_1}x, \ |y|^2 \leq \frac{(1-p)\nu_2}{p\nu_1}x_2.$$

Proof. Follows directly from the fact that $\Delta(x)$ is negative for $x \in (0, x_2)$ (see Appendix A).

5.1   A Boundary Value Problem for $\Pi_0(0, y)$

For $x \in C_x$, $y = Y_0(x)$ we obtain

$$(1 - p)\Pi_0(x, 0) - p\Pi_0(0, Y_0(x)) + \frac{C(x, Y_0(x))}{F(x, Y_0(x))}\Pi_0(0, 0) = 0, \qquad (21)$$

where $F(x, y) = \frac{A(x,y)}{1-p} = -\frac{B(x,y)}{p}$. For $y \in C_y$, $x = X_0(y)$

$$(1 - p)\Pi_0(X_0(y), 0) - p\Pi_0(0, y) + \frac{C(X_0(y), y)}{F(X_0(y), y)}\Pi_0(0, 0) = 0. \qquad (22)$$

Equation (22) implies that

$$\Pi(0, y) = \frac{1 - p}{p}\Pi_0(X_0(y), 0) - \frac{C(X_0(y), y)}{F(X_0(y), y)}\Pi_0(0, 0),$$

which is a meromorphic function[5]. For $y \in \mathcal{L}^+ \cap C_y^-$, $|X_0(y)| < 1$, and thus, we can construct analytic continuation of $\Pi_0(0, y)$ for all $y \in \zeta_y := \mathcal{L}^+ \cap C_y^-$.[6]

Moreover, as $Y_0(x)$ is analytic in $\mathcal{D}_x - [0, x_2]$ [21], $\Pi_0(0, Y_0(x))$ is meromorphic in $\mathcal{D}_x - [0, x_2]$, and from (21)

$$\Pi_0(x, 0) = \frac{p}{1 - p}\Pi_0(0, Y_0(x)) + \frac{C(x, Y_0(x))}{F(x, Y_0(x))}\Pi_0(0, 0) = 0, \; x \in \mathcal{D}_x - [0, x_2]. \qquad (23)$$

We therefore have the relation (21), not only for $x \in C_x$ but also for $x \in \mathcal{D}_x - [0, x_2]$, and by continuity, for $x \in \mathcal{D}_x - [0, x_2]$ too. Since $\Pi_0(x, 0)$ is real for $x \in [0, x_2]$, we obtain

$$Re[i\Pi(0, Y_0(x))] = Im[\Pi_0(0, 0)\frac{C(x, Y_0(x))}{F(x, Y_0(x))}], \; x \in [0, x_2],$$

or equivalently,

$$Re[i\Pi(0, y)] = c(y) := Im[\Pi_0(0, 0)\frac{C(|y|^2 p\nu_1/(1 - p)\nu_2, y)}{F(|y|^2 p\nu_1/(1 - p)\nu_2, y)}], \; y \in \mathcal{L}. \qquad (24)$$

Thus our problem is reduced to the determination of a function, which is regular for $y \in \mathcal{L}^+$, continuous in $\mathcal{L}^+ \cup \mathcal{L}$ satisfying the boundary condition (24). A standard way to solve this Riemann-Hilbert boundary value problem is to conformally transform it on the unit circle [21, 26] by introducing the conformal mappings $z = \gamma(y) : \mathcal{L}^+ \to C_y^+$, and its inverse $y = \gamma_0(z) : C_y^+ \to \mathcal{L}^+$.[7]

Then, we have the following problem: Find a function $T(z) = \Pi_0(0, \gamma_0(z))$ regular for $z \in C_z^+$, and continuous for $z \in C_z \cup C_z^+$ such that, $Re(iT(z)) = c(\gamma_0(z))$, $z \in C$. Its solution (see [22]) is given by

$$\Pi_0(0, y) = -\frac{1}{2\pi}\int_{C_z} c(\gamma_0(z))\frac{z + \gamma(y)}{z - \gamma(y)}\frac{dz}{z} + K, \; y \in C_y \cup C_y^+, \qquad (25)$$

---

[5] Its possible poles are the zeros of $F(X_0(y), y)$ in $\mathcal{L}^+ \cap C_y^-$.

[6] Note that ζ_y is an empty set when $(1 - p)\nu_2 \leq p\nu_1$, since in such a case $\mathcal{L}$ lies entirely inside the unit circle.

[7] See Appendix B for details on the numerical derivation of the conformal mappings.

where K is a constant. Then, (21) can be used to obtain $\Pi_0(x,0)$, $x \in \mathcal{C}_x{}^8$. For $x \in \mathcal{C}_x^+$, $\Pi_0(x,0)$ can be derived by using the Cauchy's formula, yielding

$$\Pi_0(x,0) = \frac{1}{2\pi} \int_{\mathcal{C}_y} \frac{V(y)}{y-x} dy, \ x \in \mathcal{C}_x^+,$$

where

$$V(y) = \frac{1-p}{p} \Pi_0(X_0(y),0) - \frac{C(X_0(y),y)}{F(X_0(y),y)} \Pi_0(0,0), \ y \in \mathcal{C}_y.$$

Using (6) we obtain $\Pi_0(x,y)$. Then, using the second in (4) we obtain $\Pi_1(x,y)$ and all unknowns are fully specified.

6 Numerical Results

6.1 Numerical Validation

We now compare the PSA to the exact results derived by the BVP approach and investigate the influence of some parameters on the mean queue lengths.

Figure 1 depicts the approximations (19) as a function of p for increasing values of M. Set $\lambda_0 = 1$, $\lambda_1 = 0.5$, $\tau = 4$, $\gamma = 2$, $\nu_1 = 4$, $\nu_2 = 5$. The horizontal lines ($M = 0$) equal the values for the tandem system with priority for the second queue. Figure 1 confirms that the PSA approximations are accurate for p close to 0, and clearly, more terms provide larger regions for p where the accuracy is good.

In Fig. 2 ($\lambda_0 = 1$, $\lambda_1 = 0.5$, $M = 3$, $\nu_1 = 4$, $\nu_2 = 5$), we can observe that the increase in γ will definitely increase $E(Q_2)$, since the system switches to the set-up mode more frequently.

Figure 3 shows the total expected number of customers in the system as a function of λ_0 and γ. We can observe that by increasing both λ_0 and γ, $E(Q_1 + Q_2)$ increases too. An interesting observation arises when we increase τ and reveals the importance of controlling the arrivals. In particular, it is seen that when we increase τ, and thus, decrease the duration of the set-up period, $E(Q_1 + Q_2)$ increases faster. This is because of the reduced arrival rate when the network is in the set-up mode compared with the arrival rate at the operating mode.

6.2 Influence of System Parameters as $p \to 0$

Here on we focus on the influence of system parameters on the mean content for the near priority tandem queue as $p \to 0$ when $\tau = 4$, $\nu_1 = 4$, $\nu_2 = 5$. Recall that as $p \to 0$, our system behaves as a single server tandem queue, in which station 2 has preemptive priority over station 1.

---

[8] Note that a similar analysis can be also performed to obtain $\Pi_0(x,0)$ in terms of another Riemann-Hilbert boundary value problem.

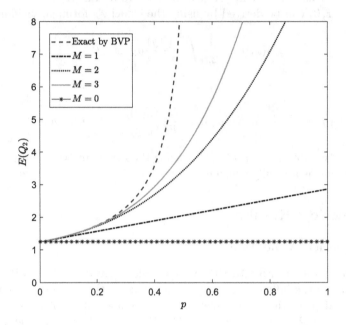

Fig. 1. Truncation approximations for $\lambda_0 = 1$, $\lambda_1 = 0.5$, $\tau = 4$, $\gamma = 2$, $\nu_1 = 4$, $\nu_2 = 5$.

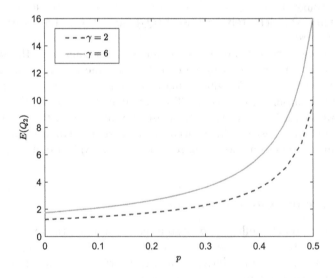

Fig. 2. Truncation approximations for $\lambda_0 = 1$, $\lambda_1 = 0.5$, $\tau = 4$, $M = 3$, $\nu_1 = 4$, $\nu_2 = 5$.

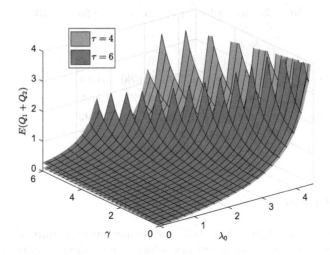

Fig. 3. Truncation approximations for $\lambda_1 = 0.5$, $M = 3$, $\nu_1 = 4$, $\nu_2 = 5$, $p = 0.2$.

Table 1 displays some values of the first-order correction term (i.e., the first derivative of the station 2 queue content as $p = 0$) for increasing values λ_0. It is seen that as γ increases the mean queue content in station 2 increases, and that increase becomes more apparent as λ_0 increases too. This is expected since by increasing the load in system, the station 1 has always customers. Thus, sharing the server with that station, even for a small percentage of the time can have a large influence, especially when the rate of failures increases too.

Table 1. First-order correction as $p \to 0$ for $\lambda_0 = 0.5$.

λ_0	γ			
	0.1	0.3	0.5	0.7
0.7	0.0173	0.1064	0.1931	0.2775
0.8	0.3683	0.4482	0.5265	0.6033
0.9	0.6844	0.7582	0.8309	0.9025
1	0.9771	1.0466	1.1154	1.1835
1.1	1.2534	1.3200	1.3861	1.4516

In Table 2 we set $\alpha = \frac{\lambda_1}{\lambda_0 + \lambda_1}$ and derive first-order correction term for station 2 queue content for increasing values of λ_0, α. It is interesting to note that for fixed λ_0, as α increases, we particularly increase the percentage of customers that may arrive when the network is down. Since we have also assumed low rate of failures the mean queue content in station 2 decreases. However, by increasing λ_0 we clearly increase the mean queue content in station 2.

Table 2. First-order correction as $p \to 0$ for $\gamma = 0.3$.

λ_0	α			
	0.1	0.2	0.3	0.4
0.7	0.8998	0.6773	0.4090	0.0796
0.8	1.0786	0.8699	0.6205	0.3175
0.9	1.2610	1.0666	0.8362	0.5592
1	1.4471	1.2672	1.0559	0.8047
1.1	1.6368	1.4717	1.2797	1.0540

7 Conclusion

In this work, we investigated the stationary behaviour of an unreliable two-node tandem queue with coupled processors, which is described by a Markov modulated RWQP. Based on the generating function approach, we applied the PSA method and obtained power series expansions of the pgfs of the stationary joint queue length distributions for each state of the network. With this result we shown the flexibility of the PSA approach to be applied in more complicated models. Moreover, we also obtained the corresponding pgfs with the aid of the theory of Riemann-Hilbert boundary value problems. By truncating the power series, we find good approximations for the expected number of customers especially when p is close to 0, by comparing them with the "exact" derivations through BVP.

In the future, we plan to expand our results to networks with more than two nodes, where the theory of BVPs cannot be applied, as well as to consider general routing among the nodes. Moreover, it could be also interesting to compare PSA method with other approximation techniques developed so far [8,31,35, 39]. Moreover, it would be interesting to derive asymptotic estimates for the occurrence of large queue lengths due to the presence of failures.

Acknowledgements. The author would like to thank the the anonymous Reviewers for the careful reading of the manuscript and the insightful remarks and input, which helped to improve the original exposition.

Appendix

A   Analysis of the Kernel

Proof of Lemma 1. Let $u(x,y) = \lambda_0(1-x) + \nu_1 p(1-\frac{y}{x}) + \nu_2(1-p)(1-\frac{1}{y}) + \gamma$. Note that

$$H(x,y) = 0 \Leftrightarrow xy\{D(x)u(x,y) - \gamma\tau\} = 0.$$

Using the principal value argument it is seen that the number of zeros of $H(x,y)$ in $\mathcal{D}_x$ equals the number of zeros of $D(x)u(x,y)$ in $\mathcal{D}_x$, which is equal to one. □

Proof of Lemma 2. The branch points of the two-valued function $Y(x)$ are the zeros of the discriminant $\Delta(x) = x(f(x) + g(x))$ of $H(x, y) = 0$, where

$$f(x) = D^2(x)[x(\lambda_0(1 - x) + \gamma + p\nu_1 + (1 - p)\nu_2)^2 - 4(p\nu_1 + (1 - p)\nu_2)],$$
$$g(x) = x\lambda_1(1 - x)\gamma[\lambda_1(1 - x)\gamma + 2D(x)(\lambda_0(1 - x) + p\nu_1 + (1 - p)\nu_2)].$$

Note $g(x) = 0$ if and only if $x = 0$, $x = 1$, and

$$2\lambda_0\lambda_1 x^2 - x[2\lambda_0(2\lambda_1 + \tau) + \lambda_1(\gamma + 2(p\nu_1 + (1 - p)\nu_2))] \\ + 2(\tau + \lambda_1)(\lambda_0 + p\nu_1 + (1 - p)\nu_2) + \lambda_1\gamma = 0. \tag{26}$$

Let x_1^*, x_2^* the zeros of (26). Then,

$$x_1^* x_2^* = 1 + \frac{2(\tau+\lambda_1)(p\nu_1+(1-p)\nu_2)+\lambda_1\gamma+2\lambda_0\tau}{2\lambda_0\lambda_1} > 1,$$
$$x_1^* + x_2^* = 2 + \frac{\lambda_1\gamma+2\lambda_1(p\nu_1+(1-p)\nu_2)+2\lambda_0\tau}{2\lambda_0\lambda_1} > 2,$$

which means that $x_1^*, x_2^* > 1$. Thus $g(x) = 0$ has exactly two roots in $[0, 1]$. By using Rouche's theorem, we can show that $D(x) = 0$, i.e., $x(f(x) + g(x)) = 0$ has exactly two zeros in $[0, 1]$, and one of them equals $x_1 = 0$.

B   On the Derivation of Conformal Mappings

A detailed approach on how we can numerically obtain the conformal mappings is given in [10,26]. We summarized the basic steps. First, we need to represent $\mathcal{L}$ in polar coordinates, i.e., $\mathcal{L} = \{y : y = \rho(\phi)\exp(i\phi), \phi \in [0, 2\pi]\}$. Since $0 \in \mathcal{L}^+$, for each $y \in \mathcal{L}$, we can have a relation between its absolute value and its real part, i.e., $|y|^2 = m(Re(y))$. Given the angle ϕ of some point on $\mathcal{L}$, the real part of this point, say $\delta(\phi)$, is the solution of $\delta - \cos(\phi)\sqrt{m(\delta)}$, $\phi \in [0, 2\pi]$. Since $\mathcal{L}$ is a smooth, egg-shaped contour, the solution is unique. Clearly, $\rho(\phi) = \frac{\delta(\phi)}{\cos(\phi)}$, and the parametrization of $\mathcal{L}$ is fully specified. Then, the mapping from $z \in \mathcal{C}_z^+$ to $y \in \mathcal{L}^+$, where $z = e^{i\phi}$ and $y = \rho(\psi(\phi))e^{i\psi(\phi)}$, satisfying $\gamma_0(0) = 0$, $\gamma_0(z) = \overline{\gamma_0(\bar{z})}$ is uniquely determined by,

$$\gamma_0(z) = z\exp[\frac{1}{2\pi}\int_0^{2\pi}\log\{\rho(\psi(\omega))\}\frac{e^{i\omega}+z}{e^{i\omega}-z}d\omega], \ |z| < 1,$$
$$\psi(\phi) = \phi - \int_0^{2\pi}\log\{\rho(\psi(\omega))\}\cot(\frac{\omega-\phi}{2})d\omega, \ 0 \leq \phi \leq 2\pi, \tag{27}$$

i.e., $\psi(.)$ is uniquely determined as the solution of a Theodorsen integral equation with $\psi(\phi) = 2\pi - \psi(2\pi - \phi)$. Due to the correspondence-boundaries theorem, $\gamma_0(z)$ is continuous in $\mathcal{C}_z \cup \mathcal{C}_z^+$.

C   On the Analyticity of $\Pi_j(x, y)$ Close to $p = 0$

We focus only on the analyticity of $\Pi_0(x, y)$ in a neighborhood of $p = 0$ by using a variant of the implicit function theorem on the functional Eq. (6). The analyticity of $\Pi_1(x, y)$, follows directly by the analyticity of $\Pi_0(x, y)$ from the second in (4). We follow the lines in [18, 41], and use the implicit function theorem for Banach spaces (see Theorem 10.2.3, p. 272 in [12]). Define the mapping $f : S \subset \mathbb{C} \times B_2 \to B_3 \times \mathbb{C}$,

$$f(p, \Pi_0) = [\Pi_0(x, y)H(x, y) - D(x)\{A(x, y)\Pi_0(x, 0) \\ + B(x, y)\Pi_0(0, y) + C(x, y)\Pi_0(0, 0)\}, \Pi_0(1, 1) - \tfrac{\tau}{\tau + \gamma}],$$

where S contains the point $(0, V_0^{(0)})$, H, A, B, C, are as in (20) and (5) respectively, B_2 be the Banach space comprising all bivariate analytic bounded functions in $\mathbb{D}^2$, with $\mathbb{D}$ the open complex unit disk, and B_3 be the Banach space comprising all trivariate analytic bounded functions in $\mathbb{D}^3$ that have a limit of 0 for the first two arguments going to 1.

Since H, A, B, C are bounded analytic functions in $\mathbb{D}^3$, and since f is affine in Π_0 and p, it is easily seen that f is r-times continuously differentiable for all r. Note also that $f(0, V_0^{(0)}) = [0, 0]$. Then, the (Banach space) derivative of f at the point $(0, V_0^{(0)})$ [12] equals

$$df(0, V_0^{(0)}) = [\Pi_0(x, y)H(x, y) - D(x)\{A(x, y)\Pi_0(x, 0) \\ + C(x, y)\Pi_0(0, 0)\}, \Pi_0(1, 1)].$$

We need to show that this mapping is a homeomorphism. Indeed,

1. $df(0, V_0^{(0)})$ is a continuous mapping for the same reasons that the mapping f itself is continuous.
2. For given $\Pi_0^{(1)}$, $\Pi_0^{(2)}$, let $df(0, V_0^{(0)})(\Pi_0^{(1)}) = df(0, V_0^{(0)})(\Pi_0^{(1)})$. Then,

$$[\Pi_0^{(1)}(x, y) - \Pi_0^{(2)}(x, y)]H(x, y) \\ - D(x)\{A(x, y)(\Pi_0^{(1)}(x, 0) - \Pi_0^{(2)}(x, 0)) + C(x, y)(\Pi_0^{(1)}(0, 0) - \Pi_0^{(2)}(0, 0))\} = 0, \\ \Pi_0^{(1)}(1, 1) - \Pi_0^{(2)}(1, 1) = 0.$$

or equivalently $f(0, \Pi_0^{(1)} - \Pi_0^{(2)}) = (0, -\tfrac{\tau}{\tau + \gamma})$, which in turn has the zero solution as a unique solution [3], and thus $\Pi_1^{(0)} = \Pi_2^{(0)}$ so that $df(0, V_0^{(0)})$ is injective.

3. To show that $df(0, V_0^{(0)})$ is surjective, we solve the $df(0, V_0^{(0)})(\Pi_0) = (g, c)$ with g a bivariate analytic bounded function in $\mathbb{D}^2$ with limit 0 for its arguments going to 1, and c a complex number. The solution is

$$\Pi_0(x, y) \\ = \frac{g(x, y)A(x, Y_0(x)) - g(x, Y_0(x))A(x, y) + \Pi^{(0)}(0, 0)[C(x, y)A(x, Y_0(x)) - C(x, Y_0(x))A(x, y)]}{H(x, y)A(x, Y_0(x))}.$$

4. The Π_0 obtained previously equals $(df(0, V_0^{(0)}))^{-1}$, which is readily seen that it is continuous.

Thus, $\Pi_0 \to df(0, V_0^{(0)})(\Pi^{(0)})$ is a linear homeomorphism and using Theorem 10.2.3 in [12], $\Pi_0(x, y)$ is r-times differentiable at $p = 0$. Having this result, and using the second in (4), $\Pi_1(x, y)$ is also r-times differentiable at $p = 0$.

References

1. Adan, I., Wessels, J., Zijm, W.: A compensation approach for two-dimensional Markov processes. Adv. Appl. Probab. **25**(4), 783–817 (1993)
2. Andradóttir, S., Ayhan, H., Down, D.G.: Server assignment policies for maximizing the steady-state throughput of finite queueing systems. Manage. Sci. **47**(10), 1421–1439 (2001)
3. Asmussen, S.: Applied Probability and Queues. Wiley, New York (1987)
4. Avrachenkov, K., Nain, P., Yechiali, U.: A retrial system with two input streams and two orbit queues. Queueing Syst. **77**(1), 1–31 (2014)
5. Baskett, F., Chandy, K.M., Muntz, R.R., Palacios, F.G.: Open, closed, and mixed networks of queues with different classes of customers. J. ACM **22**(2), 248–260 (1975)
6. Blanc, J.: A numerical study of a coupled processor model. In: Iazeolla, G., Courtois, P.-J., Boxma, O.J. (ed) Computer Performance and Reliability, pp. 289–303 (1988)
7. Borst, S.: User-level performance of channel-aware scheduling algorithms in wireless data networks. In: IEEE INFOCOM 2003. Twenty-second Annual Joint Conference of the IEEE Computer and Communications Societies (IEEE Cat. No. 03CH37428), vol. 1, pp. 321–331, March 2003
8. Chakka, R., Mitrani, I.: Approximate Solutions for Open Networks with Breakdowns and Repairs, vol. 4, pp. 267–280. Oxford University Press, Oxford (1996)
9. Cohen, J.: Boundary value problems in queueing theory. Queueing Syst. **3**, 97–128 (1988)
10. Cohen, J., Boxma, O.: Boundary Value Problems in Queueing Systems Analysis. North Holland Publishing Company, Amsterdam (1983)
11. Devos, A., Walraevens, J., Bruneel, H.: A priority retrial queue with constant retrial policy. In: Takahashi, Y., Phung-Duc, T., Wittevrongel, S., Yue, W. (eds.) QTNA 2018. LNCS, vol. 10932, pp. 3–21. Springer, Cham (2018). https://doi.org/10.1007/978-3-319-93736-6_1
12. Dieudonné, J.: Foundations of Modern Analysis. Academic Press, NY (1969)
13. Dimitriou, I.: A queueing system for modeling cooperative wireless networks with coupled relay nodes and synchronized packet arrivals. Perform. Eval. **114**, 16–31 (2017)
14. Dimitriou, I.: A two class retrial system with coupled orbit queues. Probab. Eng. Inf. Sci. **31**(2), 139–179 (2017)
15. Dimitriou, I., Pappas, N.: Stability and delay analysis of an adaptive channel-aware random access wireless network. In: Thomas, N., Forshaw, M. (eds.) ASMTA 2017. LNCS, vol. 10378, pp. 63–80. Springer, Cham (2017). https://doi.org/10.1007/978-3-319-61428-1_5
16. Dimitriou, I., Pappas, N.: Performance analysis of an adaptive queue-aware random access scheme with random traffic. In: 2018 IEEE International Conference on Communications (ICC), pp. 1–6, May 2018
17. Dimitriou, I., Pappas, N.: Stable throughput and delay analysis of a random access network with queue-aware transmission. IEEE Trans. Wirel. Commun. **17**(5), 3170–3184 (2018)
18. Dimitriou, I.: On the power series approximations of a structured batch arrival two-class retrial system with weighted fair orbit queues. Perform. Eval. **132**, 38–56 (2019)

19. Dimitriou, I., Pappas, N.: Performance analysis of a cooperative wireless network with adaptive relays. Ad Hoc Netw. **87**, 157–173 (2019)
20. Fayolle, G., Iasnogorodski, R.: Two coupled processors: the reduction to a Riemann-Hilbert problem. Z. Wahrscheinlichkeitstheorie Verw. Geb. **47**(3), 325–351 (1979)
21. Fayolle, G., Iasnogorodski, R., Malyshev, V.: Random Walks in the Quarter-Plane: Algebraic Methods, Boundary Value Problems, Applications to Queueing Systems & Analytic Combinatorics. Springer, Berlin (2017). https://doi.org/10.1007/978-3-319-50930-3
22. Gakhov, F.: Boundary Value Problems. Pergamon Press, Oxford, UK (1966)
23. Guillemin, F., van Leeuwaarden, J.S.H.: Rare event asymptotics for a random walk in the quarter plane. Queueing Syst. **67**(1), 1–32 (2011)
24. Guillemin, F., Pinchon, D.: Analysis of generalized processor-sharing systems with two classes of customers and exponential services. J. Appl. Probab. **41**(3), 832–858 (2004)
25. van den Hout, W., Blanc, J.P.C.: The power-series algorithm for Markovian queueing networks. In: Stewart, W.J. (eds)Computations with Markov Chains, pp. 321–338. Springer, Boston (1995). https://doi.org/10.1007/978-1-4615-2241-6_19
26. van Leeuwaarden, J.S.H., Resing, J.A.C.: A tandem queue with coupled processors: computational issues. Queueing Syst. **51**(1), 29–52 (2005)
27. Mikou, N., Idrissi-Kacimi, O., Saadi, S.: Two processes interacting only during breakdown: the case where the load is not lost. Queueing Syst. **19**(3), 301–317 (1995)
28. Mitrany, I.L., Avi-Itzhak, B.: A many-server queue with service interruptions. Oper. Res. **16**(3), 628–638 (1968)
29. Noufissa, M.: A two-node jackson's network subject to breakdowns. Commun. Stat. Stochast. Models **4**(3), 523–552 (1988)
30. Ozawa, T.: Asymptotics for the stationary distribution in a discrete-time two-dimensional quasi-birth-and-death process. Queueing Syst. **74**(2), 109–149 (2013)
31. Reiser, M.: A queueing network analysis of computer communication networks with window flow control. IEEE Trans. Commun. **27**(8), 1199–1209 (1979)
32. Resing, J., Órmeci, L.: A tandem queueing model with coupled processors. Oper. Res. Lett. **31**(5), 383–389 (2003)
33. Song, Y., Liu, Z., Zhao, Y.Q.: Exact tail asymptotics: revisit of a retrial queue with two input streams and two orbits. Ann. Oper. Res. **247**(1), 97–120 (2016)
34. Takine, T., Sengupta, B.: A single server queue with service interruptions. Queueing Syst. Theory Appl. **26**(3–4), 285–300 (1997)
35. Thomas, N., Mitrani, I.: Routing among different nodes where servers break down without losing jobs. In: Proceedings of 1995 IEEE International Computer Performance and Dependability Symposium, pp. 246–255, April 1995
36. Vanlerberghe, J., Walraevens, J., Maertens, T., Bruneel, H.: Approximating the optimal weights for discrete-time generalized processor sharing. In: 2014 IFIP Networking Conference, pp. 1–9, June 2014
37. Vanlerberghe, J.: Analysis and optimization of discrete-time generalized processor sharing queues. Ph.D. thesis, Ghent University (2018)
38. Vanlerberghe, J., Walraevens, J., Maertens, T., Bruneel, H.: On the influence of high priority customers on a generalized processor sharing queue. In: Gribaudo, M., Manini, D., Remke, A. (eds.) ASMTA 2015. LNCS, vol. 9081, pp. 203–216. Springer, Cham (2015). https://doi.org/10.1007/978-3-319-18579-8_15
39. Vinod, B., Altiok, T.: Approximating unreliable queueing networks under the assumption of exponentiality. J. Oper. Res. Soc. **37**(3), 309–316 (1986)

40. Vitale, C., Mancuso, V., Rizzo, G.: Modelling D2D communications in cellular access networks via coupled processors. In: 2015 7th International Conference on Communication Systems and Networks (COMSNETS), pp. 1–8, January 2015
41. Walraevens, J., van Leeuwaarden, J.S.H., Boxma, O.J.: Power series approximations for two-class generalized processor sharing systems. Queueing Syst. **66**(2), 107–130 (2010)

Scheduling Policies

Diffusion Limits for SRPT and LRPT Queues via EDF Approximations

Łukasz Kruk[(✉)] [iD]

Institute of Mathematics, Maria Curie-Skłodowska University, Lublin, Poland
lkruk@hektor.umcs.lublin.pl

Abstract. We derive a heavy traffic analysis for a G/G/1 queue in which the server uses the Shortest Remaining Processing Time (SRPT) policy from diffusion limits for G/G/1 Earliest Deadline First (EDF) systems. Our approach yields simple, concise justifications and new insights for SRPT heavy traffic limit theorems of Gromoll, Kruk and Puha [9]. Corresponding results for the longest remaining processing time (LRPT) policy are also provided.

Keywords: Heavy traffic · Queueing ·
Shortest Remaining Processing Time · Earliest Deadline First ·
Diffusion limit

1 Introduction

The Shortest Remaining Processing Time (SRPT) service protocol gives preemptive priority to the job with the shortest residual service time. It is of great theoretical interest, because it is known to minimize the queue length in a single-server system at any point of time (Schrage [22]). In spite of this, SRPT is not often implemented in practice, because it is believed to unfairly penalize large jobs (see, e.g., Bender, Chakrabarti and Muthukrishnan [2]), although this objection has largely been dismissed, e.g., by Bansal and Harchol-Balter [1], Wierman and Harchol-Balter [24].

Expressions for the mean response time for an M/G/1 SRPT queue were developed by Schrage and Miller [23], and extended later in Schassberger [21] and Perera [19]. Pavlov [17] and Pechinkin [18] characterized the heavy traffic limit of the steady state distributions for the queue length of an M/G/1 SRPT queue. The tail behavior of single server queues under SRPT was investigated, e.g., by Núñez Queija [15] and Nuyens and Zwart [16], who discussed the advisability of implementing SRPT using large deviations.

There has also been a growing body of work on functional limit theorems for SRPT systems. Down, Gromoll and Puha [4] proposed a fluid model and obtained fluid limits for G/G/1 SRPT queues. Gromoll and Keutel [8] obtained the same fluid limits in the case of a non-preemptive variant of SRPT, called Shortest Job First (SJF). Recently, the results of [4] were extended by Kruk

© Springer Nature Switzerland AG 2019
T. Phung-Duc et al. (Eds.): QTNA 2019, LNCS 11688, pp. 263–275, 2019.
https://doi.org/10.1007/978-3-030-27181-7_16

and Sokołowska [14] to SRPT queues with multiple inputs. Down and Wu [5] employed diffusion limits to show certain optimality properties of a multi-layered round robin routing policy for a system of parallel servers, each operating under SRPT, under the assumption of a finitely supported service time distribution. Gromoll, Kruk and Puha [9] provided diffusion limits for G/G/1 SRPT queues with general service time distributions. Not surprisingly, their limiting distributions are supported on the invariant manifolds for the corresponding fluid limit models, identified in [4]. More recently, Puha [20] provided a diffusion limit for a G/G/1 SRPT system under nonstandard spacial scaling for the queue length process.

In this paper, we show that under some technical conditions, the diffusion limits of [9] for G/G/1 SRPT queues can alternatively be derived from diffusion limits of Kruk [13] for preemptive G/G/1 Earliest Deadline First (EDF) systems with job service times correlated with their initial lead times. (Recall that under the EDF discipline, the server gives priority to the customer with the shortest lead time.) The main idea of our proofs is to approximate the SRPT system under consideration by G/G/1 queueing systems with the same stochastic primitives, but operating under the preemptive EDF protocol. The initial lead times of the customers in these auxiliary EDF queues are set to be large multiples of their service times. Since, by the above-mentioned result of Schrage [22], SRPT minimizes the number of jobs in the system at any point of time, the queue length in an EDF system is an upper bound for the queue length in the corresponding SRPT system. Using this fact, together with heavy traffic limiting distributions of [13] and obvious lower bounds, we arrive at the required results. The idea of comparing the behavior of a SRPT system with that of a corresponding EDF system goes back to Bender, Chakrabarti and Muthukrishnan [2], although it was originally used to "regularize" the SRPT protocol in order to make it more fair to large jobs, rather than to provide asymptotics for SRPT queues.

An advantage of our approach is simplicity and conciseness of the proofs. Indeed, once the limiting distributions for sufficiently general G/G/1 EDF queues are available, the derivation of heavy traffic limits for G/G/1 SRPT systems by our method is relatively easy. Moreover, as it is shown in Sect. 5, our approach carries over, with only minor modifications, to the Longest Remaining Processing Time (LRPT) queue discipline, which gives preemptive priority to the job in the system with the longest remaining processing time. While the latter protocol may seem to be of little practical interest since it maximizes the queue length, it does appear in some applications. See Kittsteiner and Moldovanu [12], where both SRPT and LRPT queue disciplines arise in equilibria for priority auctions. It is also plausible that our approach will turn out to be useful in the analysis of other service protocols with priorities dependent on the job service times.

This paper is organized as follows. Section 2 presents the models, notation and assumptions. Section 3 states limit theorems for G/G/1 SRPT queues, while Sect. 4 is devoted to the proofs of these results. Finally, in Sect. 5 we provide analogous results for G/G/1 LRPT queueing systems.

2 The Models, Assumptions and Notation

2.1 Notation

The following notation will be used throughout the paper. Let $\mathbb{N} = \{1, 2, \ldots\}$ and let $\mathbb{R}$ denote the set of real numbers. Let $\mathbb{R}_+ \overset{\Delta}{=} [0, \infty)$. Let $\overline{\mathbb{R}}_+ \overset{\Delta}{=} \mathbb{R}_+ \cup \{\infty\}$ and $\overline{\mathbb{R}} \overset{\Delta}{=} \mathbb{R} \cup \{\infty\}$ be equipped with the obvious topologies. Denote by e the identity map on $\mathbb{R}$, i.e., $e(t) = t$, $t \in \mathbb{R}$.

Denote by $\mathcal{M}$ the set of all finite, nonnegative measures on $\mathcal{B}(\mathbb{R})$, the Borel subsets of $\mathbb{R}$. Under the weak topology, $\mathcal{M}$ is a Polish space. We denote the zero measure in $\mathcal{M}$ by $\mathbf{0}$ and the Dirac delta measure with unit mass at $x \in \mathbb{R}$ by δ_x. For $x \in \mathbb{R}_+$, let δ_x^+ be δ_x if $x > 0$ and $\mathbf{0}$ otherwise. For $\xi \in \mathcal{M}$ and a Borel measurable function $f : \mathbb{R} \to \mathbb{R}$ integrable with respect to ξ, we write $\int_R f(x) d\xi(x)$ as $< f, \xi >$.

We will use the symbol $\Rightarrow$ to denote weak convergence of measures, either on $\mathbb{R}$ (in this case, the same symbol is used for convergence of the corresponding cumulative distribution functions (c.d.f.s)), or on the space $D_S[0, \infty)$ of right-continuous functions with left-hand limits (RCLL functions) from $[0, \infty)$ to a Polish space S, equipped with the Skorokhod J_1 topology. See Ethier and Kurtz [10] for details. When dealing with $D_S[0, \infty)$, we take $S = \mathbb{R}$ or $\mathbb{R}^d$, with appropriate dimension d for vector-valued functions, unless explicitly stated otherwise.

2.2 The Basic Models

We have a sequence of single-station queueing systems, each serving one class of customers. The queueing systems are indexed by superscript n.

The inter-arrival times for the customer arrival process are $\{u_j^n\}_{j=1}^{\infty}$, a sequence of strictly positive, independent, identically distributed (i.i.d.) random variables (r.v.s) with mean $1/\lambda_n$ and standard deviation α_n. The service times are $\{v_j^n\}_{j=1}^{\infty}$, another sequence of strictly positive, i.i.d. r.v.s with distribution function G^n, mean $1/\mu_n$ and standard deviation β_n. For every n, the sequences $\{u_j^n\}_{j=1}^{\infty}$ and $\{v_j^n\}_{j=1}^{\infty}$ are mutually independent, each queue is empty at time zero and

$$\lim_{n \to \infty} \lambda_n = \lim_{n \to \infty} \mu_n = \lambda > 0. \tag{1}$$

We assume that for some c.d.f. G

$$G^n \Rightarrow G \tag{2}$$

and

$$G_v^n(y) \overset{\Delta}{=} \mathbb{E}\left[v_j^n \mathbb{I}_{\{v_j^n \leq y\}}\right] \Rightarrow G_v(y), \tag{3}$$

where G_v is a c.d.f. of a finite positive measure on $\mathbb{R}_+$. In particular, by (1), G_v has total mass $1/\lambda$.

In our analysis of SRPT queueing systems, we will consider two cases, corresponding to bounded and unbounded service times, respectively. In the first one, we additionally assume that for all n,

$$v^* \stackrel{\Delta}{=} \min\{y \in \mathbb{R} : G^n(y) = 1\} = \min\{y \in \mathbb{R} : G(y) = 1\} < \infty. \quad (4)$$

Observe that the constant v^* in (4) does not depend on n. In the second case we additionally make the following assumptions. First,

$$G_v(y) < 1/\lambda, \qquad y \in \mathbb{R}, \quad (5)$$

and

$$G_{v^2}^n(y) \stackrel{\Delta}{=} \mathbb{E}\left[(v_j^n)^2 \mathbb{I}_{\{v_j^n \le y\}} \right] \Rightarrow G_{v^2}(y), \quad (6)$$

where G_{v^2} is a c.d.f of a finite positive measure on $\mathbb{R}_+$. We extend $G_{v^2}^n$ to $\overline{\mathbb{R}}_+$ by $G_{v^2}^n(\infty) \stackrel{\Delta}{=} \mathbb{E}(v_j^n)^2$. For every $x, y \in \overline{\mathbb{R}}_+$, we define a semimetric ρ on $\overline{\mathbb{R}}_+$ by the formula

$$\rho(x, y) \stackrel{\Delta}{=} \sup_{n \in \mathbb{N}} |G_{v^2}^n(x) - G_{v^2}^n(y)|.$$

We assume that $(\overline{\mathbb{R}}_+, \rho)$ is a totally bounded semimetric space, i.e., for every $\epsilon > 0$, $\overline{\mathbb{R}}_+$ may be decomposed into a finite number of sets with radius less than ϵ. This is the case if, for example, G_{v^2} is continuous or $G_{v^2}^n \equiv G_{v^2}$. However, the assumption (6) does not always imply total boundedness of $(\overline{\mathbb{R}}_+, \rho)$, a counterexample is $G_{v^2}^n(y) = (1 + 1/n)^2 \mathbb{I}_{\{1+1/n \le y\}}$, $n \in \mathbb{N}$. Finally, in the second case we assume that

$$\lim_{y \to \infty} \sup_{n \in \mathbb{N}} \int_y^\infty (1 - G^n(\eta))\, d\eta = \lim_{y \to \infty} \sup_{n \in \mathbb{N}} \int_y^\infty \left(\frac{1}{\mu_n} - G_v^n(\eta) \right) d\eta = 0. \quad (7)$$

(In the first case, (7) is an immediate consequence of (4).) Either (4), or (1)–(3), (7) and Fatou's lemma imply that

$$\int_0^\infty (1 - G(\eta))d\eta < \infty, \qquad \int_0^\infty (1 - \lambda\, G_v(\eta))d\eta < \infty. \quad (8)$$

For the asymptotic analysis of LRPT queueing systems, we do not need the assumptions (4)–(7) and total bondedness of $(\overline{\mathbb{R}}_+, \rho)$. Instead, we assume that for every $n \in \mathbb{N}$

$$\inf\{y \in \mathbb{R} : G^n(y) > 0\} = v_* \stackrel{\Delta}{=} \inf\{y \in \mathbb{R} : G(y) > 0\}. \quad (9)$$

We define the *customer arrival times*

$$S_0^n \stackrel{\Delta}{=} 0, \qquad S_k^n \stackrel{\Delta}{=} \sum_{i=1}^k u_i^n, \quad k \ge 1,$$

the *customer arrival process*

$$A^n(t) \triangleq \max\{k : S^n_k \le t\}, \quad t \ge 0,$$

and the *work arrival process*

$$V^n(t) \triangleq \sum_{j=1}^{\lfloor t \rfloor} v^n_j, \quad t \ge 0.$$

The work which has arrived to the queue by time t is then $V^n(A^n(t))$. The *netput process*

$$N^n(t) \triangleq V^n(A^n(t)) - t$$

measures the amount of work in queue at time t provided that the server is never idle up to time t. The *cumulative idleness process*

$$I^n(t) \triangleq - \inf_{0 \le s \le t} N^n(s),$$

gives the amount of time the server is idle, and adding this to the netput process, we obtain the *workload process*

$$W^n(t) = N^n(t) + I^n(t),$$

which records the amount of work in the queue, taking server idleness into account. All the above processes are independent of the queue service discipline, provided that the server is never idle when there are customers in the queue. However, the *queue length process* $Q^n(t)$, which is the number of customers in the queue at time t, depends on the queue discipline. All these processes are RCLL.

2.3 Heavy Traffic Assumptions

We assume that

$$\lim_{n \to \infty} \alpha_n = \alpha > 0, \quad \lim_{n \to \infty} \beta_n = \beta > 0. \tag{10}$$

Define the *traffic intensity* $\rho_n \triangleq \lambda_n / \mu_n$. We make the *heavy traffic assumption*

$$\lim_{n \to \infty} \sqrt{n}(1 - \rho_n) = \gamma \tag{11}$$

for some $\gamma \in \mathbb{R}$. We impose the Lindeberg condition on the inter-arrival times:

$$\lim_{n \to \infty} \mathbb{E}\left[\left(u^n_j - (\lambda_n)^{-1}\right)^2 \mathbb{I}_{\{|u^n_j - (\lambda_n)^{-1}| > c\sqrt{n}\}} \right] = 0 \quad \forall c > 0. \tag{12}$$

One may check that the analogous Lindeberg condition on the service times follows from (1)–(2) and (10). We introduce the *heavy traffic scaling* for the idleness, workload and queue length processes

$$\widehat{I}^n(t) = \frac{1}{\sqrt{n}} I^n(nt), \quad \widehat{W}^n(t) = \frac{1}{\sqrt{n}} W^n(nt), \quad \widehat{Q}^n(t) = \frac{1}{\sqrt{n}} Q^n(nt).$$

We define also

$$\widehat{N}^n(t) = \frac{1}{\sqrt{n}} \left[V^n\big(A^n(nt)\big) - nt \right].$$

Note that $\widehat{W}^n(t) = \widehat{N}^n(t) + \widehat{I}^n(t)$. It is a standard result (see Iglehart and Whitt [11]) that

$$(\widehat{N}^n, \widehat{I}^n, \widehat{W}^n) \Rightarrow (N^*, I^*, W^*), \tag{13}$$

where N^* is a Brownian motion with variance $(\alpha^2 + \beta^2)\lambda$ per unit time and drift $-\gamma$, $I^*(t) \overset{\triangle}{=} -\min_{0 \le s \le t} N^*(s)$, and $W^*(t) = N^*(t) + I^*(t)$. In other words, W^* is a reflected Brownian motion with drift, and I^* causes the reflection.

3   Main Results

In this section and in Sect. 4 we assume that customers are served using the SRPT queue discipline. The following theorems are the main results of this paper.

Theorem 1. *Assume that (1)–(4) and (10)–(12) hold. Then $\widehat{Q}^n \Rightarrow \frac{1}{v^*} W^*$ in $D[0, \infty)$ jointly with (13) as $n \to \infty$.*

Theorem 2. *Assume that (1)–(3), (5)–(7), (10)–(12) hold and that $(\overline{\mathbb{R}}_+, \rho)$ is a totally bounded semimetric space. Then $\widehat{Q}^n \Rightarrow 0$ in $D[0, \infty)$ as $n \to \infty$.*

Theorems 1 and 2 can be easily upgraded to limit theorems for measure-valued state descriptors. Let $w_j^n(t)$ be the residual service time at time t of the j-th customer to appear in the n-th SRPT queueing system. For $n \in \mathbb{N}$, $t \ge 0$ and $B \in \mathcal{B}(\mathbb{R})$, let

$$\mathcal{Q}^n(t) = \sum_{j=1}^{A^n(t)} \delta^+_{w_j^n(t)}, \qquad \widehat{\mathcal{Q}}^n(t)(B) \overset{\triangle}{=} \frac{1}{\sqrt{n}} \mathcal{Q}^n(nt)(B). \tag{14}$$

Note that $< 1, \mathcal{Q}^n(t) >= Q^n(t)$ and $< e, \mathcal{Q}^n(t) >= W^n(t)$, so $< 1, \widehat{\mathcal{Q}}^n(t) >= \widehat{Q}^n(t)$ and $< e, \widehat{\mathcal{Q}}^n(t) >= \widehat{W}^n(t)$. Theorem 2 obviously implies

Corollary 1. *Under the assumptions of Theorem 2 we have $\widehat{\mathcal{Q}}^n \Rightarrow \mathbf{0}$ in $D_{\mathcal{M}}[0, \infty)$ as $n \to \infty$.*

Theorem 1 in turn implies

Corollary 2. *Under the assumptions of Theorem 1 we have* $\widehat{\mathcal{Q}}^n \Rightarrow \frac{1}{v^*}W^*\delta_{v^*}$ *in* $D_{\mathcal{M}}[0,\infty)$ *as* $n \to \infty$.

The proofs of Theorems 1, 2 and Corollary 2 will be given in the next section.

The assumptions of Theorems 1 and 2 can be simplified considerably if the service time distribution $G^n \equiv G$ does not vary with n. In this case, we have

Corollary 3. *Asume that* $G^n \equiv G$ *for every* $n \in \mathbb{N}$,

$$\lim_{n\to\infty} \lambda_n = \lambda \overset{\triangle}{=} \frac{1}{\mathbb{E}v_j^n} \tag{15}$$

and that (10)–(12) hold. If $v^* \overset{\triangle}{=} \min\{y \in \mathbb{R} : G(y) = 1\} < \infty$, *then* $\widehat{\mathcal{Q}}^n \Rightarrow \frac{1}{v^*}W^*\delta_{v^*}$ *in* $D_{\mathcal{M}}[0,\infty)$ *and* $\widehat{Q}^n \Rightarrow \frac{1}{v^*}W^*$ *in* $D[0,\infty)$ *jointly with (13) as* $n \to \infty$. *In the opposite case,* $\widehat{\mathcal{Q}}^n \Rightarrow \mathbf{0}$ *in* $D_{\mathcal{M}}[0,\infty)$ *as* $n \to \infty$.

Corollary 3 follows directly from Theorem 1 and Corollaries 1, 2.

It is interesting that in the case of SRPT (and LRPT, see Sect. 5, to follow) the limiting behavior of the measure-valued state descriptor process can be retrieved from the limiting behavior of the one-dimensional queue length process which contains much less information. In fact, in the existing literature on heavy traffic approximations for queueing systems operating under EDF [6,13] and processor sharing [7], the opposite approach was used: first limiting distributions for the measure-valued state descriptors were found, and then heavy traffic limits for the real-valued queue length processes were derived as immediate consequences. In the case of SRPT and LRPT this order can be reversed due to simple and very special forms of the limiting distributions under consideration.

4   Proofs of the Main Results

4.1   Approximating EDF Systems

Without loss of generality we can assume that $\lambda = 1$, since this is only a matter of rescaling. Fix $M \in \mathbb{N}$. Consider a sequence of auxiliary EDF queueing systems, indexed by superscript n. The inter-arrival times for the n-th system are $\{u_j^n\}_{j=1}^\infty$ and the service times are $\{v_j^n\}_{j=1}^\infty$. The j-th customer arrives at the n-th system with an initial lead time (i.e., the time between the arrival time and the deadline for completion of service for that customer) equal to $M\sqrt{n}v_j^n$. Denote by $Q_M^n(t)$ the queue length process in the n-th EDF system and let

$$\widehat{Q}_M^n(t) = \frac{1}{\sqrt{n}}Q_M^n(nt).$$

By the above-mentioned SRPT optimality result, $Q^n(t) \le Q_M^n(t)$ for each $t \ge 0$, so

$$\widehat{Q}^n(t) \le \widehat{Q}_M^n(t), \qquad t \ge 0. \tag{16}$$

For any $y \in \mathbb{R}$, define

$$H_v(y) \overset{\triangle}{=} \int_y^\infty (1 - G_v(\eta)) \, d\eta, \qquad H(y) \overset{\triangle}{=} \int_y^\infty (1 - G(\eta)) \, d\eta.$$

By (8), H and H_v are finite on $\mathbb{R}$. Under the assumptions of Theorem 1 (Theorem 2), the function H_v maps $(-\infty, v^*]$ $(\overline{\mathbb{R}})$ onto $\mathbb{R}_+$ and is strictly decreasing and continuous on $(-\infty, v^*]$ $(\overline{\mathbb{R}})$. Therefore, there exists a continuous inverse function H_v^{-1} mapping $\mathbb{R}_+$ onto $(-\infty, v^*]$ $(\overline{\mathbb{R}})$. Let

$$H^M(y) = M \int_{\frac{y}{M}}^\infty (1 - G(\eta)) d\eta = M H \left(\frac{y}{M} \right),$$

$$H_v^M(y) = M \int_{\frac{y}{M}}^\infty (1 - G_v(\eta)) d\eta = M H_v \left(\frac{y}{M} \right),$$

and let $f_M = H^M \circ (H_v^M)^{-1}$. For $x \geq 0$ we have $f_M(x) = M H \left((H_v)^{-1}(x/M) \right)$, so f_M is a continuous, strictly increasing mapping of $\mathbb{R}_+$ onto $\mathbb{R}_+$.

Under the assumptions of Theorem 1, Theorem A2 from [13], together with elementary computations, implies that

$$\widehat{Q}_M^n \Rightarrow Q_M^* \overset{\triangle}{=} f_M(W^*), \tag{17}$$

in $D[0, \infty)$ as $n \to \infty$, jointly with (13), where W^* is as in (13). Under the assumptions of Theorem 2, we again have (17), but now by Corollary 3.2 from [13].

4.2   Proof of Theorem 1

We will first show that for each $x \geq 0$ we have

$$\lim_{M \to \infty} f_M(x) = \frac{x}{v^*}. \tag{18}$$

Since $f_M(0) = 0$ for each M, it suffices to show (18) for $x > 0$. Fix $x > 0$ and let $y_M = (H_v^M)^{-1}(x)$. We have

$$\int_{\frac{y_M}{M}}^{v^*} (1 - G_v(\eta)) d\eta = H_v \left(\frac{y_M}{M} \right) = \frac{x}{M} \to 0, \qquad M \to \infty,$$

so $y_M / M < v^*$ and

$$\lim_{M \to \infty} \frac{y_M}{M} = v^*. \tag{19}$$

Also, for $z < v^*$ being the continuity point of both G and G_v we have, by (1)–(4), the fact that $\lambda = 1$ and the Markov inequality,

$$\frac{1 - G_v(z)}{1 - G(z)} = \lim_{n \to \infty} \frac{\frac{1}{\lambda_n} - G_v^n(z)}{1 - G^n(z)} = \lim_{n \to \infty} \frac{\mathbb{E}[v_1^n \mathbb{I}_{[z < v_1^n \leq v^*]}]}{\mathbb{P}[z < v_1^n \leq v^*]} \in [z, v^*]. \tag{20}$$

If $z < v^*$ is a point of discontinuity of either G, or G_v, there exist points $z_n < v^*$, $z_n \downarrow z$ such that both G and G_v are continuous at z_n. Thus, (20) holds at z_n and right-continuity of G, G_v implies that

$$z \leq \frac{1 - G_v(z)}{1 - G(z)} \leq v^*, \qquad 0 < z < v^*. \tag{21}$$

Consequently, by (19) and (21), as $M \to \infty$,

$$\frac{H^M(y_M)}{H_v^M(y_M)} = \frac{\int_{\frac{y_M}{M}}^{v^*}(1 - G(\eta))d\eta}{\int_{\frac{y_M}{M}}^{v^*}(1 - G_v(\eta))d\eta} \to \frac{1}{v^*}.$$

Hence, as $M \to \infty$,

$$f_M(x) = H^M(y_M) = H_v^M(y_M)\frac{H^M(y_M)}{H_v^M(y_M)} = x\frac{H^M(y_M)}{H_v^M(y_M)} \to \frac{x}{v^*}.$$

We have proved (18). Since the functions $f_M(x)$, x/v^* are continuous and increasing in x, it is not hard to check that the convergence (18) is uniform on compact subsets of $\mathbb{R}_+$ (see, e.g., the proof of Proposition 3.4 in [6] for a similar argument).

Fix $T > 0$, $\epsilon > 0$. Let $N \in \mathbb{N}$ be so large that

$$\mathbb{P}\left[\max_{0 \leq t \leq T} W^*(t) \leq N\right] \geq 1 - \frac{\epsilon}{2}. \tag{22}$$

Fix M so large that $\sup_{0 \leq x \leq N} |f_M(x) - x/v^*| \leq \epsilon/2$. Thus, by (22),

$$\mathbb{P}\left[\max_{0 \leq t \leq T}\left|f_M(W^*(t)) - \frac{1}{v^*}W^*(t)\right| \leq \frac{\epsilon}{2}\right] \geq 1 - \frac{\epsilon}{2}. \tag{23}$$

Using (13), (17) and the Skorokhod representation theorem (see, e.g., [3], Theorem 6.7), we can construct the model primitives u_j^n and v_j^n for $j \in \mathbb{N}$ and $n \in \mathbb{N}$ on a common probability space $(\Omega, \mathcal{F}, \mathbb{P})$ such that the sequences of processes $\widehat{N}^n$, $\widehat{I}^n$, $\widehat{W}^n$, $\widehat{Q}_M^n$, $n \in \mathbb{N}$, and the processes N^*, I^*, W^* are defined on this space and

$$\widehat{N}^n \to N^*, \qquad \widehat{I}^n \to I^*, \qquad \widehat{W}^n \to W^*, \qquad \widehat{Q}_M^n \to f_M(W^*) \tag{24}$$

almost surely (a.s.). Here each a.s. convergence is in the J_1 topology on $D[0, \infty)$ and since the limits are continuous, this is equivalent to uniform convergence on compact intervals. Therefore, for n large enough,

$$\mathbb{P}\left[\max_{0 \leq t \leq T}\left|\widehat{Q}_M^n(t) - f_M(W^*(t))\right| \leq \frac{\epsilon}{2}\right] \geq 1 - \frac{\epsilon}{2}.$$

This, together with (23), yields

$$\mathbb{P}\left[\max_{0 \leq t \leq T}\left|\widehat{Q}_M^n(t) - \frac{1}{v^*}W^*(t)\right| \leq \epsilon\right] \geq 1 - \epsilon. \tag{25}$$

for n large enough. The relations (16) and (25) yield

$$\mathbb{P}\left[\widehat{Q}^n(t) \leq \frac{1}{v^*}W^*(t) + \epsilon \quad \forall t \in [0, T]\right] \geq 1 - \epsilon. \tag{26}$$

On the other hand, (4) clearly implies that $W^n(t) \leq v^* Q^n(t)$ (and hence $\widehat{W}^n(t) \leq v^* \widehat{Q}^n(t)$) for all $n \in \mathbb{N}$ and $t \geq 0$, so by (24) for n large enough we have

$$\mathbb{P}\left[\widehat{Q}^n(t) \geq \frac{1}{v^*}W^*(t) - \epsilon \quad \forall t \in [0, T]\right] \geq 1 - \epsilon. \tag{27}$$

The relations (26) and (27) imply Theorem 1.

4.3   Proof of Theorem 2

We will first show that for each $x \geq 0$ we have

$$\lim_{M \to \infty} f_M(x) = 0. \tag{28}$$

Since $f_M(0) = 0$ for each M, it suffices to show (28) for $x > 0$. Let $x > 0$ and let $y_M = (H_v^M)^{-1}(x)$. We have

$$\int_{\frac{y_M}{M}}^{\infty}(1 - G_v(\eta))d\eta = H_v\left(\frac{y_M}{M}\right) = \frac{x}{M} \to 0, \qquad M \to \infty,$$

so

$$\lim_{M \to \infty} \frac{y_M}{M} = \infty. \tag{29}$$

Also, for z being the continuity point of both G and G_v we have, by (2)–(3), the fact that $\lambda = 1$ and the Markov inequality,

$$\frac{1 - G_v(z)}{1 - G(z)} = \lim_{n \to \infty} \frac{\frac{1}{\lambda_n} - G_v^n(z)}{1 - G^n(z)} = \lim_{n \to \infty} \frac{\mathbb{E}[v_1^n \mathbb{I}_{[v_1^n > z]}]}{\mathbb{P}[v_1^n > z]} \geq z. \tag{30}$$

If z is a point of discontinuity of either G, or G_v, there exist points $z_n \downarrow z$ such that both G and G_v are continuous at z_n. Thus, (30) holds at z_n and right-continuity of G, G_v implies that

$$\frac{1 - G_v(z)}{1 - G(z)} \geq z, \qquad z > 0. \tag{31}$$

Consequently, by (29) and (31), as $M \to \infty$,

$$\frac{H^M(y_M)}{H_v^M(y_M)} = \frac{\int_{\frac{y_M}{M}}^{\infty}(1 - G(\eta))d\eta}{\int_{\frac{y_M}{M}}^{\infty}(1 - G_v(\eta))d\eta} \leq \frac{\int_{\frac{y_M}{M}}^{\infty}(1 - G(\eta))d\eta}{\int_{\frac{y_M}{M}}^{\infty}\eta(1 - G(\eta))d\eta} \leq \frac{M}{y_M} \to 0,$$

so as $M \to \infty$,

$$f_M(x) = H^M(y_M) = H_v^M(y_M) \frac{H^M(y_M)}{H_v^M(y_M)} = x \frac{H^M(y_M)}{H_v^M(y_M)} \to 0.$$

We have proved (28).

Let $T > 0$, $\epsilon > 0$ be arbitrary and let $N \in \mathbb{N}$ be so large that (22) holds. By (28), for M large enough, $f_M(N) \leq \epsilon/2$. Thus, by (17), (22) and the fact that f_M is increasing, for n sufficiently large, we have

$$\mathbb{P}\left[\max_{0 \leq t \leq T} \widehat{Q}_M^n(t) \leq \epsilon \right] \geq 1 - \epsilon. \tag{32}$$

This, together with (16) and the fact that $Q^n(t) \geq 0$ for all t, proves Theorem 2.

4.4   Proof of Corollary 2

Since $w_j^n(t) \leq v_j^n$, (4) implies that $\widehat{\mathcal{Q}}^n(t)(v^*, \infty) = 0$ for all $n \in \mathbb{N}$ and $t \geq 0$. Also, by Theorem 1, the total mass $\widehat{Q}^n(t)$ of the random measure $\widehat{\mathcal{Q}}^n(t)$ converges to $\frac{1}{v^*} W^*(t)$ in $D[0, \infty)$. Therefore, to prove Corollary 2 it is sufficient to show that for any $x \in (0, v^*)$,

$$\widehat{\mathcal{Q}}^n(t)[0, x) \Rightarrow 0, \qquad n \to \infty. \tag{33}$$

Fix $x \in (0, v^*)$. We have

$$\widehat{W}^n(t) = \, < e, \widehat{\mathcal{Q}}^n(t) > \, \leq x \, \widehat{\mathcal{Q}}^n(t)[0, x) + v^* \widehat{\mathcal{Q}}^n(t)[x, v^*]$$
$$= v^* \widehat{Q}^n(t) - (v^* - x) \widehat{\mathcal{Q}}^n(t)[0, x).$$

Thus,

$$\widehat{\mathcal{Q}}^n(t)[0, x) \leq \frac{1}{v^* - x} \left(v^* \widehat{Q}^n(t) - \widehat{W}^n(t) \right). \tag{34}$$

The right-hand side of (34) converges weakly to zero by Theorem 1, which proves (33).

5   Limiting Distributions for LRPT

In this section we assume that customers are served using the LRPT queue discipline. We assume that (1)–(3), (9) and (10)–(12) hold. In this case, we have the following two theorems.

Theorem 3. *If $v_* > 0$, then $\widehat{Q}^n \Rightarrow \frac{1}{v_*} W^*$ in $D[0, \infty)$ jointly with (13) as $n \to \infty$.*

Theorem 4. *If $v_* = 0$, then for every fixed $t > 0$ $\widehat{Q}^n(t) \Rightarrow \infty$ as $n \to \infty$.*

Let us note that it is not possible to generalize Theorem 4 to convergence in $D_{\overline{\mathbb{R}}}[0, \infty)$. Indeed, $Q^n(0) = 0$ a.s. for every n. Moreover, if $\rho_n < 1$, $0 < c < T$ and n is large, then a "typical" sample path of the process Q^n hits zero at some $t \in [nc, nT]$, so in fact for any $0 < c < T$ the convergence $\widehat{Q}^n \Rightarrow \infty$, $n \to \infty$, cannot hold in $D_{\overline{\mathbb{R}}}[c, T]$.

Let the measure-valued processes $\mathcal{Q}^n$ and $\widehat{\mathcal{Q}}^n$ be defined by (14). Theorem 3 has the following

Corollary 4. *Under the assumptions of Theorem 3 we have* $\widehat{\mathcal{Q}}^n \Rightarrow \frac{1}{v_*} W^* \delta_{v_*}$ *in* $D_{\mathcal{M}}[0, \infty)$ *as* $n \to \infty$.

The proofs of Theorems 3, 4 and Corollary 4 are similar to the proofs of Theorems 1, 2 and Corollary 2 presented in Sect. 4. The most notable difference is that in the case of LRPT we use approximating EDF G/G/1 systems with initial lead times equal to $-M\sqrt{n}v_j^n$ instead of $M\sqrt{n}v_j^n$. We omit the details.

As in the case of SRPT, if the service time distribution $G^n \equiv G$ does not depend on n, then the assumptions for the results of this section can be simplified. Namely, in this case only (10)–(12) and (15) need to be assumed for Theorems 3, 4 and Corollary 4 to hold.

References

1. Bansal, N., Harchol-Balter, M.: Analysis of SRPT scheduling: investigating unfairness. ACM Sigmetrics Perform. Eval. Rev. **29**, 279–290 (2001)
2. Bender, M., Chakrabarti, S., Muthukrishnan, S.: Flow and stretch metrics for scheduling continuous job streams. In: Proceedings of the 9th Annual ACM-SIAM Symposium on Discrete Algorithms (1998)
3. Billingsley, P.: Convergence of Probability Measures, 2nd edn. Wiley, New York (1999)
4. Down, D.G., Gromoll, H.C., Puha, A.L.: Fluid limits for shortest remaining processing time queues. Math. Oper. Res. **34**, 880–911 (2009)
5. Down, D.G., Wu, R.: Multi-layered round robin routing for parallel servers. Queueing Syst. **53**, 177–188 (2006)
6. Doytchinov, B., Lehoczky, J.P., Shreve, S.E.: Real-time queues in heavy traffic with earliest-deadline-first queue discipline. Ann. Appl. Probab. **11**, 332–378 (2001)
7. Gromoll, H.C.: Diffusion approximation for a processor sharing queue in heavy traffic. Ann. Appl. Probab. **14**, 555–611 (2004)
8. Gromoll, H.C., Keutel, M.: Invariance of fluid limits for the shortest remaining processing time and shortest job first policies. Queueing Syst. **70**, 145–164 (2012)
9. Gromoll, H.C., Kruk, Ł., Puha, A.L.: Diffusion limits for shortest remaining processing time queues. Stoch. Syst. **1**(1), 1–16 (2011)
10. Ethier, S.N., Kurtz, T.G.: Markov Processes: Characterization and Convergence. Wiley, New York (1985)
11. Iglehart, D., Whitt, W.: Multiple channel queues in heavy traffic I. Adv. Appl. Probab. **2**, 150–177 (1970)
12. Kittsteiner, T., Moldovanu, B.: Priority auctions and queue disciplines that depend on processing time. Manage. Sci. **51**, 236–248 (2005)
13. Kruk, Ł.: Diffusion approximation for a G/G/1 EDF queue with unbounded lead times. Ann. UMCS Math. A **61**, 51–90 (2007)

14. Kruk, Ł., Sokołowska, E.: Fluid limits for multiple-input shortest remaining processing time queues. Math. Oper. Res. **41**(3), 1055–1092 (2016)
15. Núñez Queija, R.: Queues with equally heavy sojourn time and service requirement distributions. Ann. Oper. Res. **113**, 101–117 (2002)
16. Nuyens, M., Zwart, B.: A large deviations analysis of the GI/GI/1 SRPT queue. Queueing Syst. **54**, 85–97 (2006)
17. Pavlov, A.V.: A system with Schrage servicing discipline in the case of a high load. Engrg. Cybernetics **21**, 114–121 (1984). translated from Izv: Akad. Nauk SSSR Tekhn. Kibernet. **6**, 59–66 (1983). (Russian)
18. Pechinkin, A.V.: Heavy traffic in a system with a discipline of priority servicing for the job with the shortest remaining length with interruption (Russian). Math. Issled. No. 89, Veroyatn. Anal. **97**, 85–93 (1986)
19. Perera, R.: The variance of delay time in queueing system M/G/1 with optimal strategy SRPT. Archiv für Elektronik und Übertragungstechnik **47**, 110–114 (1993)
20. Puha, A.L.: Diffusion limits for shortest remaining processing time queues under nonstandard spacial scaling. Ann. Appl. Probab. **25**(6), 3301–3404 (2015)
21. Schassberger, R.: The steady-state appearance of the M/G/1 queue under the discipline of shortest remaining processing time. Adv. Appl. Probab. **22**, 456–479 (1990)
22. Schrage, L.E.: A proof of the optimality of the shortest remaining processing time discipline. Oper. Res. **16**, 687–690 (1968)
23. Schrage, L.E., Miller, L.W.: The queue M/G/1 with the shortest remaining processing time discipline. Oper. Res. **14**, 670–684 (1966)
24. Wierman, A., Harchol-Balter, M.: Classifying scheduling policies with respect to unfairness in an M/G/1. In: Proceedings of the 2003 ACM SIGMETRICS International Conference on Measurement and Modeling of Computer Systems, pp. 238–249 (2003)

Revisiting SRPT for Job Scheduling in Computing Clusters

Huanle Xu[1(✉)], Huangting Wu[2], and Wing Cheong Lau[2]

[1] College of Computer Science and Technology, Dongguan University of Technology, Dongguan, China
xhlcuhk@gmail.com
[2] Department of Information Engineering, The Chinese University of Hong Kong, Sha Tin, Hong Kong

Abstract. As the scheduling principle of Shortest Remaining Processing Time (SRPT) has been proven to be optimal in the single-machine setting, it's a natural thought that SRPT shall also be extended to yield various scheduling algorithms with theoretical performance guarantees in distributed computing clusters which consist of multiple machines. In this paper, we revisit the SRPT scheduling principle to derive new and tight competitive performance bounds with respect to the overall job flowtime. In particular, for the transient scheduling scenario where all jobs arrive at the cluster at time zero, we study two different cases and show that the SRPT-based scheduling algorithm can achieve a constant competitive ratio of at most two, compared to the prior state-of-the-art ratio of 12 in the algorithm of Moseley et al. For online scheduling, we study a special case where each job only consists of one single task and show that the online SRPT Algorithm is $(1+\epsilon)$-speed, $(3+\frac{3}{\epsilon})$-competitive with respect to the overall job flowtime for $\epsilon > 0$, improving the recent result of Fox and Moseley which upper bounds SRPT to be $(1+\epsilon)$-speed, $\frac{4}{\epsilon}$-competitive.

Keywords: Job scheduling · SRPT ·
Competitive performance bound · Dual-fitting

1 Introduction

The recent years witnessed rapid emergence and proliferation of distributed computing. A key performance metric for such computing paradigm is the job response time, which is also referred to as the job flowtime. As demand for large-scale analytics soars, today's computing clusters usually scale out to consist of hundreds of thousands of machines. In addition, the job profiles are becoming increasingly diverse as small latency-sensitive jobs coexist with large batch processing applications that may take hours to months to complete [22]. To make things even challenging, jobs can have multiple small tasks with complicated precedence constraints. With a huge number of heterogeneous tasks sharing

© Springer Nature Switzerland AG 2019
T. Phung-Duc et al. (Eds.): QTNA 2019, LNCS 11688, pp. 276–291, 2019.
https://doi.org/10.1007/978-3-030-27181-7_17

a computing cluster with multiple machines, it becomes extremely difficult to design job schedulers to allocate computing machines efficiently which achieving fast job response time.

To provide efficient resource sharing among heterogeneous jobs/ tasks, various schedules have been designed for production clusters. It is well known that the SRPT scheduler (Shortest Remaining Processing Time) is optimal for minimizing the overall/ average job flowtime [5] on a single machine under the clairvoyant setting. As such, many works have aimed to extend the SRPT principle to yield efficient scheduling algorithms in the multiprocessor setting for different systems and programming frameworks [6,17,18,23]. Under SRPT, job sizes are known to the job scheduler upon job arrival and smaller jobs are given priority. However, most existing SRPT-based algorithms do not come with systematic analysis or reference study on how well or suboptimal they are when compared to the corresponding theoretical limits (e.g., [6]). [17,18,23] are exceptions and can provide performance guarantees in terms of overall job flowtime via a competitive-based analysis. However, these studies mainly address the online scheduling problem and the competitive bounds they derived are quite loose.

With the above observations in mind, in this paper, we revisit the SRPT scheduling principle in distributed computing clusters and derive tighter competitive performance bounds. Moreover, we investigate both the transient scheduling as well as the online scheduling scenarios. Under our study, we assume preemption is allowed, i.e., the scheduler can preempt a running task and later resume its execution.

Under the transient setting where all jobs arrive at the cluster at the same time, we analyze two cases, namely, the single-task case and the multi-task case. In the single-task case where each job only consists of one single task, we take a new approach to show that SRPT is optimal in terms of minimizing the overall job flowtime when jobs can be preempted. Our approach is novel in the sense that it decomposes the job flowtime into two parts, namely, the waiting time and the processing time. By contrast, the previous analysis for SRPT are mainly based on tuning the scheduling sequence of different jobs [19]. In the multi-task setting where each job consists of multiple small tasks, we consider the general case where the execution times of tasks within the same job are different. In this setting, our SRPT-based algorithm shall give scheduling priority to jobs with the smallest amount of workload where the workload is the sum of all task execution times. We also show that this algorithm is two-competitive with respect to the overall job flowtime.

For online setting where all jobs arrive at the cluster over time, we study the case in which each job consists of only one single task. Note that, this case has been extensively studied in the literature and the online SRPT algorithm is shown to be $O(\min\{\log P, \log N/M\})$-competitive where P is the ratio of maximum execution time to minimum execution time with N and M being the number of jobs and the number of machines in the cluster respectively [5]. In fact, [15] has shown that no online algorithm can achieve a smaller bound than $O(\min\{\log P, \log N/M\})$. Due to this negative result, previous works [12,14] have adopted a resource augmentation analysis to bound the competitive performance of online scheduling algorithms. Under such analysis, the performance of the offline optimal algorithm on M unit-speed machines is compared with that of the proposed algorithms on M δ-speed machines where $\delta > 1$. In this paper, we also follow the resource augmentation setup and apply a novel dual fitting approach to show that SRPT is $(1 + \epsilon)$-speed, $(3 + \frac{3}{\epsilon})$-competitive with respect to the overall job flowtime for $\epsilon > 0$. In summary, we have made the following technical contributions in this paper:

- In Sect. 4.1, we adopt a different approach from the analysis in the seminal work [19] to show that, SRPT is optimal in terms of overall job flowtime in the one-single-task setting. Our approach takes preemption into consideration and does not restrict to only tune the scheduling sequence of different jobs [19].
- In Sect. 4.2, we show the SRPT-based algorithm achieves a competitive ratio of 2, which is much tighter than the previous best result of 12. Notice that, [18] proposes two additional algorithms to assign priorities to tasks within each job. By contrast, in our analysis, we only consider the total workload of each job and use it to determine the scheduling priority of each job. As such, we can avoid most of the unnecessary approximations in [18].
- In Sect. 5, we apply a new dual-fitting framework to bound the competitive performance of online SRPT. We show that, the online SRPT algorithm is $(1 + \epsilon)$-speed, $(3 + \frac{3}{\epsilon})$-competitive with respect to the overall job flowtime for $\epsilon > 0$. Interestingly, when given small resource augmentation where $\epsilon \leq \frac{1}{3}$, our algorithm improves the recent result in [5], i.e., SRPT on multiple identical machines is $(1 + \epsilon)$-speed, $\frac{4}{\epsilon}$-competitive. To prove this result, we compute the difference of the overall job flowtime caused by each job arrival under SRPT. We then use this difference to set the variables of the formulated dual problem and explore the relationship between the objective of the dual problem and that of the original problem. By contrast, [5] defines a complicated potential function to track the dynamics of the overall job flowtime in the cluster achieved by the SRPT algorithm and the optimal scheme respectively. However, the construction of the required potential function is problem specific, which makes it difficult for one to generalize the approach to solve other scheduling problems.

2   Related Work

The design of job schedulers for large-scale computing clusters is currently an active research topic, e.g., [2,3,17,18,21,23]. Several works have derived performance bounds on algorithms geared at minimizing the total job completion time by formulating an approximate linear programming problem [2,3,21]. Other works, e.g., [17,18,23] derive performance bounds for algorithms with respect to the total job flowtime. Leonardia et al. show in [16] that there is a strong lower bound on any online randomized algorithm for the job scheduling problem on multiple unit-speed processors with the objective to minimize the overall job flowtime. Based on this lower bound, [17,18,23] extended the SRPT algorithm to design scheduling schemes aim at minimizing the overall flowtime of jobs, each consisting of multiple small tasks with precedence constraints. However, the proposed algorithm in [18] is non-work-conserving since each job can only be scheduled after it completes in another simulated system. One limitation of [17,23] is that their derived bounds are quite loose.

Recently, researchers begin to analyze the performance of SRPT from a queuing perspective. In particular, [7] proposed to analyze the mean and variance of occupancy for SRPT under the M/GI/1 setting. [4] analyzed the mean response time for SRPT scheduling policy for multi-sever systems.

Another group of works [2,3,17,18,21,23] study the clairvoyant setting where the job size is known once the job arrives. For the non-clairvoyant setting, [9–11] designed several multi-tasking algorithms (i.e., a server can serve multiple jobs simultaneously) in which machines are allocated to all active jobs and priority is given to the most-recently-arrived jobs.

Under the transient scheduling setting, previous work, e.g., [19] has shown the optimality of SRPT for the single-task case in the non-preemptive mode. The analysis of [19] is based on tuning the scheduling sequence of different jobs, which cannot be extended to the preemptive mode. By contrast, we take a new approach by decoupling the job flowtime, which can easily accommodate the preemptive setting.

For the analysis of the online SRPT algorithm in Sect. 5, we adopt the dual fitting approach. Dual fitting was first developed by [1,8] and is now widely used for the analysis of online algorithms [10,11]. On the one hand, [1,10] and [11] address linear objectives and use the dual-fitting approach to derive competitive bounds for traditional scheduling algorithms without redundancy. On the other hand, [8] focuses on a convex objective in the multi-tasking mode. In contrast, we include integer constraints associated with the non-multi-tasking mode. Moreover, our setting of dual variables is novel in the sense that it deals with the dynamical change of job flowtime across multiple machines where other settings of dual variables can only deal with the change of job flowtime on one single machine.

3   System Model

Consider a distributed cluster which consists of M servers[1], where the servers are indexed from 1 to M. Each machine can only hold one task at any time. Job j arrives at the cluster at time a_j and the job arrival process, $(a_1, a_2, \cdots, a_N)$, is an arbitrary deterministic time sequence, where N is the total amount of the jobs. Upon arrival, job j joins a global queue managed by a cluster scheduler, waiting to be scheduled.

---

[1] Each server can either represent a CPU core or a machine.

3.1 Job Service Process

Job j consists of m_j tasks which can be executed in parallel and we let δ_j^i denote the ith task of job j for all $i \in \{1, 2, \ldots, m_j\}$. The execution time of δ_j^i is denoted by p_j^i, i.e., it takes p_j^i units of time to complete δ_j^i on any of these M machines. We consider that tasks can be preempted, i.e, the scheduler can preempt a running task and later resume its execution. We shall use c_j^i to denote the completion time of task δ_j^i, therefore, the completion time of job j, c_j, can be formulated as:

$$c_j = \max_{i \in \{1, \cdots, m_j\}} c_j^i. \tag{1}$$

The flowtime of job j, f_j, is denoted by $f_j = c_j - a_j$. In this paper, we only focus on the overall job flowtime, i.e., $\sum_{j=1}^{N} f_j$.

3.2 Competitive Performance Metrics

In this paper, we revisit the SRPT scheduling algorithms under the offline and the online scheduling cases. In particular, we shall use the following metric to evaluate the performance of an offline algorithm.

Definition 1. *An algorithm is c-competitive if the algorithm's objective is within a factor of c of the optimal solution's objective.*

Kalyanasundaram et al. show in [13] that, no online algorithm can achieve a constant competitive performance bound even for the total flowtime of jobs with a single task on multiple machines. As such, previous work [12,13] has adopted a resource augmentation analysis. The following definition characterizes the competitive performance of an online algorithm with resource augmentation.

Definition 2. *[13] An online algorithm is δ-speed, c-competitive if the algorithm's objective is within a factor of c of the optimal solution's objective when the algorithm is given δ resource augmentation.*

In this paper, we adopt the resource augmentation setup to bound the competitive performance of the online SRPT algorithm.

4 Transient Scheduling: All the Jobs Arrive at the Cluster at Time Zero

Before studying of the online scheduling algorithm, in this section, we consider the transient scheduling case, i.e., all jobs arrive at the cluster at time zero.

4.1   Each Job Only Consists of One Single Task

In this subsection, we model a large-scale computing cluster as follows: the cluster consists of M identical machines which are indexed from 1 to M. Job j arrives at the cluster at time 0 and consists of only one single task, i.e., δ_j^1. The execution time of δ_j^1 is denoted by p_j^1, i.e., it takes p_j^1 units of time to complete δ_j^1 on any of these M machines.

[19] has shown that SRPT is optimal for minimizing the overall job flowtime when preemption is not allowed. In this section, we show a more general result in the following theorem:

Theorem 1. *Under the transient scheduling setting, SRPT is optimal with respect to the overall job flowtime when each job only consists of one single task and preemption is allowed.*

Proof. Without loss of generality, we assume that jobs have been ordered such that $c_1 \leq c_2 \leq \cdots \leq c_N$. When $N \leq M$, the theorem follows immediately since all jobs can be scheduled simultaneously and f_j is equal to p_j^1.

Let us then consider the case where $N > M$. Let $N = zM + q$ where $z \geq 1$, $0 \leq q \leq M - 1$ and z, q are non-negative integers. We first show that for all k such that $M \leq k \leq N$, the following result holds:

$$\sum_{j=k-M+1}^{k} f_j = \sum_{j=1}^{k} p_j^1 \tag{2}$$

At any time between 0 and c_1, there are $(k - M)$ jobs waiting to be processed among those k jobs which complete first. Hence, the accumulated waiting time in this period is $(k - M)f_1$. Similarly, at any time between c_1 and c_2, there are $(k-M-1)$ jobs waiting to be processed and they contribute $(k-M-1)\cdot(c_2-c_1) = (k - M - 1) \cdot (f_2 - f_1)$ waiting time. Hence, the total waiting time of the k jobs is given by:

$$\sum_{j=0}^{k-M-1} (k - M - j) \cdot (f_{j+1} - f_j) = \sum_{j=1}^{k-M} f_j. \tag{3}$$

Therefore, the total flowtime for these k jobs is given by:

$$\sum_{j=1}^{k} f_j = \sum_{j=1}^{k} p_j^1 + \sum_{j=1}^{k-M} f_j. \tag{4}$$

By shifting terms in (4), we have:

$$\sum_{j=k-M+1}^{k} f_j = \sum_{j=1}^{k} p_j^1.$$

Summing up all job flowtime, it follows that:

$$\sum_{j=1}^{N} f_j = \sum_{j=1}^{q} f_j + \sum_{k=1}^{z} \sum_{j=(k-1)M+q+1}^{kM+q} f_j$$

$$\stackrel{(i)}{=} \sum_{j=1}^{q} p_j^1 + \sum_{k=1}^{z} \sum_{j=1}^{kM+q} p_j^1 = \sum_{j=1}^{N} (\lfloor \frac{N-j}{M} \rfloor + 1) p_j^1, \tag{5}$$

where the first term on the R.H.S. of (i) is due to the fact that the flowtime of the first q jobs is equal to their task execution time and the second term follows from the equality of $\sum_{j=(k-1)M+q+1}^{kM+q} f_j = \sum_{j=1}^{kM+q} p_j^1$.

One can easily observe that the total flowtime achieved by SPRT is a lower bound of (5). This completes the proof.

4.2   Each Job Consists of Multiple Tasks

In this subsection, we consider a more general case where each job consists of multiple tasks and the execution time of all tasks within each job are different. In this case, let w_j define the workload of job j, thus, w_j is given by:

$$w_j = \sum_{i=1}^{m_j} p_j^i.$$

In this case, the SRPT-based algorithm works as follows: when there is an available machine, the scheduler assigns the unscheduled task with the largest execution time from the job which has the smallest w_j to this machine.

Our main result, characterizing the competitive performance of SRPT in this setting, is given in the following theorem:

Theorem 2. *Under the transient scheduling setting where each job consists of multiple tasks and the task execution times within a job are different, the SRPT-based algorithm is 2-competitive with respect to the overall job flowtime.*

Proof. To prove this result, we first show an upper bound of the total response time achieved by the SRPT-based algorithm and then give a lower bound of the total job flowtime achieved by any other algorithm.

Without loss of generality, assume jobs are ordered such that $w_1 \leq w_2 \leq \cdots w_N$. Let p_j define the largest task execution time of job j, i.e., $p_j = \max_{i \in \{1,2,\cdots,m_j\}} p_j^i$. Based on the scheduling policy of the SRPT-based algorithm, the whole cluster must be busy processing the tasks of jobs with index no larger than j during the time period $[0, f_j - p_j]$. Therefore, we have:

$$M \cdot (f_j - p_j) \leq \sum_{k=1}^{j} w_k, \tag{6}$$

which implies that, $f_j \leq \sum_{k=1}^{j} w_k/M + p_j$, therefore, the total job flowtime achieved by the SRPT-based algorithm is upper bounded by:

$$\sum_{j=1}^{N} f_j \leq \sum_{j=1}^{N} \left(\sum_{k=1}^{j} \frac{w_k}{M} + p_j \right) = \frac{\sum_{j=1}^{N}(N+1-j)w_j}{M} + \sum_{j=1}^{N} p_j. \qquad (7)$$

On the one hand, the total job flowtime achieved by any algorithm on M machines with unit speed is no less than that under one machine with a speed of M. Since SRPT is optimal for overall job flowtime on one machine, we have:

$$\sum_{j=1}^{N} f_j^* \geq \sum_{j=1}^{N} \sum_{k=1}^{j} \frac{w_k}{M} = \frac{\sum_{j=1}^{N}(N+1-j)w_j}{M}, \qquad (8)$$

where f_j^* is the job flowtime under the optimal algorithm. On the other hand, the flowtime of job j is no less than the maximum of the task execution time, i.e.,

$$f_j^* \geq p_j. \qquad (9)$$

Combining Eqs. (7), (8) and (9), the result immediately follows. This completes the proof.

5   SRPT Algorithm in the Online Setting

In this section, we study the SRPT algorithm in the online scheduling case, i.e., jobs arrive at the cluster at different times. Here we assume time is slotted for the sake of convenience.

5.1   Linear Programming Relaxation

We first formulate a linear optimization problem which serves as a relaxation of the online scheduling problem. Let $x_j^i(t)$ be the scheduling variable of task δ_j^i at time t such that $x_j^i(t)$ is equal to one if task δ_j^i is being executed at time t and zero otherwise. Let $\boldsymbol{x}(t) = \left(x_j^i(t) | j = 1, 2, \cdots, N, i = 1, 2, \cdots, m_j \right)$ and $\boldsymbol{x} = \left(\boldsymbol{x}(t) | t \geq 0 \right)$, our relaxed optimization problem is formulated as follows:

$$\min_{\boldsymbol{x}} \sum_{j=1}^{N} \sum_{t \geq a_j} \sum_{i=1}^{m_j} \left(\frac{t - a_j}{\sum_i p_j^i} \times x_j^i(t) + \frac{2x_j^i(t)}{m_j} \right)$$

$$s.t. \quad \sum_{t \geq a_j} x_j^i(t) \geq p_j^i, \forall j, i \qquad (10)$$

$$\sum_{j: t \geq a_j} \sum_i x_j^i(t) \leq M, \forall t$$

Here, the first constraint is due to the fact that the total amount of work delivered for each task is no smaller than its execution time. The second constraint states that there can be at most M machines busy processing tasks in the cluster at any time.

Lemma 1. *The optimal value of the optimization problem in* (10) *is upper bounded by the overall job flowtime achieved by the optimal scheduling policy with a factor of* 3.

Proof. Consider an optimal solution to the optimal scheduling policy, $\boldsymbol{x}^*$. Denote by c_j^* the corresponding job completion time for job j. For all $j = 1, 2, \cdots, N$, $\boldsymbol{x}^*$ and c_j^* satisfy:

$$\sum_{a_j}^{c_j^*} (x_j^i(t))^* = p_j^i, \quad \forall j, i. \tag{11}$$

It follows that $(x_j^i(t))^* = 0$ for all $t \geq c_j^*$, therefore, the first term in the summation of the objective function is upper bounded by the job flowtime achieved by the optimal scheme.

Since at most m_j tasks from job j can be processed in parallel, the flowtime of job j under the optimal scheduling policy is at least $\sum_{t \geq a_j} \sum_{i=1}^{m_j} \frac{(x_j^i(t))^*}{m_j}$. Thus, the second term in the summation of the objective is upper bound by twice of the job flowtime achieved by the optimal scheme. This completes the proof.

We proceed to write down the dual of the optimization problem in (10) as follows [9]:

$$\max_{\alpha, \beta} \sum_{j,i} \alpha_{j,i} \cdot p_j^i - \sum_t \beta(t) \cdot M$$

$$\text{s.t.} \quad \frac{(t - a_j)}{\sum_i p_j^i} + \frac{2}{m_j} - \alpha_{j,i} + \beta(t) \geq 0 \quad \forall j, i, t. \tag{12}$$

The dual has a variable α_j^i for every task δ_j^i corresponding to the first constraint in the primal and a variable $\beta(t)$ corresponding to the second constraint.

5.2   Each Job Only Consists of One Task

In this section, we apply the optimization framework in (12) to study a special online case where each job only consists of one single task. For this case, we characterize the performance of online SRPT using the following theorem:

Theorem 3. *SRPT is* $(1+\epsilon)$*-speed,* $(3+\frac{3}{\epsilon})$*-competitive with respect to the overall job flowtime for* $\epsilon > 0$ *when each job only consists of one task.*

Proof. We omit the notation i in α and p for convenience. We use $n(t)$ and $p_j(t)$ to denote the number of active jobs in the cluster and the remaining execution time of job j at time t respectively. Let $\Theta_j = \{k : a_k \leq a_j \leq c_k\}$, i.e., the set of jobs that are active when job j arrives and $A_j = \{k \neq j : k \in \Theta_j$ and $p_k(a_j) \leq p_j\}$, i.e., jobs whose residual workload upon job j's arrival is less than job j's processing requirement, and let $\rho_j = |A_j|$. Our setting of dual variables as follows:

$$\alpha_j = \left\{ \left(\frac{1}{p_j} \times \sum_{k=1}^{\rho_j} \left(\lfloor \frac{n(a_j) - k}{M} \rfloor - \lfloor \frac{n(a_j) - k - 1}{M} \rfloor \right) p_k(a_j) \right) \right.$$
$$\left. + \left(\lfloor \frac{n(a_j) - \rho_j - 1}{M} \rfloor + 1 \right) \right\} \times \frac{1}{1 + \epsilon} \tag{13}$$

and

$$\beta(t) = \frac{n(t)}{(1 + \epsilon) \cdot M} \tag{14}$$

Lemma 2. *The setting of dual variables above produces a feasible solution* (12).

Proof. Since α and β are both nonnegative, we only need to show

$$\alpha_j - \beta(t) \leq \frac{t - a_j}{p_j} + 2 \quad \forall j; t \geq a_j. \tag{15}$$

Suppose $n(a_j) = zM + q > M$, and notice that the multiplicative factor for $p_k(a_j)$ in (13) is nonzero only when $k = lM + q$ for some $l = 0, 1, \cdots, z$. Therefore, (13) can be rewritten as:

$$\alpha_j = \frac{\sum_{l=0}^{z} p_{lM+q}(a_j) \mathbb{1}(lM + q \leq \rho_j)}{(1 + \epsilon)p_j} + \frac{\left(\lfloor \frac{n(a_j) - \rho_j - 1}{M} \rfloor + 1 \right)}{1 + \epsilon}. \tag{16}$$

For ease of illustration, let Ω_1 and Ω_2 be the first and second term on the R.H.S of (16) respectively.

If $n(a_j) \leq M$, then $\Omega_1 = 0$. Thus, we have $\alpha_j = \frac{1}{1+\epsilon}$ and the result follows. When $n(a_j) > M$, we have the following three cases:

Case I: Suppose the jobs in Θ_j are completed at time t. If there are no job arrivals after time a_j, jobs indexed by $lM + q$ where l is a non-negative integer, are all processed on Machine q. Since the service capacity of Machine q is $(1 + \epsilon)(t - a_j)$ during $(a_j, t]$, it follows that,

$$t - a_j \geq \frac{1}{1 + \epsilon} \sum_{l=0}^{z} p_{lM+q}(a_j). \tag{17}$$

In contrast, if there are other job arrivals after time a_j, since we assume Θ_j must have been completed by time t, Machine q needs to process an amount of work which exceeds $\sum_{l=0}^{z} p_{lM+q}(a_j)$. In other words, (17) still holds. Thus, we have:

$$
\begin{aligned}
\frac{t - a_j}{p_j} &\geq \frac{1}{(1+\epsilon)p_j} \sum_{l=0}^{z} p_{lM+q}(a_j) \\
&= \frac{\sum_{l=0}^{z} p_{lM+q}(a_j)\mathbb{1}(lM + q \leq \rho_j)}{(1+\epsilon)p_j} + \frac{\sum_{l=0}^{z} p_{lM+q}(a_j)\mathbb{1}(lM + q \geq \rho_j + 1)}{(1+\epsilon)p_j} \\
&\overset{(i)}{\geq} \Omega_1 + \frac{\sum_{l=0}^{z} \mathbb{1}(lM + q \geq \rho_j + 1)}{(1+\epsilon)} \\
&= \alpha_j \geq \alpha_j - \beta(t) - 2,
\end{aligned}
\tag{18}
$$

where (i) is due to the fact that $p_{lM+q}(a_j) \geq p_j$ when $lM + q > \rho_j$.

Case II: Suppose jobs indexed from 1 to κ in Θ_j have been completed where $\kappa \leq \rho_j$. Let $\kappa = z_1 M + q_1$. An argument similar to that for Case I shows that by time t, we have:

$$
t - a_j \geq \frac{1}{1+\epsilon} \sum_{k=0}^{z_1} p_{lM+q_1}(a_j).
\tag{19}
$$

On the one hand, when $q \leq q_1$,

$$
\begin{aligned}
\Omega_1 &= \frac{1}{(1+\epsilon)p_j} \sum_{l=0}^{z_1} p_{lM+q}(a_j) + \frac{\sum_{l=z_1+1}^{z} p_{lM+q}(a_j)\mathbb{1}(lM + q \leq \rho_j)}{(1+\epsilon)p_j} \\
&\overset{(ii)}{\leq} \frac{1}{(1+\epsilon)p_j} \sum_{l=0}^{z_1} p_{lM+q_1}(a_j) + \frac{\sum_{l=z_1+1}^{z} \mathbb{1}(lM + q \leq \rho_j)}{1+\epsilon} \\
&= \frac{1}{(1+\epsilon)p_j} \sum_{l=0}^{z_1} p_{lM+q_1}(a_j) + \frac{\sum_{l=0}^{z} \mathbb{1}(\kappa < lM + q \leq \rho_j)}{1+\epsilon},
\end{aligned}
\tag{20}
$$

where (ii) is due to the fact that jobs with index smaller than ρ_j in Θ_j have remaining workload no less than p_j. On the other hand, when $q > q_1$, we have:

$$
\begin{aligned}
\Omega_1 &= \frac{\sum_{l=0}^{z_1-1} p_{lM+q}(a_j)}{(1+\epsilon)p_j} + \frac{\sum_{l=z_1}^{z} p_{lM+q}(a_j)\mathbb{1}(lM + q \leq \rho_j)}{(1+\epsilon)p_j} \\
&\leq \frac{1}{(1+\epsilon)p_j} \sum_{l=0}^{z_1} p_{lM+q_1}(a_j) + \frac{\sum_{l=z_1}^{z} \mathbb{1}(lM + q \leq \rho_j)}{1+\epsilon} \\
&= \frac{1}{(1+\epsilon)p_j} \sum_{l=0}^{z_1} p_{lM+q_1}(a_j) + \frac{\sum_{l=0}^{z} \mathbb{1}(\kappa < lM + q \leq \rho_j)}{1+\epsilon}.
\end{aligned}
\tag{21}
$$

Therefore, it follows that,

$$\Omega_1 \le \frac{1}{(1+\epsilon)p_j} \sum_{l=0}^{z_1} p_{lM+q_1}(a_j) + \frac{\sum_{l=0}^{z} \mathbb{1}(\kappa < lM + q \le \rho_j)}{1+\epsilon}$$

$$\le \frac{1}{(1+\epsilon)p_j} \sum_{l=0}^{z_1} p_{lM+q_1}(a_j) + \frac{\lceil \frac{\rho_j-\kappa}{M} \rceil}{1+\epsilon} \qquad (22)$$

$$\overset{(iii)}{\le} \frac{t-a_j}{p_j} + \frac{1}{1+\epsilon}\lceil \frac{\rho_j-\kappa}{M} \rceil,$$

where $\lceil x \rceil$ denotes the smallest integer which is no less than x and (iii) is due to (19). Based on (22), we then have:

$$\alpha_j \le \frac{t-a_j}{p_j} + \frac{1}{1+\epsilon}\lceil \frac{\rho_j-\kappa}{M} \rceil + \frac{\left(\lfloor \frac{n(a_j)-\rho_j-1}{M} \rfloor + 1\right)}{1+\epsilon}$$

$$\le \frac{t-a_j}{p_j} + \frac{1}{1+\epsilon}(\lfloor \frac{n(a_j)-\kappa}{M} \rfloor + 2) \qquad (23)$$

$$\le \frac{t-a_j}{p_j} + \beta(t) + 2,$$

where the last inequality is based on the observation that $\beta(t) \ge \frac{1}{1+\epsilon}(\lfloor \frac{n(a_j)-\kappa}{M} \rfloor)$ since the number of active jobs at time t, $n(t)$, is no less than $n(a_j) - \kappa$.

Case III: Suppose jobs indexed from 1 to κ in Θ_j have been completed where $\kappa = z_1 M + q_1 > \rho_j$. In this case, (19) still holds. Moreover, an argument similar to that of (22) shows that,

$$\Omega_1 \le \frac{1}{(1+\epsilon)p_j} \sum_{l=0}^{z_1} p_{lM+q_1}(a_j) - \lfloor \frac{\kappa-\rho_j}{M} \rfloor$$

$$\le \frac{t-a_j}{p_j} - \frac{1}{1+\epsilon}\lfloor \frac{\kappa-\rho_j}{M} \rfloor. \qquad (24)$$

Therefore, it follows that,

$$\alpha_j \le \frac{t-a_j}{p_j} - \frac{1}{1+\epsilon}\lfloor \frac{\kappa-\rho_j}{M} \rfloor + \frac{1}{1+\epsilon}\lceil \frac{n(a_j)-\rho_j}{M} \rceil$$

$$\le \frac{t-a_j}{p_j} + \frac{1}{1+\epsilon}(\lfloor \frac{n(a_j)-\kappa}{M} \rfloor + 2) \qquad (25)$$

$$\le \frac{t-a_j}{p_j} + \beta(t) + 2.$$

In summary, for all the three cases above, Inequality (15) is satisfied. This completes the proof of Lemma 2.

Let $F'_j(a_j)$ and $F_j(a_j)$ denote the overall remaining job flowtime at time a_j without and with job j respectively. Based on Theorem 1, if job j never arrives at the cluster and the subsequent jobs do not enter the cluster, the overall remaining job flowtime at time a_j under SRPT is given by:

$$F_j'(a_j) = \frac{1}{1+\epsilon} \sum_{k=1}^{n(a_j)-1} (\lfloor \frac{n(a_j)-1-k}{M} \rfloor + 1)p_k(a_j). \tag{26}$$

By contrast, when job j arrives at time a_j but the subsequent jobs do not enter the cluster, the overall remaining job flowtime at time a_j under SRPT is given by:

$$F_j(a_j) = \frac{1}{1+\epsilon} \sum_{k=1}^{\rho_j} (\lfloor \frac{n(a_j)-k}{M} \rfloor + 1)p_k(a_j) + \frac{1}{1+\epsilon} (\lfloor \frac{n(a_j)-\rho_j-1}{M} \rfloor + 1)p_j$$

$$+ \frac{1}{1+\epsilon} \sum_{k=\rho_j+1}^{n(a_j)} (\lfloor \frac{n(a_j)-k}{M} \rfloor + 1)p_k(a_j),$$

$$\tag{27}$$

therefore, one can view α_j as the difference of (27) and (26). This is also the incremental increase of the overall job flowtime caused by the arrival of job j and divided by $(1+\epsilon)p_j$. As a result, $\sum_j p_j\alpha_j$ corresponds to the overall job flowtime under SRPT, i.e., $\sum_j \alpha_j p_j = SRPT$ where $SRPT$ is the total job flowtime achieved by the SRPT algorithm.

Moreover, $(1+\epsilon)M\beta(t)$ is the number of active jobs in the cluster at time t, so, $M \int_0^\infty \beta(t)dt = \frac{1}{1+\epsilon}SRPT$. Therefore, we have $\sum_j \alpha_j p_j - M \int_0^\infty \beta(t)dt = \frac{\epsilon}{1+\epsilon}SRPT$.

Based on Lemmas 1 and 2, we conclude that $\frac{\epsilon}{1+\epsilon}SRPT \le 3OPT$ where OPT is the overall job flowtime achieved by the optimal algorithm. This completes the proof of Theorem 3.

6   Performance Evaluation

In this section, we evaluate the performance of the SRPT-based algorithm in the online setting via extensive simulations driven by Google cluster-usage traces [20]. The traces contain the information of job submission and the completion time of Google services. It also includes the number of tasks in each job as well as the duration of each task without preemption. From the traces, we extract the statistics of jobs during a 28-h period. We also exclude the unfinished jobs as well as those which have specific constraints on machine attributes. All the

Table 1. Simulation results of different performance metrics under all schemes.

Algorithm	Utilization	Makespan (s)	Average flowtime (s)	Average waiting time (s)
Our scheme	95.01%	1.97×10^5	1.92×10^4	1.80×10^4
FIFO	99.58%	1.89×10^5	7.87×10^4	7.75×10^4
LJF	99.91%	1.88×10^5	14.1×10^4	14.0×10^4

experiments are conducted on a PC with a 2.6 GHz Intel i5 Dual-core CPU. Since a job may consist of multiple small tasks in the traces, we randomly select one task as its representative. In addition, we set the number of servers to 500 under our simulations.

Baseline Algorithms: We use the following algorithms as the baselines for comparison with the SRPT-based algorithm:

- First-In First-Out (FIFO): Jobs are served according to their arrival times and the earliest arrived jobs are served first without preemption.
- Longest Job First (LJF): The jobs with longest processing times are served first.

Performance Metrics: For each of the scheduling algorithms, we run all jobs contained in the traces until they are completed. We then compare across different scheduling algorithms the largest job completion time (which is referred to as the makespan) and the sum of the job flowtime. To evaluate the fairness across jobs, we also study the waiting time of jobs under different algorithms.

Simulation Results: We depict the simulation results in Table 1, which shows that the average job flowtime under the SRPT-based scheme is 1.92×10^4 s while it is 7.89×10^4 and 1.41×10^5 s under FIFO and LJF respectively. As such, the SRPT-based scheme can reduce the job flowtime by nearly 80% compared to other two baselines. LJF has shown to be optimal for minimizing makespan in the single-sever setting. Nevertheless, Table 1 shows that the make-span achieved by SRPT is similar to that achieved by LJF. To quantify the fairness of SRPT, we also characterize the waiting time of all jobs achieved by different schemes. As illustrated in Table 1, the variance between job waiting times under SRPT is much smaller compared to that under LJF and FIFO. In this sense, the SRPT-based scheme achieves a much better fairness comparing to LJF and FIFO. Interestingly, one can note from Table 1 that the server utilization rate under SRPT is also the lowest among all three schemes, which indicates that, SRPT produces the smallest computation cost.

7 Conclusions

This paper revisits the SRPT scheduling principle and provides tighter competitive performance bounds of SRPT-based algorithms in distributed computing clusters under both the transient scheduling and online settings. Using new approaches, we focus on the competitive performance ratio of SRPT with respect to the overall job flowtime and successfully derive new bounds which improve prior state-of-the-art considerably. Our analytical approaches can also be applicable to more complicated scenarios where there is service variability among machines in a cluster. In the future, we aim to generalize the SRPT scheduling to take the multi-dimensional resource requirement into consideration and derive corresponding competitive performance bounds.

Acknowledgements. This research was supported in part by a CUHK-RGC direct grant (project #4055108).

References

1. Anand, S., Garg, N., Kumar, A.: Resource augmentation for weighted flow-time explained by dual fitting. In: Proceedings of SODA (2002)
2. Chang, H., Kodialam, M., Kompella, R.R., Lakshman, T.V., Lee, M., Mukherjee, S.: Scheduling in MapReduce-like systems for fast completion time. In: IEEE Infocom, March 2011
3. Chen, F., Kodialam, M., Lakshman, T.V.: Joint scheduling of processing and shuffle phases in MapReduce systems. In: Proceedings of IEEE Infocom, March 2012
4. Elahi, M., Williamson, C.: On saturation effects in coupled speed scaling. In: McIver, A., Horvath, A. (eds.) QEST 2018. LNCS, vol. 11024, pp. 407–422. Springer, Cham (2018). https://doi.org/10.1007/978-3-319-99154-2_25
5. Fox, K., Moseley, B.: Online scheduling on identical machines using SRPT. In: SODA, January 2011
6. Grandl, R., Ananthanarayanan, G., Kandula, S., Rao, S., Akella, A.: Multi-resource packing for cluster schedulers. In: ACM SIGCOMM, August 2014
7. Grosof, I., Scully, Z., Harchol-Balter, M.: SRPT for multiserver systems. In: ACM SIGMETRICS Performance Evaluation Review (2018)
8. Gupta, A., Krishnaswamy, R., Pruhs, K.: Online primal-dual for non-linear optimization with applications to speed scaling. In: Erlebach, T., Persiano, G. (eds.) WAOA 2012. LNCS, vol. 7846, pp. 173–186. Springer, Heidelberg (2013). https://doi.org/10.1007/978-3-642-38016-7_15
9. Im, S., Kulkarni, J., Moseley, B.: Temporal fairness of round robin: competitive analysis for lk-norms of flow time. In: SPAA (2015)
10. Im, S., Kulkarni, J., Munagala, K.: Competitive algorithms from competitive equilibria: non-clairvoyant scheduling under polyhedral constraints. In: Proceedings of STOC (2014)
11. Im, S., Kulkarni, J., Munagala, K., Pruhs, K.: Selfishmigrate: a scalable algorithm for non-clairvoyantly scheduling heterogeneous processors. In: Proceedings of FOCS, pp. 531–540 (2014)
12. Im, S., Moseley, B., Pruhs, K., Torng, E.: Competitively scheduling tasks with intermediate parallelizability. In: SPAA, June 2014
13. Kalyanasundaram, B., Pruhs, K.: Speed is as powerful as clairvoyance. In: Proceedings of FOCS, October 1995
14. Kalyanasundaram, B., Pruhs, K.: Speed is as powerful as clairvoyance. J. ACM **47**, 214–221 (2000)
15. Leonardi, S.: Approximating total flowtime on parallel machines. In: STOC (1997)
16. Leonardia, S., Raz, D.: Approximating total flow time on parallel machines. J. Comput. Syst. Sci. **73**(6), 875–891 (2007)
17. Lin, M., Zhang, L., Wierman, A., Tan, J.: Joint optimization of overlapping phases in MapReduce. In: Proceedings of IFIP Performance, September 2013
18. Moseley, B., Dasgupta, A., Kumar, R., Sarlos, T.: On scheduling in map-reduce and flow-shops. In: SPAA, pp. 289–298, June 2011
19. Conway, L.R.W., Maxwell, W.L.: Theory of scheduling (1967)
20. Reiss, C., Wilkes, J., Hellerstein, J.L.: Google cluster-usage traces, May 2011. http://code.google.com/p/googleclusterdata

21. Yuan, Y., Wang, D., Liu, J.: Joint scheduling of MapReduce jobs with servers: performance bounds and experiments. In: IEEE Infocom (2014)
22. Zaharia, M., Das, T., Li, H., Hunter, T., Shenker, S., Stoica, I.: Discretized streams: fault-tolerant streaming computation at scale. In: SOSP, pp. 423–438 (2013)
23. Zheng, Y., Shroff, N., Sinha, P.: A new analytical technique for designing provably efficient MapReduce schedulers. In: Proceedings of IEEE Infocom, Turin, Italy, April 2013

Multidimensional Systems

Sojourn Time Distribution in Fluid Queues

Eleonora Deiana[1]([✉])[iD], Guy Latouche[2], and Marie-Ange Remiche[1]

[1] Faculté d'informatique, Université de Namur, Avenue Grandgagnage, 21,
5000 Namur, Belgium
{Eleonora.Deiana,Marie-Ange.Remiche}@unamur.be
[2] Département d'informatique, Université libre de Bruxelles,
CP 212 - Boulevard du Triomphe, 1050 Bruxelles, Belgium
Guy.Latouche@ulb.ac.be

Abstract. We consider a fluid flow model with infinite buffer. We compute the Laplace-Stieltjes transform of the sojourn time proceeding in two steps. We first compute the stationary distribution of the buffer at arrival instants, using a change of clock. Secondly, we compute the transform of the time spent to empty the buffer. Numerical examples of sojourn time in a fluid flow are finally examined.

Keywords: Markov-modulated fluid flow · Sojourn time ·
Laplace-Stieljes transform

1 Introduction

A fluid flow $(X(t), S(t))_{t \geq 0}$ represents the behavior of a reservoir content $X(t)$ which evolves linearly following some rates modulated by a background Markov process $S(t)$ with finite state space $\mathcal{S}$. When the background process $S(t)$ takes a value $i \in \mathcal{S}$, the rate of evolution of $X(t)$ is r_i. Fluid flows have applications in telecommunication and computer systems, where they model the amount of data entering and being processed by a server. When studying these systems the attention is generally focused on the stationary distribution of the level, while less attention is paid to sojourn times.

In a traditional queueing system, where clients arrive and are served individually, the sojourn time of a typical client is easy to define. This is the time spent inside the system, from the instant of its arrival until its departure. There exist many results on the determination of the sojourn time distribution in such a system ([1,5,8]). When we consider a fluid flow, the "clients" have infinitesimal size and they arrive and leave the system continuously. The analysis of the sojourn time is more involved, and a different approach is required. Some work has already been done in this direction (see for example [7] and [4]).

When defining the time spent into the buffer by a unit of fluid, it is necessary to separate the *input rates* of the fluid, and the *output rates* at which the server

© Springer Nature Switzerland AG 2019
T. Phung-Duc et al. (Eds.): QTNA 2019, LNCS 11688, pp. 295–313, 2019.
https://doi.org/10.1007/978-3-030-27181-7_18

is working. The net rate of the fluid r_i is then given by the difference between the input and the output rates: $r_i = r_{in,i} - r_{out,i}$, i in $\mathcal{S}$.

Masuyama and Takine [7] derive the distribution of the sojourn time in a fluid flow where the server works at a constant rate $r_{out,j} = 1$, for all the phases j in the state space $\mathcal{S}$. Their analysis is based on a geometric approach: the sojourn time is given by a linear transformation of the original process. Their definition is simple, thanks to their assumption of a constant output rate. Moreover, the authors assume the sojourn time to be zero when the buffer is empty. On the other hand, while the buffer is empty, input rates might be strictly positive but smaller than the corresponding output rates. In this case there are units of fluid joining the system whose sojourn time is null, and it definitely contribute to the sojourn time distribution. However, in a fluid flow the buffer might be empty also when input rates are null and so there is no unit of fluid arriving. These epochs should not contribute to the sojourn time. This is illustrated with Example 7.1, where the buffer is empty during a positive amount of time, while the sojourn time is strictly positive with probability one.

Another approach is that of Horváth and Telek [4], who proceed in two steps. They first determine the distribution of the buffer content at arrival epochs, denoted as $\pi^*(x)$. They next condition on the initial state of the system, distributed following $\pi^*(x)$, to determine the sojourn time distribution. To obtain $\pi^*(x)$, the authors state that they proceed by similarity with [8], where the author analyses sojourn times in discrete level QBDs, but they do not give more details. Moreover, they prove in [4] that the asymptotic distribution of the sojourn time is phase-type of order $n \cdot n_+$, with n being the number of phases, and n_+ the number of phases corresponding to positive rates.

We follow the same two steps approach but we give a precise development of the distribution $\pi^*(x)$. This is based on the definition of a new clock $\phi(y)$, which is the epoch when a total of y units of fluid have entered the buffer. The distribution of the level at arrival instant $\pi^*(x)$ is then determined applying this change of clock to the original process. This approach is being adapted to more complex systems.

Let W be the stationary sojourn time of the fluid, with distribution function $V(t) = \Pr[W \le t]$. As it will be explained in Sect. 2, the distribution $V(t)$ is

$$V(t) = \int_0^\infty \pi^*(x) P^{out} \Gamma(x,t) \mathbf{1} \mathrm{d}x, \tag{1}$$

where $\Gamma(x,t)$ is a matrix whose entries give the distribution of the time needed to empty the buffer when the initial level is x, and P^{out} is a permutation matrix which we will define later. We determine the Laplace-Stieltjes (LS) transform of the distribution $V(t)$. This provides us with an alternative representation, using matrices of order $n \times n_+$, smaller than the order $(n \cdot n_+)^2$ of the phase-type representation in [4].

The paper is organized as follows. We analyze in Sect. 2 a simple discrete queueing system to illustrate our introduction of a new clock and to show that the sojourn time of this system has the same form as in Eq. (1). Section 3 details the fluid model we are interested in, and focus on well-known preliminary results of crucial interest in the sequel. The distribution of the buffer content at arrival epochs is completely determined in Sect. 4, while in Sect. 5 we focus on the time needed to empty the buffer. The sojourn time distribution is finally determined in Sect. 6. Section 7 exhibits some numerical examples. We briefly conclude our work in Sect. 8.

2   A Discrete Queueing System

We propose here to work with a discrete queue in order to explain the change of clock for the distribution at arrival epochs, and justify the form of Eq. (1) for the sojourn time distribution.

Let us consider a queueing system in a random environment. This is a two dimensional process $(N(t), S(t))_{t \geq 0}$, where $N(t)$ is the number of clients in the system at instant t, and $S(t)$ is the phase process, with finite state space S. Arrivals and services depend on the actual phase of the process $S(t)$. Let us tag every customers in order of arrival: the nth customer arrives at instant τ_n and has sojourn time W_n.

Our goal is to calculate the distribution of the stationary sojourn time W, defined as

$$\Pr\left[W \leq t\right] = \lim_{n \to \infty} \Pr\left[W_n \leq t\right]. \tag{2}$$

In a discrete queueing system, the sojourn time W_n of a single client is the time spent into the system starting from the arrival instant τ_n. In particular, it is a function of the number of clients in the buffer, and the phase at instant τ_n, thus we write

$$W_n = W\left(N(\tau_n), S(\tau_n)\right).$$

Let us define the function $A(t)$ counting the number of arrivals. For every instant t, $A(t)$ is the number of clients arrived in the interval of time $[0, t]$. This function makes a jump at every instant of type τ_n. We can define the reciprocal function A^{-1} such that

$$A^{-1}(n) = \tau_n \Leftrightarrow A(\tau_n) = n.$$

We can write Eq. (2) using the function A^{-1}, it gives

$$\Pr\left[W \leq t\right] = \lim_{n \to \infty} \Pr\left[W_n \leq t\right]$$
$$= \lim_{n \to \infty} \Pr\left[W\left(N(\tau_n), S(\tau_n)\right) \leq t\right]$$
$$= \lim_{n \to \infty} \Pr\left[W\left(N(A^{-1}(n)), S(A^{-1}(n))\right) \leq t\right]. \tag{3}$$

We can see $A^{-1}(\cdot)$ as a new clock for the process. Using this clock, we no longer consider the process continuously in time, but only at those instants when an arrival occurs. We define a new process,

$$(N^*(n), S^*(n))_{n \geq 0} = \left(N(A^{-1}(n)), S(A^{-1}(n))\right)_{n \geq 0},$$

and the corresponding stationary distribution $\gamma^*(n)$, as

$$\gamma_i^*(k) = \lim_{n \to \infty} \Pr\left[N^*(n) = k, S^*(n) = i\right], \qquad i \in \mathcal{S}.$$

We condition Eq. (3) on the initial level and phase, distributed following $\gamma^*(k)$, and we write the distribution of the sojourn time W as

$$\Pr\left[W \le t\right] = \lim_{n \to \infty} \Pr\left[\mathcal{W}\left(N^*(n), S^*(n)\right) \le t\right],$$
$$= \lim_{n \to \infty} \sum_{k \ge 0} \sum_{i \in \mathcal{S}} \Pr\left[N^*(n) = k, S^*(n) = i\right] \times$$
$$\Pr\left[\mathcal{W}(k, i) \le t | N^*(n) = k, S^*(n) = i\right]$$
$$= \sum_{k \ge 0} \sum_{i \in \mathcal{S}} \gamma_i^*(k) \Pr\left[\mathcal{W}(k, i) \le t | N^*(0) = k, S^*(0) = i\right]$$
$$= \gamma^* F(t) \mathbf{1},$$

where $F(t)$ is a matrix with entries $F_{k,i}(t) = \Pr\left[\mathcal{W}(k, i) \le t\right]$: the sojourn time is the product of two factors. Equation (1) for the sojourn time in a fluid flow has the same structure.

3   The Fluid Flow of Interest

We consider a classic fluid flow $(X(t), S(t))_{t \ge 0}$ in continuous time, where

- $X(t) \in \mathbb{R}^+$ is the *level* of the buffer content,
- $S(t) \in \mathcal{S} = \{1, 2, \ldots, n\}$, with $n < \infty$, is the *phase process*, an irreducible Markov process with generator matrix T and stationary distribution $\boldsymbol{\alpha}$, such that

$$\begin{cases} \boldsymbol{\alpha} T = \mathbf{0}, \\ \boldsymbol{\alpha} \mathbf{1} = 1. \end{cases}$$

For $S(t) = i$, the net rate of the fluid is $r_i = r_{in,i} - r_{out,i}$. We define the diagonal rate matrices R, R^{in} and R^{out} with elements of the diagonal r_i, $r_{in,i}$ and $r_{out,i}$ respectively. The input and output rates are always non-negative

$$r_{in,i} \ge 0, \quad r_{out,i} \ge 0, \quad \forall i \in \mathcal{S}.$$

The net rates r_i can take any real values. It is customary to partition the state space of the phases according to the sign of the net rates:

$$\mathcal{S} = \mathcal{S}_+ \cup \mathcal{S}_- \cup \mathcal{S}_0, \tag{4}$$

with

$$\begin{cases} r_i > 0 & \text{for } i \in \mathcal{S}_+, \\ r_i < 0 & \text{for } i \in \mathcal{S}_-, \\ r_i = 0 & \text{for } i \in \mathcal{S}_0. \end{cases}$$

Following this partition, the generator matrix T and the rate matrix R may be written as,

$$T = \begin{bmatrix} T_{++} & T_{+-} & T_{+0} \\ T_{-+} & T_{--} & T_{-0} \\ T_{0+} & T_{0-} & T_{00} \end{bmatrix}, \quad R = \begin{bmatrix} R_+ & & \\ & R_- & \\ & & 0 \end{bmatrix}. \tag{5}$$

Let us define the matrix Q as the generator of the process censored to the phases with non zero rates:

$$Q = \begin{bmatrix} Q_{++} & Q_{+-} \\ Q_{-+} & Q_{--} \end{bmatrix} = \begin{bmatrix} T_{++} & T_{+-} \\ T_{-+} & T_{--} \end{bmatrix} + \begin{bmatrix} T_{+0} \\ T_{-0} \end{bmatrix} (-T_{00})^{-1} \begin{bmatrix} T_{0+} & T_{0-} \end{bmatrix}.$$

Let $\Pi(x)$ be the vector of elements $\Pi_j(x)$ defined as the stationary distribution of the buffer content in phase j, for $j \in \mathcal{S}$,

$$\Pi_j(x) = \lim_{t \to \infty} P\left[X(t) \le x, S(t) = j\right],$$

and $\pi(x)$ vector of elements $\pi_j(x)$ defined as its density

$$\pi_j(x) = \frac{\partial}{\partial x} \Pi_j(x).$$

We also define the elements p_j of the vector p, as the stationary probabilities of observing a null buffer content while the phase process is j, that is

$$p_j = \lim_{t \to \infty} P\left[X(t) = 0, S(t) = j\right].$$

We recall the stationary density $\pi(x)$ of a classic fluid flow [6] in the following theorem.

Theorem 1. *The mean drift of the fluid flow is $\lambda = \alpha R \mathbf{1}$. If $\lambda < 0$, then the stationary density of the level is given by*

$$\pi(x) = \mu e^{Kx} \left[R_+^{-1} \ \Psi |R_-|^{-1} \ \Theta \right],$$

with $\Theta = \left(R_+^{-1} T_{+0} + \Psi |R_-|^{-1} T_{-0}\right)(-T_{00})^{-1}$, and where $|R_-|$ is the matrix of absolute values of the elements of R_-.
The matrix K is given by

$$K = R_+^{-1} Q_{++} + \Psi |R_-|^{-1} Q_{-+},$$

and the matrix Ψ is the unique solution, with $\Psi \mathbf{1} = \mathbf{1}$, of the following Riccati equation

$$\Psi |R_-|^{-1} Q_{-+} \Psi + \Psi |R_-|^{-1} Q_{--} + R_+^{-1} Q_{++} \Psi + R_+^{-1} Q_{+-} = 0.$$

The vector μ is defined by the system

$$[\mu \ p] \begin{bmatrix} -I & -\Psi & 0 \\ T_{-+} & T_{--} & T_{-0} \\ T_{0+} & T_{0-} & T_{00} \end{bmatrix} = 0$$

$$p\mathbf{1} + \mu(-K)^{-1} \left[R_+^{-1} \ \Psi |R_-|^{-1} \ \Theta \right] \mathbf{1} = 1.$$

□

4   Distribution at Arrival Epochs

As mentioned earlier, the distribution at arrival epochs is obtained working with a new clock $\phi(y)$. We proceed step by step and define successively the input process, the new clock, and then the distribution at arrival epochs. We use Fig. 1 to illustrate the different steps.

Input Process. We define the fluid flow $(Y(t), S(t))_{t \geq 0}$ of the accumulated fluid, as the *input process*, where $Y(t) \in \mathbb{R}^+$ represents the total amount of fluid which entered the buffer up to time t. The process $S(t)$ is the phase process of the original fluid queue.

We only consider here the input rates $r_{in,i}$ at which the fluid is entering the buffer. The rate matrix of the input process is R^{in}. Recall that the input rates are always positive or null. Depending on the sign of the input rates, we partition the state space $\mathcal{S}$ using two new subsets $\mathcal{S} = \mathcal{S}_+^{in} \cup \mathcal{S}_0^{in}$, defined as

$$\begin{cases} r_{in,i} > 0 & \text{for } i \in \mathcal{S}_+^{in}, \\ r_{in,i} = 0 & \text{for } i \in \mathcal{S}_0^{in}. \end{cases}$$

The left side of Fig. 1 shows a sample path of the fluid $X(t)$ above the corresponding input process $Y(t)$. We see that when $S(t)$ visits a phase $i \in \mathcal{S}_0^{in}$, $Y(t)$ stays at the same level during this interval of time, and $X(t)$ may decrease if $r_{out,i} < 0$.

New Clock. We define the new clock $\phi(y)$ as

$$\phi(y) = \max\{u, 0 < u < \infty : Y(u) = y\},$$

so we have $Y(\phi(y)) = y$. For a fixed value y, one can see $\phi(y)$ as the last instant when the total amount of fluid entered in the buffer is y, and new fluid is coming in at that time.

When $S(u) = i \in \mathcal{S}_+^{in}$, we have

$$\phi(y) = u \Leftrightarrow y = Y(u). \tag{6}$$

In general, however, the implication goes only in one direction

$$\phi(y) = u \Rightarrow y = Y(u).$$

In the right lower part of Fig. 1 we can see how we obtain the new clock $\phi(y)$ by taking the inverse of the process $Y(t)$. At the beginning the input rate is quite high, so the clock is slow. Then for a certain interval of time there is no input, so that time does increase but the accumulated fluid y does not, and the clock jumps when input resumes. It then continues again as the beginning.

Process with the New Clock. The process defined with the new clock is specified as

$$(X^*(y), S^*(y))_{y \geq 0} = (X(\phi(y)), S(\phi(y)))_{y \geq 0}.$$

It represents the content of the buffer seen at the arrival of the yth unit of fluid.

Finally, on the right upper part of Fig. 1, we show a representation of this new process. At the beginning, $X^*(y)$ is slower than the original process, as there is a lot of entries in that interval of time. As soon as there is no entries, $X^*(y)$ does not take any values but makes a jump. This corresponds to the interval of time when the phase process takes a value i in the subset S_0^{in}. After that, $X^*(y)$ continues again to increase as at the beginning.

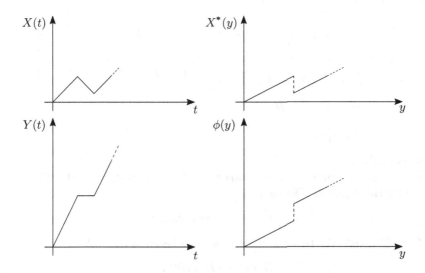

Fig. 1. An illustration of the different processes. On the upper left, a sample path of the original fluid flow $X(t)$, and below the corresponding input process $Y(t)$. On the right, the corresponding new clock $\phi(y)$ below, and the new process $X^*(y)$ above.

Stationary Distribution at Arrivals. We now determine the stationary distribution $\Pi^*(x)$ of this new process, with elements $\Pi_j^*(x)$ defined as

$$\Pi_j^*(x) = \lim_{y \to \infty} P\left[X(\phi(y)) \leq x, S(\phi(y)) = j\right],$$

and its density $\pi^*(x)$ with elements

$$\pi_j^*(x) = \frac{\partial}{\partial x} \Pi^*(x), \qquad j \in \mathcal{S}. \tag{7}$$

The next theorem explains the relation between the stationary distribution $\Pi^*(x)$ and the stationary distribution $\Pi(x)$ of the original process.

Theorem 2. *The stationary distribution of the process* $(X^*(y), S^*(y))_{y \geq 0}$ *is*

$$\Pi^*(x) = \frac{1}{\lambda} \Pi(x) R^{in},$$

where $\lambda = \alpha R^{in} \mathbf{1}$.

Proof. Let j be in $\mathcal{S}_+^{in}$, one has

$$\Pi_j^*(x) = \lim_{y \to \infty} P\left[X(\phi(y)) \leq x, S(\phi(y)) = j\right]$$

$$= \lim_{y \to \infty} \frac{1}{y} \int_0^y \mathbb{1}\{X(\phi(u)) \leq x, S(\phi(u)) = j\} du.$$

Take $t = \phi(u)$, which is equivalent to $u = Y(t)$. By definition of $Y(t)$ we have $Y'(t) = r_{in,j}$ when $S(t) = j$. Accordingly,

$$\Pi_j^*(x) = \lim_{T \to \infty} \frac{1}{Y(T)} \int_0^T \mathbb{1}\{X(t) \leq x, S(t) = j\} r_{in,j} dt$$

$$= \left(\lim_{T \to \infty} \frac{T}{Y(T)}\right) \lim_{T \to \infty} \frac{1}{T} \int_0^T \mathbb{1}\{X(t) \leq x, S(t) = j\} r_{in,j} dt$$

$$= \lim_{T \to \infty} \frac{T}{Y(T)} \Pi_j(x) r_{in,j}$$

$$= c \Pi_j(x) r_{in,j}.$$

for some scalar c.

If $j \in \mathcal{S}_0^{in}$, the process $X^*(t)$ makes a jump. The stationary distribution is then 0 for these phases. We have

$$\Pi_j^*(x) = 0, \qquad \text{with } r_{in,j} = 0.$$

As $\mathcal{S} = \mathcal{S}_+^{in} \cup \mathcal{S}_0^{in}$, the result holds for all the phases in $\mathcal{S}$, so that:

$$\Pi^*(x) = c\Pi(x) R^{in}.$$

Let us now find the constant c, defined as $c = \lim_{T \to \infty} \frac{T}{Y(T)}$. Let us remember that

$$\sum_{j \in \mathcal{S}} \lim_{x \to \infty} \Pi_j^*(x) = 1, \tag{8}$$

and

$$\lim_{x \to \infty} \Pi_j(x) = \alpha_j.$$

One may conclude from (8) that

$$c = \frac{1}{\sum_{j \in \mathcal{S}} \alpha_j r_{in,j}}$$

$$= \frac{1}{\alpha R^{in} \mathbf{1}},$$

which concludes the proof. $\qquad \square$

Corollary 1. *The stationary density of the process* $(X^*(y), S^*(y))_{y \geq 0}$ *is*

$$\pi^*(x) = \frac{1}{\lambda}\pi(x)R^{in},$$

where $\lambda = \alpha R^{in}\mathbf{1}$.

Proof. It follows from the definition of the density (7). $\square$

Remark 1. This is the same equation as in [4], but here is a proof.

5 Time to Empty the Buffer

We now proceed with the second step of our analysis, that is the computation of the distribution of the time needed to empty the buffer content. We first describe the output process that will help us to characterize the Laplace-Stieltjes transform of the time to empty the buffer.

Output Process. We define the *output process* $(Z(t), S(t))_{t \geq 0}$, where $Z(t) \in \mathbb{R}^+$ represents the amount of fluid which may leave the buffer up to time t. The process $S(t)$ is the phase process of the original fluid flow.

We take into account the fluid leaving the buffer, so the rate matrix of the fluid is $-R^{out}$. Depending on the sign of the output rates $r_{out,i}$, which are always non negative, the state space S may be partitioned using new subsets: $S = S_-^{out} \cup S_0^{out}$. These subsets are defined as

$$\begin{cases} -r_{out,i} < 0 & \text{for } i \in S_-^{out}, \\ r_{out,i} = 0 & \text{for } i \in S_0^{out}. \end{cases}$$

This leads to a different partition of the matrices. The entries of matrix T in Eq. (5) are ordered following the partition $S = S_+ \cup S_- \cup S_0$. We define an appropriate permutation matrix P^{out} to switch from the partition $S = S_+ \cup S_- \cup S_0$ to the new one $S = S_-^{out} \cup S_0^{out}$. Using the matrix P^{out} we partition the matrices T and $-R^{out}$ as

$$(P^{out})^{-1}TP^{out} = \begin{bmatrix} T_{--}^{out} & T_{-0}^{out} \\ T_{0-}^{out} & T_{00}^{out} \end{bmatrix}, \tag{9}$$

$$(P^{out})^{-1}(-R^{out})P^{out} = \begin{bmatrix} R_-^{out} & \\ & 0 \end{bmatrix}. \tag{10}$$

In particular, R_-^{out} is a diagonal matrix, whose elements $-r_{out,i}$ are all strictly negatives.

Time to Empty the Buffer. The first passage time to a given level x is $\tau(x)$, defined as

$$\tau(x) = \inf\{t \geq 0 : Z(t) = x\}.$$

The distribution of the time spent to empty the buffer, given the initial level x is

$$\Gamma_{ij}(x,t) = \Pr\left[\tau(0) \leq t, S(\tau(0)) = j | Z(0) = x, S(0) = i\right],$$

and its LS-transform is

$$\widehat{\Gamma}_{ij}(x,s) = \int_0^\infty e^{-st} d\Gamma_{ij}(x,t).$$

The matrix $\widehat{\Gamma}(x,s)$, with entries $\widehat{\Gamma}_{ij}(x,s)$, can be partitioned as

$$\widehat{\Gamma}(x,s) = \begin{bmatrix} \widehat{\Gamma}_{--}(x,s) \\ \widehat{\Gamma}_{0-}(x,s) \end{bmatrix},$$

following the partition $S = S_-^{out} \cup S_0^{out}$.

We have the following theorem.

Theorem 3. *The LS-transform $\widehat{\Gamma}(x,s)$ of the time to empty the buffer when the initial level is x, is*

$$\widehat{\Gamma}(x,s) = \begin{bmatrix} I \\ -(T_{00}^{out} - sI)^{-1} T_{0-}^{out} \end{bmatrix} e^{Q_{out}(s)x},$$

where

$$Q_{out}(s) = |R_{--}^{out}|^{-1} \left[(T_{--}^{out} - sI) - T_{-0}^{out}(T_{00}^{out} - sI)^{-1} T_{0-}^{out} \right],$$

with $\mathrm{Re}(s) \geq 0$.

Proof. This results is adapted from [2, Lemma 2]. It is proved there that if the initial phase is in S_-^{out}, then

$$\widehat{\Gamma}_{--}(x,s) = e^{Q_{out}(s)x}. \tag{11}$$

Let us now consider an initial phase $i \in S_0^{out}$. In this case the process stays an interval of time u in phase i with probability $e^{T_{ii}^{out}u}$ and then it changes to a phase k with transition rate T_{ik}^{out}. Accordingly, one has

$$\widehat{\Gamma}_{ij}(x,s) = \int_0^\infty e^{(T_{ii}^{out}-s)u} \sum_{k \neq i} T_{ik}^{out} \widehat{\Gamma}_{kj}(x,s) du$$

$$= -(T_{ii}^{out} - s)^{-1} \sum_{k \neq i} T_{ik}^{out} \widehat{\Gamma}_{kj}(x,s).$$

Using (11), we have in matrix form

$$\widehat{\Gamma}_{0-}(x,s) = -\left(T_{00}^{out} - sI\right)^{-1} T_{0-}^{out} \widehat{\Gamma}_{--}(x,s)$$
$$= -\left(T_{00}^{out} - sI\right)^{-1} T_{0-}^{out} e^{Q_{out}(s)x},$$

which gives the result.   □

Remark 2. If all output rates are strictly negative, then $\mathcal{S} = \mathcal{S}_{-}^{out}$, and Theorem 3 becomes simpler. One has

$$\widehat{\Gamma}(x,s) = e^{Q_{out}(s)x},$$

where $Q_{out}(s) = |R^{out}|^{-1}(T - sI)$.

6   Sojourn Time Distribution

We now want to obtain the stationary distribution $V(t)$ of the sojourn time W for an arriving unit of fluid. As we explained in Sect. 2, this distribution has the following form

$$V(t) = \int_0^\infty \boldsymbol{\pi}^*(x) P^{out} \Gamma(x,t) \mathbf{1} dx.$$

While determining $\boldsymbol{\pi}^*(x)$ in Corollary 1, we use the matrix of input rates R^{in} and the phases are ordered following the partition $\mathcal{S} = \mathcal{S}_{+} \cup \mathcal{S}_{-} \cup \mathcal{S}_{0}$. In Theorem 3 we compute $\Gamma(x,t)$ using the matrix of output rates R^{out}, and we partition it following $\mathcal{S} = \mathcal{S}_{-}^{out} \cup \mathcal{S}_{0}^{out}$. For this reason we need the permutation matrix P^{out} defined in (9), to multiply the two quantities $\boldsymbol{\pi}^*(x)$ and $\Gamma(x,t)$, ordered following different partitions of $\mathcal{S}$.

Using Theorem 3, the LS-transform of the sojourn time has the form

$$\widehat{V}(s) = \int_0^\infty \boldsymbol{\pi}^*(x) B_{out}(s) e^{Q_{out}(s)x} \mathbf{1} dx, \tag{12}$$

where

$$B_{out}(s) = P^{out} \left[\begin{matrix} I \\ -\left(T_{00}^{out} - sI\right)^{-1} T_{0-}^{out} \end{matrix} \right].$$

From Corollary 1 and Theorem 1, we write $\boldsymbol{\pi}^*(x)$ as

$$\boldsymbol{\pi}^*(x) = \frac{1}{\lambda} \boldsymbol{\pi}(x) R^{in}$$
$$= \frac{1}{\lambda} \left(\boldsymbol{\mu} e^{Kx} \left[R_{+}^{-1} \ \Psi R_{-}^{-1} \ \Theta \right] \right) R^{in}$$
$$= \boldsymbol{\beta} e^{Kx} \boldsymbol{\Phi}^{in},$$

with

$$\beta = \frac{1}{\lambda}\mu \quad \text{and} \quad \Phi^{in} = \left[R_+^{-1} \ \Psi R_-^{-1} \ \Theta \right] R^{in}.$$

Equation (12) becomes then

$$\widehat{V}(s) = \int_0^\infty \beta e^{Kx} \Phi^{in} B_{out}(s) e^{Q_{out}(s)x} 1 dx \tag{13}$$

$$= \beta A(s)1, \tag{14}$$

with $A(s)$ defined as

$$A(s) = \int_0^\infty e^{Kx} \Phi^{in} B_{out}(s) e^{Q_{out}(s)x} dx.$$

Let us remark that this integral is the solution of a non-singular Sylvester equation. Using integration by parts, we have

$$A(s) = \int_0^\infty e^{Kx} \Phi^{in} B_{out}(s) e^{Q_{out}(s)x} dx$$

$$= \left[(K)^{-1} e^{Kx} \Phi^{in} B_{out}(s) e^{Q_{out}(s)x} \right]_0^\infty$$

$$- \int_0^\infty (K)^{-1} e^{Kx} \Phi^{in} B_{out}(s) e^{Q_{out}(s)x} Q_{out}(s) dx$$

$$= -(K)^{-1} \Phi^{in} B_{out}(s) - (K)^{-1} \int_0^\infty e^{Kx} \Phi^{in} B_{out}(s) e^{Q_{out}(s)x} dx \ Q_{out}(s)$$

as $\lim_{x\to\infty} e^{Kx} = 0$ and $\lim_{x\to\infty} e^{Q_{out}(s)x} < \infty$. We write it as

$$KA(s) = -\Phi^{in} B_{out}(s) - \int_0^\infty e^{Kx} \Phi^{in} B_{out}(s) e^{Q_{out}(s)x} dx \ Q_{out}(s),$$

or simply

$$KA(s) + A(s)Q_{out}(s) + \Phi^{in} B_{out}(s) = 0. \tag{15}$$

The Sylvester equation can be solved numerically and the LS-transform $\widehat{V}(s)$ will be fully determined. Its numerical inversion will lead to the complete distribution determination.

Using the LS-transform of the sojourn time, it is possible to obtain the expected sojourn time.

Theorem 4. *The expected sojourn time is given by*

$$\mathbb{E}[W] = -\beta A'1.$$

Here, $A' = \lim_{s\to 0} \frac{\partial}{\partial s} A(s)$, *is solution of the Sylvester equation*

$$KA' + A'Q_{out} = -AQ'_{out} - \Phi^{in} B'_{out}, \tag{16}$$

where $A = \lim_{s \to 0} A(s)$ is solution of the Sylvester equation

$$KA + AQ_{out} + \Phi^{in} = 0,$$

and

$$Q_{out} = |R_-^{out}|^{-1} \left[T_{--}^{out} - T_{-0}^{out} \left(T_{00}^{out} \right)^{-1} T_{0-}^{out} \right],$$

$$Q'_{out} = |R_-^{out}|^{-1} \left[-I + T_{-0}^{out} \left(T_{00}^{out} \right)^{-2} T_{0-}^{out} \right],$$

$$B'_{out} = P^{out} \left[\begin{matrix} 0 \\ \left(T_{00}^{out} \right)^{-2} T_{0-}^{out} \end{matrix} \right].$$

Proof. In order to calculate the expected sojourn time, we take the first moment of the LS transform, defined in (12):

$$\mathbb{E}[W] = -\lim_{s \to 0} \frac{\partial}{\partial s} \widehat{V}(s)$$

$$= -\lim_{s \to 0} \frac{\partial}{\partial s} \beta A(s) \mathbf{1},$$

where we denote $A' = \lim_{s \to 0} \frac{\partial}{\partial s} A(s)$.

We then take the derivative and the limit for s tending to 0 in the Sylvester equation (16) to obtain the result.   □

Taking the successive derivatives in Eq. (15) allows us to obtain the successive moments of the Sojourn time.

7   Numerical Illustration

We give here some numerical examples of sojourn time distributions.

In the following, we use the notations $F(x)$ and $F^*(x)$ to indicate the stationary marginal distribution of the level for the original process $X(t)$ and the new process $X^*(t)$, respectively, without considering the phases. We have

$$F(x) = \lim_{t \to \infty} P\left[X(t) \leq x \right],$$

$$F^*(x) = \lim_{t \to \infty} P\left[X^*(t) \leq x \right].$$

7.1   Example from [4]

We start with the example in [4, Section 6]. The authors model the amount of data arriving in a buffer and being processed by a group of servers. Data arrive in the buffer at different rates, depending on the number of active sources, and the processing rates depend on the number of active servers.

The system is modeled as a fluid flow with independent input and output processes. These are controlled by two independent Markov processes $S^{in}(t)$ and $S^{out}(t)$, respectively with generator matrices Q^{in} and Q^{out}. The rate matrices for each process are R^{in} and R^{out}. The Markov process $S^{in}(t)$ represents the number of active sources. Each source switches from off to on at rate ν^{in} and from on to off at a rate μ^{in} independently from the other sources. In particular, when $S^{in}(t) = i$, there are i active sources and the total rate of data arriving in the buffer is

$$r_{i,in} = \rho_0^{in} + i\frac{\rho^{in}}{n^{in}},$$

with $0 \le i \le n^{in}$. Similarly for the number of active servers. When $S^{out}(t) = i$, there are i active servers and the total rate at which the data is processed is

$$r_{i,out} = \rho_0^{out} + i\frac{\rho^{out}}{n^{out}},$$

with $0 \le i \le n^{out}$.

Both input and output process matrices have the same structure.

For Q :
$$\begin{bmatrix} \bullet & n\nu & & & & \\ \gamma & \bullet & (n-1)\nu & & & \\ & & \ddots & & & \\ & & & (n-1)\gamma & \bullet & \nu \\ & & & & n\gamma & \bullet \end{bmatrix},$$

for R :
$$\begin{bmatrix} \rho_0 & & & & \\ & \rho_0 + \frac{\rho}{n} & & & \\ & & \rho_0 + \frac{2\rho}{n} & & \\ & & & \ddots & \\ & & & & \rho_0 + \rho \end{bmatrix},$$

where n is the number of phases in the input or in the output process respectively. The bullets $\bullet$ in the diagonal are calculated in order to have a generator matrix, with the sum of the elements in each row being equal to zero. The parameters corresponding to the two processes are given in Table 1.

Table 1. Parameters for the matrices.

Input process		Output process	
ν^{in}	1/2	ν^{out}	1/20
γ^{in}	1	γ^{out}	1/30
ρ_0^{in}	0	ρ_0^{out}	1/10
ρ^{in}	10	ρ^{out}	8
n^{in}	20	n^{out}	10

As input and output processes are independent, the total rate matrix R and the generator matrix of the whole process Q are

$$R = R^{in} \otimes I - I \otimes R^{out},$$
$$Q = Q^{in} \otimes I + I \otimes Q^{out},$$

with $\otimes$ being the Kronecker product.

Figure 2 shows the stationary distribution of the buffer $F(x)$, the distribution at the arrival epochs $F^*(x)$, calculated with the change of clock, and the distribution of the sojourn time $V(t)$. As expected, the sojourn time distribution is the same as the authors obtained in [4].

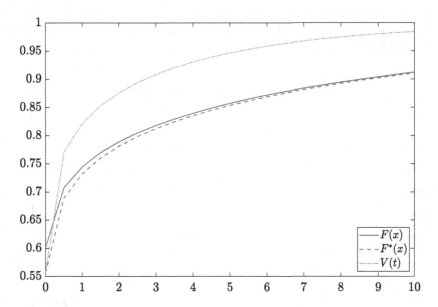

Fig. 2. Stationary distributions $F(x)$ and $F^*(x)$, sojourn time distribution $V(t)$ for Example 7.1.

Remark 3. To obtain the sojourn time distributions in Sect. 7, we have used the LS-transform inversion method from Whitt [9]. It is well-known that numerical inversion methods are prone to numerical instability, and this is considered to be a drawback of the LS-transform approach. We did not, however, encounter this issue here. In addition, we should mention recent work from Horvàth *et al.* [3] on techniques free from the risk of overshoot or undershoot.

7.2   Example with Strictly Positive Sojourn Time

We consider a system where the server works at a constant rate $r_{out,j} = 1$, as in the paper [7]. The system is simple, there is only two phases. While in the first phase there is no fluid arriving in the system, so the total rate is negative. In the second phase, the input rate is positive. We choose the values of the generator matrix T of the Markov chain so that the process spends more time in the negative phase than in the positive phase. Thus, the mean drift λ is negative. The rate matrix and the generator matrix are as follows

$$R^{in} = \begin{bmatrix} 0 & 0 \\ 0 & 2 \end{bmatrix}, \qquad R^{out} = \begin{bmatrix} 1 & 0 \\ 0 & 1 \end{bmatrix},$$

$$R = \begin{bmatrix} -1 & 0 \\ 0 & 1 \end{bmatrix}, \qquad T = \begin{bmatrix} -1 & 1 \\ 1/3 & -1/3 \end{bmatrix}.$$

Figure 3 shows the stationary distribution of the buffer $F(x)$ and the distribution at arrival epochs $F^*(x)$, calculated with the change of clock. The last one is identical to $V(t)$, the distribution of the sojourn time, as the output rate is always equal to one.

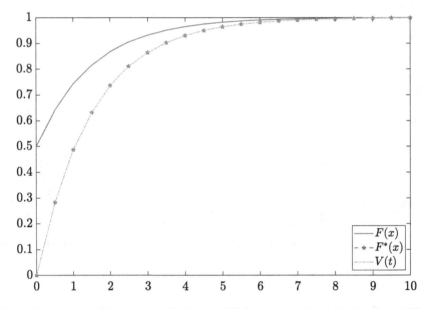

Fig. 3. Stationary distributions $F(x)$ and $F^*(x)$, sojourn time distribution $V(t)$ for Example 7.2.

The buffer is empty with a probability around 0.5. It is empty in phase 1. As soon as the phase changes to 2, the level starts to increase. In this simple example, the negative net rate corresponds to a null input rate. This means there is no fluid arriving in the system, and this period does not contribute to the sojourn time distribution. As we see in Fig. 3, the sojourn time is always strictly positive. Every time there is some input, the rate is positive, so that the arriving units of fluid are not served immediately.

7.3   A New Example

In this third example, the Markov process $S(t)$ is assumed to cycle through eight phases, remaining on average one unit of time in every phase. The generator matrix of the process is given by

$$
T = \begin{bmatrix} -1 & 1 & & \\ & -1 & 1 & \\ & & \ddots & \\ 1 & & & -1 \end{bmatrix}.
$$

We fix the mean input drift to be $\lambda_{in} = \rho$, where ρ is fixed and such that $0 < \rho < 1$. We consider three different types of environment. The first case represents a situation where during a short period (only one phase) the input rate suddenly increases. In the other two cases there is less variability in the input rates. In particular

- Case A: phase 8 by itself brings $3/4$ of the total charge arriving into the buffer;
- Case B: two phases (7 and 8) bring $3/4$ of the total charge arriving into the buffer;
- Case C: four phases (5 to 8) bring $3/4$ of the total charge arriving into the buffer.

For each environmental case, we define the vector of the input rates r_{in}^k, with $k = A, B, C$ depending on the case:

$$
\begin{aligned}
r_{in}^A &= \begin{bmatrix} 2/7, 2/7, 2/7, 2/7, 2/7, 2/7, 2/7, & 6 \end{bmatrix} \rho, \\
r_{in}^B &= \begin{bmatrix} 1/3, 1/3, 1/3, 1/3, 1/3, 1/3, & 3, & 3 \end{bmatrix} \rho, \\
r_{in}^C &= \begin{bmatrix} 1/2, 1/2, 1/2, 1/2, 3/2, 3/2, 3/2, 3/2 \end{bmatrix} \rho.
\end{aligned}
$$

We fix the output rates to be all constants and equal to one, so that the mean output drift λ_{out} is also equal to one. We also fix $\rho = 0.8$.

Figure 4 shows the stationary distribution of the level and the stationary distribution at the arrival epochs for the three cases. As the output rate is constant and equal to one, this is also the distribution of the sojourn time. We can see the influence of the variability of the input rates in the distributions. Even if the peak input rates lasts only one unit of time, its effects last longer. The sojourn time in case A is quite high, due to the unpredictability of the input rates. In case C it is much lower, thanks to the input rates more uniforms.

Let us now take the case A for the input rate, and we compare the effect of the constant output rates with some adaptive output rates. The output drift is fixed, $\lambda_{out} = 1$, and again $\rho = 0.8$. We consider three cases:

- Case a: the server works at a constant rate 1;
- Case b: during two phases the server does half of the total work;
- Case c: in phase 8 the server does half of the total work.

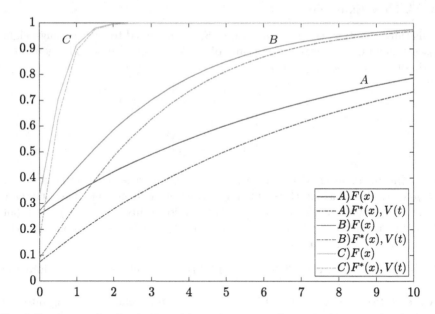

Fig. 4. Stationary distribution at arbitrary instants and at arrival instants for the input environments A, B and C.

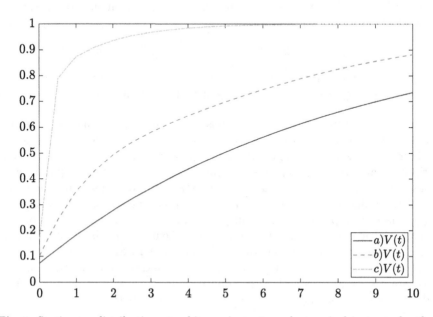

Fig. 5. Stationary distribution at arbitrary instants and at arrival instants for three different input environment, corresponding to cases a, b and c.

For each output environment, we define the vector of the output rates r_{out}^k, with $k = a, b, c$ depending on the case:

$$r_{out}^a = \begin{bmatrix} 1, & 1, & 1, & 1, & 1, & 1, & 1, & 1 \end{bmatrix}$$
$$r_{out}^b = \begin{bmatrix} 2/3, 2/3, 2/3, 2/3, 2/3, 2/3, & 2, & 2 \end{bmatrix}$$
$$r_{out}^c = \begin{bmatrix} 4/7, 4/7, 4/7, 4/7, 4/7, 4/7, 4/7, & 4 \end{bmatrix}$$

The sojourn time distribution for these output cases is shown in Fig. 5. As we make the output rates more adapted to the input rate, the peak input rate does not influence so much the sojourn time, which becomes lower. Note that the case a here correspond to the case A in Fig. 4.

8   Conclusion and Further Works

We have computed the LS-transform of the sojourn time distribution for a fluid flow with infinite buffer. The most important contribution is the introduction of the change of clock to calculate the stationary distribution at arrival epochs. The focus on the input rates when constructing the clock, allows us to adapt it to more general case. The first extension of the work will be to consider a fluid flow with finite buffer. With our clock it is possible to clearly consider the fluid which actually enters the buffer, separate from the fluid lost due to the finite buffer.

References

1. Asmussen, S.: Applied Probability and Queues. Springer, New York (2003). https://doi.org/10.1007/b97236
2. Bean, N.G., O'Reilly, M., Taylor, P.: Algorithms for return probabilities for stochastic fluid flows. Stoch. Models **21**(1), 149–184 (2005)
3. Horváth, G., Horváth, I., Almousa, S.A.D., Telek, M.: Numerical inverse Laplace transformation by concentrated matrix exponential distributions. In: Hautphenne, S., O'Reilly, M., Poloni, F. (eds.) Proceedings of the Tenth International Conference on Matrix-Analytic Methods in Stochastic Models, pp. 37–40 (2019). iSBN 978-0-646-99825-1 (Electronic version)
4. Horváth, G., Telek, M.: Sojourn times in fluid queues with independent and dependent input and output processes. Perform. Eval. **79**, 160–181 (2014)
5. Kobayashi, H., Mark, B.L.: System Modeling and Analysis: Foundations of System Performance Evaluation, 1st edn. Prentice Hall Press, Upper Saddle River (2008)
6. Latouche, G., Nguyen, G.: Analysis of fluid flow models. Queueing Models Serv. Manage. **1**(2), 1–29 (2018). arXiv:1802.04355
7. Masuyama, H., Takine, T.: Multiclass markovian fluid queues. Queueing Syst. **56**(3), 143–155 (2007). https://doi.org/10.1007/s11134-007-9019-8
8. Ozawa, T.: Sojourn time distributions in the queue defined by a general QBD process. Queueing Syst. **53**(4), 203–211 (2006)
9. Whitt, W.: A unified framework for numerically inverting laplace transforms. INFORMS J. Comput. **18**, 408–421 (2006)

An Approximate Analysis of a Bernoulli Alternating Service Model

Arnaud Devos[(✉)][iD], Dieter Fiems[iD], Joris Walraevens[iD], and Herwig Bruneel[iD]

SMACS Research Group,
Department of Telecommunications and Information Processing (EA07),
Ghent University - UGent, Sint-Pietersnieuwstraat 41, 9000 Gent, Belgium
{arnaud.devos,dieter.fiems,joris.walraevens,herwig.bruneel}@ugent.be

Abstract. We consider a discrete-time queueing system with one server and two types of customers, say type-1 and type-2 customers. The server serves customers of either type alternately according to a Bernoulli process. The service times of the customers are deterministically equal to 1 time slot. For this queueing system, we derive a functional equation for the joint probability generating function of the number of type-1 and type-2 customers. The functional equation contains two unknown partial generating functions which complicates the analysis. We investigate the dominant singularity of these two unknown functions and propose an approximation for the coefficients of the Maclaurin series expansion of these functions. This approximation provides a fast method to compute approximations of various performance measures of interest.

Keywords: Discrete-time · Alternating service · Functional equation · Approximation

1 Introduction

Two-queue queueing systems, queueing systems with two types of customers or, more generally, Markov processes with a two-dimensional state space are often harder to analyse than one-queue systems, or again more generally, Markov processes with a one-dimensional state space. Explicit expressions for the joint probability distribution or the joint probability generating function of the number of customers in the queues are usually hard to obtain. The probability generating function approach often leads to a functional equation for the joint probability generating function, which is not readily solved. In queueing systems where two types of customers share one server according to some sharing discipline, the following particular type of functional equation for the joint probability generating function of the numbers of customers of both types, $U(z_1, z_2)$, frequently occurs [5,10],

$$K(z_1, z_2)U(z_1, z_2) = f(z_1, z_2)U(z_1, 0) + g(z_1, z_2)U(0, z_2) + h(z_1, z_2)U(0, 0) , \quad (1)$$

© Springer Nature Switzerland AG 2019
T. Phung-Duc et al. (Eds.): QTNA 2019, LNCS 11688, pp. 314–329, 2019.
https://doi.org/10.1007/978-3-030-27181-7_19

where K, f, g and h are known functions. In the expression above, there are three quantities yet to be determined in the right-hand side, namely the functions $U(z_1, 0)$ and $U(0, z_2)$ and the constant $U(0, 0)$. In some fortunate circumstances where not both $U(z_1, 0)$ and $U(0, z_2)$ are present in the above equation, the kernel usually has the convenient property that it has one zero inside the unit disk for each z_2 with $|z_2| < 1$ or vice versa. This property yields an expression for the remaining unknown function since the right-hand side of the functional equation must vanish for these zero tuples. A well known example is a two-class priority queueing system, see e.g. [17].

The occurrence of both $U(z_1, 0)$ and $U(0, z_2)$ complicates the analysis, and no trivial zeroes of K seem to exist in general to determine one of the functions. Pioneering work on this problem can be found in [5,9] and [15]. In [5], the authors have developed an analytical technique such that the functions $U(z, 0)$ and $U(0, z)$ are the solutions of a so-called (Riemann-)Hilbert problem. Therefore, functional equations of the type (1) are in queueing theory sometimes referred to as boundary value problems. To obtain performance measures, the method in [5] requires the numerical evaluation of conformal mappings and (singular) integral equations. In [6], a problem where the function K in the functional equation is quadratic in both z_1, z_2 and $U(z, 0) = U(0, z)$ is studied. Instead of formulating a boundary value problem, the author proved that $U(z, 0)$ is a meromorphic function using analytic continuation and determined all it poles and zeros. This analysis was later extended in [7] to an asymmetric model, where the unknown functions are no longer equal. A profound analysis for general K that is quadratic in both variables is done in [10], where the authors either propose an algebraic method or the reduction to a boundary value problem in the complex plane. The former requires some knowledge on Galois automorphisms and elliptic functions, while the latter is similar to the approach in [5]. The function K is crucial in all these approaches and is referred to as the kernel. In [9,10,15] and most of the other literature, the kernel K is quadratic in z_1 and z_2.

The applicability of the analytical results obtained by solving these boundary value problems depends on the possibility to evaluate these results numerically [5]. Because of this drawback of the boundary value approach, some approximations for two- as well as multi-class queues have been investigated in the past. The most prominent approach is perhaps the Power Series Approximation approach [18], where the output distributions, their transforms or their moments are expressed as power series in a specified parameter and the terms of these power series are calculated iteratively (either numerically or analytically). Truncation or any other approximation based on the knowledge of a finite number of these coefficients are the result of this technique. Another approach is the so-called compensation technique [1], although this one only works for a specific class of problems. Yet another direction of analysis is obtaining partial information of the model. For instance, in some cases the asymptotics of (the decay of) the distributions of interest can be found, without the determination of the unknown functions $U(z, 0)$ and $U(0, z)$ is required [12,13].

In this paper, we analyze a discrete-time single server queueing system where the server is (alternately) responsible for the service of two types of customers. Such a queueing system is related to systems where the server or processor does not have information on the queue lengths, possibly due to technical limitations. The functional equation for the joint probability generating function of the number of type-1 and type-2 customers belongs to the class of functional equations (1). This queueing model has a simple description and is interesting because the marginal probability generating functions can be seperately determined, in contrast to the joint generating function. This is because the analysis for a single customer type is equivalent to that of a queue with server interruptions. This kind of queueing systems is well-studied [2–4]. Moreover with the concept of effective service times, see for example [8], the seperate queues are in fact equivalent to the $G - Geom - 1$ buffer system, where the parameters of the geometric service times are related: the effective service times are geometrically distributed with parameter α and parameter $1 - \alpha$, for the type-1 and type-2 queue, respectively. While we can obtain the marginal distributions, the determination of the *joint* probability generating function remains a challenging task. This becomes an issue if one is for instance interested in the distribution of the total number of customers. A closely related paper is [16] which studies a continuous-time single server two-queue polling model with random residing time service discipline. A parametric perturbation is proposed for the computation of the joint queue length distribution. However this approach also leads to a boundary value problem. In our paper we propose an approximation method that requires less complex numerical work and that can be generalized to multi-class systems where asymptotics can be obtained [13]. In contrast to [5], we approximate the unknown functions $U(z, 0)$ and $U(0, z)$ directly, partially based on information of the dominant singularities and zeros of the function K. This method is fairly simple and provides a fast method to approximate performance measures related to the joint distribution for different parameter values.

The remainder of this paper is outlined as follows. In the next section we provide a more detailed description of the queueing model under consideration. In Sect. 3, we derive the functional equation of the joint probability generating function of the number of type-1 and type-2 customers in the system. Moreover, we recapitulate the marginal distributions and their dominant singularities, as these will be helpful later on. Next, we obtain the dominant singularities of $U(z, 0)$ and $U(0, z)$ in Sect. 4. In Sect. 5, we propose a method to approximate *all* the coefficients of the Maclaurin series of these two functions. Finally, we compare this approximation with simulation results in Sect. 6, for some numerical examples.

2   Mathematical Model

We investigate a discrete-time queueing model with two infinite waiting rooms and one server. As in all discrete-time models, the time axis is divided into fixed-length intervals referred to as (time) slots. New customers may enter the system at any given (continuous) point on the time axis, but services are synchronized to (i.e. can only start and end at) slot boundaries. We further assume that the service of each customer requires exactly one slot.

Type-1 and type-2 customers arrive to the system according to two independent arrival processes. The number of type-1 and type-2 arrivals during slot k are denoted by $a_{1,k}$ and $a_{2,k}$ respectively. The sequence $a_{j,k}$ is assumed to be i.i.d. with common probability generating function (pgf) $A_j(z)$ ($j = 1, 2$). The mean number of type-j arrivals within a slot is given by

$$\lambda_j \triangleq A_j'(1), \quad j = 1, 2. \tag{2}$$

Let us define the dominant singularity[1] of $A_j(z)$ by σ_j. We assume that

$$\sigma_j > 1 \text{ and } A_j(\sigma_j) = \infty \quad j = 1, 2. \tag{3}$$

This includes all usual arrival processes, except arrival processes with a long tail [14]. We emphasize that the dominant singularity of these distributions is not necessary a pole.

As the server is (alternately) responsible for the service of two types of customers, our model basically divides the time axis into two types of time slots, referred to as "X-slots" and "Y-slots", respectively. During X-slots and Y-slots, the server can only serve customers of type 1 or 2 respectively; if no customers of the designated type are present in the system, the server remains idle. The state of the slot (X or Y) evolves independently from slot to slot: α and $1 - \alpha$ indicate the probabilities that a slot is an X-slot or Y-slot respectively.

We assume a stable system, i.e. the mean number of arrivals per slot is strictly less than the mean number of customers that can be served per slot. For the type-1 customers this yields the condition

$$\lambda_1 < \alpha,$$

while for type-2 customers we have the following constraint

$$\lambda_2 < 1 - \alpha'.$$

Throughout the remainder of this paper we use the notation $\Pr[\cdot]$ for the probability measure and $\mathrm{E}[\cdot]$ for the expectation operator.

3   The Functional Equation

We denote the system content of type-1 and type-2 customers at the beginning of slot k by $u_{1,k}$ and $u_{2,k}$, respectively. The evolution of the system content from

---

[1] Dominant singularities are singularities that lie on the boundary of the disk of convergence.

slot k to slot $k+1$ is described by the following system equations:

$$u_{1,k+1} = \begin{cases} (u_{1,k} - 1)^+ + a_{1,k}, & \text{if slot } k \text{ is an } X\text{-slot ;} \\ u_{1,k} + a_{1,k}, & \text{if slot } k \text{ is a } Y\text{-slot ;} \end{cases} \quad (4)$$

$$u_{2,k+1} = \begin{cases} u_{2,k} + a_{2,k}, & \text{if slot } k \text{ is an } X\text{-slot ;} \\ (u_{2,k} - 1)^+ + a_{2,k}, & \text{if slot } k \text{ is an } Y\text{-slot ,} \end{cases} \quad (5)$$

where $(\cdot)^+ = \max(\cdot, 0)$. Note that $u_{1,k}$ and $u_{2,k}$ are independent of the state of the system during slot k. From the system equations we obtain the following relation for the joint pgf $U_{k+1}(z_1, z_2)$ of the number of type-1 and type-2 customers at the beginning of slot $k+1$ and the joint pgf $U_k(z_1, z_2)$ of the number of type-1 and type-2 customers at the beginning of slot k:

$$\begin{aligned} U_{k+1}(z_1, z_2) &\triangleq \mathrm{E}[z_1^{u_{1,k+1}} z_2^{u_{2,k+1}}] \\ &= \frac{A_1(z_1) A_2(z_2)}{z_1 z_2} \{[(1-\alpha)z_1 + \alpha z_2] U_k(z_1, z_2) \\ &\quad + (1-\alpha)(z_2 - 1)z_1 U_k(z_1, 0) + \alpha(z_1 - 1)z_2 U_k(0, z_2)\}. \end{aligned} \quad (6)$$

Notice that

$$U_k(z_1, 0) = \sum_{n=0}^{\infty} \Pr[u_{1,k} = n, u_{2,k} = 0] z_1^n, \quad (7)$$

$$U_k(0, z_2) = \sum_{n=0}^{\infty} \Pr[u_{1,k} = 0, u_{2,k} = n] z_2^n, \quad (8)$$

by definition.

Since we are interested in the joint steady state distribution of $u_{1,k}$ and $u_{2,k}$, we define $U(z_1, z_2)$ as

$$U(z_1, z_2) \triangleq \lim_{k \to \infty} \mathrm{E}[z_1^{u_{1,k}} z_2^{u_{2,k}}].$$

Finally, applying this definition in Eq. (6) we find the following functional equation for $U(z_1, z_2)$,

$$\begin{aligned} K(z_1, z_2) U(z_1, z_2) &= A_1(z_1) A_2(z_2) \\ &\quad \times [(1-\alpha)(z_2 - 1)z_1 U(z_1, 0) + \alpha(z_1 - 1)z_2 U(0, z_2)], \quad (9) \end{aligned}$$

where we defined

$$K(z_1, z_2) = z_1 z_2 - [(1-\alpha)z_1 + \alpha z_2] A_1(z_1) A_2(z_2). \quad (10)$$

There are two unknown functions yet to be determined in the right-hand side of (9), namely the functions $U(z, 0)$ and $U(0, z)$.

3.1   The Marginal Pgfs and Their Radius of Convergence

From the functional equation (9), we easily obtain expressions for the marginal pgfs $U_1(z)$ and $U_2(z)$, describing the numbers of type-1 customers and type-2 customers respectively at the beginning of random time slot. The pgf $U_1(z)$ is given by

$$
\begin{aligned}
U_1(z) &\triangleq \lim_{k \to \infty} \mathrm{E}[z^{u_{1,k}}] \\
&= U(z, 1) \\
&= \frac{(z - 1)A_1(z)(\alpha - \lambda_1)}{z - [(1 - \alpha)z + \alpha]A_1(z)} .
\end{aligned}
\tag{11}
$$

This result can be obtained as a special case of the result(s) obtained in [2].

Let us denote the radius of convergence of $U_1(z)$ by τ_1. Since $U_1(z)$ is a pgf, we have that $\tau_1 \geq 1$. According to Pringsheim's Theorem [11, Th. IV.6], τ_1 is a singularity of $U_1(z)$. This is one of the so-called dominant singularities. The singularities of $U_1(z)$ are those of $A_1(z)$ and the possible zeros of the denominator in (11). Because $A_1(z)$ is a strictly increasing, convex function on the positive real axis and A_1 becomes infinite in its dominant singularity (3), the denominator of (11) has a simple zero in the interval $(1, \sigma_1)$. Hence, the unique dominant singularity τ_1 is this zero, which is a simple pole of $U_1(z)$ and satisfies

$$
\tau_1 = [(1 - \alpha)\tau_1 + \alpha]A_1(\tau_1), \quad 1 < \tau_1 < \sigma_1 .
\tag{12}
$$

The residue of $U_1(z)$ in τ_1 is given by

$$
\begin{aligned}
\operatorname*{res}_{z = \tau_1} U_1(z) &= \lim_{z \to \tau_1} (z - \tau_1)U_1(z) \\
&= \frac{(\tau_1 - 1)A_1(\tau_1)(\alpha - \lambda_1)}{1 - (1 - \alpha)A_1(\tau_1) - [(1 - \alpha)\tau_1 + \alpha]A_1'(\tau_1)} .
\end{aligned}
\tag{13}
$$

We furthermore calculate the pgf $U_2(z)$ of the number of type-2 customers as follows

$$
\begin{aligned}
U_2(z) &\triangleq \lim_{k \to \infty} \mathrm{E}[z^{u_{2,k}}] \\
&= U(1, z) \\
&= \frac{(z - 1)A_2(z)(1 - \alpha - \lambda_2)}{z - (\alpha z + 1 - \alpha)A_2(z)} .
\end{aligned}
\tag{14}
$$

As for $U_1(z)$, similar remarks hold for $U_2(z)$ concerning the radius of convergence. Let τ_2 be the unique dominant singularity of $U_2(z)$. It holds that τ_2 is a simple pole of $U_2(z)$ and

$$
\tau_2 = (\alpha\tau_2 + 1 - \alpha)A_2(\tau_2), \quad 1 < \tau_2 < \sigma_2 .
\tag{15}
$$

The residue of $U_2(z)$ in τ_2 is given by

$$
\begin{aligned}
\operatorname*{res}_{z = \tau_2} U_2(z) &= \lim_{z \to \tau_2} (z - \tau_2)U_2(z) \\
&= \frac{(\tau_2 - 1)A_2(\tau_2)(1 - \alpha - \lambda_2)}{1 - \alpha A_2(\tau_2) - (\alpha\tau_2 + 1 - \alpha)A_2'(\tau_2)} .
\end{aligned}
\tag{16}
$$

3.2   The Functional Equation Revisited

First of all, we want to emphasize that u_1 and u_2 are not independent. This is easily proven by the fact that the function $U_1(z_1)U_2(z_2)$ does not satisfy the functional equation (9), not even for $\alpha = 1/2$. Simulation results suggests that the correlation coefficient between type-1 and type-2 customers is negative. We now give a possible intuitive explanation for this. If u_1 is exceptionally large, then either there were a lot of type-1 arrivals in the previous time slot, or type-1 customers are not served often during the last couple of slots. In the latter case, it is likely that u_2 is small.

Secondly, we note that $(z_2 - 1)z_1 U(z_1, 0)$ in the functional equation (9) vanishes for $z_2 = 1$, if $U(z_1, 0)$ is bounded. The resulting formula (11) is therefore a priori only valid for those z values where $U(z, 0)$ is bounded. This implies that the radius of convergence of $U_1(z)$ cannot be greater than the radius of convergence of $U(z_1, 0)$. This is not surprising since the coefficients in the Maclaurin series expansion of $U(z, 0)$ are bounded by those in the Maclaurin series expansion of $U_1(z)$, i.e. $\Pr[u_1 = n, u_2 = 0] \leq \Pr[u_1 = n]$, $\forall n \geq 0$. Hence $U(z, 0)$ converges in any disk centred at the origin, where $U_1(z)$ converges. The same conclusion can be drawn for $U(0, z)$ and $U_2(z)$. Consequently, it follows that for fixed z_2 with $|z_2| \leq 1$, the joint pgf $U(z_1, z_2)$ is analytic in $|z_1| < \tau_1$. Similar, we have that for fixed z_1 with $|z_1| \leq 1$, $U(z_1, z_2)$ is analytic in $|z_2| < \tau_2$. Hence $U(z_1, z_2)$ is finite for these tuples (z_1, z_2). This implies that

$$(1 - \alpha)(z_2 - 1)z_1 U(z_1, 0) + \alpha(z_1 - 1)z_2 U(0, z_2) = 0$$

for those tuples (z_1, z_2) where $K(z_1, z_2) = 0$ and either $|z_1| < \tau_1, |z_2| \leq 1$ or $|z_1| \leq 1, |z_2| < \tau_2$.

As in [5], we can consider tuples of the form $(ze^{i\varphi}, ze^{-i\varphi})$, with $\varphi \in [0, 2\pi[$.

Theorem 1. *The kernel $K(ze^{i\varphi}, ze^{-i\varphi})$, $\varphi \in]0, 2\pi[$ has exactly two zeros in $|z| < 1$; one of them is always equal to zero. We define $z = f(e^{i\varphi})$ as the non-trivial zero. Further we have that $\lim_{\varphi \to 0} f(e^{i\varphi}) = 1$.*

Proof. See [5].

Hence, from (9) we have the following relationship for $w \triangleq e^{i\varphi}$,

$$(1 - \alpha)(f(w) - w)U(f(w)w, 0) + \alpha(f(w) - w^{-1})U(0, f(w)w^{-1}) = 0 . \quad (17)$$

In order to obtain a so called (Riemann-) Hilbert problem, one has to further transform the problem to obtain a relation between two unknown functions on a specific contour such that one of two functions is analytic in the interior of the contour, while the other is analytic outside of the contour.

In Sect. 5, we propose a method to approximate the function $U(z, 0)$ and $U(0, z)$, based on their dominant singularities and using Eq. (17).

4   Dominant Singularities of $U(z,0)$ and $U(0,z)$

In this section we show that the dominant singularity of $U(z,0)$ is τ_1 and that it is also a simple pole. Likewise, we show that the dominant singularity of $U(0,z)$ is τ_2 and that it is a simple pole as well. We start this section with the dominant singularity of $U(z,0)$.

Proposition 1. *The dominant singularity of $U(z,0)$ is τ_1. Moreover, this singularity is a simple pole with residue*

$$\operatorname*{res}_{z=\tau_1} U(z,0) = \left(\operatorname*{res}_{z=\tau_1} U_1(z) \right) \left(1 - \frac{\lambda_2}{(1-\alpha)A_1(\tau_1)} \right). \tag{18}$$

Proof. We have that $K(z_1, z_2)$, as defined in (10), is jointly analytic near $z_1 = \tau_1$, $z_2 = 1$. By the definition of τ_1, we further have that $K(\tau_1, 1) = 0$. Moreover, we have $\frac{\partial}{\partial z_2} K(\tau_1, 1) \neq 0$ since,

$$\frac{\partial}{\partial z_2} K(\tau_1, 1) = \tau_1 - \alpha A_1(\tau_1) - ((1-\alpha)\tau_1 + \alpha)A_1(\tau_1)\lambda_2$$

$$= \tau_1(1 - \lambda_2) - \alpha A_1(\tau_1)$$

$$> \tau_1(1 - \alpha - \lambda_2)$$

$$> 0. \tag{19}$$

Here we used (12) in the second step, the fact that $A_1(\tau_1) < \tau_1$ in the third step and part of the stability condition in the last step. By the implicit function theorem for analytic functions,[2] there exists a unique function $Y(z)$ and a radius $r > 0$ such that $Y(z)$ is analytic in a neighbourhood $\{z \in \mathbb{C} : |z - \tau_1| < r\}$ of τ_1, $Y(\tau_1) = 1$ and $K(z, Y(z)) = 0$ for $z \in \{z \in \mathbb{C} : |z - \tau_1| < r\}$.

Notice that

$$Y'(\tau_1) = \frac{(1-\alpha)A_1(\tau_1) + ((1-\alpha)\tau_1 + \alpha)A_1'(\tau_1) - 1}{\tau_1 - \alpha A_1(\tau_1) - ((1-\alpha)\tau_1 + \alpha)A_1(\tau_1)\lambda_2} > 0. \tag{20}$$

Indeed, the denominator of (20) is strictly positive by (19). Now consider the numerator of (11) and notice that the derivative of this function evaluated at one is positive by the stability condition. Since τ_1 is the smallest real zero bigger than one, the derivative of the numerator of (11) evaluated at τ_1 is strictly[3] negative. The numerator of (20) is equal to the value of this derivative, but multiplied by minus one. Therefore the numerator of (20) is strictly positive. Because $Y'(\tau_1) > 0$, we have that $Y(x) < 1$, for real values $x < \tau_1$ close enough to τ_1. Moreover since Y is analytic in τ_1 and $Y'(\tau_1) \neq 0$, Y is injective in a neighbourhood of τ_1.

Define $\mathcal{D}$ as the subset of the open ball $\{z \in \mathbb{C} : |z - \tau_1| < r\}$ wherefore $|z| < \tau_1$ and $|Y(z)| \leq 1$. This subset is non-empty in view of the previous

---

[2] For a reference, see e.g. [11, Th B.4].
[3] Because τ_1 is a zero of multiplicity one.

reasoning. Since for $z \in \mathcal{D} \setminus \{\tau_1\}$, $U(z, Y(z))$ remains bounded, we have that

$$U(z,0) = -\frac{\alpha(z-1)Y(z)}{(1-\alpha)z(Y(z)-1)}U(0,Y(z)), \quad z \in \mathcal{D} \setminus \{\tau_1\}. \tag{21}$$

For $0 < r' \leq \min(r, \tau_1)$ sufficiently small such that $|Y(z)| < \tau_2$ and such that Y is injective in $\{z \in \mathbb{C} : |z - \tau_1| < r'\}$, the right-hand side of this equation is analytic in this open ball, except at the point $z = \tau_1$, since $Y(\tau_1) = 1$. Therefore $U(z,0)$ can be analytically continued into this punctured disk, which proves that $z = \tau_1$ is an isolated singularity of $U(z,0)$.

Using that $U(0,1) = 1 - \lambda_1/\alpha$, we find that

$$\lim_{z \to \tau_1} (z - \tau_1)U(z,0) = -\frac{1}{Y'(\tau_1)}\frac{\alpha - \lambda_1}{1 - \alpha}\frac{\tau_1 - 1}{\tau_1} < 0.$$

Hence, τ_1 is a simple pole of $U(z,0)$. Substituting (20) into the above expression yields (18).

If we define

$$b_1 \triangleq -\operatorname*{res}_{z=\tau_1} U_1(z), \tag{22}$$

then we can write

$$\Pr[u_1 = n] \sim \frac{b_1}{\tau_1^{n+1}},$$

and

$$\Pr[u_1 = n, u_2 = 0] \sim \frac{b_1}{\tau_1^{n+1}}\left(1 - \frac{\lambda_2}{(1-\alpha)A_1(\tau_1)}\right), \tag{23}$$

where we write $f_n \sim g_n$ for $n \to \infty$ if $\lim_{n\to\infty} g_n/f_n = 1$. From this we can conclude that

$$\Pr[u_1 = n, u_2 = 0] \sim \Pr[u_1 = n]\left(1 - \frac{\lambda_2}{(1-\alpha)A_1(\tau_1)}\right).$$

This identity shows that u_1 and u_2 are non-independent. Indeed, if u_1 and u_2 were independent, then it should be the case that $\Pr[u_1 = n, u_2 = 0] = \Pr[u_1 = n]\left(1 - \frac{\lambda_2}{1-\alpha}\right)$, which is certainly not the case since $A(\tau_1) > 1$ is a constant.

Using identical arguments, we have the following similar result for $U(0,z)$, for which we omit the proof.

Proposition 2. *The dominant singularity of $U(0,z)$ is τ_2. Moreover, this singularity is a simple pole with residue*

$$\operatorname*{res}_{z=\tau_2} U(0,z) = \left(\operatorname*{res}_{z=\tau_2} U_2(z)\right)\left(1 - \frac{\lambda_1}{\alpha A_2(\tau_2)}\right). \tag{24}$$

Let us define

$$b_2 \triangleq -\operatorname*{res}_{z=\tau_2} U_2(z), \tag{25}$$

then we have the following asymptotic behaviour

$$\Pr[u_1 = 0, u_2 = n] \sim \frac{b_2}{\tau_2^{n+1}}\left(1 - \frac{\lambda_1}{\alpha A_2(\tau_2)}\right). \tag{26}$$

5 Approximation of $U(z, 0)$ and $U(0, z)$

In this section we propose an approximation for the functions $U(z, 0)$ and $U(0, z)$, which can then be used to approximate several other performance measures, using the functional equation (9). The functions $U(z, 0)$ and $U(0, z)$ are characterized by a countable infinite number of (unknown) coefficients.[4] From now on, if we speak about the coefficients of the functions $U(z, 0)$ and $U(0, z)$ we are referring to the coefficients of their Maclaurin series expansion. The asymptotics (23) and (26) approximates these coefficients accurately, except for small values of n. Let b_1^* be given by $- \operatorname*{res}_{z=\tau_1} U(z, 0)$ and b_2^* given by $- \operatorname*{res}_{z=\tau_2} U(0, z)$. We propose the following approximations $U^*(z, 0)$ and $U^*(0, z)$ for $U(z, 0)$ and $U(0, z)$, respectively:

$$U^*(z, 0) = p_0 + p_1 z + \ldots + p_{m-1} z^{m-1} + \frac{b_1^*}{\tau_1^m} \frac{z^m}{\tau_1 - z}, \tag{27}$$

$$U^*(0, z) = p_0 + p_{-1} z + \ldots p_{-(m-1)} z^{m-1} + \frac{b_2^*}{\tau_2^m} \frac{z^m}{\tau_2 - z}. \tag{28}$$

Apart for the first m coefficients, we thus approximate the coefficients by the obtained asymptotics (23), (26). We therefore reduce the problem to that of finding $N \triangleq 2m - 1$ unknowns, namely $\mathbf{p} := (p_0, p_1, \ldots, p_{m-1}, p_{-1}, p_{-2}, \ldots, p_{-m-1})$. In Sect. 3.2 we showed how the function $K(z_1, z_2)$ plays a crucial role in the determination of $U(z, 0)$ and $U(0, z)$. We therefore transfer equation (17) to the approximated functions, i.e. replace $U(z, 0)$ and $U(0, z)$ in (17) by $U^*(z, 0)$ and $U^*(0, z)$ respectively. Furthermore we can sample N values on the complex unit circle, and plug them into equation (17), resulting in N linear equations between the unknowns $\mathbf{p}$. To this end, the following lemma will be useful.

Lemma 1. *For the function f defined in Theorem 1, we have that*

1. $f(e^{-i\varphi}) = \overline{f(e^{i\varphi})}$,
2. $f(-e^{i\varphi}) = -f(e^{i\varphi})$,
3. $f(e^{i\varphi})e^{i\varphi} = f(e^{i(\varphi+\pi)})e^{i(\varphi+\pi)}$,
4. $f(e^{i\varphi})e^{-i\varphi} = f(e^{i(\varphi+\pi)})e^{-i(\varphi+\pi)}$.

Proof. See [5].

From Lemma 1 we conclude that it is no use to sample both $e^{i\varphi}$ and $e^{i(\varphi+\pi)}$ because this results into the same equation. The same holds true for $e^{i\varphi}$ and $-e^{i\varphi}$. Let $\omega \triangleq \exp\left(\frac{-2\pi i}{N}\right)$, such that $\omega^N = 1$. We propose to use $\omega, \omega^2, \ldots, \omega^{N-1}$ as values for $e^{i\varphi}$ in (17). Since N is odd, the two previously mentioned problems do not occur. Moreover, it suffices to compute $f(\omega), \ldots, f(\omega^{m-1})$ because of the first statement in Lemma 1.

---

[4] We refer to the steady-state versions of Eqs. (7) and (8).

Using the definition of $U^*(z,0)$ and $U^*(0,z)$ (17) translates to

$$(1-\alpha)(f(\omega^k)-\omega^k)\sum_{j=1}^{m-1}f(\omega^k)^j\omega^{kj}p_j + \alpha(f(\omega^k)-\omega^{-k})\sum_{j=1}^{m-1}f(\omega^k)^j\omega^{kj}p_{-j}$$

$$+(f(\omega^k)-(1-\alpha)\omega^k-\alpha\omega^{-k})p_0 = \frac{(1-\alpha)(\omega^k-f(\omega^k))f(\omega^k)^m\omega^{km}b_1^*}{\tau_1^m(\tau_1-f(\omega^k)\omega^k)}$$

$$+\frac{\alpha(\omega^{-k}-f(\omega^k))f(\omega^k)^m\omega^{-km}b_2^*}{\tau_2^m(\tau_2-f(\omega^k)\omega^{-k})}\ ,\ (29)$$

for $k=1,2,\ldots,N-1$. Hence we have $N-1$ linear equations in the unknowns **p**. We cannot use the equation for $k=0$ since $f(1)=1$ and the right-hand side of (9) vanishes trivially in that case. Instead, we propose to use the equation $(1-\alpha)U^*(1,0)+\alpha U^*(0,1)=1-\lambda_1-\lambda_2$, which originates from the normalization condition $\lim_{z\to1}U(z,z)=1$. For the approximate functions this yields

$$(1-\alpha)\sum_{j=1}^{m-1}p_j+\alpha\sum_{j=1}^{m-1}p_{-j}+p_0 = 1-\lambda_1-\lambda_2-\frac{b_1^*}{\tau_1^m(\tau_1-1)}-\frac{b_2^*}{\tau_2^m(\tau_2-1)}\ .\ (30)$$

The system of equations (29), (30) constitute a system of linear equations, i.e. we can write it as $A\mathbf{p}=\mathbf{v}$. Unfortunately, the matrix A cannot be inverted analytically. We want to remark that the matrix A is highly ill-conditioned for large N (which is typical for polynomial fitting). Since the approximation of the tail coefficients, i.e. (23) and (26) are more accurate for large N, there is an obvious tradeoff. We need sufficient terms for accurately determining the partial generating functions for lower order terms. Increasing the size however may lead to numerical problems while the tail approximation is already accurate. There are several heuristics to observe if the obtained solution is close to the real solution. For example, one can

1. compare $U^*(1,0)$ with $1-\frac{\lambda_2}{1-\alpha}$,
2. compare $U^*(0,1)$ with $1-\frac{\lambda_1}{\alpha}$,
3. compare p_{m-1} with $\frac{b_1^*}{\tau_1^m}$, or
4. compare $p_{-(m-1)}$ with $\frac{b_2^*}{\tau_2^m}$.

6   Validation and Numerical Results

In the previous section we obtained an approximation for the functions $U(z,0)$ and $U(0,z)$. In this section we compare our approximation with simulation results. To this end, we assume a Binomial arrival process for both customer types, the pgf of the number of type-j arrivals taking the form

$$A_j(z)=\left(1-\frac{\lambda_j}{16}+\frac{\lambda_j}{16}z\right)^{16},\quad j=1,2\,.\tag{31}$$

Fig. 1 shows the partial probability mass functions $\Pr[u_1 = k, u_2 = 0]$, for $k \in \{0, 1, \ldots, 9\}$, with $\lambda_1 = \lambda_2 = 0.3$. The figure shows the simulation results and the results from the approximation method of Sect. 5 for the threshold values $m = 3$, $m = 6$ and $m = 10$. These figures roughly justify the approximation method, at least for this choice of parameters. The figures indicate that a greater threshold value m does not necessarily leads to better results. In contrast, for $m = 10$ we even get negative values for the coefficients. This is likely due to the numerical instability, which increases for increasing m. By the choice of λ_1 and λ_2, type-1 and type-2 customers have the same arrival process. Hence, changing α to $1 - \alpha$ will yield the same result, but with type-1 and type-2 customers interchanged. Therefore we omitted the results for $\Pr[u_1 = 0, u_2 = k]$.

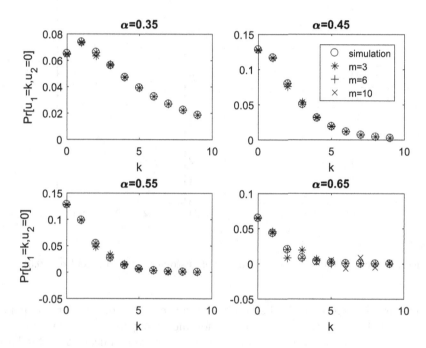

Fig. 1. The partial pmf $\Pr[u_1 = k, u_2 = 0]$, $k \in \{0, \ldots, 9\}$: comparison between simulation and approximation method for various values of m ($\lambda_1 = \lambda_2 = 0.3$).

Next, we show the influence of varying λ_2 on the approximation. In Figs. 2 and 3 we show the mean and the variance respectively of the total buffer occupancy versus λ_2, with $\lambda_1 = 0.3$ and $\alpha = 0.5$ kept fixed. Remark the constraint $\lambda_2 < 1 - \alpha$, see also Sect. 2. The right most value of λ_2 in Figs. 2 and 3 equals $\lambda_2 = 0.46$. As we can see from Fig. 2, the approximation method gives accurate results for the mean total buffer occupancy. Remark that we did not need simulation results to validate the mean total buffer occupancy, because the expectation operator is linear and we have explicit expressions for the marginal pgfs. The variance of the total buffer occupancy is an example of a performance measure that cannot be obtained analytically. As illustrated in Fig. 3, the approximation of this performance measure is accurate w.r.t. simulation results. In the left subfigure of Fig. 3 ($m = 3$), we see that the approximation becomes worse when λ_2 is closer to the stability border. Increasing the threshold value m seems to overcome this problem, as shown in the right subfigure of Fig. 3 ($m = 6$). However there is an upperbound for increasing m as shown in Fig. 4.

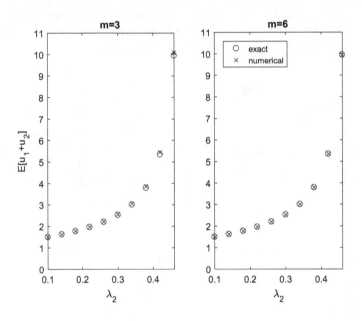

Fig. 2. Mean value of the total number of customers ($\lambda_1 = 0.3$ and $\alpha = 0.5$).

The first subfigure in Fig. 4 shows the variance of the total buffer occupancy for $\lambda_2 = 0.46$, $\lambda = 0.3$ and $\alpha = 0.5$, for increasing m. As we can see, the approximation improves until $m \approx 12$. We do however indicate that for large values of m, the vector **p** contains some negative values. The second subfigure in Fig. 4 shows the variance of the total buffer occupancy for $\lambda_2 = 0.48$, i.e. even closer to the stability bound. For visual reference, this point was not shown in Fig. 3. Also, the approximation is bad in this particular case. We indicate that even for $m = 3$, we already obtain a negative value in the vector **p**.

Finally, Fig. 5 shows the variance of the total buffer occupancy versus α, with $\lambda_1 = \lambda_2 = 0.3$ kept fixed. The smallest and largest value of α that we used are $\alpha = 0.34$ and $\alpha = 0.66$, respectively. The same remarks as above hold for Fig. 5 as well. Here we also observe that the shape of the exact curve of the variance of the total buffer occupancy in terms of α is very well approximated by our technique. More precisely, we observe a minimal variance for $\alpha = 0.5$ and a rapidly increasing variance for α close to the stability border.

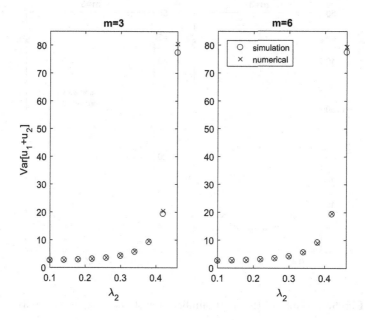

Fig. 3. Variance of the total number of customers ($\lambda_1 = 0.3$ and $\alpha = 0.5$).

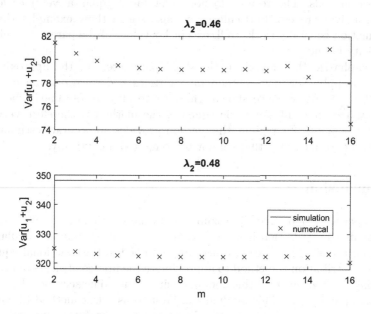

Fig. 4. Variance of the total number of customers versus threshold parameter m ($\lambda_1 = 0.3$ and $\alpha = 0.5$).

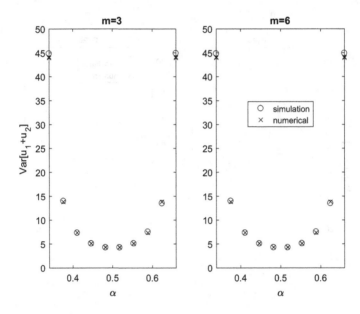

Fig. 5. Variance of the total number of customers ($\lambda_1 = \lambda_2 = 0.3$).

The previous experiments in this section were also executed for Poisson and Geometric arrivals. The results do not differ much (qualitatively) as for the binomial arrival process. Particularly, we also see in these examples that our approximation generally works well, except for values of the parameters close to the stability bound.

We emphasize that for given threshold parameter m, the approximation requires to apply Newton's method $m - 1$ times and solve one linear system of size $2m - 1$. As we have shown, the best results are obtained when m is small, say $3 < m < 14$. As a consequence, the method is therefore very fast. Except for heavy traffic regime, the method provides an accurate estimation of the performance measures that cannot be computed analytically.

7   Conclusion

In this paper we proposed an approximation to the solution of a functional equation. The first contribution is the determination of the dominant singularity of the unknown partial generating functions contained in the functional equation. A second contribution is method to approximate the first coefficients of the functions using the knowledge of this dominant singularity. The accuracy depends on the system parameters. This method can be used as a fast method to compute performance measures from the functional equation. In future, we will try to refine the method in order to obtain more accurate results, especially for high loads.

Acknowledgments. The authors wish to thank Bart Steyaert and Hans Vernaeve for preliminary discussions.

References

1. Adan, I.J.B.F., Wessels, J., Zijm, W.H.M.: A compensation approach for two-dimensional Markov processes. Adv. Appl. Probab. **25**(4), 783–817 (1993)
2. Bruneel, H.: A general model for the behaviour of infinite buffers with periodic service opportunities. Eur. J. Oper. Res. **16**(1), 98–106 (1984)
3. Bruneel, H.: A general treatment of discrete-time buffers with one randomly interrupted output line. Eur. J. Oper. Res. **27**(1), 67–81 (1986)
4. Bruneel, H., Kim, B.G.: Discrete-time Models for Communication Systems Including ATM, vol. 205. Springer, Heidelberg (2012). https://doi.org/10.1007/978-1-4615-3130-2
5. Cohen, J.W., Boxma, O.J.: Boundary Value Problems in Queueing System Analysis. North-Holland, Amsterdam (1983)
6. Cohen, J.W.: On the determination of the stationary distribution of a symmetric clocked buffered switch. In: Teletraffic Science and Engineering, vol. 2, pp. 297–307. Elsevier (1997)
7. Cohen, J.W.: On the asymmetric clocked buffered switch. Queue. Syst. **30**(3–4), 385–404 (1998)
8. Fiems, D., Steyaert, B., Bruneel, H.: Randomly interrupted GI-G-1 queues: service strategies and stability issues. Ann. Oper. Res. **112**(1–4), 171–183 (2002)
9. Fayolle, G., Iasnogorodski, R.: Two coupled processors: the reduction to a Riemann-Hilbert problem. Zeitschrift für Wahrscheinlichkeitstheorie verwandte Gebiete **47**(3), 325–351 (1979)
10. Fayolle, G., Malyshev, V.A., Iasnogorodski, R.: Random Walks in The Quarterplane. Springer, Heidelberg (1999). https://doi.org/10.1007/978-3-642-60001-2
11. Flajolet, P., Sedgewick, R.: Analytic Combinatorics. Cambridge University Press, Cambridge (2009)
12. Li, H., Zhao, Y.Q.: Tail asymptotics for a generalized two-demand queueing model - a kernel method. Queue. Syst. **69**(1), 77–100 (2011)
13. Li, H., Zhao, Y.Q.: A kernel method for exact tail asymptotics - random walks in the quarter plane. arXiv preprint arXiv:1505.04425 (2015)
14. Maertens, T., Walraevens, J., Bruneel, H.: Priority queueing systems: from probability generating functions to tail probabilities. Queue. Syst. **55**(1), 27–39 (2007)
15. Malyshev, V.A.: An analytical method in the theory of two-dimensional positive random walks. Siberian Math. J. **13**(6), 917–929 (1972)
16. Mayank, M., Boxma, O.J., Kapodistria, S., Queija, N.R.: Two queues with random time-limited polling. arXiv preprint arXiv:1701.06834 (2017)
17. Walraevens, J., Steyaert, B., Bruneel, H.: Performance analysis of a single-server ATM queue with a priority scheduling. Comput. Oper. Res. **30**(12), 1807–1829 (2003)
18. Walraevens, J., van Leeuwaarden, J.S., Boxma, O.J.: Power series approximations for two-class generalized processor sharing systems. Queue. Syst. **66**(2), 107–130 (2010)

A Discrete-Time Queueing Model in a Random Environment

Rein Nobel$^{(\boxtimes)}$ and Annette Rondaij

Department of Econometrics and Operations Research,
Vrije Universiteit Amsterdam, Amsterdam, The Netherlands
r.d.nobel@vu.nl, arondaij@gmail.com

Abstract. To study the effect of burstiness in arrival streams on the congestion in queueing systems this paper presents a one-server queueing model in a random environment in discrete time. The environment can be in two states. The number of time slots between two consecutive transitions of the environment follows a geometric distribution with a transition-dependent parameter. In every slot customers arrive in batches, and the batch-size distribution depends on the environment. Each customer requires a generally distributed service time. Arriving customers are put in a queue which is served in FIFO order. Arrivals have precedence over departures and departures have precedence over a change of the environment. The generating functions of the number of customers in the queue and the individual waiting time will be derived. Numerical results will show the effect of the burstiness in the arrival stream on the waiting-time and the queue-size distribution by calculating in parallel the corresponding results for the standard discrete-time model with a mixed batch-size distribution, *ceteris paribus*.

Keywords: Discrete-time queue · Random environment ·
Generating functions

1 Introduction

Burstiness is a well-known and nasty phenomenon in the dynamics of complex systems (see Goh and Baharabási [2] for a nice general introduction to the concept of 'burstiness'). E.g. in telecommunication systems the burstiness of the incoming traffic leads to more congestion, visualized in longer queues and larger waiting times, than in comparable non-bursty systems. To get a better insight into burstiness, queueing theorists have studied systems with bursty traffic by considering queueing models with 'correlated input streams' or enriched with an exogenous 'random environment process'. Yechiali and Naor [14] is one of the first papers in which a queueing model in a random environment is considered. In this paper an $M/M/1$ queue is discussed of which both the arrival rate and the service rate undergo Poissonian jumps. In van Hoorn and Seelen [5]

This paper is based on the second author's Master thesis [11].

© Springer Nature Switzerland AG 2019
T. Phung-Duc et al. (Eds.): QTNA 2019, LNCS 11688, pp. 330–348, 2019.
https://doi.org/10.1007/978-3-030-27181-7_20

an $SPP/G/1$ queue is studied, i.e. a single-server queue with a switched Poisson process as input process. For a much more recent paper we refer to Jiang et al. [6] and the references therein. In Jiang et al. an $M/G/1$ queueing model is discussed in a multi-phase random environment with disasters. These papers are just three examples of the many studies on continuous-time queueing models enriched with an exogenously changing environment, i.e. the system non-intermittently changes environment after an exponentially distributed time with a parameter only depending on the actual environment state. A different way to model burstiness in queueing systems can be found in Nain [9] who discusses a discrete-time single-server queue fed by the number of active servers in an $M/G/\infty$ queue. This arrival stream to the discrete-time queue shows several forms of dependencies and becomes more bursty as the tail of the service-time distribution G of the input process becomes heavier. Other authors have studied discrete-time queueing models with an on/off source for the arrival stream: in the off-state there are no arrivals (see e.g. Zhou and Wang [15]). In Hashida et al. [4] the authors study a discrete-time single-server queue with a switched batch Bernoulli process (SBBP) as input process. They present a thorough analysis of the statistical characteristics of the SBBP, showing among others the autocorrelation between the sizes of the subsequent arriving batches and they derive the probability generating function of the queue size and the individual waiting time of a customer. Also discrete-time models with Markov modulated (batch) input or with working vacations (creating a non-exogenous randomly changing environment) have been studied. We only mention Li and Liu [7] and Li et al. [8] for two studies of discrete-time models with (working) vacations in a random environment. A discrete-time model with Markov modulated (batch) Bernoulli input has been studied in Tsuchiya and Takahashi [13], but contrary to the other papers mentioned before, in this paper the buffer size is taken finite. So, in this paper apart from the steady-state queue-size distribution also the loss-probability is an important performance measure to be studied. As a side remark we notice that often the steady-state distribution of the infinite-buffer model can be used to approximate the loss probability of the equivalent finite-buffer model (see e.g. Gouweleeuw and Tijms [3]), which is another motivation to study infinite-buffer models, and specifically the discrete-time model discussed in this paper.

After this short and necessarily incomplete overview we will sketch the main objective of this paper. To get a deeper insight into burstiness, it is interesting to compare numerically the performance measures (queue length, waiting time, et cetera) of a bursty model with the same performance measures of the equivalent non-bursty model, e.g. by taking the *average arrival rate* of the bursty model as the *constant* arrival rate for the equivalent non-bursty model, *ceteris paribus*. And that is exactly what we want to pursue in this paper. We take the discrete-time single-server queue discussed in Bruneel [1] as a starting point and enrich this model with a two-state exogenous environment process. Then we get a discrete-time model similar to the model studied in Hashida et al. [4], but our setup is slightly different (e.g. we work with Delayed Access (see below)), and for that reason we will give a complete steady-state analysis of this random environment model from scratch. Due to the Delayed Access setup our analysis becomes more complicated and we claim that this analysis has its own merits. In Sect. 2 the description of this discrete-time single-server model will be presented in full detail. Here we only give a short self-contained preview of our model.

As said, we will discuss a one-server queueing model in a random environment in discrete time, i.e. time is counted only in slots. The environment can be in two states, say green and orange. The environment switches, non-intermittently and independently from any other event, from green to orange and vice versa. The number of slots between two consecutive transitions of the environment follows a geometric distribution with a transition-dependent parameter. In every slot a generally distributed number (batch) of customers arrives, and the probability distribution of the batch size depends on the actual state of the environment. The different numbers of arrivals in consecutive slots are mutually independent. Each customer requires a generally distributed service time, also counted in slots. Customers arriving in a slot can start their service only at the beginning of the next slot at the earliest (i.e. Delayed Access!). When upon arrival customers find the server busy, they join a queue and wait for their service. When upon arrival customers find the server idle, then one of the incoming customers (randomly chosen) starts his service at the beginning of the *next* slot, whereas the other incoming customers, if any, join the queue. The customers in the queue are served in the order of arrival (within a batch in random order). Arrivals have precedence over departures (the so-called Late Arrivals Setup), and departures have precedence over a change of the environment. We coin this model as the discrete-time single server LAS-DA-RE model (Late Arrival System with Delayed Access and a Random Environment).

To analyze this model we will derive the probability generating function (p.g.f.) of the joint equilibrium distribution of the number of customers in the queue, the residual service time of the customer in service, and the state of the environment. From this p.g.f. several performance measures will be deduced, like the average queue size, and using the *discrete Fast Fourier Transform (FFT) method* the tail probabilities of the queue-size distribution. We refer to Tijms [12] for a nice tutorial on the discrete FFT method. Also for the individual waiting time of a customer we will derive the p.g.f. from which the average individual waiting time and the tail probabilities (using the discrete FFT method again) will be calculated.

In Sect. 3 we will present the full steady-state analysis from scratch. In Sect. 4 we discuss the steady-state distribution of the queue length, and in Sect. 5 the individual waiting-time distribution of a customer. In Sect. 6 we will present some numerical results which will illustrate the discrepancies between the LAS-DA-RE model studied in this paper and the analogous results for the *standard discrete-time single-server model* as discussed in Bruneel [1] (and here coined as the standard LAS-DA model), by taking in this LAS-DA model the batch-size arrival distribution equal to the mixture of the two different batch-size distributions chosen for the numerical results of the LAS-DA-RE model. Our results show large differences between the results of the LAS-DA-RE model and the analogous results of the standard LAS-DA model, which illustrate anew that in bursty-traffic models the congestion should not be underestimated by using results from non-bursty equivalent models by simply averaging the different arrival streams of the bursty model.

2 Description of the Model

We consider a discrete-time queueing model with one server and an infinite waiting space in a random environment. So, time is only counted in slots and in every time slot the system finds itself in one of two possible *environment states* (e-states), say green (0) or orange (1). When in a time slot the system is in e-state i the environment changes to e-state $1 - i$ in the next slot with probability p_i and stays unaltered in the next slot with probability $1 - p_i$ $(i = 0, 1)$. In other words, the system stays in e-state i a geometrically distributed number of slots with parameter p_i. For technical reasons we assume that $p_0 + p_1 < 1$. In every time slot customers arrive in batches and the batch-size distribution depends on the environment state. To be precise, let $a_k^{(i)}$ be the probability that an arriving batch consists of k customers, given that the system is in e-state i $(k = 0, 1, \ldots;$ $i = 0, 1)$. We introduce the p.g.f.'s

$$\mathcal{A}_i(z) := \sum_{k=0}^{\infty} a_k^{(i)} z^k.$$

All batch sizes are mutually independent. Arriving batches are put in a queue and every customer is served individually by the single server in the order of arrival (within the same batch in random order). Each individual customer requires a generally distributed service time which is independent of the environment state. Let b_j be the probability that a customer requires j time slots for his service. We introduce the p.g.f.

$$\mathcal{B}(w) := \sum_{j=1}^{\infty} b_j w^j.$$

All service times are mutually independent and the service times are also independent of the batch sizes of the arriving customers. Because we discuss a discrete-time system, it is crucial to establish the precedences among the different types of events which can take place in a time slot, i.e. a (batch-)arrival, a service completion (departure) and/or a change in the environment. We have chosen for the so-called *Late Arrival Setup* (LAS) with *Delayed Access* (DA), i.e. arrivals have precedence over departures and an arriving customer can start his service in the next time slot at the earliest. Further we postulate that departures have precedence over a possible change in the environment. We coin this choice for the precedence relations among the different types of events as LAS-DA-RE setup, and in Fig. 1 these precedence relations are illustrated. In this figure we also show the observation point of the system: just after the (possible) start of a service, but before any possible arrivals. For a more detailed description of the discrete-time setup with late arrivals and delayed access (LAS-DA) we refer to Nobel and Moreno [10]. Recall that after a departure which leaves the system empty *the server always stays idle for at least one slot* due to LAS-DA setup and this characteristic makes our model more complicated than the discrete-time model discussed in Hashida et al. [4], which will become apparent in the steady-state analysis below, which is also much more involved than the steady-state analysis in Nobel and Moreno [10].

To analyze this discrete-time delay queueing model in a random environment, we define a discrete-time Markov chain (DTMC) by observing the system at the observation points $k-$, that is at the start of the time slots k just after, possibly, a service of a customer has started, but before the arrivals during time slot k have occurred (see again Fig. 1). We define the following random variables,

$$E_k = \text{the environment state at time } k-,$$
$$H_k = \text{the residual service time of the ongoing service at time } k-,$$
$$Q_k = \text{the number of customers in the queue at time } k-.$$

We define $H_k = 0$ when at epoch $k-$ the server is idle.

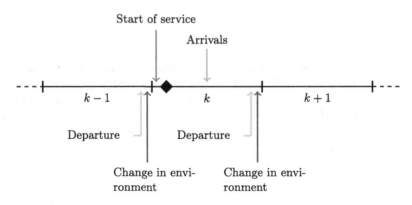

$\blacklozenge$: Observation point $k-$ of the system.

Fig. 1. Order of events taking place in the late arrival queueing system with delayed access and a random environment (LAS-DA-RE).

Then, $\{(E_k, H_k, Q_k) : k = 0, 1, 2, \ldots\}$ is an *irreducible aperiodic DTMC* which is positive recurrent under the stability condition

$$\varrho := \frac{p_0 \mathcal{A}_1'(1) + p_1 \mathcal{A}_0'(1)}{p_0 + p_1} \mathcal{B}'(1) < 1.$$

Notice that ϱ is just the offered load. We skip the formal proof that $\varrho < 1$ is the stability condition and just refer to Nobel and Moreno [10] where a similar stability proof is spelled out in detail.

As said before, we want to study the steady-state behaviour of this DTMC. Define the following limiting joint distribution,

$$\pi_i(j, n) = \lim_{k \to \infty} \mathbf{P}(E_k = i; H_k = j; Q_k = n),$$
$$i = 0, 1; \quad j = 0, 1, 2, \ldots; \quad n = 0, 1, 2, \ldots,$$

and introduce the partial two-dimensional generating functions,

$$\Pi_i(w, z) = \sum_{j=0}^{\infty} \sum_{n=0}^{\infty} \pi_i(j, n) w^j z^n, \quad i = 0, 1,$$

and the partial one-dimensional generating functions,

$$\Pi_{ij}(z) = \sum_{n=0}^{\infty} \pi_i(j, n) z^n, \quad i = 0, 1 \quad j = 0, 1, 2, \ldots.$$

Our first main goal is to find expressions for the two two-dimensional partial p.g.f.'s $\Pi_0(w, z)$ and $\Pi_1(w, z)$. Once we have obtained these expressions we can write down almost immediately the p.g.f. of the steady-state queue-length distribution (Sect. 4) and it is also rather easy to find the p.g.f. of the individual waiting-time distribution using these expressions (Sect. 5).

3   The Joint Distribution of Queue Length, Residual Service Time and Environment State

In this section we will derive expressions for the two (partial) joint probability generating functions $\Pi_i(w, z)$ of the steady-state distribution of the DTMC $\{(E_k, H_k, Q_k) : k = 0, 1, 2, \ldots\}$. To find the p.g.f. $\Pi_i(w, z)$ we write down the following system of balance equations $(i = 0, 1)$,

$$\pi_i(0, 0) = (1 - p_i) a_0^{(i)} [\pi_i(0, 0) + \pi_i(1, 0)] + p_{1-i} a_0^{(1-i)} [\pi_{1-i}(0, 0) + \pi_{1-i}(1, 0)],$$

and for $j = 1, 2, \ldots; \quad n = 0, 1, 2, \ldots$

$$\pi_i(j, n) = (1 - p_i) \sum_{k=0}^{n} a_k^{(i)} \pi_i(j + 1, n - k) + p_{1-i} \sum_{k=0}^{n} a_k^{(1-i)} \pi_{1-i}(j + 1, n - k)$$

$$+ b_j \left((1 - p_i) \left[\sum_{k=0}^{n+1} a_k^{(i)} \pi_i(1, n + 1 - k) + a_{n+1}^{(i)} \pi_i(0, 0) \right] \right.$$

$$\left. + p_{1-i} \left[\sum_{k=0}^{n+1} a_k^{(1-i)} \pi_{1-i}(1, n + 1 - k) + a_{n+1}^{(1-i)} \pi_{1-i}(0, 0) \right] \right).$$

By multiplying both sides with z^n and summing over n we get

$$\Pi_{ij}(z) = (1 - p_i)\mathcal{A}_i(z)\Pi_{i,j+1}(z) + p_{1-i}\mathcal{A}_{1-i}(z)\Pi_{1-i,j+1}(z)$$
$$+ \frac{b_j}{z}\left\{(1 - p_i)\left(\mathcal{A}_i(z)\Pi_{i1}(z) - a_0^{(i)}\pi_i(1,0)\right)\right.$$
$$+ \left[\mathcal{A}_i(z) - a_0^{(i)}\right]\pi_i(0,0)\Big)$$
$$+ p_{1-i}\left(\mathcal{A}_{1-i}(z)\Pi_{1-i,1}(z) - a_0^{(1-i)}\pi_{1-i}(1,0)\right.$$
$$+ \left.\left[\mathcal{A}_{1-i}(z) - a_0^{(1-i)}\right]\pi_{1-i}(0,0)\Big)\right\}.$$

We can get rid of the $\pi_i(1,0)$ and the $a_0^{(i)}$,

$$\Pi_{ij}(z) = (1 - p_i)\mathcal{A}_i(z)\Pi_{i,j+1}(z) + p_{1-i}\mathcal{A}_{1-i}(z)\Pi_{1-i,j+1}(z)$$
$$+ \frac{b_j}{z}\left\{(1 - p_i)\mathcal{A}_i(z)\left[\Pi_{i1}(z) + \pi_i(0,0)\right]\right.$$
$$+ p_{1-i}\mathcal{A}_{1-i}(z)\left[\Pi_{1-i,1}(z) + \pi_{1-i}(0,0)\right] - \pi_i(0,0)\right\}.$$

Multiplying by w^j and summing over $j = 1, 2, \ldots$ gives after rearranging terms two equations ($i = 0, 1$),

$$z\left[w - (1 - p_i)\mathcal{A}_i(z)\right]\Pi_i(w, z) = p_{1-i}z\mathcal{A}_{1-i}(z)\Pi_{1-i}(w, z)$$
$$+ \pi_i(0,0)\left[(1 - p_i)\mathcal{A}_i(z)(w\mathcal{B}(w) - z) + w(z - \mathcal{B}(w))\right]$$
$$+ \pi_{1-i}(0,0)\left[p_{1-i}\mathcal{A}_{1-i}(z)(w\mathcal{B}(w) - z)\right]$$
$$+ \Pi_{i1}(z)\left[(1 - p_i)\mathcal{A}_i(z)w(\mathcal{B}(w) - z)\right]$$
$$+ \Pi_{1-i,1}(z)\left[p_{1-i}\mathcal{A}_{1-i}(z)w(\mathcal{B}(w) - z)\right]. \tag{1}$$

So, we have to specify four unknown quantities/functions:

- the probabilities $\pi_0(0,0)$ and $\pi_1(0,0)$ (we know of course that $\pi_0(0,0) + \pi_1(0,0) = 1 - \varrho$),
- the partial generating functions $\Pi_{01}(z)$ and $\Pi_{11}(z)$.

From the two Eq. (1) for $\Pi_0(w, z)$ and $\Pi_1(w, z)$ we get after eliminating $\Pi_1(w, z)$

$$\left[(w - (1 - p_0)\mathcal{A}_0(z))(w - (1 - p_1)\mathcal{A}_1(z)) - p_0p_1\mathcal{A}_0(z)\mathcal{A}_1(z)\right]z\Pi_0(w, z)$$
$$= \pi_0(0,0)[w(w - (1 - p_1)\mathcal{A}_1(z))(z - \mathcal{B}(w))$$
$$+ \mathcal{A}_0(z)(w\mathcal{B}(w) - z)((1 - p_0)w - (1 - p_0 - p_1)\mathcal{A}_1(z))]$$
$$+ \pi_1(0,0)p_1w(w - 1)\mathcal{A}_1(z)\mathcal{B}(w)$$
$$+ \Pi_{01}(z)w\mathcal{A}_0(z)(\mathcal{B}(w) - z)((1 - p_0)w - (1 - p_0 - p_1)\mathcal{A}_1(z))$$
$$+ \Pi_{11}(z)p_1w^2\mathcal{A}_1(z)(\mathcal{B}(w) - z). \tag{2}$$

Notice that the first factor of the left-hand side is a *quadratic expression* in w. So, consider the roots of the following quadratic equation

$$w^2 - [(1 - p_0)\mathcal{A}_0(z) + (1 - p_1)\mathcal{A}_1(z)]w + (1 - p_0 - p_1)\mathcal{A}_0(z)\mathcal{A}_1(z) = 0.$$

These are given by

$$w^{\pm}(z) = \frac{1}{2}\left\{(1 - p_0)\mathcal{A}_0(z) + (1 - p_1)\mathcal{A}_1(z) \pm \sqrt{\mathcal{D}(z)}\right\},$$

where $\mathcal{D}(z)$ is the discriminant of this quadratic equation. So,

$$\mathcal{D}(z) = (1 - p_0)^2 \mathcal{A}_0^2(z) + (1 - p_1)^2 \mathcal{A}_1^2(z) - 2(1 - p_0 - p_1 - p_0 p_1)\mathcal{A}_0(z)\mathcal{A}_1(z).$$

Notice that it is easy to see that for all real $z \in [0, 1]$

$$\mathcal{D}(z) \geq (p_0 + p_1)^2 \left(\min\left\{\mathcal{A}_0(z), \mathcal{A}_1(z)\right\}\right)^2 > 0.$$

Hence, for all real $z \in [0, 1]$ the roots $w^+(z)$ and $w^-(z)$ are different. Next, choosing $w = w^+(z)$ or $w = w^-(z)$ the left-hand side in (2) vanishes, which leads to two equations in the unknowns

$$\pi_0(0, 0), \quad \pi_1(0, 0), \quad \Pi_{01}(z), \quad \Pi_{11}(z).$$

To enhance readability we introduce the following abbreviations

$$\mathcal{L}_i(w, z) := w(w - (1 - p_i)\mathcal{A}_i(z))(z - \mathcal{B}(w))$$
$$+ \mathcal{A}_{1-i}(z)(w\mathcal{B}(w) - z)((1 - p_{1-i})w - (1 - p_0 - p_1)\mathcal{A}_i(z)),$$
$$\mathcal{K}_i(w, z) := p_i w(w - 1)\mathcal{A}_i(z)\mathcal{B}(w),$$
$$\mathcal{M}_i(w, z) := w\mathcal{A}_{1-i}(z)(\mathcal{B}(w) - z)((1 - p_{1-i})w - (1 - p_0 - p_1)\mathcal{A}_i(z)),$$
$$\mathcal{N}_i(w, z) := p_i w^2 \mathcal{A}_i(z)(\mathcal{B}(w) - z).$$

Plugging in $w = w^+(z)$ and $w = w^-(z)$, respectively in the right-hand side of (2) and using the above notations we get the following two equations in z,

$$\pi_0(0, 0)\mathcal{L}_1\left(w^+(z), z\right) + \pi_1(0, 0)\mathcal{K}_1\left(w^+(z), z\right)$$
$$+ \Pi_{01}(z)\mathcal{M}_1\left(w^+(z), z\right) + \Pi_{11}(z)\mathcal{N}_1\left(w^+(z), z\right) = 0, \qquad (3)$$
$$\pi_0(0, 0)\mathcal{L}_1\left(w^-(z), z\right) + \pi_1(0, 0)\mathcal{K}_1\left(w^-(z), z\right)$$
$$+ \Pi_{01}(z)\mathcal{M}_1\left(w^-(z), z\right) + \Pi_{11}(z)\mathcal{N}_1\left(w^-(z), z\right) = 0. \qquad (4)$$

We emphasize that for real $0 \leq z < 1$ ($z \neq 1$!) this is a system of two *independent* linear equations, due the fact that (i) $w^+(z) \neq w^-(z)$, (ii) all the coefficients of the four unknowns are non-zero for these z-values, and (iii) for all real $z \in (0, 1)$

$$\frac{\mathcal{L}_1(w^+(z), z)}{\mathcal{L}_1(w^-(z), z)}, \frac{\mathcal{K}_1(w^+(z), z)}{\mathcal{K}_1(w^-(z), z)}, \frac{\mathcal{M}_1(w^+(z), z)}{\mathcal{M}_1(w^-(z), z)} \neq \frac{\mathcal{N}_1(w^+(z), z)}{\mathcal{N}_1(w^-(z), z)}.$$

Eliminating $\Pi_{11}(z)$ from (3) and (4) we get (z in $w^{\pm}(z)$ will be suppressed)

$$\Pi_{01}(z) = \frac{\begin{aligned}&\pi_0(0, 0)\left[\mathcal{L}_1\left(w^+, z\right)\mathcal{N}_1\left(w^-, z\right) - \mathcal{L}_1\left(w^-, z\right)\mathcal{N}_1\left(w^+, z\right)\right]\\&+\\&\pi_1(0, 0)\left[\mathcal{K}_1\left(w^+, z\right)\mathcal{N}_1\left(w^-, z\right) - \mathcal{K}_1\left(w^-, z\right)\mathcal{N}_1\left(w^+, z\right)\right]\end{aligned}}{\mathcal{N}_1\left(w^+, z\right)\mathcal{M}_1\left(w^-, z\right) - \mathcal{N}_1\left(w^-, z\right)\mathcal{M}_1\left(w^+, z\right)}.$$

Evaluating the denominator gives

$$
\begin{aligned}
\mathcal{N}_1 &\left(w^+(z), z\right) \mathcal{M}_1\left(w^-(z), z\right) - \mathcal{N}_1\left(w^-(z), z\right) \mathcal{M}_1\left(w^+(z), z\right) \\
&= w^+(z) w^-(z) p_1(1 - p_0 - p_1) \mathcal{A}_0(z) \mathcal{A}_1^2(z) \\
&\quad \times \left[\mathcal{B}\left(w^+(z)\right) - z\right]\left[\mathcal{B}\left(w^-(z)\right) - z\right]\left(w^-(z) - w^+(z)\right) \\
&= -p_1(1 - p_0 - p_1)^2 \mathcal{A}_0^2(z) \mathcal{A}_1^3(z)\left[\mathcal{B}\left(w^+(z)\right) - z\right] \\
&\quad \times \left[\mathcal{B}\left(w^-(z)\right) - z\right] \sqrt{\mathcal{D}(z)}.
\end{aligned}
\tag{5}
$$

Evaluating the two terms in the numerator gives (argument z in $w^{\pm}(z)$ suppressed again)

$$
\begin{aligned}
\mathcal{L}_1 &\left(w^+, z\right) \mathcal{N}_1\left(w^-, z\right) - \mathcal{L}_1\left(w^-, z\right) \mathcal{N}_1\left(w^+, z\right) \\
&= p_1 \mathcal{A}_1^2(z) w^+ w^-\left\{\mathcal{B}\left(w^+\right) \mathcal{B}\left(w^-\right)\left(w^+ - w^-\right)\left((1 - p_0 - p_1) \mathcal{A}_0(z) - (1 - p_1)\right)\right. \\
&\quad + z\left[(1 - p_0 - p_1) \mathcal{A}_0(z)(1 - (1 - p_0) \mathcal{A}_0(z))\left[\mathcal{B}\left(w^+\right) - \mathcal{B}\left(w^-\right)\right]\right. \\
&\quad \left.\left. + (1 - p_1 - (1 - p_0 - p_1) \mathcal{A}_0(z))\left[w^+ \mathcal{B}\left(w^-\right) - w^- \mathcal{B}\left(w^+\right)\right]\right]\right\},
\end{aligned}
$$

and

$$
\begin{aligned}
\mathcal{K}_1 &\left(w^+, z\right) \mathcal{N}_1\left(w^-, z\right) - \mathcal{K}_1\left(w^-, z\right) \mathcal{N}_1\left(w^+, z\right) \\
&= p_1^2 \mathcal{A}_1^2(z) w^+ w^-\left\{\mathcal{B}\left(w^+\right) \mathcal{B}\left(w^-\right)\left(w^+ - w^-\right)\right. \\
&\quad \left. + z\left[w^+\left(w^- - 1\right) \mathcal{B}\left(w^-\right) - w^-\left(w^+ - 1\right) \mathcal{B}\left(w^+\right)\right]\right\}.
\end{aligned}
\tag{6}
$$

Taking everything together gives after cancelation of the common factor

$$
p_1(1 - p_0 - p_1) \mathcal{A}_0(z) \mathcal{A}_1^3(z)
$$

in the numerator and the denominator,

$$
\Pi_{01}(z) =
$$

$$
\frac{\left[\begin{array}{l} \mathcal{B}\left(w^+\right) \mathcal{B}\left(w^-\right)\left(w^+ - w^-\right)\left((1 - p_0 - p_1) \mathcal{A}_0(z) - (1 - p_1)\right) \\ + z\left\{(1 - p_0 - p_1) \mathcal{A}_0(z)(1 - (1 - p_0) \mathcal{A}_0(z))\left[\mathcal{B}\left(w^+\right) - \mathcal{B}\left(w^-\right)\right]\right. \\ \left. + (1 - p_1 - (1 - p_0 - p_1) \mathcal{A}_0(z))\left[w^+ \mathcal{B}\left(w^-\right) - w^- \mathcal{B}\left(w^+\right)\right]\right\} \end{array}\right] \pi_0(0,0) + \left[\begin{array}{l} p_1\left\{\mathcal{B}\left(w^+\right) \mathcal{B}\left(w^-\right)\left(w^+ - w^-\right)\right. \\ \left. + z\left[w^+\left(w^- - 1\right) \mathcal{B}\left(w^-\right) - w^-\left(w^+ - 1\right) \mathcal{B}\left(w^+\right)\right]\right\} \end{array}\right] \pi_1(0,0)}{-(1 - p_0 - p_1) \mathcal{A}_0(z)\left[\mathcal{B}\left(w^+(z)\right) - z\right]\left[\mathcal{B}\left(w^-(z)\right) - z\right] \sqrt{\mathcal{D}(z)}}.
\tag{7}
$$

Next, we can make the following observations (for all straightforward calculus-type proofs we refer to Rondaij [11])

- $w^+(1) = 1$ and $w^-(1) = 1 - p_0 - p_1$.
- For $z = 1$ both the numerator and the denominator vanish.
- Both $w^+(0)$ and $w^-(0)$ are positive. [We have assumed $1 - p_0 - p_1 > 0$!]
- The denominator is negative for $z = 0$.
- The factor $\left[\mathcal{B}\left(w^+(z)\right) - z\right]$ is positive on the real interval $(0, 1)$.
- The factor $\left[\mathcal{B}\left(w^-(z)\right) - z\right]$ is *positive* for $z = 0$ and *negative* for $z = 1$.

Hence, the denominator has a unique zero in the real interval $(0, 1)$, because $\mathcal{D}(z) > 0$ for $z \in [0, 1]$ and this zero is the unique solution in $(0, 1)$ (for a detailed proof see Rondaij [11] again), say $z = z^\bullet$, of the equation

$$\mathcal{B}\left(w^-(z)\right) - z = 0.$$

So, for $z = z^\bullet$ also the numerator must vanish. Plugging in $z = z^\bullet$ in the numerator of (7) gives after some simplifications

$$\left\{z^\bullet \left(\mathcal{B}\left(w^+(z^\bullet)\right) - z^\bullet\right)\left(1 - w^-(z^\bullet)\right)\right.$$
$$\times \left.\left[(1 - p_0 - p_1)\mathcal{A}_0(z^\bullet) - (1 - p_1)w^+(z^\bullet)\right]\right\} \pi_0(0, 0)$$
$$+ \left\{p_1 z^\bullet w^+(z^\bullet)\left(1 - w^-(z^\bullet)\right)\left(\mathcal{B}\left(w^+(z^\bullet)\right) - z^\bullet\right)\right\} \pi_1(0, 0).$$

So, we get the following equation for the unknown probabilities $\pi_0(0, 0)$ and $\pi_1(0, 0)$,

$$\left[(1 - p_0 - p_1)\mathcal{A}_0\left(z^\bullet\right) - (1 - p_1)w^+\left(z^\bullet\right)\right]\pi_0(0, 0) + p_1 w^+\left(z^\bullet\right)\pi_1(0, 0) = 0. \quad (8)$$

Notice that because for all $z \in (0, 1)$ we have

$$w^+(z) > \max\left\{(1 - p_0)\mathcal{A}_0(z), (1 - p_1)\mathcal{A}_1(z)\right\},$$

the coefficient of $\pi_0(0, 0)$ in (8) is negative! From (8) we get

$$\pi_1(0, 0) = \frac{(1 - p_1)w^+\left(z^\bullet\right) - (1 - p_0 - p_1)\mathcal{A}_0\left(z^\bullet\right)}{p_1 w^+\left(z^\bullet\right)}\pi_0(0, 0),$$

and using

$$\pi_0(0, 0) + \pi_1(0, 0) = 1 - \varrho$$

we find

$$\pi_0(0, 0) = (1 - \varrho)\frac{p_1 w^+\left(z^\bullet\right)}{w^+\left(z^\bullet\right) - (1 - p_0 - p_1)\mathcal{A}_0\left(z^\bullet\right)}$$
$$\pi_1(0, 0) = (1 - \varrho)\frac{(1 - p_1)w^+\left(z^\bullet\right) - (1 - p_0 - p_1)\mathcal{A}_0\left(z^\bullet\right)}{w^+\left(z^\bullet\right) - (1 - p_0 - p_1)\mathcal{A}_0\left(z^\bullet\right)},$$

where

$$w^+\left(z^\bullet\right) = \frac{1}{2}\left\{(1 - p_0)\mathcal{A}_0\left(z^\bullet\right) + (1 - p_1)\mathcal{A}_1\left(z^\bullet\right) + \sqrt{\mathcal{D}\left(z^\bullet\right)}\right\},$$

with $\mathcal{D}\left(z^\bullet\right) = \left[(1 - p_0)\mathcal{A}_0\left(z^\bullet\right) - (1 - p_1)\mathcal{A}_1\left(z^\bullet\right)\right]^2 + 4p_0 p_1$.

Due to the fact that all formulae are symmetric with respect to the environment-index $i = 0, 1$ we have the alternative expressions

$$\pi_1(0, 0) = (1 - \varrho)\frac{p_0 w^+\left(z^\bullet\right)}{w^+\left(z^\bullet\right) - (1 - p_0 - p_1)\mathcal{A}_1\left(z^\bullet\right)}$$
$$\pi_0(0, 0) = (1 - \varrho)\frac{(1 - p_0)w^+\left(z^\bullet\right) - (1 - p_0 - p_1)\mathcal{A}_1\left(z^\bullet\right)}{w^+\left(z^\bullet\right) - (1 - p_0 - p_1)\mathcal{A}_1\left(z^\bullet\right)},$$

which can also be easily checked algebraically using the elementary facts

$$w^+(z) + w^-(z) = (1 - p_0)\mathcal{A}_0(z) + (1 - p_1)\mathcal{A}(z),$$
$$w^+(z)w^-(z) = (1 - p_0 - p_1)\mathcal{A}_0(z)\mathcal{A}_1(z).$$

By interchanging the environment-index i in (7) throughout we immediately find

$$\Pi_{11}(z) =$$

$$\frac{
\left[\begin{array}{l}
\mathcal{B}(w^+)\,\mathcal{B}(w^-)\,(w^+ - w^-)\,((1 - p_0 - p_1)\mathcal{A}_1(z) - (1 - p_0)) \\
+z\left\{(1 - p_0 - p_1)\mathcal{A}_1(z)(1 - (1 - p_1)\mathcal{A}_1(z))\left[\mathcal{B}(w^+) - \mathcal{B}(w^-)\right]\right. \\
\left. +(1 - p_0 - (1 - p_0 - p_1)\mathcal{A}_1(z))\left[w^+\mathcal{B}(w^-) - w^-\mathcal{B}(w^+)\right]\right\}
\end{array}\right]\pi_1(0,0)
+ \left[\begin{array}{l}
p_0\left\{\mathcal{B}(w^+)\,\mathcal{B}(w^-)\,(w^+ - w^-)\right. \\
\left. +z\left[w^+(w^- - 1)\mathcal{B}(w^-) - w^-(w^+ - 1)\mathcal{B}(w^+)\right]\right\}
\end{array}\right]\pi_0(0,0)
}{
-(1 - p_0 - p_1)\mathcal{A}_1(z)\left[\mathcal{B}(w^+(z)) - z\right]\left[\mathcal{B}(w^-(z)) - z\right]\sqrt{\mathcal{D}(z)}
}. \qquad (9)$$

Once we have found the probabilities $\pi_0(0,0)$ and $\pi_1(0,0)$ and the partial generating functions $\Pi_{01}(z)$ and $\Pi_{11}(z)$ we get from (2)

$$\Pi_0(w,z) =$$

$$\frac{
\begin{array}{c}
\pi_0(0,0)[w(w - (1 - p_1)\mathcal{A}_1(z))(z - \mathcal{B}(w)) \\
+\mathcal{A}_0(z)(w\mathcal{B}(w) - z)((1 - p_0)w - (1 - p_0 - p_1)\mathcal{A}_1(z))] \\
+\pi_1(0,0)p_1 w(w - 1)\mathcal{A}_1(z)\mathcal{B}(w) \\
+\Pi_{01}(z)w\mathcal{A}_0(z)(\mathcal{B}(w) - z)((1 - p_0)w - (1 - p_0 - p_1)\mathcal{A}_1(z)) \\
+\Pi_{11}(z)p_1 w^2 \mathcal{A}_1(z)(\mathcal{B}(w) - z)
\end{array}
}{
[(w - (1 - p_0)\mathcal{A}_0(z))(w - (1 - p_1)\mathcal{A}_1(z)) - p_0 p_1 \mathcal{A}_0(z)\mathcal{A}_1(z)]\,z
},$$

and a completely similar formula for $\Pi_1(w,z)$ by interchanging the environment-index from 0 to 1 and vice versa throughout the expression above.

4   The Queue-Length Distribution

Using the explicit expressions for the two-dimensional p.g.f.'s $\Pi_0(w,z)$ and $\Pi_1(w,z)$ we immediately get the p.g.f. $\mathcal{Q}(z)$ of the queue-length distribution,

$$\mathcal{Q}(z) = \Pi_0(1,z) + \Pi_1(1,z).$$

So, using the results of the previous section we find

$$\mathcal{Q}(z) = \frac{1 - z}{[(1 - (1 - p_0)\mathcal{A}_0(z))(1 - (1 - p_1)\mathcal{A}_1(z)) - p_0 p_1 \mathcal{A}_0(z)\mathcal{A}_1(z)]\,z}$$
$$\times \left(\begin{array}{l}
\pi_0(0,0)[\mathcal{A}_0(z)(1 - p_0 - (1 - p_0 - p_1)\mathcal{A}_1(z)) - (1 - (1 - p_1)\mathcal{A}_1(z))] \\
+ \pi_1(0,0)[\mathcal{A}_1(z)(1 - p_1 - (1 - p_0 - p_1)\mathcal{A}_0(z)) - (1 - (1 - p_0)\mathcal{A}_0(z))] \\
+ \Pi_{01}(z)\mathcal{A}_0(z)(1 - (1 - p_0 - p_1)\mathcal{A}_1(z)) \\
+ \Pi_{11}(z)\mathcal{A}_1(z)(1 - (1 - p_0 - p_1)\mathcal{A}_0(z))
\end{array}\right).$$

Due to the complexity of this expression, we will not differentiate the p.g.f. $Q(z)$ to derive e.g. the long-run average queue length, say $\overline{Q}$, by setting $\overline{Q} = Q'(1)$. Instead we will use the discrete FFT method to calculate the tail probabilities $\mathbf{P}(Q > n)$ from the p.g.f. $Q(z)$, where Q stands for an artefact random variable having the steady-state queue-length distribution. Subsequently, we get the long-run average queue length using the well-known formula

$$\overline{Q} = \mathbf{E}[Q] = \sum_{n=0}^{\infty} \mathbf{P}(Q > n).$$

Of course, to get numerical results we can only use a finite number of terms in this series. See Sect. 6 for more details.

5   The Individual Waiting-Time Distribution

In this section we will derive the p.g.f. $\mathcal{W}(z)$ of the steady-state individual waiting time, say W, of a tagged customer. So,

$$\mathcal{W}(z) := \sum_{r=0}^{\infty} \mathbf{P}\{W = r\}z^r.$$

Let

$$\gamma_m^{(i)}(j, n) = \text{the probability that a tagged customer occupies the}$$
$$(m + 1)\text{-th position in his batch and arrives in e-state } i,$$
$$\text{finds a residual time of } j \text{ slots of the ongoing service,}$$
$$\text{and } n \text{ customers waiting in queue.}$$

Standard *ergodicity arguments* give

$$\gamma_m^{(i)}(j, n) = \frac{p_0 + p_1}{p_0 \mathcal{A}_1'(1) + p_1 \mathcal{A}_0'(1)} \sum_{k>m} a_k^{(i)} \pi_i(j, n).$$

Remark that for the waiting time of an individual customer we do not count the slot in which the customer arrives, although we use the setup of Delayed Access.

Take $r = 0, 1, 2, \ldots$. Then we have (an empty sum is zero w.p. 1)

$$\mathbf{P}\{W = r\} = \sum_{m=0}^{r} \sum_{j=1}^{r-m+1} \sum_{i=0}^{1} \sum_{n=0}^{r-m+1-j} \mathbf{P}\left\{\sum_{k=1}^{n+m} B_k = r - j + 1\right\} \gamma_m^{(i)}(j, n)$$

$$+ \sum_{m=0}^{r} \sum_{i=0}^{1} \mathbf{P}\left\{\sum_{k=1}^{m} B_k = r\right\} \gamma_m^{(i)}(0, 0).$$

In the usual way we find the p.g.f.

$$W(z) = \sum_{r=0}^{\infty} \mathbf{P}(W = r)z^r = \frac{p_0 + p_1}{p_0 \mathcal{A}_1'(1) + p_1 \mathcal{A}_0'(1)}$$

$$\times \left\{ \frac{1 - \mathcal{A}_0(\mathcal{B}(z))}{1 - \mathcal{B}(z)} \left[\frac{\Pi_0(z, \mathcal{B}(z)) - \pi_0(0, 0)}{z} + \pi_0(0, 0) \right] \right.$$

$$\left. + \frac{1 - \mathcal{A}_1(\mathcal{B}(z))}{1 - \mathcal{B}(z)} \left[\frac{\Pi_1(z, \mathcal{B}(z)) - \pi_1(0, 0)}{z} + \pi_1(0, 0) \right] \right\}.$$

As for the long-run average queue length, also for the long-run average individual waiting time, say $\overline{W}$ we will not differentiate the p.g.f. $W(z)$, but use the tail probabilities $\mathbf{P}(W > r)$ instead, which will be calculated using the discrete FFT method again. See Sect. 6 for further details.

6   Numerical Results: Comparisons Between the Random Environment Model and the Standard LAS-DA Model

In this section we want to compare numerical results of the random environment LAS-DA-RE model with the analogous results for the standard LAS-DA model with a mixed batch-size distribution, *ceteris paribus*. So, we begin with a short overview of results for the standard model.

6.1   Queue-Size and Waiting-Time Distribution for the Standard LAS-DA Model

In case there is no change in the environment [so only *one* batch-size distribution $\mathcal{A}(z)$, *ceteris paribus*] we have the following well-known results (see e.g. Bruneel [1] or Rondaij [11])

$$\mathcal{Q}(z) = (1 - \mathcal{A}'(1)\mathcal{B}'(1)) \frac{1 - z}{\mathcal{B}(\mathcal{A}(z)) - z},$$

$$\overline{Q} = \mathcal{Q}'(1) = \frac{\mathcal{A}''(1)\mathcal{B}'(1) + [\mathcal{A}'(1)]^2 \mathcal{B}''(1)}{2[1 - \mathcal{A}'(1)\mathcal{B}'(1)]},$$

$$W(z) = \left[\frac{\Pi(z, \mathcal{B}(z)) - \pi(0, 0)}{z} + \pi(0, 0) \right] \frac{1 - \mathcal{A}(\mathcal{B}(z))}{\mathcal{A}'(1)(1 - \mathcal{B}(z))},$$

$$\overline{W} = W'(1) = \frac{\mathcal{A}''(1)\mathcal{B}'(1) + [\mathcal{A}'(1)]^2 \mathcal{B}''(1)}{2\mathcal{A}'(1)[1 - \mathcal{A}'(1)\mathcal{B}'(1)]}.$$

Notice *Little's Law*: $\overline{Q} = \mathcal{A}'(1)\overline{W}$.

To compare our numerical results for the random environment LAS-DA-RE model with the equivalent results for the standard LAS-DA model we take the p.g.f. of the batch-size distribution in the standard LAS-DA model equal to

$$\mathcal{A}(z) := \frac{p_0 \mathcal{A}_1(z) + p_1 \mathcal{A}_0(z)}{p_0 + p_1},$$

Then we have for this model

$$\overline{Q} = \frac{(p_0 + p_1)[p_1 \mathcal{A}_0''(1) + p_0 \mathcal{A}_1''(1)]\mathcal{B}'(1) + [p_1 \mathcal{A}_0'(1) + p_0 \mathcal{A}_1'(1)]^2 \mathcal{B}''(1)}{2(p_0 + p_1)\{p_0 + p_1 - [p_1 \mathcal{A}_0'(1) + p_0 \mathcal{A}_1'(1)]\mathcal{B}'(1)\}},$$

and

$$\overline{W} = \frac{(p_0 + p_1)[p_1 \mathcal{A}_0''(1) + p_0 \mathcal{A}_1''(1)]\mathcal{B}'(1) + [p_1 \mathcal{A}_0'(1) + p_0 \mathcal{A}_1'(1)]^2 \mathcal{B}''(1)}{2[p_1 \mathcal{A}_0'(1) + p_0 \mathcal{A}_1'(1)]\{p_0 + p_1 - [p_1 \mathcal{A}_0'(1) + p_0 \mathcal{A}_1'(1)]\mathcal{B}'(1)\}}.$$

6.2 How to Calculate the Performance Measures for the Random Environment LAS-DA-RE Model?

As said before, differentiating the p.g.f.'s $\mathcal{Q}(z)$ and $\mathcal{W}(z)$ is too cumbersome, so we use the *discrete Fast Fourier Transform [FFT] method* to calculate the *long-run average mean queue length* $\overline{Q}$ and the *long-run average mean individual waiting time* $\overline{W}$ as follows (see Tijms [12] for more details on the discrete FFT method)

- use the discrete FFT method to calculate the tail probabilities $\mathbf{P}(Q > n)$ and $\mathbf{P}(W > r)$ for $n, r = 0, 1, 2, \ldots, 2^{15} - 1$,
- calculate subsequently $\overline{Q} = \sum_{n>0} \mathbf{P}(Q > n)$ and $\overline{W} = \sum_{r>0} \mathbf{P}(W > r)$,
- finally, we checked the quality of the numerical results by using *Little's Law*.

6.3 Comparison of the Numerical Results

In our first examples we take the *offered load*

$$\varrho = \frac{p_0 \mathcal{A}_1'(1) + p_1 \mathcal{A}_0'(1)}{p_0 + p_1} \mathcal{B}'(1) = 0.96, \quad p_0 = 0.9, \quad p_1 = 0.05, \quad \mathcal{B}'(1) = 3$$

and so the *overall mean batch size*

$$\frac{p_0 \mathcal{A}_1'(1) + p_1 \mathcal{A}_0'(1)}{p_0 + p_1} = 0.32.$$

We want to compare results for the *mean queue length* $\overline{Q}$ and the *mean waiting time* $\overline{W}$ between the standard LAS-DA model with a mixed batch-size distribution and the random environment LAS-DA-RE model.

We vary $\mathcal{A}_0'(1)$ and $\mathcal{A}_1'(1)$ keeping $\mathcal{A}'(1) = 0.32, p_0 = 0.9, p_1 = 0.05$, constant for the moment. In e-state 0 we choose a Poisson distribution for the batch size and in e-state 1 the batch size follows a geometric distribution (shifted to zero). In Table 1 we show the mean queue lengths and the mean waiting times for both models for several choices of $\mathcal{A}_0'(1)$ and $\mathcal{A}_1'(1)$. We see that the differences between the standard model and the random environment model become more apparent as the gap between the two mean batch sizes increases. In Figs. 2 and 3 we show the tail probabilities of the queue length for the standard model and the random environment model, respectively. Notice the fatter tails in Figs. 3.

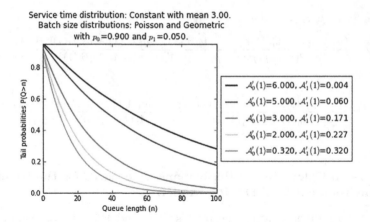

Fig. 2. Tail probabilities of the queue length for the standard LAS-DA model with mixed batch-size distribution: poisson, geometric.

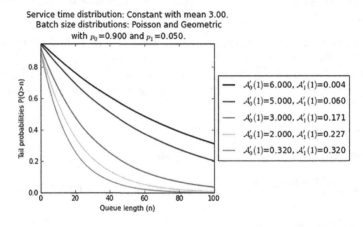

Fig. 3. Tail probabilities of the queue length for the random environment LAS-DA-RE model: poisson, geometric.

In Figs. 4 and 5 we show the tail probabilities of the waiting time, and again the fatter tails can be seen in Fig. 5.

In Table 2 and the Figs. 6, 7, 8 and 9 we present the analogous results for the case that in e-state 0 the batch-size is taken two-point 0 or 8 and the batch-size distribution in e-state 1 is kept unchanged, i.e. Poisson.

All these results clearly illustrate the phenomenon that in the random environment LAS-DA-RE model the performance measures (means and tail probabilities) are significantly larger than the equivalent performance measures in the standard LAS-DA model. For more numerical results we refer to Rondaij [11]. Finally, we remark that many of the numerical results have been checked by simulations.

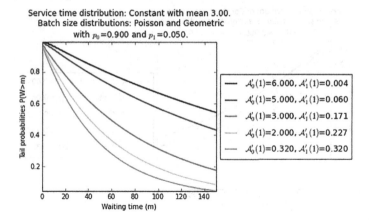

Fig. 4. Tail probabilities of the waiting-time for the standard LAS-DA model with mixed batch-size distribution: poisson, geometric.

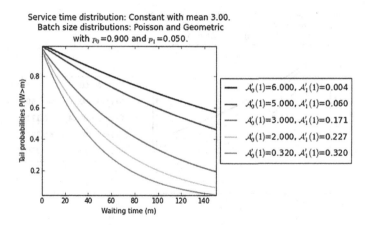

Fig. 5. Tail probabilities of the waiting-time for the random environment LAS-DA-RE model: poisson, geometric.

Table 1. Batch size distributions are poisson and geometric for environment 0 and 1, respectively. Furthermore, $p_0 = 0.9, p_1 = 0.05$. Overall mean batch size is 0.32. Service time follows a constant distribution with mean 3.

	Mean queue length		Mean waiting-time	
	Random env.	Mixed standard	Random env.	Mixed standard
$\mathcal{A}_0'(1) = 6, \mathcal{A}_1'(1) = 0.00\overline{44}$	85.7926	78.7340	268.1017	246.0439
$\mathcal{A}_0'(1) = 5, \mathcal{A}_1'(1) = 0.060$	62.0675	57.2779	193.9608	178.9934
$\mathcal{A}_0'(1) = 3, \mathcal{A}_1'(1) = 0.17\overline{11}$	29.0912	27.5235	90.9101	86.0110
$\mathcal{A}_0'(1) = 2, \mathcal{A}_1'(1) = 0.2267$	19.8401	19.2253	62.0004	60.0789
$\mathcal{A}_0'(1) = 0.32, \mathcal{A}_1'(1) = 0.32$	15.1579	15.1579	47.3684	47.3684

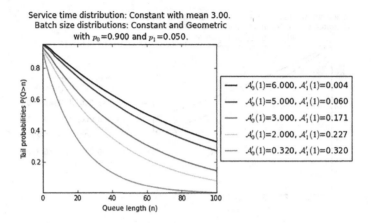

Fig. 6. Tail probabilities of the queue length for the standard LAS-DA model with mixed batch-size distribution: two-point, geometric.

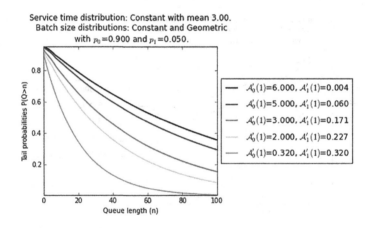

Fig. 7. Tail probabilities of the queue length for the random environment LAS-DA-RE model: two-point, geometric.

Table 2. Batch size distributions are two-point and Geometric for environment 0 and 1, respectively. Furthermore, $p_0 = 0.9, p_1 = 0.05$. Overall mean batch size is 0.32. Service time follows a constant distribution with mean 3.

	Mean queue length		Mean waiting-time	
	Random env.	Mixed standard	Random env.	Mixed standard
$\mathcal{A}_0'(1) = 6, \mathcal{A}_1'(1) = 0.00\overline{44}$	97.6388	90.5761	305.1212	283.0504
$\mathcal{A}_0'(1) = 5, \mathcal{A}_1'(1) = 0.060$	81.8097	77.0147	255.6552	240.6711
$\mathcal{A}_0'(1) = 3, \mathcal{A}_1'(1) = 0.17\overline{11}$	52.7807	51.2077	164.9397	160.0241
$\mathcal{A}_0'(1) = 2, \mathcal{A}_1'(1) = 0.2267$	39.5807	38.9621	123.6896	121.7566
$\mathcal{A}_0'(1) = 0.32, \mathcal{A}_1'(1) = 0.32$	19.3768	19.3768	60.5526	60.5526

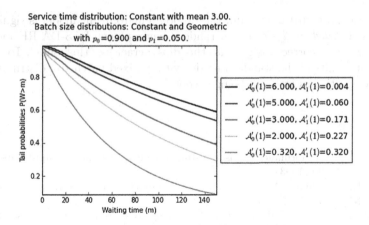

Fig. 8. Tail probabilities of the waiting-time for the standard LAS-DA model with mixed batch-size distribution: two-point, geometric.

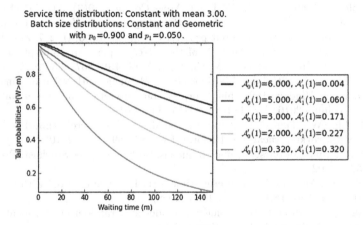

Fig. 9. Tail probabilities of the waiting-time for the random environment LAS-DA-RE model: two-point, geometric.

7 Conclusions

For the random environment LAS-DA-RE model explicit expressions have been derived for the p.g.f. $\mathcal{Q}(z)$ of the *steady-state queue-length distribution*, and for the p.g.f. $\mathcal{W}(z)$ of the *steady-state waiting-time distribution of an individual customer*. Using the discrete FFT method the *tail probabilities* $\mathbf{P}(Q > n)$ and $\mathbf{P}(W > m)$ have been calculated, and using these tail probabilities approximate values for the *mean queue length* $\overline{Q}$ and the *mean individual waiting time* $\overline{W}$ have been presented. We compared the numerical results of the standard LAS-DA model with a mixed batch-size distribution with the numerical results of the random environment LAS-DA-RE model. Especially in case the mean batch sizes are very different for the two e-states in the LAS-DA-RE model the discrepancy between the standard model and the random environment model is significant. Hence, a *detailed analysis* of the random environment LAS-DA-RE model is required to get a correct insight into the idiosyncracies of this model. To use the equivalent results of the standard model with a mixed batch-size distribution as an approximation is far from being acceptable.

References

1. Bruneel, H.: Performance of discrete-time queueing systems. Comput. Oper. Res. **20**(3), 303–320 (1993)
2. Goh, K.I., Barabasi, A.L.: Burstiness and memory in complex systems. Europhys. Lett. **81**(4), 48002 (2008)
3. Gouweleeuw, F.N., Tijms, H.C.: Computing loss probabilities in discrete-time queues. Oper. Res. **46**(1), 149–154 (1998)
4. Hashida, O., Takahashi, Y., Shimogawa, S.: Switched batch Bernoulli process (SBBP) and the discrete time SBBP/G/1 queue with applications to statistical multiplexer performance. IEEE J. Sel. Areas Commun. **9**(3), 394–401 (1991)
5. van Hoorn, M.H., Seelen, L.P.: the $SPP/G/1$ queue: a single server queue with a switched poisson process as input process. OR Spektrum **5**(4), 207–218 (1983)
6. Jiang, T., Liu, L., Li, J.: Analysis of the $M/G/1$ queue in multi-phase random envirionment with disasters. J. Math. Anal. Appl. **430**, 857–873 (2015)
7. Li, J., Liu, L.: On the discrete-time $Geo/G/1$ queue with vacations in random environment. Hindawi Publishing Corporation, Discrete Dynamics in Nature and Society Volume 2016, Article ID 4029415, 9 p. (2016)
8. Li, J., Liu, W., Tian, N.: Steady-state analysis of a discrete-time batch arrival queue with working vacations. Perform. Eval. **67**, 897–912 (2010)
9. Nain, P.: Impact of bursty traffic on queues. Stat. Inferences Stochast. Process. **5**, 307–320 (2002)
10. Nobel, R.D., Moreno, P.: A discrete-time retrial queueing model with one server. Eur. J. Oper. Res. **189**(3), 1088–1103 (2008)
11. Rondaij, A.: A discrete-time queueing model in a random environment. Master thesis, Vrije Universiteit Amsterdam (2018)
12. Tijms, H.C.: A First Course in Stochastic Models. Wiley, New York (2003)
13. Tsuchiya, T., Takahashi, Y.: On discrete-time single-server queues with Markov modulated batch Bernoulli input and finite capacity. J. Oper. Res. Soc. Jpn. **36**(1), 29–45 (1993)
14. Yechiali, U., Naor, P.: Queueing problems with heterogeneous arrivals and service. Oper. Res. **19**(3), 722–734 (1971)
15. Zhou, W.-H., Wang, A.-H.: Discrete-time queue with Bernoulli bursty source arrival and generally distributed service times. Appl. Math. Model. **32**, 2233–2240 (2008)

Queueing Models in Applications

Queueing Analysis of Home Delivery Services with Parcel Lockers

Shinto Hideyama[1], Tuan Phung-Duc[2(✉)], and Yukihiko Okada[2]

[1] Graduate School of Systems and Information Engineering, University of Tsukuba,
Tsukuba, Ibaraki 305-8577, Japan
`s1820487@s.tsukuba.ac.jp`
[2] Faculty of Engineering, Information and Systems, University of Tsukuba,
Tsukuba, Ibaraki 305-8577, Japan
`{tuan,okayu}@sk.tsukuba.ac.jp`

Abstract. Nowadays, Parcel Delivery Service (PDS) has become popular due to the popularization of Electric Commerce (EC). The redelivery problem has also become serious due to the absence of receivers and the courier must deliver again at a later time. The increase in the number of redeliveries causes social losses such as extra workload of delivery, carbon dioxide emissions and so on. To improve this problem, Japanese government recommends the utilization of the parcel locker which is expected to reduce the number of redeliveries. While some empirical researches reveal the effectiveness of the parcel locker, quantitative evaluation using stochastic models has not been carried out yet. In this paper, we model the PDS including the parcel locker service as a queueing model and analyze it using a Quasi-Birth-and-Death (QBD) process. Furthermore, we derive the stability condition which must be satisfied to ensure the proper service and we derive some performance measures such as blocking probability, the mean number of parcels in queues. Our numerical experiments show the influence of the collaborative behavior of users on the efficiency of the system.

Keywords: Parcel Delivery Service (PDS) ·
Parcel locker · Quasi-Birth-and-Death (QBD) process

1 Introduction

Nowadays, many people use the Parcel Delivery Service (PDS) in Japan because of the expansion of e-commerce business. According to MLIT (Ministry of Land, Infrastructure, Transport and Tourism, Japan), the handling number of home delivery parcels in Japan nearly approached 4.25 billion items in FY2017 of Japan [1]. On the other hand, redelivery due to the absence of receivers has become a serious social problem. The number of redeliveries increases with the number of delivery parcels. Redelivery causes some social losses such as extra workload of delivery, carbon dioxide emissions, and so on. According to MLIT, about 20% of parcels are being redelivered. In order to deal with the problem,

© Springer Nature Switzerland AG 2019
T. Phung-Duc et al. (Eds.): QTNA 2019, LNCS 11688, pp. 351–368, 2019.
https://doi.org/10.1007/978-3-030-27181-7_21

MLIT advocates the utilization of the parcel locker as a way of receipt of parcels. Parcel locker is the facility which enables us to receive parcels 24 hours a day. The parcel locker is located in various places and receivers could choose the most convenient one when using the service. To evaluate the parcel locker service, some studies are conducted. Focusing on the configuration of the parcel locker, Iwan et al. [2] assess the parcel locker as one of the solutions for the last mile delivery in Szczecin, Poland. Lachapelle et al. [3] evaluate the place and regional location characteristics of parcel lockers in five South East Queensland (SEQ) cities, Australia. From these results, the parcel locker can be considered as an effective solution for the redelivery problem.

On the other hand, these studies have not paid attention to the fact that the number of parcels in the parcel locker changes with time. Actually, since the size of the parcel locker is limited, services cannot be used if the parcel locker is already full. This is commonly referred to as blocking and we call the probability that it occurs the blocking probability. It is worth analyzing how often blocking occurs.

From these backgrounds, our research focuses on the dynamic analysis using queueing theory for PDS including the parcel locker. Queueing theory is one of effective and famous methods for analyzing systems with dynamic changes and it is applied to modeling service systems in various fields including transportation [4,5]. In our queueing model, the parcels which fail to home delivery because of the receiver's absence will be stored in the parcel locker under certain conditions. Such a system can be expressed by a queueing model with feedback for which several related studies are available [6,7].

The structure of this paper is organized as follows. In Sect. 2, PDS including the parcel locker is modeled by a feed-back queue which is analyzed using a Quasi-Birth-and-Death process. In Sect. 3, we present the analysis of the model and derive the stability condition and some performance measures. In Sect. 4, we conduct numerical experiments on the stability condition and performance measures obtained in Sect. 3. Finally, in Sect. 5, we conclude this paper and discuss future works.

2   Modelling

In this section, we describe the system of PDS including the parcel locker and model as a queueing model, and formulate it as a Quasi-Birth-and-Death (QBD) process. There are various kinds of actual courier services, but here we consider the PDS including the parcel locker as shown in Fig. 1 and propose two slightly different queueing systems, System 1 and System 2. System 1 considers losses and System 2 does not take the loss into consideration. We will describe details of each system.

2.1   Parcel Delivery Service

System 1. We define the system of PDS including the parcel locker with some assumptions as System 1. In this study, it is assumed that the courier has two delivery methods: delivery by truck and delivery by the parcel locker. In the case of delivery by truck, the courier goes to the receiver's home by truck. On the other hand, in the case of delivery by the parcel locker, the courier stores the parcel in the locker, then the receiver will go to collect the parcel.

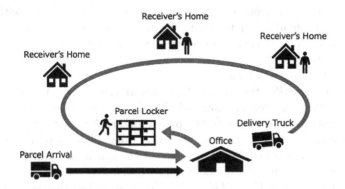

Fig. 1. PDS including the parcel locker.

First of all, we describe the arrival of parcels for delivery to the courier's office. In this system, there are two types of parcels to arrive and we will call them as Parcel A and Parcel B, respectively. Parcel A is the parcel that the receiver wishes to be delivered by delivery truck. Parcel A arrives at the office according to a Poisson process with rate $(1-r)\lambda$. On the other hand, Parcel B is the parcel that the receiver wishes to be delivered by the parcel locker. Parcel B arrives at the office according to a Poisson process with rate $r\lambda$. Then, it should be noted that the combined arrival rate of Parcel A and Parcel B is λ. The office prepares two kinds of queues: a queue of parcels for delivery truck and a queue of parcels for the parcel locker. The former has infinite size and the latter has finite size. Parcel A is always allotted to the queue of parcels for delivery truck. Parcel B is allocated to the queue of parcels for the parcel locker if there is room in the queue. However, if the queue is full, it is handled as a loss and handled out of the system. In practice, it means that the parcel is not delivered by the truck again, but by another method not presented here.

Next, we describe the queue of parcels for delivery truck. Parcels in the queue are loaded to the delivery truck in order of arrivals and delivered one by one to each receiver's home. However, in this paper, we will not consider the number of parcels that can be loaded in the delivery truck and the loading time, and the office owns just one delivery truck. The time it takes for the delivery truck to move to the next delivery destination follows the exponential distribution with rate μ_1. When the courier arrives at the receiver's home, the courier succeeds in delivery with probability p which means that they handed over a parcel to the receiver. On the other hand, the courier fails in delivery with probability $1 - p$ which means that the courier could not hand over a parcel to the receiver because of the receiver's absence. At that time, if there is still room in the queue of parcels for the parcel locker, they go to the locker before returning to the office and they distribute the parcel to the queue. However, if there is no vacancy, they bring the parcel back to the office. In fact, there is a time lag between delivery failure and the placement of the parcel at the locker. However, in this study, we assume that this time lag is negligible if $\mu_1 >> \mu_2$ (μ_2 will be described later). From the above, this queue can be seen as a single server queue with infinite waiting room. It should be noted that the parcel in service refers to the parcel to the next delivery destination and the waiting parcel refers to the parcel waiting for delivery within the delivery truck or the office. We define $N_1(t)$ as the number of parcels in the queue at time $t \geq 0$.

Finally, we describe the queue of parcels at the locker. The parcel locker is located within the delivery area of the office and we do not consider the movement time from the office to the parcel locker. We assume that the parcel locker can store up to c parcels at the same time. This queue is prepared on the side of the parcel locker, and parcels in the queue are stored in order of arrivals. Here, we define the queue as that the number of servers is c and the size of waiting room is $K - c$ ($K \geq c$). Parcels stored in the parcel locker are collected independently by receivers, and the time until collection of each parcel follows the exponential distribution with rate μ_2. We define $N_2(t)$ as the number of parcels in the queue at time $t \geq 0$.

Based on these assumptions, the queueing model diagram of System 1 can be represented as shown in Fig. 2. Also, the parameters appearing above are summarized in Table 1.

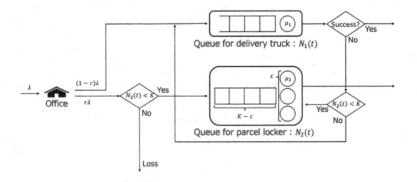

Fig. 2. The queueing model of System 1.

System 2. Next, we also describe System 2. System 2 has the same properties as System 1 except for one point described below. In System 1, Parcel B arriving when the queue of parcels for the parcel locker is full is treated as loss. However, in System 2, loss should be avoided by distributing it to the queue of parcel for the delivery truck. Thus the queueing model diagram of System 2 can be represented as shown in Fig. 3.

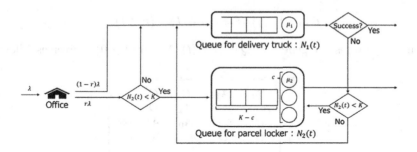

Fig. 3. The queueing model of System 2.

Table 1. The list of parameters.

Parameters	Definitions
λ	The arrival rate of parcels
r	The probability of the parcel locker service
μ_1	The service rate of delivery truck
μ_2	The service rate of the parcel locker
p	The success probability of home delivery
c	The number of service windows of the parcel locker service
K	The buffer size of the parcel locker service

2.2 Markov Chain

System 1. We modeled the PDS including the parcel locker in Sect. 2.1. Now we define S as follows.

$$S = \{(i,j) \mid i \in I, j \in J\},$$

where,

$$I = \{0, 1, 2, \ldots\}, \quad J = \{0, 1, 2, \ldots, K\}.$$

Then, it is clear that $\{(N_1(t), N_2(t)) \mid t \geq 0\}$ forms a Continuous-Time Markov Chain (CTMC) in the state space S. Furthermore, in a jump, the value of $N_1(t)$ changes at most 1, $\{(N_1(t), N_2(t)) \mid t \geq 0\}$ forms a Quasi-Birth-and-Death (QBD) process. Moreover, we separate S as follows.

$$S = \bigcup_{k=0}^{\infty} \mathcal{L}_k,$$

where,

$$\mathcal{L}_k := \{(i,j) \mid i = k, j \in J\} \quad (k \in I).$$

Then, the infinitesimal generator $\boldsymbol{Q}$ of $\{(N_1(t), N_2(t)) \mid t \geq 0\}$ is given as follows.

$$
\boldsymbol{Q} =
\begin{array}{c}
\phantom{\mathcal{L}_0} \\
\mathcal{L}_0 \\
\mathcal{L}_1 \\
\mathcal{L}_2 \\
\mathcal{L}_3 \\

\end{array}
\begin{array}{c}
\begin{array}{cccccc}
\mathcal{L}_0 & \mathcal{L}_1 & \mathcal{L}_2 & \mathcal{L}_3 & \cdots
\end{array} \\
\left(
\begin{array}{ccccc}
\boldsymbol{B_0} & \boldsymbol{A_1} & \boldsymbol{O} & \boldsymbol{O} & \cdots \\
\boldsymbol{A_{-1}} & \boldsymbol{A_0} & \boldsymbol{A_1} & \boldsymbol{O} & \cdots \\
\boldsymbol{O} & \boldsymbol{A_{-1}} & \boldsymbol{A_0} & \boldsymbol{A_1} & \cdots \\
\boldsymbol{O} & \boldsymbol{O} & \boldsymbol{A_{-1}} & \boldsymbol{A_0} & \cdots \\
\vdots & \vdots & \vdots & \vdots & \ddots
\end{array}
\right),
\end{array}
$$

where,

$$
\boldsymbol{B_0} =
\begin{array}{c}
0 \\ 1 \\ \vdots \\ c-1 \\ c \\ c+1 \\ \vdots \\ K-1 \\ K
\end{array}
\begin{array}{c}
\begin{array}{cccccccc}
0 & 1 & \cdots & c & \cdots & K-1 & & K
\end{array} \\
\left(
\begin{array}{cccccccc}
-\lambda & r\lambda & \cdots & 0 & \cdots & 0 & & 0 \\
\mu_2 & -(\mu_2+\lambda) & \ddots & 0 & \ddots & 0 & & 0 \\
\vdots & \ddots & \ddots & \ddots & \ddots & \ddots & & \vdots \\
0 & 0 & \ddots & r\lambda & \ddots & 0 & & 0 \\
0 & 0 & \ddots & -(c\mu_2+\lambda) & \ddots & 0 & & 0 \\
0 & 0 & \ddots & c\mu_2 & \ddots & 0 & & 0 \\
\vdots & \ddots & \ddots & \ddots & \ddots & \ddots & & \vdots \\
0 & 0 & \ddots & 0 & \ddots & -(c\mu_2+\lambda) & & r\lambda \\
0 & 0 & \cdots & 0 & \cdots & c\mu_2 & & -\{c\mu_2+(1-r)\lambda\}
\end{array}
\right),
\end{array}
$$

$$
\boldsymbol{A_{-1}} =
\begin{array}{c}
0 \\ 1 \\ \vdots \\ K-1 \\ K
\end{array}
\begin{array}{c}
\begin{array}{ccccc}
0 & 1 & \cdots & K-1 & K
\end{array} \\
\left(
\begin{array}{ccccc}
p\mu_1 & (1-p)\mu_1 & \cdots & 0 & 0 \\
0 & p\mu_1 & \ddots & 0 & 0 \\
\vdots & \ddots & \ddots & \ddots & \vdots \\
0 & 0 & \ddots & p\mu_1 & (1-p)\mu_1 \\
0 & 0 & \cdots & 0 & p\mu_1
\end{array}
\right),
\end{array}
$$

$$
\boldsymbol{A_0} = \boldsymbol{B_0} -
\begin{array}{c}
0 \\ 1 \\ \vdots \\ K-1 \\ K
\end{array}
\begin{array}{c}
\begin{array}{ccccc}
0 & 1 & \cdots & K-1 & K
\end{array} \\
\left(
\begin{array}{ccccc}
\mu_1 & 0 & \cdots & 0 & 0 \\
0 & \mu_1 & \ddots & 0 & 0 \\
\vdots & \ddots & \ddots & \ddots & \vdots \\
0 & 0 & \ddots & \mu_1 & 0 \\
0 & 0 & \cdots & 0 & p\mu_1
\end{array}
\right),
\end{array}
$$

$$A_1 = \begin{array}{c} \\ 0 \\ 1 \\ \vdots \\ K-1 \\ K \end{array} \begin{array}{ccccc} 0 & 1 & \cdots & K-1 & K \\ \begin{pmatrix} (1-r)\lambda & 0 & \cdots & 0 & 0 \\ 0 & (1-r)\lambda & \ddots & 0 & 0 \\ \vdots & & \ddots & \ddots & \vdots \\ 0 & 0 & \ddots & (1-r)\lambda & 0 \\ 0 & 0 & \cdots & 0 & (1-r)\lambda \end{pmatrix} \end{array},$$

and O is the zero matrix with an appropriate dimension.

System 2. System 2 can be formulated in the same way as System 1 by replacing the (K, K) element of B_0 with $-(c\mu_2 + \lambda)$ and the (K, K) element of A_1 with λ. These changes are due to the fact that parcels handled as losses in System 1 are allocated to the queue of parcels for delivery truck in System 2.

3 Analysis

In this section, we present the analysis of our model. We derive the stability condition, the stationary distribution and performance measures. Because the analysis of both systems is the same, we present the same description and indicate the difference when necessary.

3.1 Stability Condition

We derive the stability condition which must be satisfied to have the stationary distribution. We consider the column vector η which satisfies (1) as follows.

$$\eta A = 0, \quad \eta e = 1, \tag{1}$$

where,

$$A = A_{-1} + A_0 + A_1.$$

Then, let 0 and e denote the row vector which all elements are 0 with an appropriate dimension and the column vector which all elements are 1 with an appropriate dimension, respectively. In this study, we obtain η as follows.

$$\eta = \left(\eta_0, \quad \eta_0\rho, \quad \eta_0\frac{\rho^2}{2!}, \quad \cdots, \quad \eta_0\frac{\rho^c}{c!}, \quad \eta_0\frac{\rho^c}{c!}\left(\frac{\rho}{c}\right), \quad \cdots, \quad \eta_0\frac{\rho^c}{c!}\left(\frac{\rho}{c}\right)^{K-c} \right),$$

where,

$$\eta_0 = \frac{1}{\sum_{k=0}^{c}\frac{\rho^k}{k!} + \frac{\rho^c}{c!}\sum_{k=1}^{K-c}\left(\frac{\rho}{c}\right)^k}, \qquad \rho = \frac{(1-p)\mu_1 + r\lambda}{\mu_2}.$$

It should be noted that η corresponds to the stationary distribution of M/M/c/K queues which the arrival rate is $(1-p)\mu_1 + r\lambda$ and the service rate is μ_2. According to Takine [8], QBD's stability condition is given as follows.

$$\eta A_1 e - \eta A_{-1} e < 0. \tag{2}$$

Symplifying (2), we can obtain the stability condition of each system. In this study, the stability condition of System 1 is (3) and the stability condition of System 2 is (4).

$$(1-r)\lambda + \{(1-p)\mu_1\} \frac{\frac{\rho^c}{c!}\left(\frac{\rho}{c}\right)^{K-c}}{\sum_{k=0}^{c}\frac{\rho^k}{k!} + \frac{\rho^c}{c!}\sum_{k=1}^{K-c}\left(\frac{\rho}{c}\right)^k} < \mu_1, \tag{3}$$

$$(1-r)\lambda + \{r\lambda + (1-p)\mu_1\} \frac{\frac{\rho^c}{c!}\left(\frac{\rho}{c}\right)^{K-c}}{\sum_{k=0}^{c}\frac{\rho^k}{k!} + \frac{\rho^c}{c!}\sum_{k=1}^{K-c}\left(\frac{\rho}{c}\right)^k} < \mu_1. \tag{4}$$

These stability conditions mean the conditions for the queue of parcels for delivery truck does not diverge within each system. In both (3) and (4), the first term on the left side represents the arrival rate of Parcel A and the second term on the left side expresses the arrival rate of parcels which are blocked by the queue for parcel locker. The sum of these arrival rates should not exceed the service rate of the delivery truck.

3.2   Stationary State Probability

Since our Markov chains are irreducible, the stationary distribution exists under the stability condition. Then, stationary state probability $\pi_{i,j}$ is defined as follows.

$$\pi_{i,j} = \lim_{t\to\infty} P\{N_1(t) = i, N_2(t) = j\}.$$

Moreover, we define π_i and π as follows.

$$\pi_i = (\pi_{i,0}, \pi_{i,1}, \pi_{i,2}, \cdots, \pi_{i,K}), \quad (i \in I),$$
$$\pi = (\pi_0, \pi_1, \pi_2, \cdots).$$

From the property of Markov chain, π is the unique solution of the following equations.

$$\pi Q = 0, \tag{5}$$
$$\pi e = 1. \tag{6}$$

These equations are obtained by the global balance equations and the normalization condition. Equation (5) is rewritten as follows.

$$\pi_0 B_0 + \pi_1 A_{-1} = 0, \tag{7}$$
$$\pi_{k-1} A_1 + \pi_k A_0 + \pi_{k+1} A_{-1} = 0, \quad k = 1, 2, \ldots. \tag{8}$$

From Takine [8], we obtain the solution for $\pi_{i,j}$ as follows.

$$\pi_k = \pi_1 R^{k-1}, \quad k = 1, 2, \ldots, \tag{9}$$

where $\boldsymbol{R}$ is the minimal nonnegative solution of (10).

$$A_1 + \boldsymbol{R}A_0 + \boldsymbol{R}^2 A_{-1} = \boldsymbol{0}. \tag{10}$$

Matrix $\boldsymbol{R}$ is numerically computed using the method in Takine [8]. Moreover, we also obtain (11) if stability condition is satisfied.

$$\sum_{k=1}^{\infty} \pi_k e = \pi_1 (\boldsymbol{I} - \boldsymbol{R})^{-1} e < \infty, \tag{11}$$

where $\boldsymbol{I}$ is an identity matrix with an appropriate dimension. Next, we will derive π_0 and π_1. From (7) and (8) with $k = 1$, we obtain the matrix equation as follows.

$$(\pi_0, \pi_1) \begin{pmatrix} \boldsymbol{B}_0 & A_1 \\ A_{-1} & A_0 + \boldsymbol{R}A_{-1} \end{pmatrix} = (\boldsymbol{0}, \boldsymbol{0}). \tag{12}$$

Also, (6), (9) and (11) yield

$$\pi_0 e + \pi_1 (\boldsymbol{I} - \boldsymbol{R})^{-1} e = 1. \tag{13}$$

From the above, $\boldsymbol{R}$ can be calculated numerically by solving (10) by Takine [8], π_0 and π_1 can be calculated by solving (12) and (13) and π_k $(k = 2, 3, \dots)$ can be calculated by using (9) recursively.

3.3 Performance Measures

In order to analyze our model, we define several performance measures summarized in Table 2. Here, $\mathbb{E}[\cdot]$ denotes the expectation. The derivations of these performance measures will be described in sequel. First, we define P_b. It must be described as follows.

$$P_b = \frac{\sum_{i \in I} (\text{arrival rate for the parcel locker service in state } (i, K)) \times \pi_{i,K}}{\sum_{(i,j) \in S} (\text{arrival rate for the parcel locker service in state } (i, j)) \times \pi_{i,j}} \tag{14}$$

We could rewrite (14) as follows.

$$P_b = \frac{r\lambda\pi_0 e_1 + \{(1-p)\mu_1 + r\lambda\} \sum_{i=1}^{\infty} \pi_i e_1}{r\lambda\pi_0 e + \{(1-p)\mu_1 + r\lambda\} \sum_{i=1}^{\infty} \pi_i e}, \tag{15}$$

where,

$$e_1 = (0, 0, \cdots, 0, 1)^{\mathrm{T}}.$$

From (9) and (15), we rewrite P_b as follows.

$$P_b = \frac{r\lambda\pi_0 e_1 + \{(1-p)\mu_1 + r\lambda\}\pi_1 (\boldsymbol{I} - \boldsymbol{R})^{-1} e_1}{r\lambda\pi_0 e + \{(1-p)\mu_1 + r\lambda\}\pi_1 (\boldsymbol{I} - \boldsymbol{R})^{-1} e}.$$

Table 2. The list of performance measures.

Indicators	Definitions
P_b	The blocking probability of the queue of parcels for the parcel locker
$\mathbb{E}[N_1(t)]$	The mean number of parcels in the queue for delivery truck
$\mathbb{E}[N_2(t)]$	The mean number of parcels in the queue for the parcel locker
T_1	The number of parcels delivered by delivery truck per a unit time
T_2	The number of parcels delivered by the parcel locker per a unit time
T_{Loss}	The number of parcels treated as loss per a unit time (defined only for System 1)
TC	The total of social costs

Next, we define $\mathbb{E}[N_1(t)]$ and $\mathbb{E}[N_2(t)]$ as follows.

$$\mathbb{E}[N_1(t)] = \sum_{i=1}^{\infty} i\pi_i e, \quad \mathbb{E}[N_2(t)] = \sum_{i=0}^{\infty} \pi_i e_2, \tag{16}$$

where,

$$e_2 = \left(0, 1, \cdots, K-1, K\right)^{\mathrm{T}}.$$

For $\mathbb{E}[N_1(t)]$, (17) can be described from the nature of the expectation,

$$\mathbb{E}[N_1(t)] = \sum_{i=1}^{\infty} P\{N_1(t) \geq i\} = \sum_{i=1}^{\infty} \sum_{k=i}^{\infty} \pi_k e. \tag{17}$$

By (11), (16) and (17), $\mathbb{E}[N_1(t)]$ and $\mathbb{E}[N_2(t)]$ can be transformed as follows.

$$\mathbb{E}[N_1(t)] = \pi_1(I - R)^{-2}e, \quad \mathbb{E}[N_2(t)] = \pi_0 e_2 + \pi_1(I - R)^{-1}e_2.$$

Next, we derive T_1, T_2 and T_{Loss}. These are given by

$$T_1 = \mu_1 \sum_{i=1}^{\infty} \pi_i e, \quad T_2 = \mu_2 \sum_{i=0}^{\infty} \pi_i e_3, \quad T_{Loss} = r\lambda \sum_{i=0}^{\infty} \pi_i e_1. \tag{18}$$

From (11) and (18), we derive T_1, T_2 and T_{Loss} as follows.

$$T_1 = \mu_1(1 - \pi_0 e),$$
$$T_2 = \mu_2 \left\{\pi_0 e_3 + \pi_1(I - R)^{-1}e_3\right\},$$
$$T_{Loss} = r\lambda \left\{\pi_0 e_1 + \pi_1(I - R)^{-1}e_1\right\},$$

where,

$$e_3 = \left(\min(0,c), \min(1,c), \cdots, \min(K-1,c), \min(K,c) \right)^{\mathrm{T}}.$$

It should be noted that T_{Loss} is defined only for System 1, hence $T_{Loss} = 0$ in System 2. Finally, we define TC as follows.

$$TC = C_{Truck} + C_{Locker} + w_3 T_{Loss},$$

where,

$$C_{Truck} = w_{11}\mathbb{E}[N_1(t)] + w_{12}T_1,$$
$$C_{Locker} = w_{21}\mathbb{E}[N_2(t)] + w_{22}T_2.$$

C_{Truck} denotes the social costs about the delivery of truck. w_{11} is the management cost per a unit time for a parcel for delivery truck; w_{12} is the delivery cost of delivery truck per a parcel. On the other hand, we define C_{Locker} as the social cost which is mainly borne by the receiver. w_{21} is the management cost per a unit time for a parcel for the parcel locker and w_{22} is the delivery cost of the parcel locker per a parcel. Also, w_3 is the processing cost per a parcel treated as loss. It should be noted that $T_{Loss} = 0$ in System 2. By defining such an TC, it is possible to compare the two systems in terms of cost.

4   Numerical Examples

In this section, we plot the numerical examples of stability region and performance measures by using computer. It should be noted that stability region is the range where stability condition is satisfied.

4.1   Stability Region

In order to evaluate the increment of stability region by the introduction of the parcel locker service, we need to look at the boundary of the stability region of each of the following three situations.

- (a): The situation which the parcel locker service is introduced and $p = 1.0$.
- (b): The situation which the parcel locker service is introduced.
- (c): The situation which the parcel locker service is not introduced.

By comparing the stability region of (b) and the stability region of (c), we can see the increment of the stability region by introducing the parcel locker service. Therefore, the increment of the stability region is the area sandwiched between the boundaries of these stability regions. Furthermore, by comparing the boundary of the stability region of (a) and the boundary of the stability region of (b), it is possible to see the difference of the stability region. Actually, (b) is a situation given by parameters at the time of analysis, (c) is a situation where $c = K = 0$ are changed among the parameters of (b), and (a) is a situation where $p = 1.0$ is changed among the parameters of (b). For this time we set $p = 0.8$,

$r = 0.1$, $c = K = 50$ and see the increment of stability region in each of $\mu_2 = 0.1$, 1.0 and 5.0 for System 1 and System 2. Results of numerical experiments are as shown in Fig. 4. It should be noted that the left column shows the result of System 1, and the right column shows the result of System 2. From these results, it can be seen that the larger the μ_2, the larger the increment of the stability region. It is easy to understand intuitively. In addition, it can be seen that the increment of the stability region in System 2 is larger than that in System 1. This is because System 2 tends to accumulate parcels in the system more than System 1 due to the non-existence of lost parcels in System 2. Also, when $\mu_2 = 5.0$, both System 1 and System 2 show that the boundary of the stability region of (b) is considerably close to the boundary of the stability region of (a). Therefore, when μ_2 is large, it is possible to obtain a stability region close to the situation where redelivery does not exist in both systems.

4.2   Performance Measures

In this section, we conduct the numerical experiments on performance measures which are defined in Sect. 3.3 by changing r from 0 to 1 and discuss the results. We fix $\lambda = 80$, $\mu_1 = 100$, $\mu_2 = 1.0$, $p = 0.8$, and $K = c$. Also, for each value of $c = 5, 10, 20$ and 30, numerical experiments are conducted on System 1 and System 2. First, for P_b, $\mathbb{E}[N_1(t)]$ and $\mathbb{E}[N_2(t)]$, numerical experiment results are as shown in Fig. 5. From these results, P_b and $\mathbb{E}[N_2(t)]$ increase in both System 1 and System 2 as r increases. The increase in P_b can be easily understood from the increase in the amount of parcels arriving at the parcel locker due to the increase in r. Also, $\mathbb{E}[N_2(t)]$ is asymptotically approaching c as r increases. This is due to the fact that the parcel locker is always close to full since the maximum service rate of the parcel locker will be lower than the arrival rate of parcels to the parcel locker in this parameter setting. On the other hand, $\mathbb{E}[N_1(t)]$ tends to decrease roughly. Therefore, roughly speaking, the number of parcels for the delivery truck decreases as the use of the parcel locker becomes active. However, looking closely at Fig. 5d, the graph of $c = 20$ takes the minimum value $\mathbb{E}[N_1(t)] = 3.68711994$ when $r = 0.24$ and the graph of $c = 30$ takes the minimum value $\mathbb{E}[N_1(t)] = 2.15232469$ when $r = 0.31$. In other words, increasing the utilization rate of the parcel locker service to reduce parcels for the delivery truck can be counterproductive depending on the situation.

Next, for T_1, T_2 and T_{Loss}, numerical experiment results are as shown in Fig. 6. As shown in Fig. 6a, T_1 of System 1 decreases monotonically with the increase of r. On the other hand, from Fig. 6b, each graph decreases asymptotically to a certain value, respectively. With the increase of r, more parcels which are desired to be received at the parcel locker arrive. If the parcel locker is full when the parcel arrives, it will be delivered by the delivery truck in System 2 but will leave the system by treating it as a loss in System 1. The difference in T_1 is considered to be due to the difference in these systems. Also, from Fig. 6c and d, the trend is almost the same. As for T_2, it can be seen that each graph changes asymptotically to the maximum service rate $c\mu_2$. T_{Loss} is defined only for System 1 and shown in Fig. 6e, it is monotonically increasing with the increase of r.

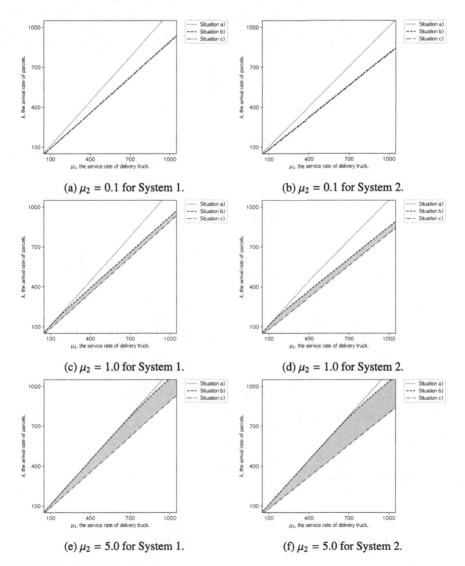

(a) $\mu_2 = 0.1$ for System 1.

(b) $\mu_2 = 0.1$ for System 2.

(c) $\mu_2 = 1.0$ for System 1.

(d) $\mu_2 = 1.0$ for System 2.

(e) $\mu_2 = 5.0$ for System 1.

(f) $\mu_2 = 5.0$ for System 2.

Fig. 4. Numerical results for the increment of stability region.

Finally, we compare TC of System 1 and System 2. As mentioned above, the difference between System 1 and System 2 is whether or not the parcel blocked by the parcel locker upon arrival is treated as a loss. When considering the delivery cost, we want to know under what conditions System 1 (or System 2) will be better. In Fig. 7, we set the values of w_{11}, w_{12}, w_{21} and w_{22} and conduct the numerical calculation on w_3 which TC of System 1 becomes equal to TC of System 2. Therefore, in Fig. 7, System 1 is better than System 2 when the

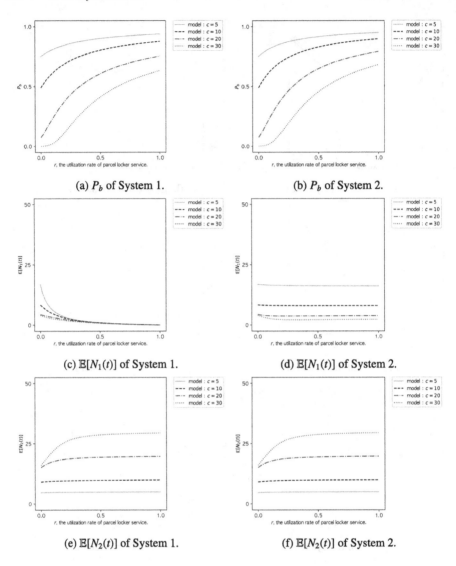

Fig. 5. Numerical results of P_b, $\mathbb{E}[N_1(t)]$ and $\mathbb{E}[N_2(t)]$.

value of w_3 is smaller than that in the curve and reverse. We use the numerical experiment results of each of the performance measures that is already shown in Figs. 5 and 6 and the five cost unit setting patterns described below.

- Setting 1: $w_{11} = 1.0, w_{12} = 1.0, w_{21} = 1.0$ and $w_{22} = 1.0$.
- Setting 2: $w_{11} = 5.0, w_{12} = 1.0, w_{21} = 1.0$ and $w_{22} = 1.0$.
- Setting 3: $w_{11} = 1.0, w_{12} = 5.0, w_{21} = 1.0$ and $w_{22} = 1.0$.
- Setting 4: $w_{11} = 1.0, w_{12} = 1.0, w_{21} = 5.0$ and $w_{22} = 1.0$.
- Setting 5: $w_{11} = 1.0, w_{12} = 1.0, w_{21} = 1.0$ and $w_{22} = 5.0$.

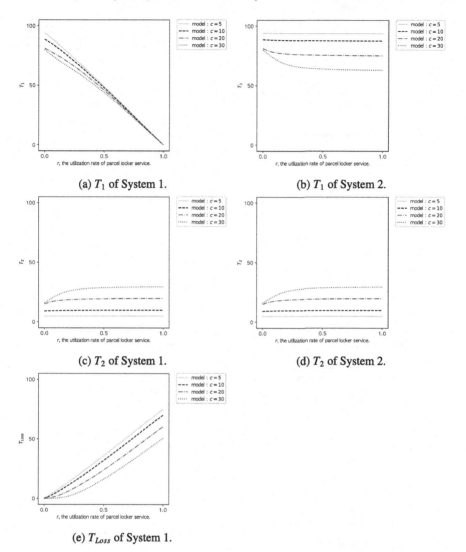

(a) T_1 of System 1.

(b) T_1 of System 2.

(c) T_2 of System 1.

(d) T_2 of System 2.

(e) T_{Loss} of System 1.

Fig. 6. Numerical results of T_1, T_2 and T_{Loss}.

The results of numerical experiments under these settings are as shown in Fig. 7. It should be noted that Fig. 7a assumes $w_{11} = w_{12} = w_{21} = w_{22} = 1.0$. Each graph decreases monotonically with the increase of r, and approaches a certain value. Also, we observe that the border value of w_3 decreases with the increase in c. Figure 7b shows the case $w_{11} = 5$ while other costs are the same as in Fig. 7a and we observe that the tendency is more pronounced. This difference might be due to the difference in $\mathbb{E}[N_1(t)]$ among the two systems. Figure 7c is for the case $w_{12} = 5$ and other costs are the same as in Setting 1. From the result, it shows a monotonically increasing tendency with respect to the increase of r when $c = 30$.

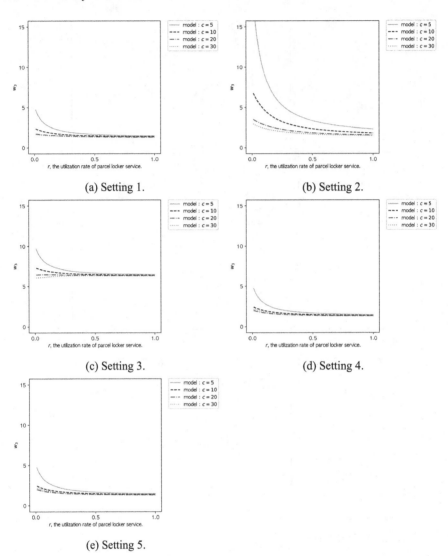

Fig. 7. Numerical results of w_3.

In this regard, when c is large, the difference in T_1 between the two systems is small when r is small, but the difference enlarges when r is large. This is the reason while the curve for $c = 30$ increases with the increase in r. Also, Fig. 7d and e give almost the same results as Fig. 7a. This is because the difference of $\mathbb{E}[N_2(t)]$ and T_2 between the two systems is small.

5 Conclusions

We have modeled the PDS system by two queueing systems. In System 1, blocked parcels are lost and thus are proceeded outside the system while in System 2, blocked parcels are redelivered by the truck. Each system contains the parcel locker that is considered as one of the solutions for the redelivery problem. By analyzing these systems, we have evaluated the impact of the use of the parcel locker on the performance measures such as the blocking probability of the parcel locker and the mean number of parcels. Moreover, based on a cost function, we have found that if the cost of lost parcels is smaller (or bigger) than a certain value, System 1 (or System 2) is better.

Finally, we consider the future works. In this study, we have modeled the PDS which is composed by some simple elements. However, the real PDS is more complicated so that we should consider more elements so as to be closer to the reality. For instance, we have not considered the loading time and the number of parcels on the truck, but these expansions may allow more detailed analysis. Furthermore, we have assumed an exponential distribution of delivery times and arrival intervals of parcels, but parcel arrivals are often batch arrivals and delivery time do not always follow an exponential distribution. Hence, we may be able to make a more general model by considering batch arrival of parcels or general distribution of delivery time intervals. It is also worth using simulations to verify the differences that arise from changing these assumptions.

Acknowledgements. This study was supported by the SamuRAI (Strategic Advancement of Multi-Purpose Ultra-Human Robot and Artificial Intelligence Technologies) project commissioned by the New Energy and Industrial Technology Development Organization (NEDO), Japan, the joint program of the University of Tsukuba and Toyota Motor Corporation, titled "Research on the next generation social systems and mobilities." and JSPS KAKENHI Grant Number 18K18006. Also, in order to conduct this study, we interviewed Yamato Transport Co., Ltd., a major Japanese courier company, in Saitama Prefecture on November 20, 2018. We thank the company for giving us various insights in the interview.

References

1. Ministry of Land, Infrastructure, Transport and Tourism, Japan: Toward the reduction of redelivery. http://www.mlit.go.jp/seisakutokatsu/freight/re_delivery_reduce. html. Accessed 17 Mar 2019, in Japanese
2. Iwan, S., Kijewska, K., Lemke, J.: Analysis of parcel lockers' efficiency as the last mile delivery solution-the results of the research in Poland. Transp. Res. Proc. **12**, 644–655 (2016)
3. Lachapelle, U., Burke, M., Brotherton, A., Leung, A.: Parcel locker systems in a car dominant city: location, characterisation and potential impacts on city planning and consumer travel access. J. Transp. Geogr. **71**, 1–14 (2018)
4. Wang, Y., Guo, J., Ceder, A.A., Currie, G., Dong, W., Yuan, H.: Waiting for public transport services: queueing analysis with balking and reneging behaviors of impatient passengers. Transp. Res. Part B: Methodol. **63**, 53–76 (2014)

5. Phung-Duc, T.: Multiserver queues with finite capacity and setup time. In: Gribaudo, M., Manini, D., Remke, A. (eds.) ASMTA 2015. LNCS, vol. 9081, pp. 173–187. Springer, Cham (2015). https://doi.org/10.1007/978-3-319-18579-8_13

6. Schrage, L.E.: The queue M/G/1 with feedback to lower priority queues. Manag. Sci. **13**(7), 466–474 (1967)

7. Coffman, E.G., Kleinrock, L.: Feedback queueing models for time-shared systems. J. ACM (JACM) **15**(4), 549–576 (1968)

8. Takine, T.: Beyond M/M/1 – invitation to quasi-birth-and-death processes. Commun. Oper. Res. Soc. Jpn. **59**(4), 179–184 (2014). In Japanese

A MAP-Based Performance Analysis on an Energy-Saving Mechanism in Cloud Computing

Xuena Yan[1,2], Shunfu Jin[1(✉)], Wuyi Yue[3], and Yutaka Takahashi[4]

[1] School of Information Science and Engineering, Yanshan University,
Qinhuangdao 066004, China
yanxuena@126.com, jsf@ysu.edu.cn
[2] Langfang Yanjing Vocational Technical College,
Langfang 065200, China
[3] Department of Intelligence and Informatics, Konan University,
Kobe 658-8501, Japan
yue@konan-u.ac.jp
[4] Graduate School of Informatics, Kyoto University,
Kyoto 606-8501, Japan
takahashi@i.kyoto-u.ac.jp

Abstract. For the purpose of satisfying the service level agreement of cloud users while at the same time reducing the energy consumption in cloud computing, we analyze the system performance of an energy-saving mechanism with a synchronous sleep mode. Considering the correlation of traffic in cloud computing, the arrival of the data requests is described as a Markovian Arrival Process (MAP), and a MAP/M/N/$N+K$ queue with synchronous multiple-vacations is established to evaluate the energy-saving mechanism. Taking into account the random cloud environment, we construct a state transition rate matrix and analyze the queueing model in the steady state. Accordingly, we derive the energy conservation level and the average latency of tasks to evaluate the energy-saving mechanism. Finally, numerical experiments are provided to illustrate the impact of the sleep parameter on the system performance.

Keywords: Cloud computing · Energy-saving mechanism ·
Markovian Arrival Process · Synchronous vacation

1 Introduction

Cloud computing is a state-of-the-art paradigm, and the cloud service market is estimated to be valued at \$383.4 billion US dollars by 2020 [1]. Construction size

This work was supported in part by National Science Foundations (Nos. 61872311, 61472342) and Natural Science Foundation of Hebei Province (F2017203141), China, and was supported in part by MEXT and JSPS KAKENHI Grant (No. JP17H01825), Japan.

T. Phung-Duc et al. (Eds.): QTNA 2019, LNCS 11688, pp. 369–378, 2019.
https://doi.org/10.1007/978-3-030-27181-7_22

and power consumption of the cloud data centers worldwide will dramatically increase, and a considerable amount of greenhouse gases will be emitted into the environment [2,3]. Constructing and evaluating green cloud computing has become an extremely challenging research issues.

Queueing theory has made significant contributions to performance studies on cloud computing. In [4], an M/G/c queueing model was constructed for analyzing cloud computing centers, and the mean response time of a task for different offered loads was shown with analytical and simulation results. In [5], aiming to determine and measure the Quality of Service (QoS) guarantees of cloud users, an open Jackson network was built to provide cloud users with the best service option. In [6], Cheng et al. used a vacation queue with exhaustive service to model the task schedule of a heterogeneous cloud computing system and proposed a task scheduling algorithm with similar traffic to reduce the energy consumption. In [7], a task scheduling strategy with a sleep-delay timer and a waking-up threshold was proposed in cloud computing. The task scheduling strategy was modeled as an M/M/c synchronous vacation queue with a vacation-delay and a N-policy. The energy conservation level of this proposed strategy was evaluated through numerical experiments. In [8], a second optional service queue was proposed to model the service provided by cloud vendors. The parameters of the cloud service queue was optimized by establishing a two-dimensional Markov chain. In all the literature mentioned above, the arrival of customers was assumed to follow a Poisson process. However, using a Markovian Arrival Process (MAP) would be more appropriate for capturing the stochastic behavior of correlated data requests in cloud computing.

There have been a lot of studies on MAP-based queues in recent years since a MAP is more universal than a Poisson distribution. In [9], a queueing model with MAP arrivals and negative customers was studied. In the model, there were two classes of removal rules: the arrival of a negative customer removed all the customers in the system; the arrival of a negative customer removed only one customer from the head of the system. In [10], a MAP/M/c queue was studied. Numerical results illustrated the relationship of the loss probability, the mean waiting time and the mean queue size to different impatience times. In [11], a MAP/M/N retrial queueing model with asynchronous single-vacations was constructed to calculate the loss probability, the blocking probability, the expected queue length and the actual arrival rate of customers entering the servers. Numerical examples illustrated the relationship of the mean number of customers in the orbit and the mean number of customers under service with different fundamental arrival rates of MAP and different arrival processes. Although there is a body of literature dedicated to MAP-based queues, there have been few works on the MAP queueing model in cloud computing. Unlike the research mentioned above, in [12], Dudina et al. proposed a MAP/M/N/N+K queueing model in a random environment to evaluate a call center. The Laplace-Stieltjes transform of the sojourn time was derived by using the method of collective marks.

As far as we know, MAP-based queueing models have not been applied to cloud computing. In this paper, we investigate the energy-saving mechanism in cloud computing by building a type of MAP-based queueing model. The rest structure of this paper is organized as follows. In Sect. 2, for application in an energy-saving mechanism in a random cloud environment, we construct a MAP-based queueing model with multiple servers and a synchronous multiple-vacations. In Sect. 3, we analyze the queueing model through constructing the infinitesimal generator of the four-dimensional Markov chain. In Sect. 4, the performance measures for the energy-saving mechanism in cloud computing are derived. Then, in Sect. 5, we carry out experiments and elaborate to evaluate the energy-saving mechanism. Finally, we make our conclusions in Sect. 6.

2   System Model

Cloud computing deploys multiple Virtual Machines (VMs) on one Physical Machine (PM) with the help of virtualization technology and provides services to users in the form of a resource pool, so as to allocate cloud resources more efficiently [13]. Considering the energy conservation of PMs in cloud computing, we investigate an energy-saving mechanism with a sleep mode in a random cloud environment. In this energy-saving mechanism, if all the VMs on a PM are idle, all the VMs hosted on a PM together with the PM itself (abbreviated to PM) will simultaneously switch to a sleep period controlled by a sleep timer with a random length. If no task arrives before the sleep timer expires, the PM will switch to another sleep period with a new sleep timer. Otherwise, the PM will be awakened after the sleep timer expires.

We regard the tasks to be submitted to the cloud data center as customers, the VMs as servers, the buffer as a waiting space and the sleep period as a vacation. A queueing model with multiple servers and a synchronous multiple-vacations is established.

Let N be the number of VMs and K be the buffer size. At an arbitrary task arrival epoch: (i) if the VMs are asleep and the buffer is not full, the arriving task will queue in the buffer; (ii) if the VMs are awake and there is at least one idle VM, the arriving task will join the system and occupy one of the idle VMs immediately; (iii) if the VMs are awake and there are i, $i \in \{N, N + 1, \ldots, N + K - 1\}$ tasks in the system, the arriving task has to queue in the buffer; (iv) if the buffer is full, the arriving task will be balked by the system.

The operation of the queueing model is dependent on the state of the random cloud environment. The random cloud environment is described as a stochastic process $r_t, t \geq 0$, which is a homogeneous irreducible continuous-time Markov chain. The state space of r_t is $E_r = \{1, 2, \ldots, R\}$ and the infinitesimal generator of r_t is $\boldsymbol{H}$.

The arrival of tasks is supposed to follow a MAP. The arrival of tasks is directed by the stochastic process $\nu_t, t \geq 0$ with state space $E_\nu = \{0, 1, \ldots, W\}$. Under a fixed environment state r, $r \in E_r$, the stochastic process ν_t is an irreducible continuous-time Markov chain. The time duration for the Markov chain sojourning in state ν, $\nu \in E_\nu$ is supposed to follow an exponential distribution with a positive parameter $\lambda_\nu^{(r)}$. At the probability $p_0^{(r)}(\nu, \nu')$, $\nu, \nu' \in E_\nu, \nu \neq \nu', r \in E_r$, the Markov chain ν_t jumps to state ν' from state ν without any task arrival, and at the probability $p_1^{(r)}(\nu, \nu')$, $\nu, \nu' \in E_\nu, r \in E_r$, the Markov chain ν_t jumps to state ν' from state ν with one task arrival. The arrival pattern of tasks is characterized by the matrices $\boldsymbol{D}_0^{(r)}$ and $\boldsymbol{D}_1^{(r)}$, which are given by

$$
\boldsymbol{D}_0^{(r)} = \begin{bmatrix} -\lambda_0^{(r)} & \lambda_0^{(r)} p_0^{(r)}(0,1) & \cdots & \lambda_0^{(r)} p_0^{(r)}(0,W) \\ \lambda_1^{(r)} p_0^{(r)}(1,0) & -\lambda_1^{(r)} & \cdots & \lambda_1^{(r)} p_0^{(r)}(1,W) \\ \vdots & \vdots & \ddots & \vdots \\ \lambda_W^{(r)} p_0^{(r)}(W,0) & \lambda_W^{(r)} p_0^{(r)}(W,1) \cdots & & -\lambda_W^{(r)} \end{bmatrix},
$$

$$
\boldsymbol{D}_1^{(r)} = \begin{bmatrix} \lambda_0^{(r)} p_1^{(r)}(0,0) & \lambda_0^{(r)} p_1^{(r)}(0,1) & \cdots & \lambda_0^{(r)} p_1^{(r)}(0,W) \\ \lambda_1^{(r)} p_1^{(r)}(1,0) & \lambda_1^{(r)} p_1^{(r)}(1,1) & \cdots & \lambda_1^{(r)} p_1^{(r)}(1,W) \\ \vdots & \vdots & \ddots & \vdots \\ \lambda_W^{(r)} p_1^{(r)}(W,0) & \lambda_W^{(r)} p_1^{(r)}(W,1) \cdots & & \lambda_W^{(r)} p_1^{(r)}(W,W) \end{bmatrix}.
$$

The matrix $\boldsymbol{D}^{(r)} = \boldsymbol{D}_0^{(r)} + \boldsymbol{D}_1^{(r)}$ represents the infinitesimal generator of Markov chain ν_t. The average arrival rate $\lambda^{(r)}$ of tasks can be calculated by

$$
\lambda^{(r)} = \boldsymbol{\theta}^{(r)} \boldsymbol{D}_1^{(r)} e
$$

where $\boldsymbol{\theta}^{(r)}$ is the invariant vector of the Markov chain ν_t with generator matrix $\boldsymbol{D}^{(r)}$. The vector $\boldsymbol{\theta}^{(r)}$ can be calculated by $\boldsymbol{\theta}^{(r)} \boldsymbol{D}^{(r)} = \boldsymbol{0}$ and $\boldsymbol{\theta}^{(r)} e = 1$. Here e is a column vector of appropriate size consisting of 1 and $\boldsymbol{0}$ is a row vector of appropriate size consisting of 0.

Under a fixed environment state r, $r \in E_r$, the service times of tasks are supposed to follow an exponential distribution with a positive parameter μ_r. The time durations of sleep periods are supposed to be exponentially distributed with a positive parameter α. The service time of a task, the arrival epoch of a task and the time duration of a sleep period are supposed to be independent of each other.

3   Model Analysis

Let x_t, $x_t \in \{0,1\}$ be the VM states at epoch $t, t \geq 0$. $x_t = 0$ means that the VMs are asleep, and $x_t = 1$ means that the VMs are awake. Let y_t, $y_t \in \{0,1,\ldots,N+K\}$ be the number of tasks in the queueing system, including the tasks being served on VMs and the tasks waiting in the buffer, at epoch $t, t \geq 0$. We call y_t the system level for convenience. Let h_t, $h_t \in E_r$ be the state of the random cloud environment at epoch $t, t \geq 0$, and z_t, $z_t \in E_\nu$ be the state of MAP at epoch $t, t \geq 0$. Thus, the system model under consideration can be described in terms of the regular irreducible continuous-time four-dimensional Markov chain $\xi_t = \{x_t, y_t, h_t, z_t\}$, $t \geq 0$ with infinitesimal generator matrix $\boldsymbol{Q}$.

The stationary distribution of the four-dimensional Markov chain ξ_t is defined as follows:

$$\pi_{j,i,r,\nu} = \lim_{t\to\infty} P\{x_t = j, y_t = i, h_t = r, z_t = \nu\},$$
$$j = 0, 1, i \in \{0, 1, \ldots, N+K\}, r \in E_r, \nu \in E_\nu.$$

Let $\boldsymbol{\pi}_j$ be the stationary probability vector for the VMs being at state j. When $j = 0$:

$$\boldsymbol{\pi}_0 = (\boldsymbol{\pi}_{0,0}, \boldsymbol{\pi}_{0,1}, \ldots, \boldsymbol{\pi}_{0,K})$$

where $\boldsymbol{\pi}_{0,i}$ is the stationary probability vector for the VMs being at state $j = 0$ and the system being at level i, $i = 0, 1, \ldots, K$. $\boldsymbol{\pi}_{0,i}$ is given as

$$\boldsymbol{\pi}_{0,i} = (\pi_{0,i,1,0}, \pi_{0,i,1,1}, \ldots, \pi_{0,i,1,W}, \pi_{0,i,2,0}, \pi_{0,i,2,1}, \ldots, \pi_{0,i,2,W}, \ldots,$$
$$\pi_{0,i,R,0}, \pi_{0,i,R,1}, \ldots, \pi_{0,i,R,W}), \quad i = 0, 1, \ldots, K.$$

When $j = 1$:

$$\boldsymbol{\pi}_1 = (\boldsymbol{\pi}_{1,1}, \boldsymbol{\pi}_{1,2}, \ldots, \boldsymbol{\pi}_{1,N+K})$$

where $\boldsymbol{\pi}_{1,i}$ is the stationary probability vector for the VMs being at state $j = 1$ and the system being at level i, $i = 1, 2, \ldots, N+K$. $\boldsymbol{\pi}_{1,i}$ is given as

$$\boldsymbol{\pi}_{1,i} = (\pi_{1,i,1,0}, \pi_{1,i,1,1}, \ldots, \pi_{1,i,1,W}, \pi_{1,i,2,0}, \pi_{1,i,2,1}, \ldots, \pi_{1,i,2,W}, \ldots,$$
$$\pi_{1,i,R,0}, \pi_{1,i,R,1}, \ldots, \pi_{1,i,R,W}), \quad i = 1, 2, \ldots, N+K.$$

Then, the stationary probability vector $\boldsymbol{\Pi}$ of the four-dimensional Markov chain ξ_t is shown as follows:

$$\boldsymbol{\Pi} = (\boldsymbol{\pi}_0, \boldsymbol{\pi}_1).$$

The stationary distribution $\boldsymbol{\Pi}$ can be obtained by solving the system of linear equations

$$\boldsymbol{\Pi Q} = \mathbf{0} \tag{1}$$

subject to the condition of $\boldsymbol{\Pi e} = 1$.

The infinitesimal generator matrix $\boldsymbol{Q}$ can be given in a 2×2 block-structure form as follows:

$$Q = \begin{bmatrix} Q_{0,0} & Q_{0,1} \\ Q_{1,0} & Q_{1,1} \end{bmatrix}$$

where $\boldsymbol{Q}_{b,c}$, $b, c = 0, 1$ indicates that the VMs change to state c from state b.

In order to analyze the non-zero sub-blocks of $\boldsymbol{Q}_{b,c}$, we introduce some notations as follows:

$\boldsymbol{I}_{\overline{W}}$: an identity matrix of $W + 1$ dimension;
$\boldsymbol{I}_R$: an identity matrix of R dimension;
$\boldsymbol{I}_{R \times \overline{W}}$: an identity matrix of $R \times (W+1)$ dimension;
$\boldsymbol{O}$: a zero matrix of appropriate dimension;
$\otimes$: the symbol Kroneckers product;

$$\tilde{D}_l = diag\left\{D_l^{(r)}, r \in E_r\right\}, l = 0, 1;$$
$$A = diag\{\mu_r, r \in E_r\}.$$

Submatrix $Q_{0,0}$ can be given in a $(K+1) \times (K+1)$ block-structure form as follows:

$$Q_{0,0} = \begin{bmatrix} L_{0,0} & L_{0,1} & & & \\ & L_{1,1} & L_{1,2} & & \\ & & \ddots & & \ddots & \\ & & & L_{K-1,K-1} & L_{K-1,K} \\ & & & & L_{K,K} \end{bmatrix}$$

where $L_{s,d}$, $s, d \in \{0, 1, \ldots, K\}$ represents the transition rate submatrix from the system level s to d when the VMs stay in the sleep state. $L_{s,d}$ is given by

$$L_{0,0} = \tilde{D}_0 + H \otimes I_{\overline{W}},$$
$$L_{s,s} = \tilde{D}_0 + (H - \alpha I_R) \otimes I_{\overline{W}}, \quad s \in \{1, 2, \ldots, K-1\},$$
$$L_{K,K} = \tilde{D}_0 + \tilde{D}_1 + (H - \alpha I_R) \otimes I_{\overline{W}},$$
$$L_{s,s+1} = \tilde{D}_1, \quad s \in \{0, 1, \ldots, K-1\}.$$

Submatrix $Q_{1,1}$ can be given in a $(N+K) \times (N+K)$ block-structure form as follows:

$$Q_{1,1} = \begin{bmatrix} U_{1,1} & U_{1,2} & & & & \\ U_{2,1} & U_{2,2} & U_{2,3} & & & \\ & \ddots & \ddots & & \ddots & \\ & & U_{N+K-1,N+K-2} & U_{N+K-1,N+K-1} & U_{N+K-1,N+K} \\ & & & U_{N+K,N+K-1} & U_{N+K,N+K} \end{bmatrix}$$

where $U_{s,d}$, $s, d \in \{1, 2, \ldots, N+K\}$ represents the transition rate submatrix from the system level s to d when the VMs remain awake. $U_{s,d}$ can be calculated as follows:

$$U_{s,s} = \tilde{D}_0 + (H - sA) \otimes I_{\overline{W}}, \quad s \in \{1, 2, \ldots, N\},$$
$$U_{s,s} = \tilde{D}_0 + (H - NA) \otimes I_{\overline{W}}, \quad s \in \{N+1, N+2, \ldots, N+K-1\},$$
$$U_{N+K,N+K} = \tilde{D}_0 + \tilde{D}_1 + (H - NA) \otimes I_{\overline{W}},$$
$$U_{s,s-1} = sA \otimes I_{\overline{W}}, \quad s \in \{2, 3, \ldots, N\},$$
$$U_{s,s-1} = (NA) \otimes I_{\overline{W}}, \quad s \in \{N+1, \ldots, N+K\},$$
$$U_{s,s+1} = \tilde{D}_1, \quad s \in \{1, 2, \ldots, N+K-1\}.$$

Submatrix $Q_{0,1}$ represents the transition rate submatrix where the system is awakened from a sleep period. $Q_{0,1}$ can be given in a $(K+1) \times (N+K)$ block-structure form as follows:

$$Q_{0,1} = \begin{bmatrix} O & & & & \\ & \alpha I_{R \times \overline{W}} & & & \\ & & \ddots & & \\ & & & \alpha I_{R \times \overline{W}} & \\ & & & & \alpha I_{R \times \overline{W}} \end{bmatrix}.$$

Submatrix $Q_{1,0}$ represents the transition rate submatrix where the system goes to sleep from an awake state. $Q_{1,0}$ can be given in a $(N+K) \times (K+1)$ block-structure form as follows:

$$Q_{1,0} = \begin{bmatrix} A \otimes I_{\overline{W}} & & & \\ & O & & \\ & & \ddots & \\ & & & O \\ & & & O \end{bmatrix}.$$

There are mainly two methods for computing the stationary distribution of finite state Markov chains: the direct method and the iterative method. If the dimension of the infinitesimal generator Q is small, it can be easily solved using the direct method. Otherwise, it can be solved using the iterative method. In this paper, we use the Gauss-Seidel method, one of the iterative methods, to calculate the stationary distribution Π of the four-dimensional Markov chain ξ_t.

4 Performance Measures

In order to evaluate the system performance, we derive some performance measures in terms of the energy conservation level and the average latency of tasks in the system.

We define the energy conservation level as the energy conservation per unit time for the VMs in cloud computing with an energy-saving mechanism. Let C_v be the energy consumption per time unit when the system is in a sleep period, and C_b be the energy consumption per time unit when the system is in the awake state. The energy conservation level ω can be calculated by the criterion

$$\omega = (C_b - C_v) \, \pi_0 e \tag{2}$$

where π_0 is the probability that the VMs are asleep.

We define the average latency of tasks as the sum of the average waiting time of tasks in the buffer and the average service time of tasks on the VMs. By Little's law, the average latency σ of tasks in the system can be calculated as follows:

$$\sigma = \left(\sum_{i=1}^{K} i\pi_{0,i} e + \sum_{i=1}^{N+K} i\pi_{1,i} e \right) / \lambda^{out} \tag{3}$$

where λ^{out} is the effective arrival rate of the tasks. λ^{out} is calculated as

$$\lambda^{out} = \sum_{i=1}^{N+K} \min\{i, N\} \, \boldsymbol{\pi}_{1,i} \left(\boldsymbol{A} \otimes \boldsymbol{I}_{\overline{W}}\right) \boldsymbol{e}.$$

5   Numerical Results

In order to evaluate the energy-saving mechanism, we provide experiments to demonstrate the relationship between the performance measures and the sleep parameter.

Figure 1 demonstrates the relationship between the energy conservation level ω and the sleep parameters α.

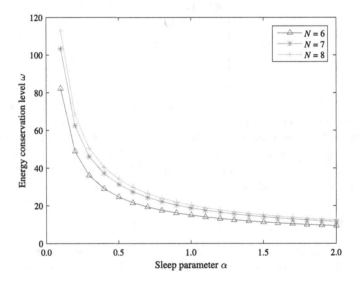

Fig. 1. Change trend of energy conservation level ω.

From Fig. 1, we observe that for the same number N of VMs, the energy conservation level ω will decrease as the sleep parameter α increases. When the sleep parameter increases, the VMs change to the awake state from a sleep period earlier, so the VMs are less likely to be asleep. Thus, the energy consumption of VMs will increase, and the energy conservation of the system will decrease.

In addition, it appears also from Fig. 1 that for the same sleep parameter α, the energy conservation level ω will increase as the VM number N increases. The larger the VM number is, the stronger the system service ability is, so the quicker the system will empty. Consequently, the system will be more likely to be asleep, and the energy conservation of the system will increase.

Figure 2 shows the relationship between the average latency σ of tasks and the sleep parameter α.

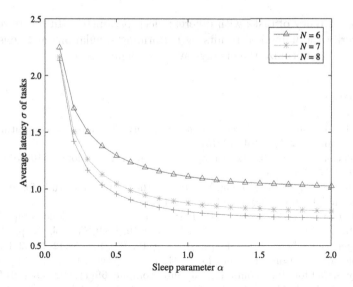

Fig. 2. Change trend of average latency σ.

From Fig. 2, we find that for the same number N of VMs, the average latency σ of tasks will decrease as the sleep parameter α increases. We note that the tasks arriving while the system is in the sleep state have to wait for the expiration of the sleep timer before getting service. The bigger the sleep parameter is, the earlier the VMs are awakened. Thus, the shorter time the tasks have to wait during sleep state, and the average latency of arrival tasks will decrease.

On the other hand, from Fig. 2, we also observe that for the same sleep parameter α, the average latency σ of tasks will decrease as the VM number N increases. This is because the larger the VM number is, the stronger the system service ability is. Consequently, the shorter time a task will wait in the buffer, and the average latency of tasks will decrease.

Comparing Figs. 1 and 2, we can see that there is a trade-off between the energy conservation level and the average latency of tasks when setting the sleep parameter in the energy-saving mechanism.

6   Conclusions

Considering the correlation of cloud traffic, we evaluated the energy-saving mechanism in random cloud computing. We assumed that the arrival of cloud data requests follows a MAP, and we established a $MAP/M/N/N+K$ queue with synchronous multiple-vacations to evaluate the energy-saving mechanism. Based on the model analysis, we derived performance measures in terms of the energy conservation level and the average latency of tasks for the energy-saving mechanism. With correlated traffic in a random cloud environment, we carried out numerical experiments to illustrate the change trends of the energy conservation level and the average latency of tasks with different sleep parameters. In future research, we will verify the numerical results by performing simulations and construct a cost function to optimize the energy-saving mechanism.

References

1. Mastelić, T., Brandić, I.: Recent trends in energy-efficient cloud computing. IEEE Cloud Comput. **2**(1), 40–47 (2015)
2. Publications Office of the European Union. https://doi.org/10.2760/358256. Accessed 17 Mar 2019
3. Jing, S., Ali, S., She, K., Zhong, Y.: State-of-the-art research study for green cloud computing. J. Supercomput. **65**(1), 445–468 (2013)
4. Khazaei, H., Mišić, J., Mišić, V.B.: Performance analysis of cloud computing centers. In: Zhang, X., Qiao, D. (eds.) QShine 2010. LNICST, vol. 74, pp. 251–264. Springer, Heidelberg (2012). https://doi.org/10.1007/978-3-642-29222-4_18
5. Vilaplana, J., Solsona, F., Teixidó, I., Mateo, J., Abella, F., Rius, J.: A queuing theory model for cloud computing. J. Supercomput. **69**(1), 492–507 (2014)
6. Cheng, C., Li, J., Wang, Y.: An energy-saving task scheduling strategy based on vacation queuing theory in cloud computing. Tsinghua Sci. Technol. **20**(1), 28–39 (2015)
7. Jin, S., Wang, X., Yue, W.: A task scheduling strategy with a sleep-delay timer and a waking-up threshold in cloud computing. In: Takahashi, Y., Phung-Duc, T., Wittevrongel, S., Yue, W. (eds.) QTNA 2018. LNCS, vol. 10932, pp. 115–123. Springer, Cham (2018). https://doi.org/10.1007/978-3-319-93736-6_8
8. Jin, S., Wu, H., Yue, W.: Pricing policy for a cloud registration service with a novel cloud architecture. Cluster Comput. **22**(1), 271–283 (2019)
9. Li, Q., Zhao, Y.: A MAP/G/1 queue with negative customers. Queueing Syst. **47**(1–2), 5–43 (2004)
10. Choi, B.D., Kim, B., Zhu, D.: MAP/M/c queue with constant impatient time. Math. Oper. Res. **29**(2), 309–325 (2004)
11. Zhou, Z., Zhou, Z.: A MAP/M/N retrial queueing model with asynchronous single vacations. In: 2018 International Conference on Virtual Reality and Intelligent Systems, Changsha, pp. 245–249. IEEE (2018)
12. Dudina, O., Dudin, S.: Queueing system MAP/M/N/$N+K$ operating in random environment as a model of call center. In: Dudin, A., Klimenok, V., Tsarenkov, G., Dudin, S. (eds.) BWWQT 2013. CCIS, vol. 356, pp. 83–92. Springer, Heidelberg (2013). https://doi.org/10.1007/978-3-642-35980-4_10
13. Jin, S., Hao, S., Yue, W.: Energy-efficient strategy with a speed switch and a multiple-sleep mode in cloud data centers. In: Yue, W., Li, Q.-L., Jin, S., Ma, Z. (eds.) QTNA 2017. LNCS, vol. 10591, pp. 143–154. Springer, Cham (2017). https://doi.org/10.1007/978-3-319-68520-5_9

Analysis of the Average Confirmation Time of Transactions in a Blockchain System

Wenjuan Zhao[1], Shunfu Jin[1($\boxtimes$)], and Wuyi Yue[2]

[1] School of Information Science and Engineering, Yanshan University,
Qinhuangdao 066004, China
zwj8569@163.com, jsf@ysu.edu.cn
[2] Department of Intelligence and Informatics, Konan University,
Kobe 658-8501, Japan
yue@konan-u.ac.jp

Abstract. Blockchain is well known as a database technology support-
ing digital currencies, such as Bitcoin, Ether, and Ripple. Regarding
the mining process as a vacation, and the block-verification process as
a service, we establish a type of non-exhaustive queueing model with a
limited batch service and a possible zero-transaction service. By selecting the beginning instant of a block-verification process as a regeneration
point and using the method of a generating function, we obtain the stationary probability distribution for the number of transactions in the
system at the regeneration points. Then, we derive the average number
of transactions and the average confirmation time of transactions in the
blockchain system. Finally, we provide numerical results with analysis
and simulation to demonstrate how the average number of transactions
and the average confirmation time of transactions in the blockchain system change with the mining parameter.

Keywords: Blockchain · Regeneration point · Generating function ·
Average confirmation time

1 Introduction

Blockchain is a decentralized distributed ledger that does not allow deletion of
data [1]. Compared with traditional accounting techniques, blockchain has many
obvious advantages, such as irreversibility, anonymity, and autonomy [2]. Since
the terminology of blockchain was first presented by Nakamoto in [3], blockchain
has continued to develop. In recent years, considerable efforts have been devoted
to the study of the blockchain system.

This work was supported in part by National Science Foundations (Nos. 61872311,
61472342) and Natural Science Foundation of Hebei Province (F2017203141), China,
and was supported in part by MEXT and JSPS KAKENHI Grant (No. JP17H01825),
Japan.

T. Phung-Duc et al. (Eds.): QTNA 2019, LNCS 11688, pp. 379–388, 2019.
https://doi.org/10.1007/978-3-030-27181-7_23

One of the research topics in the field of the blockchain system is the application technology. In [4], Konstantinos et al. investigated how blockchain and smart contracts work on the Internet of Things (IoT), and pointed out certain issues that should be considered before the deployment of a blockchain network. In [5], Huang et al. presented a decentralized security model, called the lightning network and smart contracts (LNSC). With this model, the security of trades between electric vehicles (EVs) and charging piles can be effectively enhanced. In [6], with the help of the blockchain technology, Muhamed et al. proposed a global higher education credit platform, named EduCTX. The establishment of this platform was the first step toward a higher education system with transparency and technical advancement. However, one important issue that has been overlooked by the afore-mentioned research is how to evaluate and improve the performance of the blockchain system.

From a queueing theory point of view, some works have appeared concerning the performance evaluation of the blockchain system. In [7] and [8], Kasahara et al. established a single-server queue with a batch service and a priority mechanism based on the Bitcoin system. By using the method of a supplementary variable, they derived the average transaction-confirmation time. With numerical experiment results, they were able to quantitatively evaluate the effects of the block size on the transaction-confirmation time. However, in the research above, neither the process of solving the puzzle-like problem, nor the implementation of the coinbase in the blockchain system was taken into account.

In [9], Li et al. built a Markovian batch-service queueing system with two different service stages and derived the stationary probability vector of the system. They obtained formulas for the average number of transactions in the queue, the average number of transactions in a block and the average confirmation time of transactions, yet the coinbase transaction was also omitted in this model, and the model analysis was short of generality due to the assumption of exponentially distributed service time.

In this paper, by considering the process for solving the puzzle-like problem and the implementation of the coinbase in the blockchain system, we present a modeling approach to analyze the average number of transactions and the average confirmation time of transactions in the blockchain system.

The rest of this paper is organized as follows. In Sect. 2, based on the mining cycle in the blockchain system, we establish a type of non-exhaustive queueing model with a limited batch service and a possible zero-transaction service. In Sect. 3, we carry out an analysis of the system model and derive the average number of transactions and the average confirmation time of transactions in the blockchain system. In Sect. 4, we provide numerical results with analysis and simulation. In Sect. 5, we summarize the conclusions.

2   System Model

In this section, we establish a type of non-exhaustive queueing model with a limited batch service and a possible zero-transaction service to model the mining cycle in a blockchain system.

Transactions, nodes and blocks are the basic components of a blockchain system. In order to quantitatively evaluate the response performance of a blockchain system, we need to establish a mathematical model to more realistically capture the stochastic behaviors of transactions. Note that the size of a block is limited and the block can not be confirmed during the mining process in a blockchain system. A non-exhaustive queue is naturally suitable to capture the working flow of a mining cycle in a blockchain system.

Once a newly generated block is connected to the blockchain, i.e. a block-verification process ends, a mining process begins no matter whether there are transactions waiting in the system. Moreover, when a puzzle-like problem is solved, i.e. a mining process ends, a zero-transaction begins if the newly generated block is empty, otherwise, a normal service period begins.

Considering that transactions in the newly generated block are simultaneously validated and the newly generated block is possibly empty, we establish a type of non-exhaustive queueing model with a limited batch service and a possible zero-transaction service. In this queueing model, the mining process is regarded as a vacation, the block-verification process is regarded as a service.

We assume that the arrivals of transactions follow a Poisson process with the parameter λ ($\lambda > 0$).

We assume that the time duration V for a mining process is an independent and identically distributed (i.i.d) random variable and follows a general distribution with a distribution function $V(t)$. The Laplace-Stieltjes Transform (LST) $V^*(s)$, the mean value $E[V]$ and the second moment $E[V^2]$ of the time duration V for a mining process are given as follows:

$$V^*(s) = \int_0^\infty e^{-st}\, dV(t), \quad E[V] = \frac{1}{\theta} = \int_0^\infty t\, dV(t), \quad E[V^2] = \int_0^\infty t^2\, dV(t).$$

We assume that the time duration S for a block-verification process is an i.i.d random variable and follows a general distribution with a distribution function $S(t)$. The LST $S^*(s)$, the mean value $E[S]$ and the second moment $E[S^2]$ of the time duration S for a block-verification process are given as follows:

$$S^*(s) = \int_0^\infty e^{-st}\, dS(t), \quad E[S] = \frac{1}{\mu} = \int_0^\infty t\, dS(t), \quad E[S^2] = \int_0^\infty t^2\, dS(t).$$

3   Model Analysis

3.1   Number of Transactions at the Regeneration Points

Let $Q^{(n)}$ be the number of transactions in the system at the beginning instant of the nth block-verification process. We select the beginning instant of a block-verification process as a regeneration point. The numbers of transactions in the system at the regeneration points constitute a Markov chain $\{Q^{(n)}, n \geq 1\}$.

The stability condition of $\{Q^{(n)}, n \geq 1\}$ is given as follows:

$$\lambda(E[S] + E[V]) < b$$

where b is the maximum block capacity.

The transition probability of this Markov chain is given by

$$p_{jk} = P\left\{Q^{(n+1)} = k \mid Q^{(n)} = j\right\}$$

$$= \begin{cases} \int_0^\infty \dfrac{(\lambda t)^k}{k!} e^{-\lambda t}\, dS * V(t), & j < b \\ \int_0^\infty \dfrac{(\lambda t)^{k-j+b}}{(k-j+b)!} e^{-\lambda t}\, dS * V(t), & b < j < b+k \\ 0, & j \geq b+k \end{cases} \tag{1}$$

where $S * V(t)$ is the distribution function for the convolution of S with V.

Let q_k be the probability distribution for the number of transactions at the regeneration point in the steady state. It follows that

$$q_k = \sum_{j=0}^{b-1} q_j \int_0^\infty \frac{(\lambda t)^k}{k!} e^{-\lambda t}\, dS * V(t)$$

$$+ \sum_{j=b}^{k+b} q_j \int_0^\infty \frac{(\lambda t)^{k-j+b}}{(k-j+b)!} e^{-\lambda t}\, dS * V(t). \tag{2}$$

The probability generating function $Q(z)$ for the number of transactions at the regeneration point in the steady state is given as follows:

$$Q(z) = \frac{1}{z^b} S^*(\lambda(1-z)) V^*(\lambda(1-z)) \left(\sum_{j=0}^{b-1} q_j z^b + Q(z) - Q_b(z) \right) \tag{3}$$

where

$$Q_b(z) = \sum_{k=0}^{b-1} q_k z^k.$$

Simplifying Eq. (3), we get

$$Q(z) = \frac{S^*(\lambda(1-z)) V^*(\lambda(1-z))(Q_b(1)z^b - Q_b(z))}{z^b - S^*(\lambda(1-z)) V^*(\lambda(1-z))}. \tag{4}$$

In the denominator on the right-hand side (r.h.s.) of Eq. (4), we introduce some notations as follow:

$$f(z) = z^b, \quad g(z) = -S^*(\lambda(1-z)) V^*(\lambda(1-z)).$$

Using Rouche's theorem [10] and Lagrange's theorem [11], for $\varepsilon > 0$, it can be proved that $|f(z)| > |g(z)|$ on the circle $|z| = 1+\varepsilon$, and that $f(z)$ and $f(z)+g(z)$ have the same number of zeros inside $|z| = 1+\varepsilon$. Therefore, the denominator on the r.h.s. of Eq. (4) has b roots inside $|z| = 1 + \varepsilon$. One of these roots is $z = 1$, and the other $b-1$ roots are given as follows:

$$z_r = \sum_{n=1}^\infty \frac{e^{\frac{2\pi r n}{b} i}}{n!} \frac{d^{n-1}}{dz^{n-1}} (S^*(\lambda(1-z)) V^*(\lambda(1-z)))^{n/b} \Big|_{z=0}, \quad r = 1, 2, \ldots, b-1$$

where $i = \sqrt{-1}$.

Since $Q(z)$ is analytic in $|z| \leq 1$, the numerator on the r.h.s. of Eq. (4) must also be zero at $z = z_r$ for $r = 1, 2, \ldots, b - 1$. Therefore, we have $b - 1$ equations as follows:

$$\sum_{k=0}^{b-1} q_k z_r^b - \sum_{k=0}^{b-1} q_k z_r^k = 0, \quad r = 1, 2, \ldots, b - 1. \tag{5}$$

Letting $z \to 1$ and using L'Hospital rule in Eq. (4), we get

$$1 = \frac{b Q_b(1) - Q_b'(1)}{b - \lambda E[S] - \lambda E[V]} \tag{6}$$

where $Q_b'(1)$ is the first derivative of $Q_b(z)$ at $z = 1$.

Rearranging Eq. (6) yields

$$\sum_{k=0}^{b-1} (b - k) q_k = b - \lambda E[S] - \lambda E[V]. \tag{7}$$

Based on Eqs. (5) and (7), we can numerically compute the coefficients $\{q_0, q_1, \ldots, q_{b-1}\}$ of $Q_b(z)$. Furthermore, we obtain the probability generating function $Q(z)$.

3.2   Average Confirmation Time $E[T]$ of Transactions

A mining process and the subsequent block-verification process combine to constitute a mining cycle. Let C be the time duration of a mining cycle. The LST $C^*(s)$ and the mean value $E[C]$ for the time duration C of a mining cycle are given as follows:

$$C^*(s) = S^*(s)V^*(s), \quad E[C] = E[S] + E[V].$$

The probability generating function $A_C(z)$ for the number of transactions arriving during a mining cycle is given as follows:

$$A_C(z) = S^*(\lambda(1 - z))V^*(\lambda(1 - z)). \tag{8}$$

Let D be the elapsed time of a mining cycle. Referencing to [12], the probability density function $h(t)$ of D is given as follows:

$$h(t) = \frac{1}{E[C]}(1 - C(t)) \tag{9}$$

where $C(t)$ is the distribution function for the time duration C of a mining cycle.

The probability generating function $A_D(z)$ for the number of transactions arriving during the elapsed time D of a mining cycle is given as follows:

$$A_D(z) = \sum_{i=0}^{\infty} z^i \int_0^{\infty} \frac{(\lambda t)^i}{i!} e^{-\lambda t} h(t) dt$$

$$= \frac{1 - S^*(\lambda(1 - z))V^*(\lambda(1 - z))}{\lambda(1 - z)(E[S] + E[V])}. \tag{10}$$

The probability generating function $L_s(z)$ for the number of transactions at the beginning instant of a mining cycle is given as follows:

$$
\begin{aligned}
L_s(z) &= \frac{Q(z)}{V^*(\lambda(1-z))} \\
&= \frac{S^*(\lambda(1-z))(Q_b(1)z^b - Q_b(z))}{z^b - S^*(\lambda(1-z))V^*(\lambda(1-z))}.
\end{aligned}
\tag{11}
$$

We note that the number of transactions at any moment within a mining cycle is the sum of the number of transactions at the beginning instant of a mining cycle and the number of transactions arriving during the elapsed time of the same mining cycle. The probability generating function $L(z)$ for the number of transactions at any moment is then obtained as follows:

$$
L(z) = L_s(z)A_D(z).
\tag{12}
$$

Substituting Eqs. (10) and (11) into Eq. (12), we give that

$$
L(z) = \frac{S^*(\lambda(1-z))(Q_b(1)z^b - Q_b(z))}{z^b - S^*(\lambda(1-z))V^*(\lambda(1-z))} \times \frac{1 - S^*(\lambda(1-z))V^*(\lambda(1-z))}{\lambda(1-z)(E[S] + E[V])}.
\tag{13}
$$

Taking the derivative of z, letting $z \to 1$ and using L'Hospital rule in Eq. (13), the average number $E[L]$ of transactions in the blockchain system is given as follows:

$$
\begin{aligned}
E[L] = \lambda E[S] &+ \frac{\lambda(E[S^2] + 2E[S]E[V] + E[V^2])}{2(E[S] + E[V])} \\
&+ \frac{b(b-1)(Q_b(1)-1) - Q_b''(1) + \lambda^2(E[S^2] + 2E[S]E[V] + E[V^2])}{2(b - \lambda E[S] - \lambda E[V])}
\end{aligned}
\tag{14}
$$

where $Q_b''(1)$ is the second derivative of $Q_b(z)$ at $z = 1$.

Following Little's law [13], the average confirmation time $E[T]$ of transactions is then given as follows:

$$
\begin{aligned}
E[T] = E[S] &+ \frac{E[S^2] + 2E[S]E[V] + E[V^2]}{2(E[S] + E[V])} \\
&+ \frac{b(b-1)(1 - Q_b(1)) + Q_b''(1)}{2\lambda(b - \lambda E[S] - \lambda E[V])} + \frac{\lambda(E[S^2] + 2E[S]E[V] + E[V^2])}{2(b - \lambda E[S] - \lambda E[V])}.
\end{aligned}
\tag{15}
$$

4 Numerical Experiments

In order to quantitatively evaluate the response performance of a blockchain system, we provide numerical experiments with analysis and simulation.

Experiments are carried out on Intel(R) Core(TM) i7-4790 CPU @ 3.60 GHz 3.60 GHz, 8.00 GB RAM. Analysis results are obtained in Matlab 2016a, simulation results are obtained using MyEclipse 2014. For simulation, we create a

TRANSACTION class with attributes in terms of UNARRIVE, WAIT, VALI-DATE and CONNECT to record the transaction state. We also create a BLOCK class with attributes in terms of MINE and VERIFY to record the state of a blockchain system.

By setting the arrival rate $\lambda = 0.5$, the mining parameter $\theta = 0.2, 0.4, 0.8$, the maximum block capacity $b = 4, 8$ and the block-verification rate $\mu \in [0.4, 1.4]$ as an example, we carry out experiments with analysis and simulation to investigate the change for the average number $E[L]$ of transactions in the blockchain system versus the block-verification rate μ for different mining parameter θ and different maximum block capacity b in Fig. 1.

From Fig. 1, we observe that for all the mining parameter θ and the max-imum block capacity b, as the block-verification rate μ increases, the duration of the block-verification process becomes shorter, the number of transactions arrived during the block-verification process decreases. So the average number of transactions in the blockchain system will decrease.

We also observe that for all the block-verification rate μ and the maximum block capacity b, as the mining parameter θ increases, the mining process lasts a shorter time, the number of transactions arrived during the mining process decreases. So the average number of transactions in the blockchain system will decrease.

Comparing Figs. 1(a) and (b), we find that for the same mining parameter θ and the same block-verification rate μ, as the maximum block capacity b increases, fewer transactions will sojourn in the Transaction Memory Pool. So the average number of transactions in the blockchain system will decrease.

By setting the arrival rate $\lambda = 0.5$, the block-verification rate $\mu = 0.5, 1.0, 1.5$, the maximum block capacity $b = 4, 8$ and the mining parameter $\theta \in [0.2, 1.2]$ as an example, we carry out experiments with analysis and simulation to inves-tigate the change trend for the average confirmation time $E[T]$ of transactions versus the mining parameter θ for different block-verification rate μ and different maximum block capacity b in Fig. 2.

From Fig. 2, we notice that for all the block-verification rate μ and the max-imum block capacity b, as the mining parameter θ increases, the mining process lasts a shorter time, the new block including transactions is generated earlier and get confirmed earlier. So the average confirmation time of transactions will decrease.

We also notice that for all the mining parameter θ and the maximum block capacity b, as block-verification rate μ increases, the duration of the block-verification process becomes shorter, the newly generated block including trans-actions is also connected earlier to blockchain. So the average confirmation time of transactions will decrease.

Comparing Figs. 2(a) and (b), we find that for the same mining parameter θ and the same block-verification rate μ, as the maximum block capacity b increases, transactions in the Transaction Memory Pool are put earlier into a newly generated block and get confirmed earlier. So the average confirmation time of transactions will decrease.

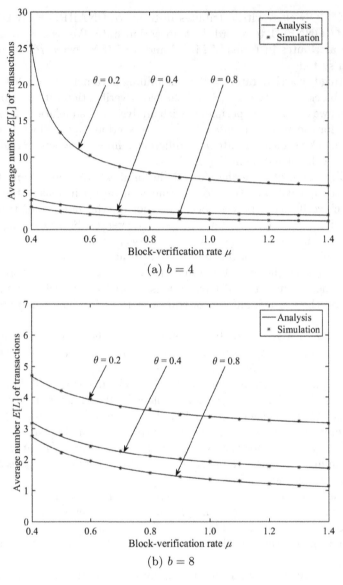

Fig. 1. Change trend for the average number $E[L]$ of transactions in the blockchain system.

The experiment results above show that when the arrival rate of transactions is given, the performance of a blockchain system is sensitive to several system parameters, such as the maximum block capacity, the block-verification rate and the mining parameter.

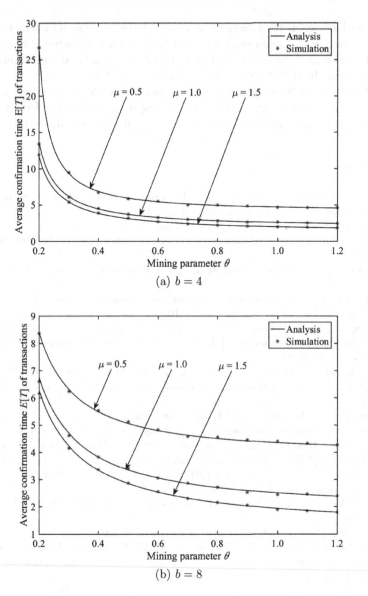

Fig. 2. Change trend for the average confirmation time $E[T]$ of transactions.

5   Conclusions

Based on the working flow of a blockchain system, in this paper, we established a type of non-exhaustive queueing model with a limited batch service and a possible zero-transaction service. By employing the methods of an embedded Markov chain and a generating function, we obtained the stationary probability for the number of transactions in the system at the regeneration points.

We derived performance measures in terms of the average number of transactions and the average confirmation time of transactions in the blockchain system. Combing analysis results and simulation results, we validated the system model and evaluated the performance of the blockchain system.

In future research, we plan to extend our study to investigate Nash equilibrium of transactions and present an appropriate remittance fee charged to transactions for maximizing the overall revenue of a blockchain system.

References

1. Wang, L., Shen, X., Li, J., Shao, J., Yang, Y.: Cryptographic primitives in blockchains. J. Netw. Comput. Appl. **127**, 43–58 (2019)
2. Qian, W., Shao, Q., Zhu, Y., Jin, C., Zhou, A.: Research problems and methods in blockchain and trusted data management. J. Softw. **29**(1), 150–159 (2018). (in Chinese)
3. Nakamoto, S.: Institute. https://nakamotoinstitute.org/bitcoin/. Accessed 21 Feb 2019
4. Christidis, K., Devetsikiotis, M.: Blockchains and smart contracts for the internet of things. IEEE Access **4**, 2292–2303 (2016)
5. Huang, X., Xu, C., Wang, P., Liu, H.: LNSC: a security model for electric vehicle and charging pile management based on blockchain ecosystem. IEEE Access **6**, 13565–13574 (2018)
6. Turkanović, M., Holbl, M., Kosic, K., Hericko, M., Kamisalic, A.: EduCTX: a blockchain-based higher education credit platform. IEEE Access **6**, 5112–5127 (2018)
7. Kasahara, S., Kawahara, J.: Priority mechanism of bitcoin and its effect on transaction-confirmation process. J. Ind. Manag. Optim. **15**(1), 365–386 (2019)
8. Kawase, Y., Kasahara, S.: Transaction-confirmation time for bitcoin: a queueing analytical approach to blockchain mechanism. In: Yue, W., Li, Q.-L., Jin, S., Ma, Z. (eds.) QTNA 2017. LNCS, vol. 10591, pp. 75–88. Springer, Cham (2017). https://doi.org/10.1007/978-3-319-68520-5_5
9. Li, Q., Ma, J., Chang, Y.: Cornell University. arXiv:abs/1808.01795. Accessed 21 Feb 2019
10. Rupp, R.: The symmetric versions of Rouché's theorem via-calculus. J. Complex Anal. **2014**, 9 (2014). https://doi.org/10.1155/2014/260953. Article ID 260953
11. Aguiar, M., Lauve, A.: Lagrange's theorem for Hopf monoids in species. Can. J. Math. **65**(2), 241–265 (2013)
12. Karr, A.: Stochastic processes (Sheldon M. Ross). SIAM Rev. **26**(3), 448–449 (1984)
13. Wolff, R., Yao, Y.: Little's law when the average waiting time is infinite. Queueing Syst. **76**(3), 267–281 (2014)

Correction to: Class Aggregation for Multi-class Queueing Networks with FCFS Multi-server Stations

Pasquale Legato [ID] and Rina Mary Mazza [ID]

Correction to:
Chapter "Class Aggregation for Multi-class Queueing Networks with FCFS Multi-server Stations"
in: T. Phung-Duc et al. (Eds.): *Queueing Theory and Network Applications*, **LNCS 11688,**
https://doi.org/10.1007/978-3-030-27181-7_14

Unfortunately a few formulas were not displayed correctly in the contribution.

The correct formulas are given in this table:

Page & formula n°	Errata	Corrige
pg. 3, formula (5)	$P_j(0\|\mathbf{n}) =$ $1 - 1/m_j\left[\sum_{c=1}^{c} R_j V_{jc} T_c(\mathbf{n}) + \sum_{j=1}^{m_j-1}(m_j - l)P_j(l\|\mathbf{n})\right]$	$P_j(0\|\mathbf{n}) =$ $1 - 1/m_j\left[\sum_{c=1}^{c} R_j V_{jc} T_c(\mathbf{n}) + \sum_{l=1}^{m_j-1}(m_j - l)P_j(l\|\mathbf{n})\right]$
pg. 8	**Until** convergence upon $D_{jc}(\mathbf{N}) = R_j, \ c = 1,\ldots,C; j = 1,\ldots,M$	**Until** convergence upon $D_{jc}(\mathbf{N}), \ c = 1,\ldots,C; j = 1,\ldots,M$
pg. 8, formula (23)	$O\left(\left(M + \sum_{j=1}^{M} m_j\right) \cdot C\right)$ per iteration	$O\left(\sum_{j=1}^{M} m_j \cdot C\right)$ per iteration
pg. 10, formula (26)	$O\left(\left(M + \sum_{j=1}^{M} m_j\right) \cdot (C+N)\right)$ per iteration	$O\left(\sum_{j=1}^{M} m_j \cdot (C+N)\right)$ per iteration
pg. 11	**Until** convergence upon $P_j(l - 1\|N - 1),$ $l = 1,\ldots,m_j - 1; j = 1,\ldots,M$	**Until** convergence upon $P_j(l - 1\|N - 2)$ and $P_j(l - 1\|N - 1),$ $l = 1,\ldots,m_j - 1; j = 1,\ldots,M$

The updated version of this chapter can be found at
https://doi.org/10.1007/978-3-030-27181-7_14

Author Index